International Conference on Industrial and Systems Engineering (IConISE 2017)

Enhancing Industrial Resilience Through Disruptive Innovation

IOP Conference Series: Materials Science and Engineering
Volume 337

Denpasar, Indonesia
29 – 30 August 2017

Editors:

Nani Kurniati
Ratna Sari Dewi
Dyah Santhi Dewi
Dody Hartanto
Niniet Indah Arvitrida
Putu Dana Karningsih

ISBN: 978-1-5108-6158-9
ISSN: 1757-8981

Printed from e-media with permission by:

Curran Associates, Inc.
57 Morehouse Lane
Red Hook, NY 12571

Some format issues inherent in the e-media version may also appear in this print version.

Printed by Curran Associates, Inc. (2018)

For permission requests, please contact the Institute of Physics
at the address below.

Institute of Physics
Dirac House, Temple Back
Bristol BS1 6BE UK

Phone: 44 1 17 929 7481
Fax: 44 1 17 920 0979

techtracking@iop.org

Additional copies of this publication are available from:

Curran Associates, Inc.
57 Morehouse Lane
Red Hook, NY 12571 USA
Phone: 845-758-0400
Fax: 845-758-2633
Email: curran@proceedings.com
Web: www.proceedings.com

Table of contents

Volume 337

International Conference on Industrial and Systems Engineering (IConISE) 2017

29–30 August 2017, Denpasar, Bali, Indonesia

Accepted papers received: 12 March 2018
Published online: 10 April 2018

Preface

Papers

Parallel Session 1, 29 August 2017, 13:00-15:00 Room 1

Parallel Session 1, 29 August 2017, 13:00-15:00 Room 3

Parallel Session 1, 29 August 2017, 13:00-15:00 Room 4

Parallel Session 2, 29 August 2017, 15:30-17:15 Room 2

Parallel Session 2, 29 August 2017, 15:30-17:15 Room 3

Parallel Session 2, 29 August 2017, 15:30-17:15 Room 4

Parallel Session 3, 30 August 2017, 13:00-15:00 Room 1

Parallel Session 3, 30 August 2017, 13:00-15:00 Room 2

Parallel Session 3, 30 August 2017, 13:00-15:00 Room 3

Parallel Session 3, 30 August 2017, 13:00-15:00 Room 4

Parallel Session 4, 30 August 2017, 15:15-17:00 Room 1

Parallel Session 4, 30 August 2017, 15:15-17:00 Room 2

Parallel Session 4, 30 August 2017, 15:15-17:00 Room 3

IOP Conf. Series: Materials Science and Engineering **337** (2018) 011001 doi:10.1088/1757-899X/337/1/011001

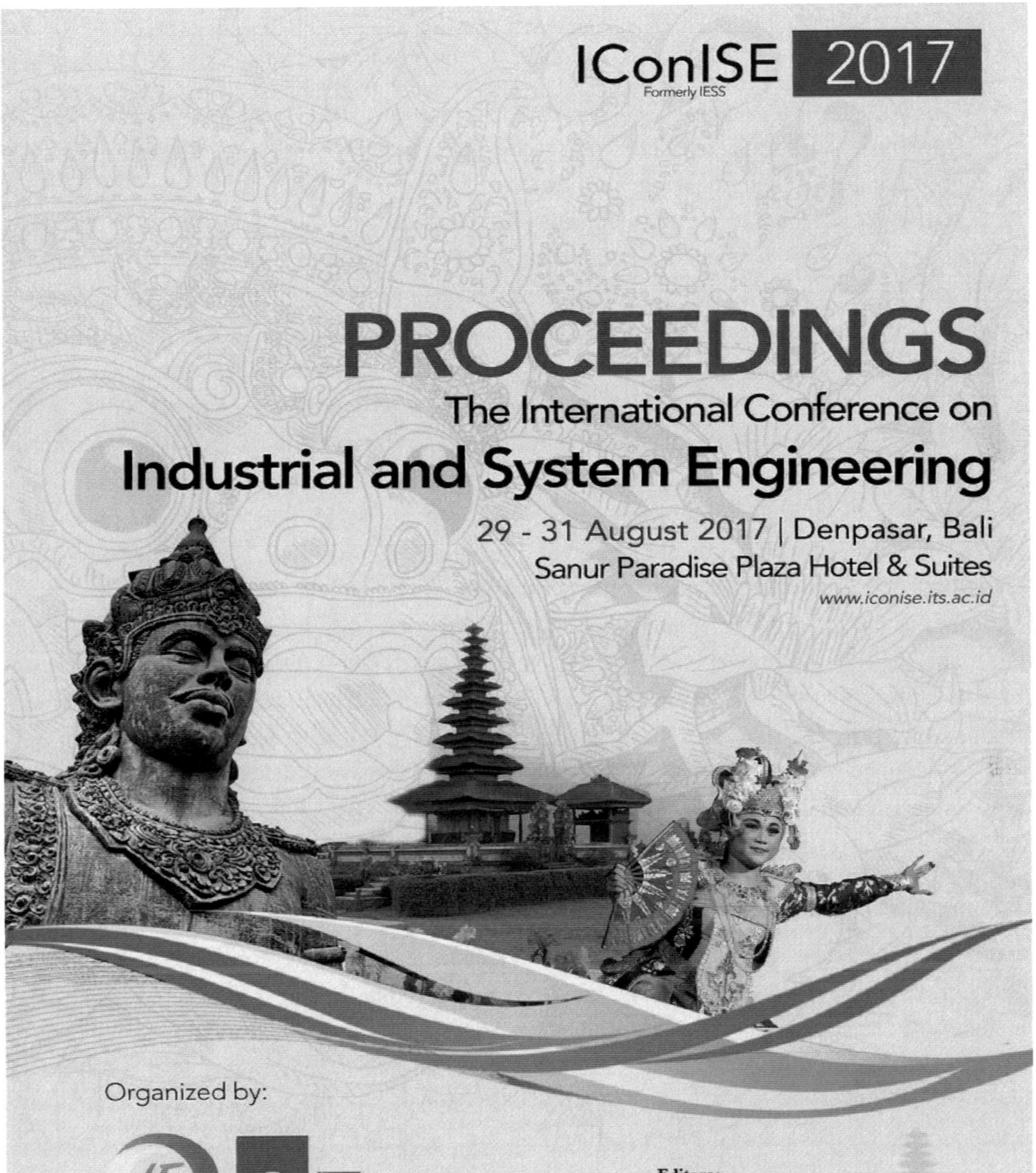

IOP Conf. Series: Materials Science and Engineering **337** (2018) 011001 doi:10.1088/1757-899X/337/1/011001

Proceeding of International Conference on Industrial and System Engineering (IConISE) 2017

Preface

The 1st International Conference on Industrial and System Engineering was hosted by Department of Industrial Engineering ITS, Surabaya, Indonesia The conference was held in Sanur Paradise Plaza Hotel & Suites, Denpasar, Bali, Indonesia, on 29-30th August 2017. The conference's theme was "***Enhancing Industrial Resilience through Disruptive Innovation***". Under this theme, we explored various theories and methods related to value creation and innovation in product and service design, business process, industrial technology, information, and management concerning disruptive innovation. We also discussed approaches to collect, process, manage, and use any information efficiently and effectively, thus the result will be able to improve industrial competitiveness and value in facing the global challenges in industrial environment.

Total abstracts submissions is approximately 180 from about 15 countries. The review is conducted in two stages: abstract submission and full paper submission. The number of papers scheduled for presentation is 94 which represent authors from 7 countries.

I would like to thank all parties that have contributed to this conference. First of all, I would like to thank the four keynote speakers: Professor Shuo-Yan Chou, Director of the Center for Internet of Things Innovation, National Taiwan University of Science and Technology, Taiwan; Professor Keith Goffin, Director of The Centre for Innovative Products and Services- Cranfield School of Management, United Kingdom; Dr.Ir.Patdono Suwignjo, M.Eng.Sc, Directorate General of Institutional Affairs for Science,Technology and Higher Education - Ministry of Research, Technology and Higher Education, Indonesia ; Fadjar Hutomo, S.T., M.MT., CFP Deputy Chairman of Financial Access – BEKRAF.

I would also like to thank for the great support from our institution, Institut Teknologi Sepuluh Nopember (ITS), Surabaya, and also Universitas Internasional Semen Indonesia (UISI), Gresik; Institut Teknologi Del, North Sumatera; and Badan Kerjasama Teknik Industri (BKSTI) East Java region. Moreover, we are also indebted to Semen Indonesia, Jakarta Industrial Estate Pulogadung (JIEP), and Unilever. In addition, we would also like to thank the advisory committee for the support and review committee who have played fundamental roles in making sure the accepted papers, are of high quality. Last but not least, I would like to thank all the committee members from Industrial Engineering Department and Business Management Departement of ITS, for the hard works to the organisation of this conference.

Surabaya, August 2017

Dr. Putu Dana Karningsih
IConISE 2017 Chair

IOP Conf. Series: Materials Science and Engineering **337** (2018) 011001 doi:10.1088/1757-899X/337/1/011001

International Conference on Industrial and System Engineering (IConISE) 2017

- ***Conference Chair*** :
  Putu Dana Karningsih, ITS, Indonesia.
- ***Secretary :***
  Dewanti Anggrahini, ITS, Indonesia.
  Mar'atus Sholihah, ITS, Indonesia.
- ***Paper Chair :***
  Nani Kurniati, ITS, Indonesia.
  Ratna Sari Dewi, ITS, Indonesia.
  Dyah Santhi Dewi, ITS, Indoneisa.
  Dody Hartanto, ITS, Indonesia.
  Niniet Indah Arvitrida, ITS, Indonesia.
- ***Program Chair :***
  Effi Latiffianti, ITS, Indonesia.
  Erwin Widodo, ITS, Indonesia.
  Mokh. Suef, ITS, Indonesia.
  Budi Santosa, ITS, Indonesia.
  M. Saiful Hakim, ITS, Indonesia.
- ***Treasury :***
  Anny Maryani, ITS, Indonesia.
  Naning Aranti Wessiani, ITS, Indonesia.
- ***Publication and Publicity :***
  Yudha Prasetyawan, ITS, Indonesia.
  Arief Rahman, ITS, Indonesia.
- ***Sponsorship :***
  Adithya Sudiarno, ITS, Indonesia.
  Berto Mulia Wibawa, ITS, Indonesia.
  Aang Kunaifi, ITS, Indonesia.
  Diesta Iva Maftuhah, ITS, Indonesia.
- ***Advisory Committee :***
  Prof. Ir. Budi Santosa, M.S., Ph.D., ITS, Indonesia.
  Prof. Ir. I Nyoman Pujawan, M.Eng., Ph.D., ITS, Indonesia.
  Prof. Dr. Ir. Udisubakti Ciptomulyono, M.Eng.Sc., ITS, Indonesia.
  Prof. Ir. Moses L. Singgih, M.Eng.Sc., ITS, Indonesia.
  Prof Iwan Vanany, S.T., M.T., Ph.D., ITS, Indonesia.
  Dr. Ir. Patdono Suwignjo, M.Eng.Sc., ITS, Indonesia.
  Dr. Ir. Sri Gunani Partiwi, M.T., ITS, Indonesia.
  Ir. Sritomo Wignjosoebroto, M.Sc., ITS, Indonesia.

IOP Conf. Series: Materials Science and Engineering **337** (2018) 011001 doi:10.1088/1757-899X/337/1/011001

Dr. Ir. I Ketut Gunarta, M.T., ITS, Indonesia.
Maria Anityasari, S.T., M.E., Ph.D., ITS, Indonesia.

- ***Reviewer :***

Ahmad Syamil, Bina Nusantara University, Indonesia.
Benny Tjahjono, Cranfield University, UK.
Chia Fen Chi, NTUST, Taiwan.
Hui Ming Wee, Chung Yuan Christian University, Taiwan.
Huynh Trung Luong, AIT, Thailand.
Jatinder N.D. Gupta, University of Alabama, USA.
Kung-Jeng Wang, NTUST, Taiwan.
Morikawa Katsumi, Hiroshima University, Japan.
Sha'ri Mohd Yusof, University of Technology, Malaysia.
Shuo Yan Chou, NTUST, Taiwan.
Thunyarat Amornpetchkul, NIDA Business School, Thailand.
Togar Simatupang, ITB, Indonesia and Del Institute of Technology.
Yudi Fernando, Universitas Sains Malaysia, Malaysia.
Yosef Manik, Del Institute of Technology.

Peer review statement

All papers published in this volume of *IOP Conference Series: Materials Science and Engineering* have been peer reviewed through processes administered by the proceedings Editors. Reviews were conducted by expert referees to the professional and scientific standards expected of a proceedings journal published by IOP Publishing.

IOP Conf. Series: Materials Science and Engineering **337** (2018) 012001 doi:10.1088/1757-899X/337/1/012001

Developing cross entropy genetic algorithm for solving Two-Dimensional Loading Heterogeneous Fleet Vehicle Routing Problem (2L-HFVRP)

D L Paramestha[1*] and B Santosa[1]**
[1]Department of Industrial Engineering, Institut Teknologi Sepuluh Nopember, Kampus ITS Sukolilo-Surabaya 60111, Indonesia

*dlaksma@gmail.com, **budi_s@ie.its.ac.id

Abstract. Two-dimensional Loading Heterogeneous Fleet Vehicle Routing Problem (2L-HFVRP) is a combination of Heterogeneous Fleet VRP and a packing problem well-known as Two-Dimensional Bin Packing Problem (BPP). 2L-HFVRP is a Heterogeneous Fleet VRP in which these costumer demands are formed by a set of two-dimensional rectangular weighted item. These demands must be served by a heterogeneous fleet of vehicles with a fix and variable cost from the depot. The objective function 2L-HFVRP is to minimize the total transportation cost. All formed routes must be consistent with the capacity and loading process of the vehicle. Sequential and unrestricted scenarios are considered in this paper. We propose a metaheuristic which is a combination of the Genetic Algorithm (GA) and the Cross Entropy (CE) named Cross Entropy Genetic Algorithm (CEGA) to solve the 2L-HFVRP. The mutation concept on GA is used to speed up the algorithm CE to find the optimal solution. The mutation mechanism was based on local improvement (2-opt, 1-1 Exchange, and 1-0 Exchange). The probability transition matrix mechanism on CE is used to avoid getting stuck in the local optimum. The effectiveness of CEGA was tested on benchmark instance based 2L-HFVRP. The result of experiments shows a competitive result compared with the other algorithm.

1. Introduction

Packing and distribution processes are among important activities in a manufacturing company. Packing and distribution processes in the combinatorial optimization field have been studied intensively but separately. This combination has been reviewed by a number of researchers in the recent years. In this paper, we address the combination of packing and distribution problem by adding some constraints to bring the model closer to the reality.

There are several types of distribution problems which are popular as the object of study by researchers. Vehicle Routing Problem (VRP) is a popular distribution problem as a research object. A great amount of research about VRP have been discussed since 1959 when Danzig and Ramser have described the Traveling Salesman Problem (TSP) [1]. In contrast with to VRP, in the TSP there are no costumer demand and capacity. In this paper, we discuss the variant of Two-dimensional Loading Capacitated Vehicle Routing Problem (2L-CVRP). 2L-CVRP is a generalization of the Capacitated Vehicle Routing Problem, in which customer demand is formed by a set of two dimensional, rectangular, weighted items [2].

This paper addresses a variant of 2L-CVRP, called Two-dimensional Loading Heterogeneous Fleet Vehicle Routing Problem (2L-HFVRP). 2L-HFVRP is a combination of Heterogeneous Fleet Vehicle Routing Problem (HFVRP) and 2L-CVRP. HFVRP is a VRP with heterogeneous fleet of vehicle which have various capacities. Since most enterprises own a heterogeneous fleet of vehicle or hire different type of vehicles to serve their customers, it is therefore crucial to study VRP with a fleet of heterogeneous vehicles [3]. As the NP-hard problem, this paper used metaheuristic approach as the solver. Cross Entropy Genetic Algorithm (CEGA) is a metaheuristic that is in this paper. CEGA algorithm was first used by Santosa et al. [4] and successfully solved a scheduling problem.

2. Problem description

The 2L-HFVRP is defined on an undirected connected graph $G = (V, E)$, where $V = \{0,1,\ldots,n\}$ is a vertex set corresponding to the depot (vertex 0) and the customers (vertices 1, 2,…,n) and $E = \{e_{ij}; i, j \in V\}$ is an edge set. For each $e_{ij} \in E$ is associated with distance d_{ij} and $d_{ii} = 0$. A fleet of P different type of vehicles is located at the depot, and the number of vehicles is unlimited for of each type. Every type of vehicle t $(t = 1,2,\ldots,P)$ is associated with capacity Q_t, fix cost F_t,variable cost V_t, length L_t, and width W_t. The loading surface of vehicle of type t is $A_t = L_t \times W_t$. On the basis that a vehicle with larger capacity usually has higher cost and greater fuel consumption, we assume that $Q_1 \leq Q_1 \leq \cdots \leq Q_p$, $F_1 \leq F_1 \leq \cdots \leq F_p$, and $V_1 \leq V_1 \leq \cdots \leq V_p$. The traveling cost of each edge $e_{ij} \in E$ by a vehicle type t is $c_{ijt} = V_t \times d_{ij}$. The transportation cost of route for vehicle type t is $C_R = F_t + \sum_{i=1}^{i<|R|} V_t \times d_{R(i),R(i+1)}$, where R is the route which starts and ends in the same point (depot). 2L-HFVRP has an objective function to minimize total transportation cost that will be calculated by the sum of C_R for all route R.

For each customer i $(i = 1,2,\ldots,n)$ demands a set of m_i rectangular items, denoted as IT_i, and the total weight of IT_i equal to D_i. Each item IT_i $(r = 1,2,\ldots,m_i)$ has a specific length l_{ir} and width w_{ir}. Total area of the items of customers i is denote as $a_i = \sum_{r=1}^{m_i} w_{ir} \times l_{ir}$. In 2L_HFVRP, a feasible loading must satisfy the following constraints:

- All items of given customer must be loaded on the same vehicle and split deliveries are not allowed.
- All items can be rotated 90^{0} and must be loaded with their sides parallel to the sides of the loading surface.
- Each vehicle must start and finish at the depot.
- Each customer can only be served once.
- The capacity, length and width of the vehicle cannot be exceeded.
- No two items can overlap in the same route.

3. Packing heuristic

Packing heuristic is composed by 5-heuristics. 5-heuristics are used to solve packing problemand to generating a feasible loading position of an inserted item. 5-heuristics are based on the research by Zachariadis et al. [2]. 5-heuristics are used in sequence, and start from heur1. If heur1 fails to produce a feasible solution, then algorithm will be called next $heur_i$ $(i = 1,2,..,5)$. Heur2 is more complex than heur1 and heur5 is the most complex heuristic. If a feasible solution is found, then the packing heuristic stops and the solution is stored.

PosList is a list of possible loading positions for the items. At the beginning, the only available loading position is PosList = {(0, 0)}. PosList = {(0, 0)} is at front left corner of the container. Whenever an item is inserted, a selected loading position will be removed from the PosList. New loading positions are generated and added into the PosList whenever an item inserted.

The loading process the item to the container divided into two scenarios. Both the sequential and the unrestricted scenarios are considered on this paper. The sequential scenario is a scenario that deals with loading and unloading item into vehicle. The sequence of the loading process is affected by the sequence of the route. Whereas the unrestricted scenario is a scenario that only deals with feasible loading of item into the vehicle. In this paper, the loading process is allowed to 90° rotation position to maximize the utility of container surface.

4. Theoretical basis

4.1. Cross entropy

Cross Entropy was originally used to estimate the probability of rare events with the implementation of adaptive algorithms in complex stochastic events, by minimizing the variation [5]. Basically, on CE algorithm, a number of samples are generated randomly. In the next iteration, a number of sample are generated randomly using the parameters that generated from the previous iteration. CE algorithm is executed by the iterative procedure, where each iteration is divided into two parts, as follows:

- Generate a number of sample randomly using a specific mechanism
- Update the parameter. This parameter will be used to generate the new samples in the next iteration. The updated parameter is based on the elite sample in the previous iteration. Elite sample is a percentage of sample that will be used to update the parameter.

4.2. Genetic algorithm

GA algorithm is an algorithm that is based on the principles of genetics and natural selection. The elements that exist in genetics and natural selection adopted in GA to solve optimization problems. The elements are cross over, mutation and elitism. Besides being able to solve the problems of continuous optimization, GA is also widely used to solve combinatorial problems such as TSP, VRP, and crew scheduling for the airline. The procedure of finding solutions in GA, is based only on the value of the objective function, not gradient or calculus techniques [5].

4.3. Local improvement

2-opt, 1-1 exchange and 1-0 exchange procedure is a mutation mechanism that is used to solve VRP. These procedures were based on Ai and Kachitvichyanukul [6]. These procedures are used to improve the quality of VRP solution. The 2-opt procedure works in single route by systematically exchanging the route direction between two pairs of consecutive customers in the route and evaluating whether the routing cost of route is improved or not. The 1-1 exchange procedure systematically interchanges one customer from the first route with another customer from the second route. The 1-0 exchange procedure systematically moves one customer from the first route to the second route [6].

5. Proposed algorithm

CEGA used in this study is based on research by Santosa et al. [4] with some modifications. Crossover mechanism is not used in this study because it would be difficult to generate the feasible solution of 2L-HFVRP. The crossover mechanism was replaced with transition probability matrix mechanism. This mechanism is only performed when the sample did not change over three iterations. This mechanism aims to escape from the local optimum. Transition probability matrix mechanism has a similar concept with crossover mechanism because they both generate solutions based on elite sample (parent).

There are two criteria for termination of iterations. They are the maximum number of iterations and terminating criterion ε. Terminating criterion ε is the difference of mutation rate on the current iteration $p_{m(it)}$ with the previous iteration $p_{m(it-1)}$. The value of ε is usually set near 0. This means that the objective function of the elite sample has the same value between iterations because they indicated already reached optimum. If one of the terminating criterion is met, then stop the iterations. The steps above are described in more detail in algorithm 1.

Algorithm 1: CEGA Algorithm
1. Generate a population of *N*
2. Calculate the objective function of the population in actual iteration.
3. Set a number of the elite sample ρ x N (N_{elite})

4. If there are a number of samples that have not changed in three iterations, then generate a new sample to replace them. It uses the transition probability matrix mechanism. Transition probability matrix P_M will be calculated for each iteration by formula 1.1.

$$P_{M\,(it)} = \alpha\omega + (1-\alpha)P_{M\,(it-1)} \tag{1.1}$$

5. Replace the sample that have not changed in three iterations by the new sample from step 4.
6. Run the mechanism of mutation for each sample
   a. Generate random numbers r
   b. If $r \leq 0.33$ then select 2-opt procedure
   c. If $0.33 < r \leq 0.66$ then select 1-0 Exchange procedure
   d. If $r > 0.66$ then select 1-1 Exchange procedure
   e. Repeat the step 6a until all samples finished mutated.
7. Update G_{best}, if the best sample in the actual iteration has a better value of objective function than the previous G_{best}.
8. Update the mutation rate p_m using elite sample with equation (1.2).

$$p_{m(it)} = \frac{A_{(it)}}{2} \tag{1.2}$$

With the value of $A_{(it)}$ is obtained from equation (1.3) and equation (1.4)

$$A_{(it)} = (1-\alpha)u + A_{(it-1)}\alpha \tag{1.3}$$

$$u = \frac{\overline{z_e}}{2z_{best}} \tag{1.4}$$

Repeat the step 2 if the maximum number of iterations or termination criterion ε unmet

$$\varepsilon = \left|p_{m(it)} - p_{m(t-1)}\right| \tag{1.5}$$

If the maximum iteration or the value of ε is met, then the iteration will stop and the best solution in the last iteration will be recorded as output of the algorithm CEGA.
Note:
$\overline{z_e}$: The average objective function of elite sample.
z_{best} : The best solution for all iterations.

To determine the type of vehicle, the algorithm used a mechanism as described in figure 1. The first step is calculating the total area, weight and maximum length of the items that has not entered into the route. After that, compare them with total area, capacity (weight) and the length of container.

Figure 1. Pseudocode of determining the types of vehicles.

```
Allnode = {1, 2,...,n}; Route ⊆ Allnode ; nonRoute ⊆ Allnode \ {Route}
Vehicle = {P,P-1,....1}; VehicleP>VehicleP-1>......>Vehicle1
TotArea = sum(Vol(areaRoute))
TotWeight= sum(Weight(nonRoute))
MaxDim=max(Dimension(nonRoute))

for i = P : 1 step -1
if AND(TotVol>= AreaVehicle P, TotWeight>= CapVehicle P, MaxDim>= LVehicle P)
   Select = Vehicle P
     elseif AND(TotArea>= AreaVehicle (i), TotWeight>= CapVehicle (i), MaxDim>= LVehicle (i))
   Select = Vehicle (i+1)
     elseif AND(TotArea < AreaVehicle 1, TotWeight < CapVehicle 1, MaxDim < LVehicle 1)
   Select = Vehicle 1
      end
end
```

6. Experiment and analysis
Experiments were performed with three different cases which are CVRP, 2L-CVRP and 2L-HFVRP. Each case of 2L-CVRP and 2L-HFVRP is divided into two versions which are the sequential and the unrestricted version. CEGA algorithms were run on MATLAB program. All experiments were run on a CPU core i3 5015U 2.1GHz with 4 GB ram in the operating system Windows 10.

6.1. Data set
This paper uses the data sets as used in Leung et al. [3]. This data set was originally from Zachariadis et al. [2] , with heterogeneous fleet data added onto it. Out of 36 instances only instances 1-15 are used in the data set. Each instance consists of 5 classes. Class 1 is for pure CVRP and class 2-5 is for 2L-HFVRP. On the 2L-HFVRP case the number of vehicles is unlimited for of each type. There are 4 types of vehicles A, B, C, and D. Each vehicle has the capacity (Q_i), the length of the container (L_i), the width of the container (W_i), variable cost (V_i), and Fix cost (F_i), $i = A, B, C, D$.

6.2. Parameter setting
CEGA algorithm in this study uses a number of parameters that must be determined. These parameters include:

- N: The number of samples in population (Population size); N=5
- α: Smoothing coefficient.
- ρ: Elite sample ratio. Elite sample rate (ρ) generally in range between 1% -10% [4], so in this paper the value of ρ = 2%.
- ε: Terminating criterion
- $maxiter$: Maximum number of iteration.

The value of α, ε, and $maxiter$ are determined by performing several experiments with certain combinations. Each combination performed 5 times to find the solutions. Convergence and Divergence are the indicators in determining the value of α, ε, and $maxiter$. Convergence is a condition that illustrates the ability of the algorithm to produce a near optimal solution. Convergence is obtained by dividing the difference between the average solution and the optimal solution with the optimal solution. The divergence is a condition that illustrates the deviation between the solutions from the algorithm compared to the optimal solution. The divergence is calculated by subtracting the worst solution in five replications with the optimal solution.

Based on the experiment, the combination of parameter $\alpha = 0.8$, ε = 0.00001 and *maxiter* = 30 is the best of out of 18 combinations for 2L-HFVRP. It is the best combination because it has the smallest value of the convergence and divergence. The combination of parameter $\alpha = 0.2$, ε = 0.00001 and *maxiter* = 60 is selected to be the best combination for 2L-CVRP.

6.3. Computational result
In this study, the CEGA algorithm is compared with SA_HSL [3] algorithm. This is because not many studies that discuss about 2L-HFVRP. The comparison of the two algorithms for sequential and unrestricted scenario can be seen in table 1.

The sequential version of 2L-HFVRP produce %gap = -4.78%. That means the average result of CEGA algorithm is 4.78% higher than SA_HSL algorithm. The unrestricted version of 2L-HFVRP produce %gap = -5,68%. That means the average result of CEGA algorithm is 5,68% higher than SA_HSL algorithm. Instances 7,8,10,11, and 13 in the sequential and the unrestricted scenario have higher minus value of %gap than others. *Sec h* is the time when the last best solution was updated and *Sec tot* is the overall time to run the algorithm.

6.3.1. Applied to the pure CVRP
As for the pure CVRP, the results of CEGA are compared with results of previous studies which were using the SA_HLS [3], GTS [2], ACO [7], and GRAPS x ELS [8] algorithm. Table 2 displays the results comparison from each algorithm for pure CVRP. Based on the resulting solution, CEGA algorithm is

able to compete with the other comparator algorithms. On the Instance 1,2,4,5 and 6, CEGA algorithm is able to produce the optimal solution. For all comparator algorithms, CEGA has an average% gap = -1. That means the average result of CEGA algorithm is 1% higher than compared to other algorithms.

Table 1. The average computational result of 2L-HFVRP in class 2-5.

	Sequential								Unrestricted						
Inst	Average Class 2-5		Sec h		Sec tot		%gap	Inst	Average Class 2-5		Sec h		Sec tot		%gap
	CEGA	SA_HSL	CEGA	SA_HSL	CEGA	SA_HSL			CEGA	SA_HSL	CEGA	SA_HSL	CEGA	SA_HSL	
1	607.786	603.15	75.8	5.73	121.05	31.04	-0.77%	1	607.49	600.77	79.68	4.52	118.33	29.89	-1.12%
2	719.525	705.03	52.27	6.09	104.53	31.28	-2.06%	2	711.71	699.21	78.09	5.2	113.6	32.88	-1.79%
3	777.527	771.81	189.33	10.35	232.03	36.61	-0.74%	3	768.79	770.12	159.69	10.37	274.36	33.84	0.17%
4	710.593	704.87	198.91	8.71	339.1	35.19	-0.81%	4	706.37	698.19	201.65	7.73	344.5	30.04	-1.17%
5	799.536	802.56	410.03	9.35	546.85	27.62	0.38%	5	789.95	786.84	504.02	7.82	563.99	27.68	-0.39%
6	841.512	834.76	311.91	9.92	393.03	43.9	-0.81%	6	831.9	831.32	329.04	8.9	429.64	42.78	-0.07%
7	6539.45	5770.83	392.79	1.95	718.87	31.4	-13.32%	7	6488.88	5630.02	688.04	1.15	948.8	31.08	-15.25%
8	6976.225	5633.04	353.24	4.77	547.51	37.42	-23.84%	8	6713.35	5602.6	835.47	1.92	875.24	30.22	-19.83%
9	1055.125	1047.6	383.43	15.74	481.92	68.84	-0.72%	9	1032.85	1035.62	687.18	7.88	1051.53	58.75	0.27%
10	8477.2	7730.73	976.95	5.87	1363.35	51.78	-9.66%	10	8620.8	7625.05	1618.99	6.15	1813.64	43.27	-13.06%
11	9243.8	8491.94	917.17	10.07	1540.98	58.88	-8.85%	11	9235.1	8329.69	1325.6	4.91	1628.83	52.61	-10.87%
12	1668.625	1681.61	134.86	32.9	323.48	158.45	0.77%	12	1678.9	1681.07	259.82	34.91	306.83	167.21	0.13%
13	29007.75	26761.4	1824.07	6.71	2070.17	65.34	-8.39%	13	29508.75	25978.7	1983.75	9.35	2339.05	69.95	-13.59%
14	11411.25	11120.3	1053.74	11.29	1530.83	55.62	-2.62%	14	11196.75	10869.1	1015.75	16.31	1613.35	78.1	-3.01%
15	11949.5	11916.3	1455.51	24.53	1686.65	148.5	-0.28%	15	12136	11490.1	1174.91	8.88	1647.61	91.53	-5.62%
					Average %Gap		-4.78%						Average %Gap		-5.68%

Table 2. Average comparison results for the pure CVRP.

Inst	CEGA	SA_HLS	GTS	ACO	G x E	SA_HLS *%gap*	GTS *%gap*	ACO *%gap*	G x E *%gap*
1	278.73	278.73	278.73	278.73	278.73	0.0%	0.0%	0.0%	0.0%
2	334.96	334.96	334.96	334.96	334.96	0.0%	0.0%	0.0%	0.0%
3	360.09	358.4	358.4	358.4	358.4	-0.5%	-0.5%	-0.5%	-0.5%
4	430.88	430.89	430.89	430.89	430.89	0.0%	0.0%	0.0%	0.0%
5	375.28	375.28	375.28	375.28	375.28	0.0%	0.0%	0.0%	0.0%
6	495.85	495.85	495.85	495.85	495.85	0.0%	0.0%	0.0%	0.0%
7	570.15	568.56	568.56	568.56	568.56	-0.3%	-0.3%	-0.3%	-0.3%
8	574.95	568.56	568.56	568.56	568.56	-1.1%	-1.1%	-1.1%	-1.1%
9	608.89	607.65	607.65	607.65	607.65	-0.2%	-0.2%	-0.2%	-0.2%
10	536.86	535.8	535.8	535.8	535.8	-0.2%	-0.2%	-0.2%	-0.2%
11	516.24	505.01	505.01	505.01	505.01	-2.2%	-2.2%	-2.2%	-2.2%
12	618.71	610	610	610	610	-1.4%	-1.4%	-1.4%	-1.4%
13	2081.20	2006.34	2006.34	2006.34	2006.34	-3.7%	-3.7%	-3.7%	-3.7%
14	846.73	837.67	837.67	837.67	837.67	-1.1%	-1.1%	-1.1%	-1.1%
15	867.59	837.67	837.67	837.67	837.67	-3.6%	-3.6%	-3.6%	-3.6%
				Rata-rata %gap		-1.0%	-1.0%	-1.0%	-1.0%

6.3.2. Applied to the 2L-CVRP

As for the sequential scenario of 2L-CVRP, the results of CEGA are compared with results of previous studies which were using the SA_HLS [3], GTS [2], and ACO [7] algorithm. For unrestricted scenario, the algorithm GRAPS x ELS [8] was added.

Table 3 displays the results comparison from each algorithm for 2L-CVRP. In table 3, the average %gap of sequential 2L-CVRP generated by SA_HSL, GTS, and the ACO algorithm are -0.22%; 0.55%; and -2.04%, respectively. As for sequential 2L-CVRP, based on the average of %gap, the solutions produced by CEGA has a better average than the GTS algorithm. Meanwhile in instance 6, 7, 9, and 10, CEGA algorithm is able to generate a better average solution than the SA_HLS algorithm. However, CEGA algorithm outperforms ACO algorithm only in instance 6. Whereas, the average %gap of unrestricted 2L-CVRP generated by SA_HSL, GTS, ACO and the G × E algorithm are -1,95%, -1,01%, -3,02%, and -3,87% respectively. Meanwhile in instance 1, 3, 4, 8 and 9, CEGA algorithm is able to generate a better average solution than the GTS.

Table 3. The average computational result of 2L-CVRP in class 2-5.

	Sequential								Unrestricted								
	Average Class 2-5				% gap				Average Class 2-5					% gap			
Inst	CEGA	SA_HSL	GTS	ACO	SA_HSL	GTS	ACO	Inst	CEGA	SA_HSL	GTS	ACO	G x E	SA_HSL	GTS	ACO	G x E
1	299.05	298.34	304.22	294.48	-0.24%	1.70%	-1.55%	1	291.8	290.37	295.74	285.77	282.65	-0.49%	1.33%	-2.11%	-3.24%
2	347.71	345.23	346.71	345.24	-0.72%	-0.29%	-0.72%	2	344.49	341.35	341.89	341.02	339.26	-0.92%	-0.76%	-1.02%	-1.54%
3	392.23	390.78	393.35	381.4	-0.37%	0.28%	-2.84%	3	380.2	377.37	384.49	376.32	376.32	-0.75%	1.11%	-1.03%	-1.03%
4	444.28	444.21	444.62	441.11	-0.02%	0.08%	-0.72%	4	439.31	435.01	441.45	438.65	435.01	-0.99%	0.49%	-0.15%	-0.99%
5	389.73	387.59	396.36	382.39	-0.55%	1.67%	-1.92%	5	386.45	379.49	382.22	379.03	379.03	-1.83%	-1.11%	-1.96%	-1.96%
6	498.86	503.66	505.04	499.49	0.95%	1.22%	0.13%	6	502.53	501.02	499.47	497.27	497.04	-0.30%	-0.61%	-1.06%	-1.11%
7	712.71	719.95	723.83	702.27	1.01%	1.54%	-1.49%	7	705.74	698.65	703.49	696.91	691.11	-1.02%	-0.32%	-1.27%	-2.12%
8	723.21	716.62	715.72	711.65	-0.92%	-1.05%	-1.62%	8	696.77	703.13	705.6	691.14	678.84	0.90%	1.25%	-0.82%	-2.64%
9	618.26	621.23	622.2	614.54	0.48%	0.63%	-0.60%	9	613.17	612.02	615.65	612.02	612.01	-0.19%	0.40%	-0.19%	-0.19%
10	718.42	727.73	727.86	697.2	1.28%	1.30%	-3.04%	10	716.68	701.61	713	682.53	675.79	-2.15%	-0.52%	-5.00%	-6.05%
11	765.38	761.25	768.15	728.61	-0.54%	0.36%	-5.05%	11	765.83	736.3	740.04	721.82	705.95	-4.01%	-3.49%	-6.10%	-8.48%
12	623.68	618.98	628.62	615.76	-0.76%	0.79%	-1.29%	12	636.95	613.94	616.83	611.26	611.26	-3.75%	-3.26%	-4.20%	-4.20%
13	2664.35	2641.36	2679.34	2591.77	-0.87%	0.56%	-2.80%	13	2667.53	2561.4	2599.4	2520.73	2490.62	-4.14%	-2.62%	-5.82%	-7.10%
14	1093.85	1079.47	1092.78	1042.33	-1.33%	-0.10%	-4.94%	14	1081.9	1015.81	1036.77	991.26	984.42	-6.51%	-4.35%	-9.14%	-9.90%
15	1239.35	1230.28	1234.21	1212.93	-0.74%	-0.42%	-2.18%	15	1230.38	1193.97	1197.83	1167.28	1144.69	-3.05%	-2.72%	-5.41%	-7.49%
				Average %Gap	-0.22%	0.55%	-2.04%						Average %Gap	-1.95%	-1.01%	-3.02%	-3.87%

7. Analysis

CEGA algorithm used on this paper in finding the solutions is affected by two factors: the value of α and the number of iterations. The number of iterations is limited by $maxiter$ and the value of ε. The value of ε will make the iterations stop when its value has reached its limit even if a solution is not optimal. In general, the smaller value of α will produce a better solution but needs more iterations. However, extending the iteration becomes ineffective when a higher value of α is used. This is because the algorithm has a tendency to get stuck in a local optimum when using a high value of alpha. Thus, if uses a high value of α, then the algorithm is possible to get near optimal solution. This option also has a possibility to stuck in a local optimum. The ability to find a solution would be better to use the small value of α, but this requires a longer number of iterations.

Based on the average % gap, it can be seen that 2L-HFVRP have higher average % gap compared to other algorithms. However, CEGA was able to resolve the case CVRP and 2L-CVRP properly. 2L-HFVRP is a variant of 2L-CVRP by adding variations such as different types of vehicles. In 2L-HFVRP, there is a mechanism to determine the type and the number of vehicles used. This mechanism was observed as the factor affecting the high value of %gap in 2L-HFVRP. On the other hand, in the context of 2L-CVRP where this mechanism does not apply, the value of %gap is relatively small.

However, it should be noted that not all examples of the 2L-HFVRP produce a solution with a high value of % gap. Instance 7, 8, 10, 11, and 13 produce a high value of %gap. On the instance 7, 8, 10, 11, and 13 there are a number of nodes in which the item's weight are higher than the average item's weight of the other nodes. Some of the items are almost filling into one type of vehicle with the largest capacity. These types of problems require a various type of vehicles. Whereas, the mechanism of selecting the type of vehicle used in this paper tended to choose the highest capacity in such a way that for some instances in 2L-HFVRP high values of %gaps are produced.

8. Discussion and conclusion

Based on the experimental results and analysis, it can be concluded that the CEGA successfully developed to solve Two-Dimensional Constraint Loading Heterogeneous Fleet Vehicle Routing Problem (2L-HFVRP). CEGA algorithms are generally able to complete CVRP and-CVRP 2L and 2L-HFVRP with some notes. Therefore, we can conclude that CEGA algorithm is able to compete with the algorithms in previous research. CEGA algorithm requires many iterations when using a small value of α while it has a risk of getting stuck in a local optimum if using a high value of α. The mechanisms of determining the type of vehicles tend to choose a vehicle with a highest capacity. This makes the search of possible solutions is limited. Local improvement is proved to have a good ability in improving the solution, but the terms of mutation make this mechanism is limited in searching solutions.

9. References

[1] Yeun L C, Ismail W a N R, Omar K and Zirour M 2008 Vehicle Routing Problem : Models and Solutions *Journal of Quality Measurement and Analysis* **4** pp 205–18
[2] Zachariadis E E, Tarantilis C D and Kiranoudis C T 2009 A Guided Tabu Search for the Vehicle Routing Problem with two-dimensional loading constraints *European Journal of Operational Research* **195** pp 729–43
[3] Leung S C H, Zhang Z, Zhang D, Hua X and Lim M K 2013 A meta-heuristic algorithm for heterogeneous fleet vehicle routing problems with two-dimensional loading constraints *European Journal of Operational Research* **225** pp 199–210
[4] Santosa B, Budiman M A and Wiratno S E 2011 A Cross Entropy-Genetic Algorithm for m-Machines No-Wait Job-ShopScheduling Problem *Journal of Intelligent Learning Systems and Applications* **3** pp 171–80
[5] Santosa B and Willy P 2011 *Metoda Metaheuristik Konsep dan Implementasi* ed B Santosa (Surabaya: Guna Widya)
[6] Ai T J and Kachitvichyanukul V 2009 Particle swarm optimization and two solution representations for solving the capacitated vehicle routing problem *Computers & Industrial Engineering* **56** pp 380–7
[7] Fuellerer G, Doerner K F, Hartl R F and Iori M 2009 Ant colony optimization for the two-dimensional loading vehicle routing problem *Computers and Operations Research* **36** pp 655–73
[8] Duhamel C, Lacomme P, Quilliot A and Toussaint H 2011 A multi-start evolutionary local search for the two-dimensional loading capacitated vehicle routing problem *Computers and Operations Research* **38** pp 617–40

Solving the Container Stowage Problem (CSP) using Particle Swarm Optimization (PSO)

Matsaini[1*] and Budi Santosa[1]
[1]Department of Industrial Engineering, Institut Teknologi Sepuluh Nopember, Kampus ITS Sukolilo-Surabaya 60111, Indonesia

*matsay.abdullah@gmail.com

Abstract. Container Stowage Problem (CSP) is a problem of containers arrangement into ships by considering rules such as: total weight, weight of one stack, destination, equilibrium, and placement of containers on vessel. Container stowage problem is combinatorial problem and hard to solve with enumeration technique. It is an NP-Hard Problem. Therefore, to find a solution, metaheuristics is preferred. The objective of solving the problem is to minimize the amount of shifting such that the unloading time is minimized. Particle Swarm Optimization (PSO) is proposed to solve the problem. The implementation of PSO is combined with some steps which are stack position change rules, stack changes based on destination, and stack changes based on the weight type of the stacks (light, medium, and heavy). The proposed method was applied on five different cases. The results were compared to Bee Swarm Optimization (BSO) and heuristics method. PSO provided mean of 0.87% gap and time gap of 60 second. While BSO provided mean of 2,98% gap and 459,6 second to the heuristcs.

1. Introduction

Sea transportation is an important mode in international trade. According to the Maritime Transport Review released by the United Nations Conference on Trade and Development at 2012, about 80% of global trade is conducted through sea. Among different ship types, about 62% used containers. Since the containerization era, containers have helped to smooth the flow of goods [1]. With the rapidly growing use of containers, container stowage problem becomes a complex problem faced daily by all shipping lines. Therefore, it requires proper container containment planning. But, good planning is not easy as it depends on the experience of the planner [2]. In the container arrangement into the vessel, we must pay attention to three rules. First rule is the location of the container. The location must be in accordance with the type and size of the container. The second rule,is the heavier container should be placed under a lighter container. The third is the balance of the ship. The arrangement of thousands of containers that have dozens of destinations with regard to the size, type, weight and equilibrium of ships, raises many possibilities that are difficult to solve (combinatorial problems). Out of many combinations, there are some solutions which are infeasible because it must be re-unloaded as much as possible so container can get into the ship. Therefore, the heuristic approach to solve container arrangement problem that makes the number of shifting and unloading time minimal is required [3].

Research on CSP have been done with various of methods, such as genetic algorithm [4], bi-level genetic algorithm [5], heuristics [6], ant colony optimization [1]), tabu search [8], constraint programming and integer programming [9], survey method and classification [10] integer linear programming [11], hybrid metaheuristic and local search heuristics [10]. The algorithm used in this paper is particle swarm optimization (PSO). The reason is aside from its simplicity, PSO has not been applied for container stowage problem (CSP) with considering the five factors : type, size, weight, purpose, and balance.

2. Container Stowage Problem (CSP)

Each container location on the vessel has a position index or location information: position of bay, row, and tier. The bay represents a segment from the front to the rear of the ship. While the row indicates a segment from the center to the outside of the ship if the ship is seen from above. Tier represents a segment from the bottom to the top if the ship is viewed from the side. The description of bay, row, and tier can be seen in figure 1.

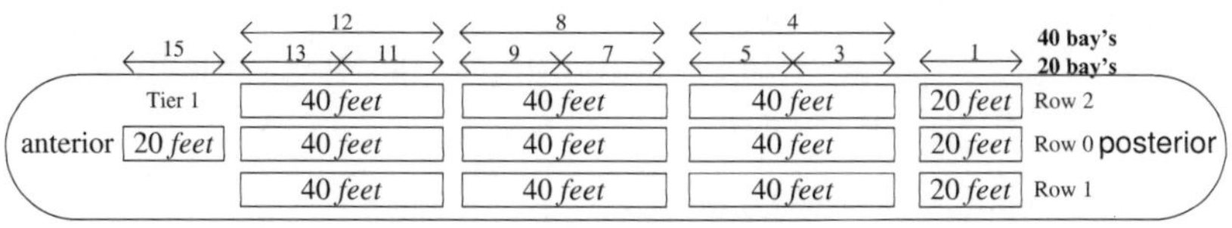

Figure 1. Description of Bay, Row, and Tier.

Figure 1 presents each location based on bay, row, and tier indices. The container size of 20 is placed on the odd bays, i.e. 01, 03, 05, 07, 09 and so on. While, container size of 40 is placed on even bays or two odd bays, i.e. 04 = 03 + 05 , 08 = 07 + 09 and so on [6].

3. Model Container Stowage Problem (CSP)

Many research on CSP use various methods and models. One of the models that considers the five factors in the stowage process (the size, type, weight, and purpose of the container, as well as, the ship's balance) was the model that proposed by Ambrosino et al. [6]. This basic model has also been used in Putamawa and Santosa [3].This model is used in this research. . The basic model is:

$$Min\ L = \sum_l \sum_c t_{lc} x_{lc} \quad (1)$$
$$\sum_l \sum_c x_{lc} = m \quad (2)$$
$$\sum_l x_{lc} \leq 1\ \forall c \quad (3)$$
$$\sum_c x_{lc} \leq 1\ \forall l \quad (4)$$
$$\sum_l \sum_c w_c x_{lc} \leq Q \quad (5)$$
$$\sum_{c\in T} x_{ijkc} = 0 \quad \forall i \in E, j, k \quad (6)$$
$$\sum_{c\in F} x_{ijkc} = 0 \quad \forall i \in O, j, k \quad (7)$$
$$\sum_{c\in T} x_{i+1jkc} + \sum_{c\in F} x_{i+1jkc} \leq 1 \quad \forall i \in E, j, k \quad (8)$$
$$\sum_{c\in T} x_{i-1jkc} + \sum_{c\in F} x_{i+1jkc} \leq 1 \quad \forall i \in E, j, k \quad (9)$$
$$\sum_{c\in T} x_{i+1jk+lc} + \sum_{c\in F} x_{i+1jkc} \leq 1 \quad \forall i \in E, j, k = 1, \ldots, |k|-1 \quad (10)$$
$$\sum_{c\in T} x_{i-1jk+lc} + \sum_{c\in F} x_{i+1jkc} \leq 1 \quad \forall i \in E, j, k = 1, \ldots, |k|-1 \quad (11)$$
$$\sum_k \sum_{c\in T} w_c x_{ijkc} \leq MT \quad \forall i, j \quad (12)$$
$$\sum_k \sum_{c\in T} w_c x_{ijkc} \leq MF \quad \forall i, j \quad (13)$$
$$\sum_{\substack{c,e\in C:\\ w_c \neq w_e}} \left(w_c x_{ijkc} - w_e x_{ijk+1e}\right) \geq 0 \quad \forall i, j, k = 1, \ldots, |k|-1 \quad (14)$$
$$\sum_{\substack{c,e\in C:\\ d_c \neq d_e}} \left(d_c x_{ijkc} - d_e x_{ijk+1e}\right) \geq 0 \quad \forall i, j, k = 1, \ldots, |k|-1 \quad (15)$$
$$-Q_2 \leq \sum_{i\in A,j,k} \sum_c w_c x_{ijkc} - \sum_{i\in P,j,k} \sum_c w_c x_{ijkc} \leq Q_2 \quad (16)$$
$$-Q_1 \leq \sum_{i,j\in L,k} \sum_c w_c x_{ijkc} - \sum_{i,j\in R,j,k} \sum_c w_c x_{ijkc} \leq Q_1 \quad (17)$$
$$x_{lc} \in \{0,1\} \quad \forall l, c \quad (18)$$

The objective function is to minimize the total unloading time (L), with the decision variable x_{lc} being the container c at location l, equals to 1 if the container c is at location l and equals 0 if not. $x_{lc} = x_{ijk}$. Other variables used in the mathematical model are:

l	= The location of the 1st placement ... n shown by i = bay, j = row, and k = tier (l = 1,2, ... n).
c	= 1st container ... m.
t_{lc}	= Time needed for unloading container c location l.
M	= Number of all containers.
w_c	= The weight of the container c.
Q	= Total capacity.
Q1	= Tolerance of anterior-posterior balance.
Q2	= Left-right balance tolerance.
E	= Bay is even. O = Bay odd. T = Container with size 20 feet.
F	= Container with size 40 feet.
MT	= Weight limit of one level container 20 feet.
MF	= Weight limit of one level container 40 feet.
K	= Tier ship.
d_c	= Destination of container.

The description of each equation is as follows. Eq. (1) objective function, minimizing the total time of unloading.Eq. (2) the total l location of the occupied container should be equal to the number of containers. Eq.(3) one container is placed in only one location. Eq.(4) one location can only be occupied one container. E. (5) the total weight of the container transported shall not exceed the capacity of the ship (Q). Eq.(6) the 20 feet (T) container will not be placed on even bay (E). In accordance with the arrangement rules based on container size is 20 feet container placed on odd bay. Eq. (7) the 40 feet (F) container will not be placed in the odd bay (O). Eq. (8) and eq.(9) container 20 and 40 feet container cannot be placed simultaneously on odd bays and successive even bays. Eq. (10) and eq.(11) container 20 feet cannot be placed on top of container 40 feet. Eq. (12) the total weight of the 20 feet container pile on one tier shall not exceed the MT limit. Eq. (13) the total weight of the 40 feet container pile on one tier should not exceed the MF limit. Eq. (14) the heavier container cannot be placed above a lighter container. Eq. (15) container with closer destination is placed on top of the container with the longest destination. Eq.(16) the total difference of the weight of the container on the anterior part with the posterior part shall not exceed the tolerance Q2. Eq. (17) the total difference of the weight of the container on the left side with the right side shall not exceed the tolerance Q1. In accordance to the rules based on the balance of the vessel is the total weight difference of the container between the axis portion of the vessel forward and the rear axis of the ship does not exceed the specified limit. Finally, Eq. (18) the decision variable is a binary (0,1).

4. Particle swarm optimization (PSO)

Particle Swarm Optimization (PSO) is developed based on the behavior of the swarm (bird or fish). The individual movement is influenced by its individual (cognitive) and social (swarm) movements [13]. Each individual (particle) is charaterized by velocity and position. Updating the velocity (v) and the position (x) of each particle based on its initial position, initial velocity, Gbest, and Pbest is done according the following formula:

$$V_t = V_{t-1} + c_1 r_1 (Pbest - X_t) + c_2 r_2 (Gbest - X_t) \quad (19)$$
$$X_{t+1} = X_t + V_t \quad (20)$$

where Pbest is the best position of each particle among iterations, Gbest is the best of Pbest. PSO is suitable to solve continuous problem, because the value of position and velocity of all of particles are continuous. The disadvantage of PSO is the possibility to be trapped at local optimal. The solution can be trapped at local optimal because of the updating solution only based on the best solution from previous iteration and the algorithm does not have a mechanism to jump out from the local optimal. The modification by changing the updating mechanism (based on some best solution) can reduce the possibility to be trapped at local optimal. To apply PSO to CSP, it is necessary to customize or modify algorithms as illustrations in figure 2.

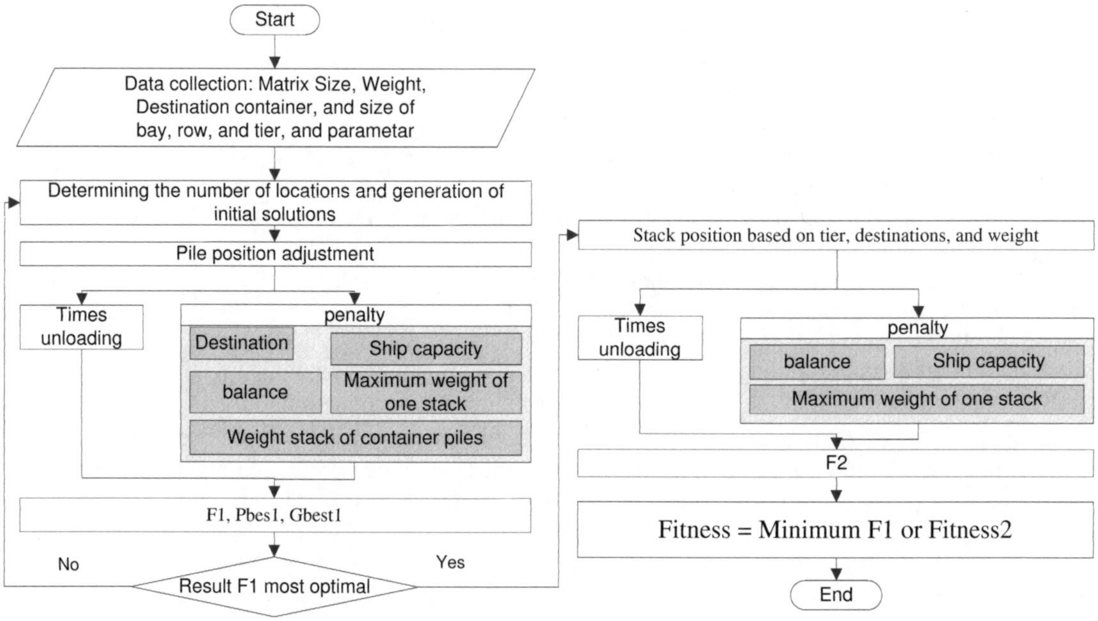

Figure 2. Flowchart of solving the CSP using PSO.

5. Validation

Validation is done by comparing the results of experment on small data with those of enumeration. The comparison of the results of both are as shown in table 1.

Table 1. Comparison of PSO and enumeration.

Method	Enumeration	PSO
Penalty	0	0
Time *Unloading*	366	366
Objective function	366	366

Based on the results, particle swarm optimization and enumeration algorithms provided the same performance. Therefore, the algorithm developed can be said to be valid, so that research can be done next step.

6. Experiment

The experiments were done on some data sets. The data used are secondary data sets from those in Putamawa and Santosa [3]. The numbers of of bay = 14, row = 4, tier = 5.The description of the data is shown in table 2.

Table 2. Data Criteria for Each Case

Case	Total Container	Size		Weight (Ton)			Destination		
		20'	40'	Light (10 - 15)	Medium (20 - 25)	Heavy (30 - 35)	d1	d2	d3
1	100	62	38	45	30	25	47	53	0
2	120	75	45	50	44	26	55	65	0
3	130	90	40	56	46	28	55	75	0
8	140	95	45	60	50	30	65	75	0
9	140	95	45	60	50	30	50	40	50

7. Analysis of performance algorithm

Particle swarm optimization (PSO) for container stowage problem (CSP) has been combined to solve CSP. CSP is the problem of arrangement of container into ship which is difficult to solve because it have many constraints and where constraints can be conflicted each other.. To overcome this issue, the PSO added a penalty to the objective function. The penalty is used to control the solution in finding the optimal solution. Container position is determined by random number, and steps added to arrange the container based on the weight of the container. It is possible to change the position of the container based on the coordinates of the position on the ship. Every position that has not been full until the upper level is possible to be placed a new container obtained from the coordinates of other positions with the least pile position. Then sorted is done by the container destination so that the container with the closer destination is always placed above the container with the further destination. Experiments were conducted on five different cases. The five cases were applied to the PSO, and the PSO modifications were compared with the BSO modifications in the previous study.

The results from five cases showed that PSO still cannot find the optimal solution. It can be seen in table 3 that shows the gap of PSO is still greater than those of BSO modification. Some steps are added to improve the performance of PSO.

$$Gap = \overline{L_{PSO}} - \overline{L_{BSO}} \tag{21}$$

$$\% \; Gap = \frac{(\overline{L_{PSO}} - \overline{L_{BSO}})}{\overline{L_{BSO}}} \; x \; 100 \tag{22}$$

Gap is the difference between the two algorithms, % gap is the percentage of the gap. To calculate gap and % gap we can use formulas (21) and (22). $\bar{L}$ = Average times unloading .

In table 3. The largest mean objective function is given by BSO. While, the results of PSO in cases 1, 2, 4, and 5 are similar to the solution obtained by the heuristic method. In case 3, PSO gives smaller result than the heuristic method.

8. Conclusion

We have proposed modified PSO to solve Container Stowage Problem. The validation process of PSO is done by comparing the results with those of enumeration technique. The experiment on small problems showed the same results. This validated the proposed algorithm. To apply PSO on CSP we added steps that changes the position of the container based on the number of tiers, sorting the container stack based on the destination container, and sorting the container stack based on the type of container weight. By adding these three steps,it guarantees that two constraints are not violated.This allows the penalty value to be smaller and the value of the objective function to be minimum. The algorithm has been carried out in five cases. The comparison between PSO and BSO provided the Gap of -2.01 percent. It means that PSO gives a smaller value of 2.01 percent or 433.86 seconds lower than the result of modified BSO. Then the comparison of PSO with heuristics provide 0.87 percent Gap. This means that PSO gives greater value of 0.87 percent or 60 seconds greater than heuristic.

For further research, it is suggested to consider not only to minimize unloading time but also loading time. Also, consider to include more then two types of container sizes.

Table 3. Comparison of Algorithms

Comparison	Case	Mean	%GAP	Best	GAP	Comparison	Case	Mean	%GAP	Best	GAP
PSO Vs BSO Modifikasi	1	14629.60	0.97	14624	135.2	PSO Modifikasi Vs Heuristik	1	14057.80	1,47	14004	150.0
	2	17294.20	0.42	17279	57.5		2	16847.00	1.59	16740	156.0
	3	18879.30	2.03	18866	362.0		3	18021.00	-0.41	17910	-186.0
	4	20304.30	2.25	20297	440.0		4	19628.00	0.97	19530	90.0
	5	20308.90	2.35	20304	462.0		5	19619.00	0.77	19560	90.0
PSO Modifikasi Vs BSO Modifikasi	1	14057.80	-2.97	14004	-484.8	BSO Modifikasi Vs Heuristik	1	14488.80	4.58	14436	582.0
	2	16847.00	-2.17	16740	-481.5		2	17221.50	3.84	17190	606.0
	3	18021.00	-2.61	17910	-594.0		3	18504.00	2.25	18456	360.0
	4	19628.00	-1.15	19530	-327.0		4	19857.00	2.15	19818	378.0
	5	19619.00	-1.12	19560	-282.0		5	19875.00	2.08	19842	372.0

9. References

[1] Wang N, Jin B and Lim A 2015 Target-guided algorithms for the container pre-marshalling problem. *Omega* (United Kingdom) **53** pp 67–77

[2] Fan L, Low M, Ying H and Jing H 2010 Stowage Planning of Large Containership with Tradeoff between Crane Work-load Balance and Ship Stability *Proceedings of the International MultiConference of Engineers and Computers Scientists* **3** pp 1537–1543

[3] Putamawa F and Santosa B 2011 *Pengembangan algoritma bee swarm optimization untuk penyelesaian container stowage problem* Institut Teknologi Sepuluh Nopember (Tugas Akhir)

[4] Dubrovsky O, Levitin G and Penn M 2002 A genetic algorithm with a compact solution encoding for the container ship stowage problem *Journal of Heuristics* **8** pp 585–599

[5] Zhang R, Jin Z, Ma Y and Luan W 2015 Optimization for two-stage double-cycle operations in container terminals *Computers and Industrial Engineering* **83** pp 316–326

[6] Ambrosino D, Sciomachen A and Tanfani E 2004 Stowing a containership: The master bay plan problem *Transportation Research Part A: Policy and Practice* **38** pp 81–99

[7] Ambrosino D, Anghinolfi D, Paolucci M and Sciomachen A 2010 *An experimental comparison of different heuristics for the master bay plan problem Lecture Notes in Computer Science (Including Subseries Lecture Notes in Artificial Intelligence and Lecture Notes in Bioinformatics)*, 6049 LNCS pp 314–325

[8] Wilson I D, Roach P A and Ware J A 2001 Container stowage pre-planning: Using search to generate solutions, a case study *Knowledge-Based Systems* **14** pp 137–145

[9] Delgado A, Jensen R M, Janstrup K, Rose T H and Andersen K H 2012 A Constraint Programming model for fast optimal stowage of container vessel bays *European Journal of Operational Research* **220** pp 251–261

[10] Lehnfeld J and Knust S 2014 Loading, unloading and premarshalling of stacks in storage areas: Survey and classification *European Journal of Operational Research* **239** pp 297–312

[11] Wang S, Liu Z and Meng Q 2015 Segment-based alteration for container liner shipping network design *Transportation Research Part B: Methodological* **72** pp 128–145.

[12] Araujo E J, Chaves, A A, De Salles Neto L L and De Azevedo A T 2016 Pareto clustering search applied for 3D container ship loading plan problem *Expert Systems with Applications* **44** pp 50–57

[13] Santosa B and Willy P 2011 *Metoda Metaheuristik Konsep dan Implementasi* (Surabaya:Guna Widya)

A meta-heuristic method for solving scheduling problem: crow search algorithm

Antono Adhi[1*], Budi Santosa[1], Nurhadi Siswanto[1]
[1] Department of Industrial Engineering, Institut Teknologi Sepuluh Nopember, Kampus ITS Sukolilo-Surabaya 60111, Indonesia

*antonoadhi@yahoo.com

Abstract. Scheduling is one of the most important processes in an industry both in manufacturingand services. The scheduling process is the process of selecting resources to perform an operation on tasks. Resources can be machines, peoples, tasks, jobs or operations.. The selection of optimum sequence of jobs from a permutation is an essential issue in every research in scheduling problem. Optimum sequence becomes optimum solution to resolve scheduling problem. Scheduling problem becomes NP-hard problem since the number of job in the sequence is more than normal number can be processed by exact algorithm. In order to obtain optimum results, it needs a method with capability to solve complex scheduling problems in an acceptable time. Meta-heuristic is a method usually used to solve scheduling problem. The recently published method called Crow Search Algorithm (CSA) is adopted in this research to solve scheduling problem. CSA is an evolutionary meta-heuristic method which is based on the behavior in flocks of crow. The calculation result of CSA for solving scheduling problem is compared with other algorithms. From the comparison, it is found that CSA has better performance in term of optimum solution and time calculation than other algorithms.

1. Introduction

Scheduling is a process associated with allocation of sources to perform task within a period of time [1]-[3]. The business field that performs scheduling activities may include anything in both productions and services. Resources may include machines in a production floor, an airline's flight path, labor on a construction work, a computer processing unit, and another [3]. The tasks assigned to the resources may be operations on the production process, for examples, the process of flying and landing the plane, the work in the construction process, the execution of computer processes, and others [3]. Machines are scheduled to run customer order jobs. The limited flight path is set to serve the plane that will take off or land. Workers in a construction project are organized to perform certain tasks that have been prepared in a certain time scale. Computer processers regularly execute commands in a programming language, either serial or parallel.

Given the role of the scheduling process, the scheduling problem is crucial in an industrial system [4]. The role, besides relating to internal production process, also concerns with service to the customer. Failure to meet customer demand will result in a great loss in customer trust. To avoid such mistakes, companies need to schedule their activities on available resources as efficient as possible [5]. The main focus is to keep the scheduled time of completion in accordance with the delivery time promised to customers [6].

To achieve good results in scheduling, the process must have targets. The main target in the scheduling is to obtain optimal value on one or more goals [3]. Optimization technology in the production scheduling process significantly improves manufacturing facilities by reducing manufacturing conflicts, processing times, work-in processes, and increasing resource utilization on the production floor [7][8]. Improvements from the optimization process of the scheduling process will impact on subsequent benefits such as organizational effectiveness and customer satisfaction [5]. Businesses will become more competitive and survive in market competition [5].

To obtain an effective scheduling result, it needs the management's role to pursue it. This is because scheduling is the result of a decision-making [1][3]. To make a decision, management will set the goal of scheduling, gathering information, and finding the best alternative. The purpose of scheduling is to look for a sequence of machine processes and jobs that pass through a route to optimize some criteria such as total completion time or makespan [9].

To find the optimum total completion time is the most popular goal in many studies [10]. The main information required in the scheduling is obtained from the customer's order. Customer orders are translated into due date [3]. The best alternative is chosen from permutation sequence of job to get the optimum decision. The literature of research on scheduling optimization from year to year is mostly to examine the best method that can be used to find the sequence of jobs that exist on a permutation.

In the manufacturing industry environment, decision making on the scheduling process will have an effect on the next process decision making. Decision making on Material Requirements Planning (MRP) systems is strongly influenced by decision making on production scheduling processes [3]. After scheduling is established, the material required in the production process must be prepared. The level of optimization generated in decision making on scheduling will affect the level of optimization of the material procurement process.

Due to the importance of scheduling problems, in this paper, scheduling problems will be solved by a method for obtaining optimum scheduling. Several methods in many studies have been done to solve the problem in order to obtain the optimum scheduling.

Many approaches can be used to obtain optimum scheduling results, either through exact, heuristic, and meta-heuristic algorithms [11]. However, exact methods such as branch and bound or dynamic programming are only suitable for small-scale scheduling [5]. This is due to scheduling is including the NP-hard problem [11]. With the NP-hard problem, in complex scheduling, the amount of computation does not increase linearly, but in a factorial manner. Since the number of possible sequences of a permutation is a factorial calculation. The exact method will fail if the problem scale is too complex because of the large memory requirements and long computational time [5]. Therefore, it is needed a method that can accomplish the computation process on the scheduling calculation at a reasonable time, and obtain an acceptable optimization results.

Possible methods for the NP-hard scheduling process with solutions closer to the nearest optimum and shorter computing times are by heuristic or meta-heuristic methods [5][12]. Both computational calculations of time from heuristic or meta-heuristic methods are acceptable. While the results obtained close to the optimum.

Many variations of heuristic and meta-heuristic algorithms are used to find the optimum solution to solve scheduling problems. Heuristic algorithms such as Dispatching Rules, Local Search, and meta-heuristic algorithms such as Taboo Search (TS), Simulated Annealing (SA), Genetic Algorithm (GA), and Particle Swarm Optimization (PSO) are able to resolve scheduling problems with approximated optimal solutions with time computations that are considered fewer [5]. New algorithms of meta-heuristic methods are continually developed to look for better algorithms for scheduling problems or other field problems [13]. A novel meta-heuristics optimization algorithm is proposed by Askarzadeh [14]. The name of the method is Crow Search Algorithm (CSA). CSA is based on the social behavior of the crow. CSA is easier to implement and has fewer parameters to adjust in comparison with some other meta-heuristic methods [14]. CSA is quite powerful to search optimization problem but it is not implemented in scheduling problem yet.

The aim of this paper is to prove the effective searching ability of CSA in order to solve scheduling problem. The proposed approach uses CSA to assign the sequence of operations of each job on available capable machine.

2. Crow Search Algorithm

Crow is the most intelligent bird. They have a relatively larger brain volume compared to their large body [14]. Compared to the ratio of head and body, their brains are slightly smaller than humans [14]. Crows are able to recognize faces and give warnings when disturbances arrive [14]. Some principles of CSA according to Askarzadeh [14] are as follows:

- crows live in groups;
- crows remember the position of the hiding place;
- crows join each other while stealing food;
- crows keep the possibility of snatching the food.

It is assumed that there is an environmental dimension d anda number of crows, N. The number of crows, N and the crow position of each iteration $iter$ is denoted by the vectors $x^{i,iter}$ (=1, 2, ..., N; $iter$ = 1, 2, ..., $iter_{max}$). $x^{i,iter} = [x_1^{i,iter}, x_2^{i,iter}, ..., x_d^{i,iter}]$ and $iter_{max}$ is the maximum iteration. Each crow has a memory of its hiding location. At each iteration, the position of the hiding place is denoted by $m^{i,iter}$. This is the best position crow i. In it neighborhood, every crow would find it best position to be a hiding place to store food.

Assuming on an iteration $iter$, a crow will visit its hiding place, $m^{j,iter}$. Then the crow i follows the crow j in the crow j's hideout location. Possible conditions are as follows:

- Crow j does not know that the crow i follows it. Therefore the crow i approaches the hiding place of crow j. The new crow i's position is as follows:

  $$x^{i,iter+1} = x^{i,iter} + r_1 \times fl^{i,iter} \times (m^{j,iter} - x^{i,iter}) \tag{1}$$

  r_1 is a random number between 0 and 1. $fl^{i,iter}$ is the crow's long fly on iteration $iter$. Figure 1 depicts the condition scheme 1 and the effect of fl on search ability. The small number fl directs the crow i to the local search (about $x^{i,iter}$) and the large number will cause it to global search (away from $x^{i,iter}$).

  Figure 1 (a) depicts the condition if fl is less than 1. The next crow position lies between $x^{i,iter}$ and $m^{i,iter}$. Figure 1 (b) indicates if fl is more than 1. The next crow i positions on the path outside $m^{i,iter}$.
- Crow j knows if crow i follow it. Therefore in order to protect the food storage as not to be ransacked, crows j will fool by moving to a new position in the search space.

This conditions can be expressed with the following calculation:

$$x^{i,iter+1} = \begin{cases} x^{i,iter} + r_i \times fl^{i,iter} \times \left(m^{j,iter} - x^{i,iter}\right) & r_j \geq AP^{j,iter} \\ a \quad random \quad position, \quad otherswise \end{cases} \tag{2}$$

IOP Conf. Series: Materials Science and Engineering **337** (2018) 012003 doi:10.1088/1757-899X/337/1/012003

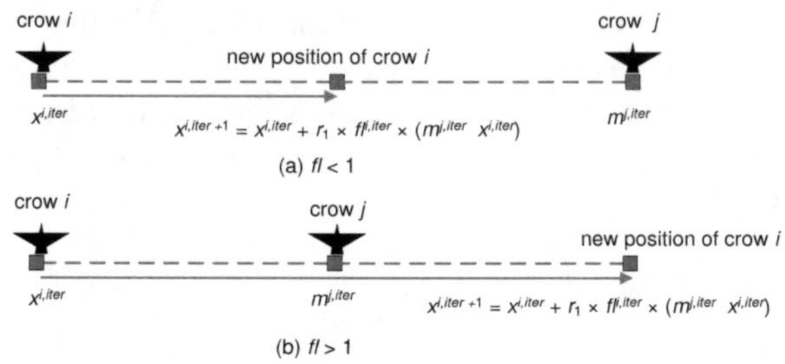

Figure 1. The schematic of CSA state and the effect of *fl* on the search capability.

3. Experiment

The computational aims to analyze the performance of CSA in order to minimize scheduling problems. The CSA algorithm is tested on a sets of problem instance from Kacem [15]. Kacem's data set contains instances ranging from 8 × 8 and 10 × 10 whose scale (n × m, n: number of jobs, m: number of machines). The type of scheduling problem obtained from Kacem's data is job shop. There is no certain pattern of operations to execute every job.

Some parameters of CSA are set to obtain global optimum. Upper bound is set to 6 while lower bound is set to -6. The number of crows is set to 10. Parameter of awareness probability (*AP*) controls intensification and diversification. By decrease of *AP*, CSA tends to conduct the search on local region in order to find current good solution. AP is set to 0.1. *fl* denotes to flight length of crow is set to 2. A set of calculation is iterated 10.000 times. Calculation of CSA for every data of Kacem is iterated for several times.

4. Result

From calculation of CSA, it is obtained optimum makespan 14 for problem size 8 × 8. For problem size 10 × 10 of Kacem's instances CSA finished it in 7 of makespan. From any calculations give different sequences of optimum makespan. The global optimum makespan obtained from greedy method are 14 for problem size 8 × 8 and 7 for problem size 10 × 10 of Kacem's instances. The results of problem size 8 × 8 sometimes are trapped in local optimum that gives solution 15 of makespan, while the results of problem size 10 × 10 always give solution 7 of makespan.

Table 1 compares the results of CSA with result of Approach by Localization + Controlled Genetic Algorithm (AL + CGA) [15], Particle Swarm Optimization + Simulated Annealing (PSO + SA) [16], Parallel Variable Neighborhood Search Algorithm (PVNS) [17], hybrid Genetic Algorithm (hGA) [18] and Quantum-behaved Particle Swarm Optimization (QPSO) [5].

Tabel 1. Makespan of Kacem Instances.

Problem Size	AL + GA	PSO + SA	PVNS	hGA	QPSO	CSA
8 × 8	15	15	14	14	14	14
10 × 10	7	7	7	7	7	7

The last column of Table 1 signifies the best makespan obtained from CSA. Problem size 8 × 8 is performed by sequence 3, 5, 4, 2, 7, 8, 6, 1 of jobs and Problem size 10 × 10 is performed by sequence 8, 9, 2, 5, 4, 6, 7, 1, 10, 3 of jobs. Gantt chart of the obtained solution by CSA method for problem 8 × 8 is illustrated in Figure 2 and problem 10 × 10 is illustrated in figure 3.

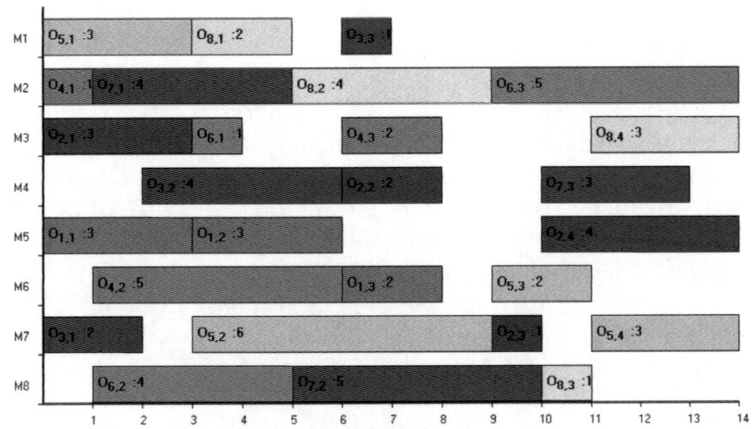

Figure 2. Gantt chart of 8 × 8 problem.

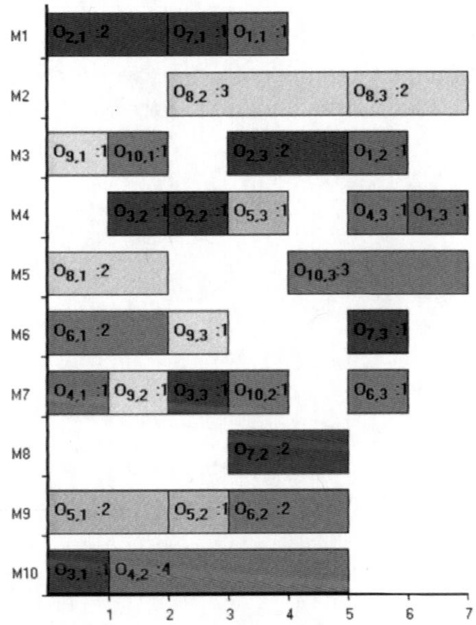

Figure 3. Gantt chart of 10 × 10 problem.

5. Conclusion

CSA is found to be a good problem solving technique for scheduling problem. The algorithm of CSA is applied sets of problem instances from Kacem [15]. The result indicates that CSA produces either better solution or same as compared to the best known solutions in the literature. CSA could be a good method for giving optimum solution of scheduling problem. Due to it's simple algorithm, CSA gives an advantage in time period of calculation.

Some calculations of CSA sometime are trapped in local optimum. For suggestion, in next research, CSA could be modified to resolve this problem. Quantum technique could be an alternative to escape from local optimum. This technique will observe the result of calculations. If there is no different result in several iterations, the position of crows will be regenerated.

CSA can be implemented to solve either continuous or discrete problems. To solve other discrete problem as scheduling problem, for suggestion, CSA could be used to solve traveling salesman problem.

6. References

[1] Baker K R 1974 *John Wiley & Sons* New York
[2] Morton T E and Pentico D W 1993 *John Wiley & Sons* New York
[3] Pinedo M L 2008 *Springer* Third Edition
[4] Kaplanoglu V 2016 *Expert Systems With Applications* **45** 71–84
[5] Singh M R and Mahapatra S S 2016 *Computers & Industrial Engineering* **93** 36–44
[6] Kuhpfahl J and Bierwirth C 2016 *Computers & Operations Research* **66** 44–57
[7] Li X and Gao L *Int. J. Production Economics* **74** 93–110
[8] Kurdi M *Computers & Operations Research* **67** 132–142
[9] Bai D, Zhang Z, and Zhang Q 2016 *Computers & Operations Research* **67** 207–215
[10] Koulamas C and Panwalkar S S 2016 *European Journal of Operational Research* **000** 1–5
[11] Lei D and Guo X 2016 *Computers & Operations Research* **65** 76–82
[12] Ding J Y, Song S, Wu C 2016 *European Journal of Operational Research* **248** 758–771
[13] Abedinnia H, Glock C H, and Brill A 2016 *Computers & Operations Research* **74** 165–174
[14] Askarzadeh A 2016 *Computers and Structures* **169** 1–12
[15] Kacem I, Hammadi S, and Borne P 2002 *IEEE Transactions on Systems, Man, and Cybernetics* **32** 1–13
[16] Xia W J and Wu Z M 2006 *Int. Journal of Advanced Manufacturing Technology* **29** 360–366
[17] Yazdani M, Amiri M, and Zandieh M *Expert Systems with Applications* **37** 678–687
[18] Gao J, Sun L Y, and Gen M 2008 *Computers and Operations Research* **35** 2892–2907.

IOP Conf. Series: Materials Science and Engineering **337** (2018) 012004 doi:10.1088/1757-899X/337/1/012004

Comparing genetic algorithm and particle swarm optimization for solving capacitated vehicle routing problem

T Iswari[1*], A M S Asih[2]
[1]Department of Industrial Engineering, Parahyangan Catholic University, Bandung, Indonesia
[2]Department of Industrial Engineering, Universitas Gadjah Mada, Yogyakarta, Indonesia

*titiiswari@unpar.ac.id

Abstract. In the logistics system, transportation plays an important role to connect every element in the supply chain, but it can produces the greatest cost. Therefore, it is important to make the transportation costs as minimum as possible. Reducing the transportation cost can be done in several ways. One of the ways to minimizing the transportation cost is by optimizing the routing of its vehicles. It refers to Vehicle Routing Problem (VRP). The most common type of VRP is Capacitated Vehicle Routing Problem (CVRP). In CVRP, the vehicles have their own capacity and the total demands from the customer should not exceed the capacity of the vehicle. CVRP belongs to the class of NP-hard problems. These NP-hard problems make it more complex to solve such that exact algorithms become highly time-consuming with the increases in problem sizes. Thus, for large-scale problem instances, as typically found in industrial applications, finding an optimal solution is not practicable. Therefore, this paper uses two kinds of metaheuristics approach to solving CVRP. Those are Genetic Algorithm and Particle Swarm Optimization. This paper compares the results of both algorithms and see the performance of each algorithm. The results show that both algorithms perform well in solving CVRP but still needs to be improved. From algorithm testing and numerical example, Genetic Algorithm yields a better solution than Particle Swarm Optimization in total distance travelled.

1. Introduction

Transportation planning, as one of the main areas of supply chain management, has become increasingly challenging for companies in terms of controlling cost because the fuel cost rises continuously [1]. Vehicle Routing Problem is a central issue in transportation planning. Since Dantzig and Ramser introduce VRP in 1959 [2], VRP has become one of the most important and widely studied topics in the area of combinatorial optimization [3][4]. VRP can be described as the process of determining optimal routes from one or several depots to a number of geographically scattered customers, subject to side constraints [5]. The most common VRP is CVRP (Capacitated Vehicle Routing Problem). The implementation of VRP in real-life applications are blood pick-up routing, waste collection, street cleaning, school bus routing, food distribution, etc. Therefore, this topic has attracted the attention of numerous researchers.

The basic VRP and basically all of its variants like CVRP, VRPTW, VRPDP, and TDVRP belong to the class of NP-hard problems. These VRP extensions make it even more complex to solve with more additional variables such that exact algorithms become high time-consuming with the increases in problem size. Thus, for large-scale problem instances typically found in industrial applications, finding an optimal solution is not practicable. Therefore, different types of heuristics, ranging from classical heuristics to sophisticated metaheuristics can produce solutions with good quality. This paper studies a comparison of two population-based metaheuristic methods : Genetic Algorithm and Particle Swarm Optimization to see which one is better to solve the Vehicle Routing Problem.

2. Literature Review

The well-known Vehicle Routing Problem was firstly introduced by Dantzig and Ramser in 1959 [2]. They tackled the optimum routing of a fleet of gasoline delivery trucks between a bulk terminal and a large number of service stations supplied by the terminal. They tried to find a way to assign stations to trucks in such a manner that the station demands are satisfied and the total mileage covered by the fleet is minimized.

Since its introduction by Dantzig and Ramser in 1959, VRP has been developed by many researchers. Lin et al [6] found that numerous types of traditional VRP were identified, including CVRP, time-dependent VRP, the pickup and delivery problem, multi-depot VRP, stochastic VRP, the location routing problem, periodic VRP (PVRP), dynamic VRP, inventory routing problem, fleet size and mix VRP, generalized VRP, multi-compartment VRP, site-dependent VRP, split delivery VRP, fuzzy VRP, OVRP, VRP with loading constraints, and multi-echelon VRP.

CVRP is the most common variant of VRP. Lin et al [7] define CVRP as a problems in which a set of routes for vehicles based at one or several depots have to be determined by the number of geographically dispersed customers. Each vehicle has the maximal loading capacity.The objective is to minimize the total cost (i.e. a weighted function of the number of vehicles and the distance traveled by the vehicles) to serve a set of customers. The customers have their known demands in which the route must be designed in such a way that each customer is visited at most once by one vehicle only.

Laporte [8] gave good overviews on solution methodologies for VRP. Those are exact algorithms, classical heuristics, and metaheuristics. From his research, given that VRP and its variants are NP-hard problems, classical heuristics and metaheuristics are considered more efficient and practical than the exact solution in solving large-scale problems. Example of classical heuristic methods is saving algorithm that developed by Clarke and Wright [9], while the metaheuristic are Ant Colony Optimization [10][11][12], Genetic Algorithm [13], Simulated Annealing [14], Particle Swarm Optimization [12][15] and Tabu Search [16].

Based on literature review, this research employs metaheuristic methods for solving the Vehicle Routing Problem. It is because VRP is a complex problem and it is not recommended to use exact method for real cases in large problems due to high computational time. This paper will compare two metaheuristic methods, those are Genetic Algorithm (GA) and Particle Swarm Optimization (PSO). GA and PSO were chosen because both methods are population-based methods. In the population-based method, the solution will move from a population to another population with some evaluation processes. GA uses a genetic scheme such as selection, mutation, and crossover, while PSO uses swarm behavior such as updating position and velocity for the evaluation process.

3. Problem Description

Laporte [5] defines the classical VRP in graph-theoretic terms as follows: Let $G = (V, A)$ be a graph where $V=\{1,\ldots,n\}$ is a set of vertices representing clients with known and deterministic demands. V_1 represents the single depot [5]. A is the a set of arcs that interconnect all clients. The weight (i.e. cost) of each arc between node i and j is denoted by c_{ij}. The objective is to minimize the total distance traveled for each of K vehicles available, subject to the following constraints:

1. Each customer is visited exactly once.
2. Each route starts and ends at the depot.
3. Other side constraints to be satisfied.

In CVRP, the vehicles have their own capacities and the total demands from the customer should not exceed the capacity of the vehicles. Below is the mathematical model of CVRP.

Notations :
N = Set of nodes $\{0,1,2,\ldots,|N|\}$
K = Set of vehicles $\{1,2,\ldots,|K|\}$
t_{ij} = Travel time from node i to node j
C = Vehicle capacity
d_{ij} = Distance from node i to node j
p_j = Demand at node j

Decision variable:
x_{ijk} = 1 if vehicle k travels directly from node i to node j; 0 otherwise.

Minimize

$$\sum_{k \in K} \sum_{i \in N, i \neq j} \sum_{j \in N, j \neq i} d_{ij} x_{ijk} \qquad (1)$$

Subject to:

$$\sum_{k \in K} \sum_{i \in N'} x_{ijk} = 1 \qquad \forall j \in N \qquad (2)$$

$$\sum_{i \in N'} x_{0ik} \leq 1 \qquad \forall k \in K \qquad (3)$$

$$\sum_{i \in N, i \neq j} x_{ilk} = \sum_{j \in N, j \neq i} x_{ljk} \qquad \forall l \in N, k \in K \qquad (4)$$

$$\sum_{i \in N, i \neq j} \sum_{j \in N, j \neq i} p_j x_{ijk} \leq C \qquad \forall k \in K \qquad (5)$$

The objective function (1) is for minimizing the total travel distance of vehicles. Each node can only be visited once defined by constraint (2), and each vehicle is used at most once as defined by constraints (3). Constraint (4) is a flow constraint. Constraint (5) guarantees that the demand picked up by a vehicle does not exceed the vehicle's capacity.

4. Genetic Algorithm and Particle Swarm Optimization

4.1 Genetic Algorithm (GA)

The first method used for solving CVRP is Genetic Algorithm (GA). GA is a metaheuristic method based on the genetic scheme in living creatures. It uses some evolutionary scheme like selection, mutation, and crossover. The flowchart of GA is shown in Figure 1. The steps of Genetic Algorithm for solving the vehicle routing is written below:

1. Determine the parameter values that are used in GA, those are population size, mutation rate, crossover rate, and the maximum number of iteration.
2. Generate random initial population with the number of chromosomes appropriate with population size. The solution represented by a chromosome consisting a permutation of n customers denoted by the set *{1,2,...,n}* and *Ndummy* zeros. Here, *Ndummy* is a variable that denotes the number of dummy depots for constructing a new route in addition to the route constraint. For example, the first chromosome is *{0,1,2,3,0}*, meaning that the route is from depot to customer 1, then customer 1 to customer 2, then customer 2 to customer 3, and customer 3 will back to the depot.
3. Calculate the fitness value of each chromosome by calculating the total distance from each route and choose the chromosome with the minimum total distance as the best chromosome.

4. After calculate fitness value, then sort the population based on their fitness value. The sorting procedure is from chromosome that has best fitness value to the chromosome that has worst fitness value.
5. Then do selection process based on truncation method. In the truncation method, with the population that has been sorted, keep some of the best chromosomes so they can enter next iteration. This mechanism called elitism. Then, for other chromosomes, choose two chromosomes as a parent and do crossover and mutation.
6. After new population is obtained, repeat step 3 to 5 until the algorithm reaches the stopping criteria. In this research, the stopping criteria is maximum number of iteration.

4.2 Particle Swarm Optimization (PSO)

The second method used for solving CVRP is Particle Swarm Optimization (PSO). PSO is a metaheuristic method based on swarm behavior by updating the position and velocity of each individual. The flowchart of PSO is shown in Figure 2. The steps for solving VRP using PSO is written below :

1. Determine the first initialization as the number of particle, initial routes, c1, c2, maximum number of iteration, initial position and initial velocity
2. Evaluate the total distance of every particle
3. Choose the particle with the smallest total distance and use this particle into Gbest. For every particle, use the initial total distance as Pbest
4. Repeat the following steps until the stopping criteria is reached. In this research, the stopping criteria is maximum number of iteration.
   a. Use Pbest and Gbest for updating the velocity of every particle with Equation (6). With the new velocity then update the position of every particle use Equation (7).

$$V_i(t) = V_i(t-1) + c_1 r_1\big(Pbest_i - X_i(t-1)\big) + c_2 r_2\big(Gbest_i - X_i(t-1)\big) \quad (6)$$

$$X_i(t) = V_i(t) + X_i(t-1) \quad (7)$$

   Where,
   X = Particle position
   V = Particle velocity
   i = *particle index*
   t = t^{th} iteration
   $Pbest_i$ = *local* best from i^{th} particle
   $Gbest$ = *global* best from all swarm
   c_1, c_2 = learning factor.
   r_1, r_2 = random number

   b. Evaluate the total distance of every particle
   c. Choose the particle with the smallest total distance and make this particle into Gbest. For every particle, use the initial total distance as Pbest
   d. Check the stopping criteria. If stopping criteria is reached then stop, otherwise go to step 'a' again.

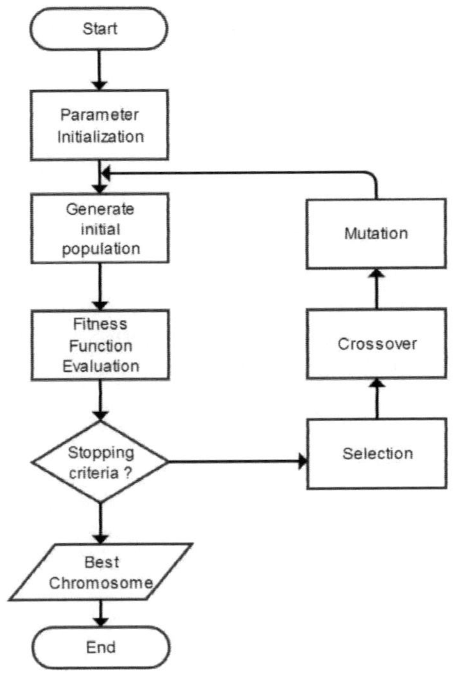

Figure 1. Genetic Algorithm Flowchart.

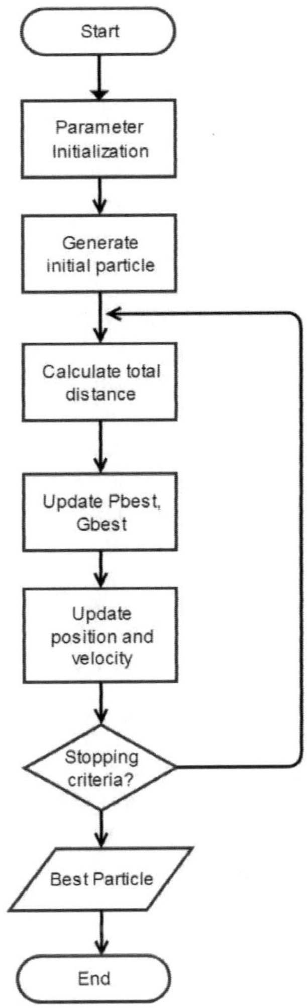

Figure 2. PSO Flowchart.

5. Computational Results

5.1 Algorithm Testing

A computational study is conducted to evaluate the performance of Genetic Algorithm and Particle Swarm Optimization to solve the Capacitated Vehicle Routing Problem. The computational study is divided to algorithm testing and numerical example on a real case. First, this research evaluates solution quality of GA and PSO by test it on a small-medium data set from Augerat [17] to see the performance of both algorithms compare to the exact solution. The data sets that are used from Augerat et al are small-medium CVRP datasets that have 31-38 instance. Parameter used for Genetic Algorithm are as follows maximum number of iteration = 50000, population size = 80, mutation rate = 0,1, crossover rate = 0,8 and for PSO are maximum number of iteration = 50000, number of particle = 80, c1=2, c2=1. Table 1 shows the comparison of Genetic Algorithm and Particle Swarm Optimization results with the optimal solution from exact solution.

Table 1. Comparison Result for Small-Medium Instances

No	Instance	Optimal Solution	GA	PSO	diff GA	diff PSO
1	31	672	699	703	0,041	0,046
2	34	788	801	835	0,016	0,060
3	35	955	998	1.081	0,045	0,132
4	38	805	829	891	0,030	0,107
mean		805	831,70	877,55	0,033	0,086

The result in Table 1 shows that the average differences between these two method and the optimal solution are 0,033 and 0,086. It means that the performance of proposed algorithm is competitive because the differences with the optimal solution are relatively small. From the comparison results, we can see that GA obtain a better solution than PSO solution in total distance traveled.

5.2 Numerical Example in a Real Case

After conducting the algorithm testing, this research compares the performance of Genetic Algorithm and Particle Swarm Optimization in a real case of Vehicle Routing Problem that belongs to a medium-large case of Vehicle Routing Problem. The case used in this research is distribution routing of Chainstore A in Yogyakarta City, Indonesia. Chainstore A has 1 depot located in West Ringroad Yogyakarta and 62 retailers scattered within Yogyakarta City region. Each retailer has its own demand and the demand is fulfilled by the depot. The vehicle for distribution has its own capacity and the demand carried by the vehicle must not exceed its capacity. The algorithm has been generated ten times, then we calculate a minimum, mean and maximum value for each algorithm. This research compares GA results PSO results in terms of both solution quality and computational time. Table 2 shows the comparison between GA and PSO in the real case.

Table 2. Comparison Result for Real Case

i	TD	Time	TD	Time	diff
1	401,54	296,98	452,64	215,91	-0,127
2	402,43	298,99	451,69	224,42	-0,122
3	418,89	298,26	437,45	222,99	-0,044
4	412,65	297,34	452,14	222,71	-0,096
5	404,54	299,18	408,59	219,04	-0,010
6	407,25	296,07	447,44	216,40	-0,099
7	413,89	300,63	452,10	216,05	-0,092
8	407,64	305,31	475,20	210,76	-0,166
9	414,05	307,99	448,25	222,27	-0,083
10	402,20	305,17	451,54	218,29	-0,123
min	401,54	296,07	408,59	210,76	-0,166
mean	408,51	300,59	447,70	218,89	-0,096
max	418,89	307,99	475,2	224,42	-0,010

Based on Table 2, GA provides a better solution than PSO in the average of total distance traveled. The average differences is -0,096, the negative sign means that GA produces smaller value than PSO. It also is shown by the minimum, mean and maximum value, the results fro GA is smaller than PSO results. It indicates that GA can produce better solutions than PSO. However, for the computational time, GA needs longer time to yields the optimal solution compared to PSO.

6. Conclusion

This study proposed two metaheuristic methods named Genetic Algorithm (GA) and Particle Swarm Optimization (PSO) to solve the Capacitated Vehicle Routing Problem (CVRP). We tested the proposed algorithm on a small-medium data set from Augerat [17] to see the performance of both algorithms compare to the exact solution. The results indicate that the performance of proposed algorithm is competitive because the gaps with the optimal solution are relatively small, but it still has room for improvement. This research also compares the performance of Genetic Algorithm and Particle Swarm Optimization in a real case of Vehicle Routing Problem that is a distribution routing of Chainstore A in Yogyakarta City, Indonesia. From the results, we can know that GA can produce better solutions than PSO. However, in term of computational time, GA needs longer time to achieve the optimal solution compared to PSO. Researchers in the future may focus on developing more effective algorithms for solving medium and large CVRP instances.

7. References

[1] Halim C and Yu V F 2016 *Minimum Cost Vertex-Disjoint Path Cover Problem Electronic Thesis and Dissertation (ETD) National Taiwan University of Science and Technology*

[2] Dantzig G B and Ramser J H 1959 *The Truck Dispatching Problem Management Science* **6**(1) pp 80-91

[3] Cordeau J -F, Laporte G, Savelsbergh M W P and Vigo D 2007 *Vehicle Routing* **14** pp 367-428

[4] Toth P and Vigo D 2002 The Vehicle Routing Problem *SIAM Monographs on Discrete Mathematics and Applications* Philadelphia

[5] Laporte G 1992 The Vehicle Routing Problem: An overview of exact and approximate algorithms *European Journal of Operational Research* **59** (3) 345-358.

[6] Lin C, Choy K L, Ho G T S, Chung S H and Lam H Y 2014 Survey of Green Vehicle Routing Problem: Past and future trends *Expert Systems with Applications* **41** pp 1118-1138

[7] Lin S -W, Ying K -C, Lee Z –J and His F -H 2006 Applying Simulated annealing Approach for Capacitated Vehicle Routing Problems *2006 IEEE Int.Conference on Systems, Man, and Cybernetics* pp 639-644

[8] Laporte G 2007 What you should know about the vehicle routing problem *Naval Research Logistics* **54** pp 811-819

[9] Clarke G and Wright J W 1964 Scheduling of Vehicles from a Central Depot to a Number of Delivery Points *Operations Research* **12** pp 568-581

[10] Bell J E and McMullen P R 2004 Ant colony optimization techniques for the vehicle routing problem *Advanced Engineering Informatics* **18** pp 41–48

[11] Bullnheimer B, Hartl R F and Strauss C 1999 An improved ant System algorithm for thevehicle Routing Problem *Annals of operations research* **89** pp 319-328

[12] Fajar A 2011 Optimasi Rute Distribusi Benda POS Berbasis TSP dengan Ant colony system dan Particle Swarm Optimization *Bachelor thesis* Program Studi Teknik Industri Jurusan Teknik Mesin dan Industri Universitas Gadjah Mada.

[13] Prins C 2004 A simple and efective evolutionary algorithm for the vehicle routing problem *Computer and Operation Research* **31** pp 1985-2002

[14] Breedam A V 1995 Improvement heuristics for the Vehicle Routing Problem based on Simulated annealing *European Journal of Operational Research* **86** pp 480-490

[15] Dewi P K 2010 Optimasi Rute Distribusi Benda POS Berbasis TSP dengan Particle Swarm Optimization *Bachelor thesis* Program Studi Teknik Industri Jurusan Teknik Mesin dan Industri Universitas Gadjah Mada.

[16] Badeau P, Guertin F, Gendreau M, Potvin J and Taillard E 1997 A Parallel Tabu Search Heuristic For The Vehicle Routing Problem With Time Windows *Transpn Res.*-C **5**(2) pp 109-122

[17] Augerat P 1995 *Capacitated VRP Instances* Available via http://neo.lcc.uma.es/vrp/vrp-instances/capacitated-vrp-instances/ (Last access 8 June, 2017)

IOP Conf. Series: Materials Science and Engineering **337** (2018) 012005 doi:10.1088/1757-899X/337/1/012005

A Hybrid Symbiotic Organisms Search Algorithm with Variable Neighbourhood Search for Solving Symmetric and Asymmetric Traveling Salesman Problem

M I H Umam[1,*] and B Santosa[1]
[1]Departement of Industrial Engineering, Institut Teknologi Sepuluh Nopember Kampus ITS Sukolilo-Surabaya 60111, Indonesia

*ISNAINI.30@gmail.com

Abstract. Combinatorial optimization has been frequently used to solve both problems in science, engineering, and commercial applications. One combinatorial problems in the field of transportation is to find a shortest travel route that can be taken from the initial point of departure to point of destination, as well as minimizing travel costs and travel time. When the distance from one (initial) node to another (destination) node is the same with the distance to travel back from destination to initial, this problems known to the Traveling Salesman Problem (TSP), otherwise it call as an Asymmetric Traveling Salesman Problem (ATSP). The most recent optimization techniques is Symbiotic Organisms Search (SOS). This paper discuss how to hybrid the SOS algorithm with variable neighborhoods search (SOS-VNS) that can be applied to solve the ATSP problem. The proposed mechanism to add the variable neighborhoods search as a local search is to generate the better initial solution and then we modify the phase of parasites with adapting mechanism of mutation. After modification, the performance of the algorithm SOS-VNS is evaluated with several data sets and then the results is compared with the best known solution and some algorithm such PSO algorithm and SOS original algorithm. The SOS-VNS algorithm shows better results based on convergence, divergence and computing time.

1. Introduction

Combinatorial optimization from the past has often been used to solve problems in science, engineering, and commercial applications. Since the Mathematical model began to be used to help solve problems in life real, algorithms have become important in recent years due to their flexibility to find a solution. One of the combinatorial problems in the field of transportation is looking for a shortest route that can be traveled from the starting point to the destination point, as well as minimize travel costs and travel time. Such problems such as this is known as Traveling Salesman Problem (TSP). Author **Error! Reference source not found.** suggests that TSP issues have become a concern over the past two decades and various approaches have been proposed for solve the problem. Then research to solve these TSP issues continue to be developed with modifications and with new approaches such as particle swarm optimization [2], Neural network for Euclidean TSP [3], 2-opt based approach on differential evolution [4]. Simulated annealing with greedy search [5], Ant colony with multidepot Multiple TSP [6], Evolutionary Algorithm [7], Lower tolerance branch and bound [8], and Genetic algorithm [9].

The asymmetric traveling salesman problem (ATSP) is a well-known combinatorial optimization problem, in which bi-directional distances between a pair of cities are not necessarily identical. An optimal solution to the ATSP minimizes the total distance traveled to visit all cities in a list [10][11]. The asymmetric traveling salesman problem is defined on a directed graph $G = (V, A)$, where $V = \{1, . . . , n\}$ is the vertex set, $A=\{(i, j) : i, j \in V\}$ is the arc set, and a non-symmetric cost matrix (cij) is defined on A [12]. For the simple equation [13]describe Let cij be the distance from city i to city j . If $cij = cji$ for all i and j , then this problem is known as the symmetric traveling salesman problem; otherwise, it is designated as an asymmetric traveling salesman problem (ATSP). According [14] and [15], the asymmetric TSP, which is more general in principle and has numerous real-world applications. In contrast to the symmetric TSP, the distance from one city to another in the ATSP may not necessarily be the same as the distance on the reverse direction. Asymmetric traveling salesman problem is one of a class of difficult problems in combinatorial optimization that is representative of a large number of scientific and engineering problems. ATSP and its variants are commonly used models for formulating many practical applications in manufacturing scheduling problem. For example, the scheduling problem in a discrete manufacturing is mainly concerned with how to determine the sequence of jobs so as to minimize the total set-up cost [16].

One of new metaheuristics is the Symbiotic Organism Search (SOS). This method was developed by Cheng and Prayogo [14] that is inspired by the biological relationship between organisms on ecological systems. Obstacles which is encountered in the application of SOS algorithm to solve the combinatorial problem is on the parasitic phase of SOS algorithm requires to create a parasitic organism using the dimensions of the objective function to be sought (Continuous function), whereas TSP and ATSP does not have dimension of purpose function for it is necessary there is another suitable mechanism to replace the mechanism in the parasitic phase. Authors [17] also suggested modifications should be made to the phase parasites by adding more specific search algorithms to overcome combinatorial problems. The proposed mechanism for replacing this parasitic phase is to use mutation mechanism. The mutation mechanism allows for bringing in new individuals who are not from the previous population with refers to the change or replacement of elements of the solution vector [18].

This paper focusses on solving TSP and ATSP using SOS algorithm by modifying the SOS algorithm to be applicable to TSP and ATSP. The modification is done by adding mutation mechanism to replace the mechanism at that parasitic phase to find the best route. SOS is hybridized with Variable Neighbourhood Search algorithm as a local search to generate the better initial leader.

2. Research Methods

2.1 *Symbiotic Organisms Search* (SOS)

SOS is one of new metaheuristics that is inspired by the interaction behavior seen among organisms in universe. The organism has the basic nature of being unable to live alone so have a dependence on other species to maintain sustainability. This dependence-based relationship is known as a symbiosis. There is some forms of symbiosis i.e symbiotic of mutualism, symbiotic of commensalism and symbiotic of parasitism. Here is the flow of SOS:

1. Initialization

$$\text{Ecosystem} = \text{rand x } ((\text{ub-lb}) + \text{lb}) \quad (1)$$

   Where :
   Rand= a number randomly between 0 to 1
   Ub = upper bound value
   Lb = lower bound value
2. REPEAT
   - The phase of mutualism
   - Commensalism Phase

- Phase Parasitism

3. UNTIL (Perform until termination criteria are met)

- Mutualism Phase

This SOS phase mimics mutualistic relationships between organisms in nature. SOS describes Xi is an organism compatible with members of the ecosystem. Other organisms Xj Then randomly selected from the ecosystem to interact with Xi. Both organisms Engage in mutualistic relationships with the goal of improving survival together with benefits in the ecosystem. New candidate solutions for Xi and Xj are calculated based on the mutualistic symbiosis between the organisms Xi and Xj, which are modeled on equations (2) and (3) follows [14]:

$$Xinew = Xi + rand(0,1) * (Xbest - Mutual\_Vector * BF1) \quad (2)$$
$$Xjnew = Xj + rand(0,1) * (Xbest - Mutual\_Vector * BF2) \quad (3)$$

$$Mutual\ Vector = \frac{Xi+Xj}{2} \quad (4)$$

- Commensalism Phase

Similar to the phase of mutualism, an organism Xj is randomly selected from the ecosystem to interact with Xi. In this case, the Xi organism tries to get advantage of interaction. However, organism Xj alone does not profit or suffer relationship. New candidate solutions from Xi are calculated according to the symbiosis between organisms commensal Xi and Xj, which are modeled in Eq. (2). Following the rules, Xi organism updated only if the new fittnes are better than the previous interaction fittnes.

$$Xinew = Xi + rand(-1,1) * (Xbest - Xj) \quad (5)$$

- Parasitism Phase

In SOS, the Xi organism is given a role similar to anopheles mosquito through the creation of an artificial parasite called " Parasite Vector ". Parasite Vector is made in space search by duplicating the Xi organism, then modify selected at random dimensions using random numbers. Organism Xj is randomly selected from the ecosystem and serves as the host for vector parasites. Parasite Vector tries to replacing Xj in the ecosystem. Both organisms are then evaluated to measure their fitness value. If Parasite Vector has a better fitness value, it will kill the organism Xj and assume its position in the ecosystem. If the fitness value of Xj is better, Xj will have immunity from parasites and Parasite Vector will no longer can live in the ecosystem.

- Mutation on TSP

Mutations are possible to elicit new non-originating individuals of cross breeding results. Mutation refers to the change of order or replacement of elements from solution vector (on the TSP problem) by generating a new value (function optimization). This process is done by choosing which individual to mutate with using random numbers. Individuals who are selected then mutate by placing a sequence of routes from the individual. Mutation can be swap, slide and flip .

2.2 *Variable Neighbourhood Search* (VNS)

According to [19], VNS's basic idea is to change the environment in searching for better solutions. VNS starts from a descent method to reach the local minimum, then investigate systematically or randomly, causing the environment to be further away from this solution. Each time, one or more points in the current environment is used as the initial solution for local descent. One jump point from the current solution as a new reference if and only if a better solution is found. VNS is not like the trajectory method on Simulated Annealing or Taboo Search and does not include prohibited steps in the process. Although simplicity is more specific.

3. Discussion

IOP Conf. Series: Materials Science and Engineering **337** (2018) 012005 doi:10.1088/1757-899X/337/1/012005

The SOS-VNS algorithm is a hybrid of SOS and VNS algorithm. The VNS is a local search algorithm widely used for solving combinatorial problem. The steps of the hybrid SOS-VNS algorithm are then described in algorithm shown in Figure 1.

Table 1. Data Set, Type, Number of city and best known solution.

Data Set	Type	Number of City	Best Known Solution
Co04	Asymetric	4	21
SP11	Symetric	11	133
UK12	Symetric	12	1733
LAU15	Symetric	15	291
GR17	Symetric	17	2085
BR17	Asymetric	17	39
WG22	Symetric	22	842

Table 2. Result of computation.

Data set	PSO			SOS			SOS-VNS		
	Convergence	divergence	comp time	Convergence	divergence	comp time	convergence	Divergence	comp time
Co04	0	0	0,0149	0	0	0,0124	0	0	0
SP11	0,00861	77	10,48	0,0015	1	13,8	**0,0015**	**1**	**0,648**
UK12	0,0547	215	17,14	0,0084	67	14,18	**0,0051**	**6**	**6,489**
LAU15	0,29	204	39,91	0,016	24	16,83	**0,0109**	**12**	**6,363**
GR17	0,1876	570	115,97	0,01813	107	63,99	**0,0162**	**73**	**8,081**
BR17	0,1743	11	49,399	0,0512	7	12,453	**0,01025**	**2**	**11,975**
WG22	0,627	621	63,35	0,114	249	33,55	**0,0762**	**165**	**14,925**

Table 2 compares the running results of Particle Swarm Optimization (PSO), original Symbiotic organisms Search (SOS) and Symbiotic organisms Search with Variable Neighbourhood Search (SOS-VNS) on a set of bechmark TSP and ATSP in TSPLIB. Convergence is use to measure the ability of the algorithm to find result near the optimal solution. The smaller of convergence value shown the algorithm is able to find the near optimal solution. While divergence is used to see how much deviation generated from the algorithm. Smaller level of divergence shown the more robust algorithm. Computation time (comp time) is used to evaluated the efficientcy of optimization algorithm. All of algorithm are written in MATLAB code and limit of iteration of each is 50000 on Intel(R) Core(TM) i3 processor 2.27GHz.

Based on the results of 10 runs are carried out for each algorithm, overall SOS-VNS algorithm can be said to be superior than PSO and SOS original. Convergence value of the SOS-VNS algorithm is superior to all data sets. This proves that the SOS-VNS algorithm is capable of generating a set of solutions closer to the optimal solution than PSO and SOS original. In terms of SOS-VNS divergence it also shows that better than PSO and SOS original, with smaller divergence values than for all data sets. For computing time SOS-VNS also shown that it more eficient than PSO and SOS original.

4. Conclusion

In this paper, a novel and hybrid SOS with VNS is proposed as a new approach for solving symetric and asymetric TSP. This new algorithm is compared with PSO and the original SOS. The result show excellent results. Using convergence, divergence, and computation time, SOS-VNS is better than PSO and original SOS. From 7 data sets tested, SOS-VNS is able to solve all data sets.

Pseudocode SOS-VNS Algorithm for TSP

```
Input  : Number of population, maximum iteration
Output : Best Route, Total distance
Begin
Generate ecosystem, calculation initial distance,
Generate neighbourhood set
        Shake.  Random solution x'of Nk (x).
        Local search.
        Improve or not. if x" better than x*, set (x ← x").
calculation distance
set IT=1
Repeat
     while it < itmax
     |  for i = 1: ecosize; update best distance and best route
     |  |  Mutualism Phase
     |  |     If distance better than before
     |  |     | eco(i; j) ← x (new1;new2)
     |  |     end
     |  |  Commensalism Phase
     |  |     If distance better than before
     |  |     |  eco(i) ← x (new1)
     |  |     end
     |  |  Parasitism Phase
     |  |   Rand
     |  |     If rand < 0,33
     |  |      | Flip
     |  |      | Elseif rand > 0,67
     |  |      |      Swap
     |  |      | Else
     |  |      |    Slide
     |  |     end
     |  |    If distance better than before
     |  |    | eco(i) ← x (new1)
     |  |   end
     |  End
     |  It=it+1
     |  If it == itmax
     |  | break
     |  end
     end
```

Figure 1. *Pseudocode* Algorithm of SOS-VNS for TSP

References

[1] Moon C, Kim J, Choi G and Seo Y 2002 An Efficient Genetic Algorithm for the Traveling Salesman Problem with Precedence constraints *European Journal of Op Res.* **140** 606-617

[2] Shi X H, Liang Y C, Lee H P, Lu C, and Wang Q X 2007 Particle Swarm Optimization-based Algorithm for TSP and Generalized TSP *Information Process Letters* **103** 169-176

[3] Creput J C and Koukam A 2009 A Memetic Neural Network for The Euclidean Traveling Salesman Problem *Neurocomputing* **72** 1250-1264

[4] Chiang C W, Lee W P and Heh J S 2010 A 2-opt Based differential Evolution for Global Optimization *Applied Soft Computing* **10** 1200-1207

[5] Geng X, Chen Z, Yang W, Shi D and Zhao K 2011 Solving the Traveling Salesman Problem Based On An Adaptive Simulated Annealing Algorithm with Greedy Search *Applied Soft Computing* **11** 3680-3689
[6] Ghaufurian S and Javadian N 2011 An Ant Colony Algorithm for Solving Fixed Destination Multi-depot Multiple Traveling Salesman Problems *Applied Soft Computing* **11** 1256-1262
[7] Liao Y F, Yau D H and Chen C L 2012 Evolutionary Algorithm to Traveling Salesman Problem *Computers and Mathematics with Applications* **64** 788-797
[8] Germs R, Goldenngorin B and Turkensteen M 2012 Lower Tolerance-based Branch and Bound Algorithms for ATSP *Computers & Operations Research* **39** 291-298
[9] Rani K and Kumar V 2014 Solving traveling salesman problem using genetic algorithm based on heuristic crossover and mutation operator *Int Journal of Research in Engineering & Technology* **2** 27-33.
[10] Nagata Yand Soler D 2012 Expert Systems with Applications A new genetic algorithm for the asymmetric traveling salesman problem. *Expert Systems With Applications* **39** 8947–8953
[11] Xing Ã L, Chen Y, Yang K, Hou F, Shen X and Cai H 2008 A hybrid approach combining an improved genetic algorithm and optimization strategies for the asymmetric traveling salesman problem *Engineering Applications of Artificial Intelligence* **21** 1370–1380
[12] Öncan T, Altınel I K and Laporte G 2009 A comparative analysis of several asymmetric traveling salesman problem formulations. *Computers & Operation Research* **36** 637–654
[13] Sherali H D, Sarin S C and Tsai P 2006 A class of lifted path and flow-based formulations for the asymmetric traveling salesman problem with and without precedence constraints, *Discrete Optimization* **3** 20–32.
[14] Cheng M Y and Prayogo 2014 Symbiotic Organisms Search : A New Metaheuristic Optimization Algorithm *Computers and Structures* **139** 98-112.
[15] Ausiello G, Bonifaci V and Laura L 2008 The on-line asymmetric traveling salesman problem. *Journal of Descrete Algorithm* **6** 290–298.
[16] Bai J, Yang G, Chen Y, Hu L and Pan C 2013 A model induced max-min ant colony optimization for asymmetric traveling salesman problem *Applied Soft Computing Journal* **13** 1365–1375
[17] Ruskartina E, Vincent F Y, Santosa B and Perwira A A N R 2015 Symbiotic Organism Search (SOS) for Solving Capacited Vehicle Routing Problem *International Journal of Mecanical, Aerospace, Industrial, Mechatronic and Manufacturing Engineering* **9** 873-877
[18] Santosa B and Willy P 2011 *Metode Metaheristik Konsep dan Implementasi,* Guna Widya, Surabaya
[19] Alba E 2005 *Parallel Metaheuristics : A New Class of Algorithms.* Hoboken, New Jersey: John Wiley & Sons

IOP Conf. Series: Materials Science and Engineering **337** (2018) 012006 doi:10.1088/1757-899X/337/1/012006

Discrete particle swarm optimization to solve multi-objective limited-wait hybrid flow shop scheduling problem

B Santosa[1*] , N Siswanto[1] and Fiqihesa[1]
[1] Industrial Engineering Department, Institut Teknologi Sepuluh Nopember Kampus ITS Sukolilo-Surabaya 60111, Indonesia

*budi_s@ie.its.ac.id

Abstract. This paper proposes a discrete Particle Swam Optimization (PSO) to solve limited-wait hybrid flowshop scheduing problem with multi objectives. Flow shop schedulimg represents the condition when several machines are arranged in series and each job must be processed at each machine with same sequence. The objective functions are minimizing completion time (makespan), total tardiness time, and total machine idle time. Flow shop scheduling model always grows to cope with the real production system accurately. Since flow shop scheduling is a NP-Hard problem then the most suitable method to solve is metaheuristics. One of metaheuristics algorithm is Particle Swarm Optimization (PSO), an algorithm which is based on the behavior of a swarm. Originally, PSO was intended to solve continuous optimization problems. Since flow shop scheduling is a discrete optimization problem, then, we need to modify PSO to fit the problem. The modification is done by using probability transition matrix mechanism. While to handle multi objectives problem, we use Pareto Optimal (MPSO). The results of MPSO is better than the PSO because the MPSO solution set produced higher probability to find the optimal solution. Besides the MPSO solution set is closer to the optimal solution

1. Introduction

Flow shop is one of scheduling problems where the sequence is the same for all jobs. The original flow shop scheduling problem is a scheduling problem at a single production line (each process only has a machine). While the problem that occurs frequently in the real world is scheduling problem at a parallel production line (there is a process with more than one machine). That problem is one of flow shop scheduling problem development that is called hybrid flow shop scheduling problem.

Flow shop scheduling model is not only grouped based on the number of production line, but also based on the characteristic of the product. There are some type of products such as a product that can wait for being processed at the next machine and a product that cannot wait. The basic model of flow shop scheduling problem does not consider the characteristic of the product (can wait or not). Therefore, that basic model is appropriate for the product that can wait for a long time (unlimited waiting time). While, the model for product that cannot wait is called no-wait flow shop scheduling problem [1]. Besides, there is a product that can wait for being processed at the next operation with limited waiting time, such as casting process that must be done before the material getting hardened. When using no-wait flow shop scheduling model, the process are forced without idle time. The production sequence can be different if there is a little idle time in the process. The model that is suitable for this characteristic is called limited wait flow shop scheduling problem. There are not many research carried out in this area.

The main purpose of scheduling model development is finding the optimal production sequence. The optimal production sequence can be seen from some performance criteria, such as production time, tardiness time, and machine utility. Production time is minimized based on total time needed, without considering the job deadline. Tardiness time is minimized based on the positive difference between the job deadline and the job finishing time, without consider the total production time needed. Then, machine utility is maximized based on the minimum total idle time, without considering the total production time and tardiness time. There is a possibility for an objective function is not optimal when the other objective functions are optimized. Therefore, there is a trade-off when there are three objective functions. When we have trade-off among the criteria then we will have a multi-objective problem.

Scheduling problem for n-job m-process at parallel line production is a combinatorial problem that classified into Non Polynomial-hard (NP-Hard) problem. The best method used to solve NP-Hard Problem is heuristic [2] . There were many research that using metaheuristic method to solve many type of flow shop scheduling problems, single objective and multi-objective problems. For example, hybrid flow shop scheduling problem solved by exact method, heuristic, and metaheuristics [3]. In this paper, it does not consider limited wait constraint. There is a few research that consider limited wait constraint, i.e. [4] , [5] and , Su (2003) and [6].

Particle Swarm Optimization (PSO) is easy to use and also efficient [7]. Originally, PSO was developed to solve continuous optimization problems. Since scheduling is a discrete problem, we need to modify PSO to fit the problem. The modifications are done by using transition probability matrix. As for handling the multiobjective, we use pareto optimal. The rest of the paper is organized as follows. Section 2 presents the problem definition, and Section 3 describes the literature review. Section 4 discusses the mathematical model of the problem. In section 5, we explain the proposed algorithm, MPSO. In section 6, we describe the experiments, results and analysis. Section 7 concludes the paper.

2. Problem Definition

This paper focuses on solving the limited wait hybrid flow shop scheduling (LWHFS) by using Modified Particle Swarm Optimization. The objectives used in this paper are minimizing total production time (makespan), minimizing tardiness time (total tardiness), and maximizing machine utility (minimizing total machine idle time). Since the original PSO is to solve continuous optimizaton problem, here we modified PSO to fit the discrete problem and also develop it to handle the multi-objectve case.

3. Literature Review

3.1 Hybrid Flow Shop

Hybrid Flow Shop Scheduling (HFS) problem is one of Flow Shop Scheduling problem extension. There are many types of HFS problems based on the characteristic of the problem. Generally HFS problems have following characteristics:

- The number of process equals 2 or more.
- The number of machine at each process equals 1 or more and the number of process that have more than 1 machine is minimum one.
- Each job processed through the same sequence, but there is a possibility that some jobs are not processed at some operations.

Research on flow shop scheduling problem usually imposes an assumption that each job is processed at all of operations. In some real production problem, sometime there is a missing operation. For example is the assembly process at stainless steel factory [8]. At that assembly process, some joint types do not need pre-processing, such that there are some jobs that do not need pre-processing. In this case, we refer to as hybrid flow shop scheduling problem because the job skip some stage without changing the others production sequence.

Limited wait is a constraint that restricts the waiting time between the operations. Waiting time means a reasonably long time when the product waits to be processed at the next operation. For example, the fabrication process of semiconductor wafer. The waiting time after processed at furnace tubes are limited to avoid particle absorption from the air [5]. That waiting time is counted from the product finished at a process until that product start being processed at the next operation. Because of this limitation, the limited wait constraint (equation 1) is adding to the main HFS model.

$$S_{(i+1,j)} - C_{ij} \leq U \quad (1)$$

where S_{ij} is a start time of job j at operation i, C_{ij} is completion time of job j at operation i, and U is a limitation of waiting time. The characteristics of HFS problem we solved in this paper are :

- The number of machine at each process is more than or equal to one.
- The waiting time for each job between the process are limited (limited wait constraints).
- Some job may not be processed at some operations, but the sequence of the other process are same (missing operations).

3.2. Particle Swarm Optimization

Particle Swarm Optimization (PSO) is developed based on the behavior of the swarm (bird or fish). The individual movement is influenced by its individual (cognitif) and social (swarm) movements [9]. Each individual (particle) is charaterized by velocity and position. Updating the velocity (v) and the position (x) of each particle based on its initial position, initial velocity, Gbest, and Pbest is done aording the following formula:

$$V_t = V_{t-1} + c_1 r_1 (Pbest - X_t) + c_2 r_2 (Gbest - X_t) \quad (2)$$

$$X_{t+1} = X_t + V_t \quad (3)$$

where Pbest is the best position of each particle among iterations, Gbest is the best of Pbest. PSO is suitable used to solve continuous problem, because the value of position and velocity of all of particles are continuous numbers. The disadvantage of PSO is the possibility to be trapped at local optimal. The solution can be trapped at local optimal because of the updating solution only based on the best solution from previous iteration and the algorithm does not have a mechanism to jump out from the local optimal. The modification by changing the updating mechanism (based on some best solution) can reduce the possibility to be trapped at local optimal. The approach that can save some best solutions is pareto optimal.

3.3. Multi Objectives

Pareto optimal is an approach based on the dominance concept. This approach used to find a solution set that closest to the optimal solution (pareto front). The number of the solution set's member depend on the number of non-dominated solution that can be found. The most suitable solution for the real condition can be chosen from the solution set. Pareto optimal insure that each criteria or objective function get a same treatment, because there is no priority from each criteria [7][. This approach have been proven effective for solving multi-objective problem which each criteria have the same weight.

Non-dominated solution is solutions that minimum have a better criteria than the others solution. When all of criterias are better than the others solution, the solution is a dominance solution. When all of criterias are worst than the others solution, the solution is a dominated solution. Non-dominated solutions will be added to the solution set, while the dominated solutions eliminated from the solution set and replace by the dominance solution. When there are some non-dominated solution at the solution set, a mechanism is needed to choose which is the best solution for generate new solutions. That mechanism is crowding distance.

Crowding distance used to measure the distance of a solution from other solutions. The bigger crowding distance will produce a better diversity [10]. The more far distance between the solution means that the solution set has better diversity and dispersion. When the solution dispersion is increasing, the probability to find the optimal solution is increasing too. Because of that, the solution with biggest crowding distance will be chosen. The steps to calculate crowding distance are shown below this [10]:

1. Sorting the solution based on each criteria value.
2. Based on the ranking at each criteria, calculate the distance for each criteria and each solution. The distance calculate based on this below equation.
$$d_{k,m} = \frac{f_{k+1,m} - f_{k-1,m}}{f_m^{max} - f_m^{min}} \tag{4}$$
where k is the solution index, n is the number of solution set member, and m is the criteria index. The distance value of the first and the last solution are infinite so that solution always chosen [11]. The first and the last solution are the limitation of the solution range. So, when one of them isn't chosen, the range of solution become smaller and the dispersion will be reduced.
3. Calculate the crowding distance by summarizing the distance at each criteria.

4. Mathematical Model

The model used in this paper is the modification of those proposed by [12]. The development were done by adding and changing constraints to suit the missing operations and limited wait characteristics. The objective functions are minimizing completion time (makespan), total tardiness time, and total machine idle time. Those three objective functions have some weight and to be minimized simultaneously.

i : operation index
j : job index
k : machine index
S_{ij} : starting time of job j at operation i
C_{ij} : completion time of job j at operation i
p_{ij} : processing time of job j at operation i
d_j : due date of job j
M : a very big number (big M)
x_{ijk} : binary variable, 1 if job j processed by machine k at operation i and 0 if not
y_{ihj} : binary variable, 1 if job h processed before job j at operation i dan 0 if not
a_{ij} : binary variable, 1 if job j processed at operation i dan 0 if not

$$MIN\ f_p(Z_1 : Z_2 : Z_3) \tag{5}$$

$$Z_1 = max_{ij}\ \{C_{ij}\} \tag{6}$$

$$Z_2 = \sum_j \left(max\ \{max_i\{C_{ij}\} - d_j, 0\}\right) \tag{7}$$

$$Z_3 = \sum_k \left(max_{ij}\ \{C_{ij}\} - min_{ij}\ \{S_{ij} + M(1 - x_{ijk})\} - \sum_i \sum_j p_{ij} x_{ijk} \right) \tag{8}$$

$$C_{ij} = \left(S_{ij} + \sum_k p_{ij} x_{ijk} \right) a_{ij}\ , \forall i \in I, j \in J \tag{9}$$

$$\sum_k x_{ijk} = a_{ij}\ , \forall i \in I, j \in J \tag{10}$$

$$S_{(i+1)j} \geq C_{ij}, \forall i \in I, j \in J \tag{11}$$

$$S_{(i+1)j} - C_{ij} \leq w_{ij}, \forall i \in I, j \in J \tag{12}$$

$$C_{ij} - S_{ih} \leq y_{ihj}M, \forall i \in I, h \in J, j \in J, h \neq j \tag{13}$$

$$C_{ih} - S_{ij} \leq (1 - y_{ihj})M, \forall i \in I, h \in J, j \in J, h \neq j \tag{14}$$

$$x_{ijk} \in \{0,1\}, \forall i \in I, j \in J, k \in K \tag{15}$$

$$y_{ihj} \in \{0,1\}, \forall i \in I, h \in J, j \in J \tag{16}$$

$$a_{ij} \in \{0,1\}, \forall i \in I, j \in J \tag{17}$$

Equation (5) is the multi-objective function to minimize Z_1, Z_2, and Z_3 simultaneously. Equation (6) is the first objective function (Z_1) to calculate makespan. Equation (7) is the second objective function (Z_2) to count total tardiness. Equation (8) is the third objective function (Z_3) to count total machine idle time. Equation (9) is to count the completion time of the job at each process. Equation (10) is to assure that each job at each operation is processed once by one machine. Equation (11) is to assure that the processing of the job at an operation is started after the predecessor process of that job finished. Equation (12) is to make sure that the waiting time between the process do not exceed the limitations. Equation (13) and (14) are to assure that each machine can process only one job at a time. Equation (15)-(17) are to make sure that the variable values are binary.

5. Proposed Algorithm

The proposed algorithm is an algorithm that combines PSO with probability transition matrix and pareto optimal. The algorithm is called Modified Particle Swarm Optimization (MPSO). The MPSO algorithm can be described as follows:

1. Generate initial position and initial velocity for particles. Each initial position is in the form of a probability transition matrix . The dimension of initial position and initial velocity respectively is N by p, where N is the number of particles and P is the number of jobs.
2. Generate solution sequence based on the probability value of the transition matrix. Normalize first the probability transition matrix such that the row sum is equal to 0.
3. Calculate the fitness function
4. Find non-dominated solution (NDS) and update the pareto archive. The updating process based on some conditions as below :
   - If the new NDS be dominated by minimum one of pareto archive member, the pareto archive does not need an updating.
   - If the new NDS dominate one of pareto archive member or more, the new NDS are adding into the pareto archive and pareto archive member that dominated by the new NDS eliminated from the pareto archive.
   - If the new NDS does not dominate and is not dominated, the new NDS are adding to the pareto archive without eliminated pareto archive member.
   - If the size of the pareto archive exceeds the limitation, the pareto archive member are sorted by its crowding distance value, from the biggest value to smallest value. Pareto archive member which has smallest crowding distance value is eliminated until the size of the pareto archive does not exceed the limitation.
5. Find Gbest and Pbest
   - Gbest are chosen from pareto archive member randomly. In the MPSO algorithm, Gbest for each particle may be different, it is not like in original PSO that each particle has same Gbest.

- Pbest is chosen by comparing the old Pbest and the new solution for each particle. When the old Pbest and the new solution are not dominated each other, the Pbest is chosen from one of them that has higher order ranking

6. Updating the velocity use equation (18).

$$v_t = v_{t-1} + c_1(Pbest - random_t) + c_2(Gbest - random_t) \tag{18}$$

7. Updating the probability value of transition matrix use equation (19).

$$random_{t+1} = random_t + v_t \tag{19}$$

8. Checking the stopping criteria. If the stopping criteria is not reached, so the iteration repeated from the step 3. If the stopping criteria is reached, so the iteration stopped.

6. Experiment

Data sets used in this experiment are from OR Library. The data sets hel1 (Heller, 1960) were modified such that we have three different data sets

1. Data set *flow shop* (FS), dimension 10x10 (10 *job* 10 *stage*) with 1 machine at each stage.
2. Data set *hybrid flow shop* 2 (HFS2), dimension 20x10 (20 *job* 10 *stage*) with 3 machines at each stage.
3. Data set *hybrid flow shop* 3 (HFS3), dimension 30x10 (30 *job* 10 *stage*) with 3 machines at each stage.

The experiments are done to evaluate performance of the algorithm. The proposed algorithm is used to solve some cases where characterized by the number of jobs and machines. Paramater used in this paper based on the results of the parameter testing in the preliminary experiments, i.e. population number (n) = 1000, maximum iterations (itmax) = 200, $c_1 = 0.3$, $c_2 = 0.7$, and the pareto archive size limitation is 10 percent of population number. There are many kind of performance metrics that can be used to know the quality of the solution set. The performance criteria that previously used in the research is shown below [13]:

- General Distance Metric (GD) or convergence metric is distance between solution set and pareto front.

$$GD = \frac{\sqrt{\sum_{i=1}^{n} d_i^2}}{n} \tag{20}$$

- Spacing Metric (S) atau divergence metric is a dispersion measure of solution set

$$S = \sqrt{\frac{1}{n-1}\sum_{i=1}^{n}(d - d_i)^2} \tag{21}$$

- Number of Non-Dominated Solution (NNDS) indicates how many non-dominated can be found.

The performance criteria used in this paper are the number of non-dominated, convergence metric, divergence metric, and computation time. The results of the experiment can be seen in Tables 1, 2, 3 and 4.

Table 1. Computation time for different cases.

Data Sets	Average Computation Time (second)		Minimum Computation Time (second)		Maximum Computation Time (second)	
	PSO	**MPSO**	**PSO**	**MPSO**	**PSO**	**MPSO**
FS1	410.6196	587.6464	396.5077	578.4517	439.0492	599.0126

FS2	453.5219	994.7531	448.7813	975.1563	458.375	1004.938
FS3	578.5492	1314.6	576.4705	1306.4	580.3237	1325.3
HFS1	371.4384	561.0381	365.7599	552.0875	377.5224	569.232
HFS2	483.7687	907.4625	481.8281	903.6406	485.7031	911.6719
HFS3	547.7273	1283.3	543.7259	1280.4	552.0719	1286
LWFS1	464.9298	687.9551	456.1313	682.9724	474.243	694.7349
LWFS2	537.7281	1041.422	534.2656	1015.328	541.0781	1055.672
LWFS3	657.825	1458.575	651.6786	1440.3	662.2866	1469.2
LWHFS1	437.9291	659.3694	434.9308	651.2886	440.656	666.1555
LWHFS2	520.5719	1042.494	516.9063	1030.813	528.2188	1055.438
LWHFS3	646.7372	1455.05	630.946	1448.8	655.7034	1465.2

The NNDS shows the number of the optimal solution that can be found. The NNDS of MPSO's solution sets are more than the NNDS of PSO's solution sets. If the number of the NNDS is increasing, the probability to find the optimal solution is increasing too. It means that MPSO algorithm has bigger probability to find the optimal solution than PSO algorithm. The convergence metric shows the distance between solution set and optimal solution. The convergence metric of MPSO's solutions sets are fewer than the convergence metric of PSO's solution sets. If the convergence metric is decreasing, the solution sets is closer to the optimal solution. It means that MPSO algorithm can produce solution sets closer to the optimal solution than those of PSO algorithm.

Table 2. The result of number of non-dominated solution (NNDS).

Data Sets	Average NNDS		Maximum NNDS		Minimum NNDS	
	PSO	**MPSO**	**PSO**	**MPSO**	**PSO**	**MPSO**
FS1	46.6	44.6	51	54	43	34
FS2	70.6	56.4	89	68	39	38
FS3	7.25	24.25	13	39	4	10
HFS1	36.2	39	39	45	35	32
HFS2	50	53	66	79	37	31
HFS3	47.75	50.75	66	76	38	23
LWFS1	39.4	42	41	47	37	39
LWFS2	71.2	62.6	81	93	46	43
LWFS3	5	26	6	32	3	20
LWHFS1	9.6	19.8	15	32	5	14
LWHFS2	17.2	28.2	29	35	4	11
LWHFS3	5.25	33	8	42	3	29

Table 3. The result of convergence metric.

Data set	Average		Minimum convergence		Maximum convergence	
	PSO	**MPSO**	**PSO**	**MPSO**	**PSO**	**MPSO**
FS1	18.69064	19.74676	17.9344	18.0728	19.1383	22.2714
FS2	15.83864	19.36146	13.9196	16.9493	20.3738	23.899
FS3	40.2946	24.75198	28.7164	19.6508	49.5226	33.0869

HFS1	20.12258	19.72883	19.4348	18.7562	20.4303	21.2668
HFS2	16.81366	18.78364	14.482	15.1776	18.5806	24.0503
HFS3	14.32188	15.58568	11.6963	11.7155	16.7184	21.7906
LWFS1	20.89856	20.2554	20.3979	18.9415	21.7414	20.9456
LWFS2	16.01526	19.08602	15.0727	15.0646	19.0826	22.2889
LWFS3	46.77843	21.79778	40.7414	19.6466	59.0604	24.5089
LWHFS1	43.23386	29.95048	32.7506	22.6481	56.9076	33.5503
LWHFS2	38.25176	29.83808	22.7312	25.5861	62.7081	41.4912
LWHFS3	45.38035	18.1212	34.1204	15.6371	54.7496	19.0556

The divergence metric shows the dispersion of solution sets. The divergence metric of MPSO's solution sets are more than those of PSO's solution sets. If the divergence metric is decreasing, the dispersion of solution sets is better. It means that MPSO algorithm cannot produce solutions sets better than the solution sets produced by PSO algorithm. The modification by adding the transition matrix to PSO make the MPSO algorithm needed much time to find the optimal solution. This is disadvantage of MPSO algorithm compared to PSO algorithm. But, MPSO algorithm also has advantage that is the quality of the solution is better than the solution produced by PSO algorithm. The solution of MPSO algorithm is better, because it has bigger probability to find optimal solution and produce solution sets that closer to the optimal solution.

Table 4. The results of divergence metric.

Data set	Average		Minimum Divergence		Maximum Divergence	
	PSO	MPSO	PSO	MPSO	PSO	MPSO
FS1	3.52396	5.21076	2.5495	3.4303	5.3137	7.6514
FS2	8.44268	8.91154	3.1566	4.9111	20.3184	12.3185
FS3	6.0442	6.797925	3.4641	2.7889	8.8918	16.9039
HFS1	6.8421	6.6502	4.6557	3.877	13.9267	8.4275
HFS2	3.94416	5.75744	2.4305	2.991	8.5563	7.9315
HFS3	5.891475	6.169175	3.0623	3.743	8.9818	7.841
LWFS1	4.04944	5.03074	3.2364	3.208	5.882	7.8089
LWFS2	9.04326	8.83724	4.6961	5.5712	12.9217	11.3803
LWFS3	10.028	7.69615	2.6833	5.2942	19.9575	10.7271
LWHFS1	15.12552	7.25384	8.8428	3.1557	25.026	15.495
LWHFS2	10.37906	9.31244	5.6195	5.3377	16.2384	19.3111
LWHFS3	7.6178	7.9582	5.0709	6.0333	9.815	11.5819

7. Conclusion

MPSO algorithm has been successfully developed to solve the multi-objective discrete problem. It can solve some type of flow shop scheduling problems, i.e. flow shop scheduling, hybrid flow shop scheduling, limited wait flow shop scheduling and limited wait hybrid flow shop scheduling problems. All of those are multi-objective problems with three objective functions (i.e minimizing makespan, total tardiness, and total machine idle time). Furthermore, this algorithm can be applied to solve other type of scheduling problems. The modification of PSO by adding transition matrix making the MPSO algorithm is more efficient than the original PSO. Although MPSO algorithm required longer computing time, the quality of solution sets are better than those of original PSO. Though the dispersion of solution sets are not better, but the distance of solution sets are closer to the optimal solution. The distance between solution sets and the optimal solution is more important than the dispersion of solution sets. MPSO algorithm still can be further developed so it has a new mechanism of generating initial solution, updating solution, and stopping criteria that can reduce the computation time. When the computation time reduced, MPSO algorithm perfectly can be better than the original PSO algorithm.

8.References

[1] Bagchi T P, Gupta J N D and Sriskandarajah C 2006 A review of TSP based approaches for flowshop scheduling *European Journal of Operational Research* **169** (3) 816–854

[2] Avriel M, Penn M and Shpirer N 2000 Container ship stowage problem : complexity and connection to the coloring of circle graphs *Discrete Applied Mathematics* **103** (1–3) 271–279.

[3] Ruiz R and Vazquez-Rodriguez J A 2010 The hybrid flow shop scheduling problem *European Journal of Operational Research* **205** (1) 1–18

[4] Yang D and Chern M 1995 A two-machine flowshop scheduling problem with limited waiting time constraints *Computers & Industrial Engineering* **28** (1) 63–70

[5] Su L 2003 A huybrid two-stage flowshop with limited waiting time constraints *Computers & Industrial Engineering* **44** (3) 409–424

[6] Gicquel C, Hege L, Minoux L and vanCanneyt W 2012 A discrete time exact solution approach for a complex hybrid flow-shop scheduling problem with limited-wait constraints *Computers & Operations Research* **39** (3) 629–636

[7] Jarraya B and Bouri A 2012 Metaheuristic Optimization Backgrounds : A Literature Review. *International Journal of Contemporary Business Studies* **3** (12)

[8] Tseng C, Liao C and Liao T 2008 A note on two-stage hybrid flowshop scheduling with missing operations *Computers & Industrial Engineering* **54** (3) 695–704

[9] Santosa B and Willy P 2011 *Metode Metaheuristik Konsep dan Implementasi* (Surabaya:Guna Widya)

[10] Rahman R A and Santosa B 2013 *Differential Evolution dan Bottleneck Heuristik untuk Menyelesaikan Permasalahan Penjadwalan Multi-Objective Hybrid Flowshop Unrelated Parallel Machines* (Surabaya: Sepuluh Nopember Institute of Technology thesis)

[11] Racquel C R and Naval Jr P C 2005 An effective use of crowding distance in multiobjective particle swarm optimization *Proc of the 2005 conference on Genetic and evolutionary computation* 257-264

[12] Liao C J, Tjandradjaja E and Cung T 2012 An approach using particle swarm optimization and bottleneck heuristic to solve hybrid flow shop scheduling problem *Applied Soft Computing* **12** (6) 1755–1764

[13] Qiu C, Wang C and Zuo X 2013 A novel multi-objective particle swarm optimization with K-means based global best selection strategy *International Journal of Computational Intelligence System* **6** (5) 822–835

Simulated annealing with restart strategy for the blood pickup routing problem

V F Yu[1*], T Iswari[1,2,3], N M E Normasari[1,2], A M S Asih[2] and H Ting[4]
[1]Department of Industrial Management, National Taiwan University of Science and Technology, Taipei, Taiwan
[2]Department of Mechanical and Industrial Engineering, Universitas Gadjah Mada, Yogyakarta, Indonesia
[3]Program Studi Teknik Industri, Universitas Katolik Parahyangan, Bandung, Indonesia
[4]Department of Information and Finance Management, National Taipei University of Technology, Taipei, Taiwan

*vincent@mail.ntust.edu.tw

Abstract. This study develops a simulated annealing heuristic with restart strategy (SA_RS) for solving the blood pickup routing problem (BPRP). BPRP minimizes the total length of the routes for blood bag collection between a blood bank and a set of donation sites, each associated with a time window constraint that must be observed. The proposed SA_RS is implemented in C++ and tested on benchmark instances of the vehicle routing problem with time windows to verify its performance. The algorithm is then tested on some newly generated BPRP instances and the results are compared with those obtained by CPLEX. Experimental results show that the proposed SA_RS heuristic effectively solves BPRP.

1. Introduction

The need for blood rapidly increases recently. However, the supply of blood steadily decreases [1] With these trends, it is necessary to find an effective and efficient way to manage blood supply chain. In the US, operation of blood service includes collection, storage and processing of blood and its components[2].

Blood collection involves two essential components: donors (suppliers) and blood bank (depot). In the collection process, to take a pint of blood (450-ml specialized package) from a donor takes about 10 minutes. After the collection, the blood samples are sent to the blood bank via vehicles equipped with temperature-controlled boxes as a temporary storage while the blood samples are in transit. Donated blood must be delivered to a blood processing center, called blood bank, within 6 hours. Thus an efficient vehicle routing to pickup the blood is needed.

The efficiency of blood collection planning/supply center and related issues have been studied [3] since 1970s, but routing aspects of the problem have rarely been studied. Therefore, this research introduces the Blood Pickup Routing Problem (BPRP), which is an extension of the well-known Vehicle Routing Problem with Time Windows (VRPTW). BPRP covers not only the typical characteristics of VRPTW but also problem rich constraints such as the spoilage time of blood and time restrictions at the depot.

Since VRPTW is NP-hard, BPRP is also NP-hard. Therefore, exact solution approaches may not be suitable for practical cases. Considering the success of Simulated Annealing (SA) on solving hard combinatorial optimization problems [4], this study develops a meta-heuristic based on Simulated Annealing (SA) for solving BPRP of practical sizes. More specifically, this research modified the Simulated Annealing with Restart Strategy (SA_RS) of Lin and Yu [11] to solve BPRP. SA has been successfully applied to many vehicle routing related problems [12],[13],[7],[8],[14],[11], including BPRP [10]. Therefore, this study improves the SA for BPRP by incorporating the restart strategy to avoid getting trapped in local optimum by exploring larger portion of search space.

The rest of this paper is organized as follows. Section 2 reviews relevant literature. Section 3 describes the problem. The proposed SA_RS is discussed in Section 4. Section 5 presents computational study. Section 6 provides conclusions and future research directions.

2. Literature Review

There are not many studies devoted to blood collection planning[15]. Blood collection planning is firstly studied by Cumming et al. [16]. They studied the forecasting of blood collection by developing population based Markovian model to decrease the gap between demand and supply of blood. Yi [2] introduced the vehicle routing aspect on blood issue. With the federal's regulation on blood goods, the study seeks an efficient way to decide which sites to shuttle blood and how these vehicles should be routed to provide a sufficient amount of platelets to meet daily production requirements at the minimum transportation cost possible. The computational results reveal significant reduction of transportation cost of nearly 60%, on average. Doerner, et al. [17] study the blood collection process of the Austrian Red Cross blood program and address multiple pickups at single site. Similar to Yi [2], they assume known and constant collection with the goal to find minimum cost tours. The resulting cost reduction ranges from 6% to 12% depending on the problem complexity.

Literature on heuristic approaches for VRPTW is extensive. Recently, Rabbani et al. [18] develop a Genetic Algorithm (GA) to solve VRP with multi-middle depots for perishable food delivery. The results prove that the GA is promising. Song and Ko [19] propose a priority-based heuristic (PBH) to find an efficient route to deliver perishable food product. The priority in PBH is computed based on distance and density of the node. The result shows that PBH can solve the problem efficiently. Rabbani, et al. [20] develop and compare GA and SA on solving VRPTW for delivering perishable products. The results reveal that SA produces better solutions than GA. Osvald and Stirn [21] develop a heuristic based on tabu seach to solve fresh vegetables distribution. The result shows that the algorithm can improve the solution up to 47%. Doerner et al. [17] propose a heuristic based on the savings and greedy procedure to solve VRPmiTW. The results show that the heuristic find solutions in a reasonable time.

3. Problem Description

BPRP is characterized by a network consisting of a blood processing facility (depot) and fixed donation sites (customers) accumulating donated blood. Donation sites to be visited are pre-determined and assumed to produce a fixed amount of blood bags. The blood bank is responsible for managing the collection process which includes dispatching of shuttles to the donation sites and bringing the donated blood back to the blood bank for further processing.

The perishability of blood must be considered for practical purpose. The requirement as to the age of the blood bags when they reach the blood bank varies from blood bank to blood bank depending on the facility's requirement. But, a standard age of at most 6 hours should be adhered [2].

The objective of the BPRP is to find the set to minimum cost routes to bring back all collected blood bags to the blood bank without waste, that is, within the shelf-life. Moreover, BPRP has the following typical VRPTW attributes and assumptions:

1. Each donation sites can be visited within a specific time window pertaining to the opening and ending time of the donation event;
2. Supply at each site is deterministic;
3. The blood bank operates within a specific time window as well;

4. The blood bank administers the scheduling of the shuttles which are assumed to have homogeneous capacity;
5. Total amount of supplies for collection in a route should not exceed the capacity of the shuttle;
6. Each donation site can be visited only once;
7. Service time is assumed to be negligible.
8. Age of the blood starts at the time of donation.

More formally, BPRP can be defined on a complete graph $G = (N, A)$ where $N = \{v_0, v_1, \ldots v_n\}$ is the node set and $A = \{(v_i, v_j): v_i, v_j \in N, i \neq j\}$ is the arc set. Node v_0 represents the depot and the remaining nodes in N represent a subset N' of supplier nodes (donation sites). Each supplier node is associated with a supply p_i and a time window $[o_i, e_i]$. Each node operates from o_i to e_i. Each vehicle route is restricted with a maximum allowable route length, T_{max}, which represents the spoilage of the blood. Each arc $(v_i, v_j: i \neq j) \in A$ *is* associated with a finite travel time t_{ij}. The goal of BPRP is to find a set of least-cost routes to collect total supply produced at each node using a fleet of homogenous vehicles.

4. SA_RS for BPRP

SA is popular heuristic for solving combinatorial optimization problems. SA is first proposed by Metropolis, et al. [22]. SA usually begins by randomly generated an initial solution. Then a new solution is taken from the neighborhood of the current solution. The new solution replaces the current solution if it is better, and then the process continues from the new current solution. SA accepts a worse solution with small probability. Thus, SA can escape to be trapped at local optima. To increase the diversity of SA to explore a larger solution space, this study develops the SA with restart strategy. Restart strategy is a well-known strategy to diversify SA procedure [11]. The following sections give more detailed explanation of the SA_RS for BPRP.

4.1. Initial Solution

The initial solution for SA_RS is generated by the Nearest Neighborhood algorithm [23] The vehicle starts from depot to visit the nearest unvisited customer, then continue to the next nearest unvisited customer until all customers are visited. The vehicle then returns to the depot. The vehicle will return to the depot when the total collected blood exceeds the capacity of the vehicle, the arrival time violates time window or the age of blood exceeds the spoilage time. The algorithm then continues to construct a new route and performs the same method until all customers have been visited.

4.2. Neighborhood

SA_RS uses a random neighborhood structure that includes swap, insert, and reverse operations. The swap operation is executed by randomly selecting two points in different routes of the current solution and then swapping their positions. The insertion operation inserts a randomly selected point from one route into a randomly selected position of another route of the current solution. The reverse operation is performed by randomly selecting two points from a route in the current solution and then reversing the sequence between them, including those two selected points. The probabilities of choosing swap, insert, and reverse operations are 0.3, 0.4, and 0.3, respectively. These probabilities are determined based on an preliminary experiment

4.3. SA_RS Parameters

The SA with restart strategy begins with five parameters I_{iter}, T_0, $N_{non\text{-}improving}$, *Max_restart*, and α. I_{iter} denotes the number of iterations the search proceeds at a particular temperature. T_0 represents the initial temperature. $N_{non\text{-}improving}$ is the maximum allowable number of reductions in temperature during which the best objective function value is not improved. The algorithm will restart at the initial temperature if $N_{non\text{-}improving}$ is reached. *Max_restart* is the maximum number of restarts. Finally, α is the coefficient controlling the cooling mechanism. Boltzmann function is used to calculate the probability of acceptance.

4.4. SA_RS Procedure
This study implements the restart strategy to prevent the solution being trapped in the local solution. The restart strategy allows the algorithm to restart from the initial temperature if the number of non-improvement solution have been reached in the temperature reductions. This mechanism helps the algorithm to explore more in the solution area. Figure 1 shows the flowchart of SA_RS algorithm.

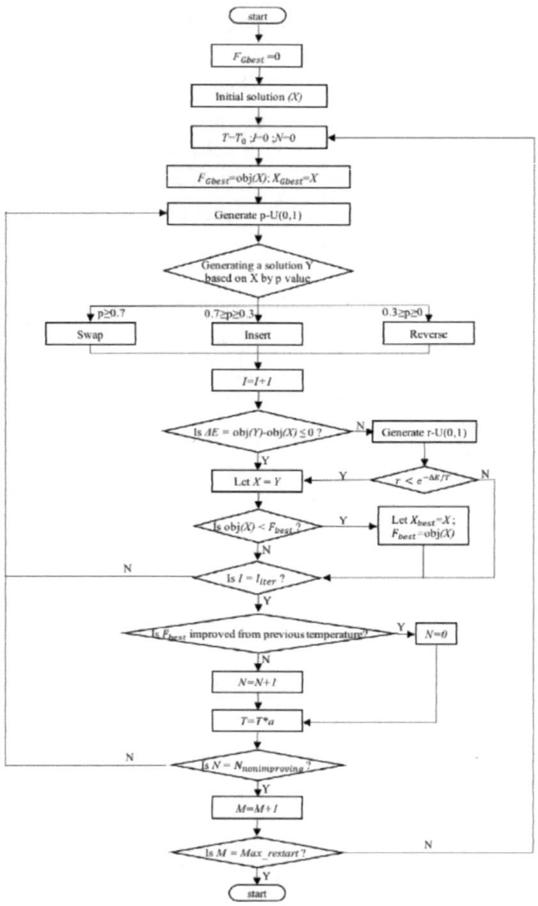

Figure 1. Flowchart of SA_RS.

5. Computational Results
The proposed SA_RS is implemented in C++ and tested on small BPRP instances to evaluate its performance.

5.1. Parameter Setting
The best parameter combination is determined by a two-level (2^k) factorial design, which has been widely used to determine parameter setting for heuristics [24]. The initial parameters used for the pilot experiment are: I_{iter} = 3000, α = 0.95, T_0= 10, $N_{non\text{-}improving}$ = 100, and *Max_restart* = 10. The best parameters for BPRP are: I_{iter} = 9000, α = 0.95, T_0= 10, $N_{non\text{-}improving}$ =200, and *Max_restart* = 10. These parameter values are used for all computational study hereafter.

5.2. Results on VRPTW Instances

Table 1 compares the performances of SA and SA_RS on solving in Solomon's VRPTW instances. SA_RS provides better solution for all instances. The average percentage gap is -1.80% and the largest gap is -11.67% for instance C203. However, SA_RS needs slightly longer computational time than SA. This shows that the proposed SA_RS can solve VRPTW effectively.

Table 1. Comparison basic SA and SA_RS in VRPTW.

Instance	SA		SA_RS		Gap (%)
	Total Distance	CPU Time (s)	Total Distance	CPU Time (s)	
C101	828.94	50.24	828.94	57.73	0.00%
C102	828.94	49.26	828.94	59.28	0.00%
C103	828.06	57.36	828.06	59.39	0.00%
C104	824.78	49.82	824.78	61.09	0.00%
C105	828.94	48.95	828.94	55.89	0.00%
C106	828.94	49.19	828.94	55.69	0.00%
C107	828.94	49.27	828.94	56.21	0.00%
C108	828.94	48.43	828.94	54.43	0.00%
C109	828.94	48.36	828.94	53.98	0.00%
C201	621.64	63.35	591.56	47.67	-4.84%
C202	621.64	62.87	591.56	54.35	-4.84%
C203	669.27	54.55	591.17	49.13	-11.67%
C204	625.06	72.63	591.17	52.37	-5.42%
C205	626.63	65.59	588.88	54.19	-6.02%
C206	621.01	68.49	588.49	53.07	-5.24%
C207	588.29	62.98	588.29	47.34	0.00%
C208	588.32	55.17	588.32	49.55	0.00%
R101	1672.08	61.27	1649.86	72.46	-1.33%
R102	1491.22	59.73	1487.32	81.52	-0.26%
R103	1243.6	68.71	1229.09	71.58	-1.17%
R104	1030.59	59.07	1014.22	70.89	-1.59%
R105	1405.99	65.12	1389.93	78.24	-1.14%
R106	1275.85	77.59	1273.43	63.53	-0.19%
R107	1118.63	80.08	1092.19	72.72	-2.36%
R108	993.45	68.31	973.49	73.74	-2.01%
R109	1186.79	60.76	1177.44	68.70	-0.79%
R110	1134.84	61.28	1113.68	76.10	-1.87%
R111	1103.16	61.04	1076.42	72.96	-2.42%
R112	985.71	61.88	971.26	73.23	-1.47%
R201	1199.58	64.64	1182.07	72.67	-1.46%
R202	1078.94	57.94	1063.38	69.14	-1.44%
R203	914.44	56.23	912.24	64.65	-0.24%
R204	756.45	74.43	743.50	68.34	-1.71%
R205	1012.95	65.66	1006.89	63.84	-0.60%
R206	931.91	70.11	911.40	62.41	-2.20%
R207	847.02	68.13	809.36	68.33	-4.45%
R208	732.74	54.75	727.52	57.36	-0.71%
R209	908.19	56.02	890.01	69.12	-2.00%
R210	920.47	56.71	920.47	71.68	0.00%
R211	795.21	63.99	769.19	65.93	-3.27%
RC101	1690.83	63.20	1673.24	70.92	-1.04%
RC102	1506.78	66.31	1501.96	73.17	-0.32%
RC103	1355.85	78.58	1297.57	69.20	-4.30%
RC104	1200.85	73.95	1172.10	80.78	-2.39%
RC105	1585.61	69.69	1547.43	79.59	-2.41%
RC106	1435.31	64.53	1396.60	72.14	-2.70%
RC107	1285.04	61.90	1258.94	71.25	-2.03%
RC108	1154.92	62.71	1147.26	70.09	-0.66%
RC201	1314	58.80	1294.99	69.85	-1.45%
RC202	1127.6	69.40	1123.55	64.22	-0.36%

Instance	SA		SA_RS		Gap (%)
	Total Distance	CPU Time (s)	Total Distance	CPU Time (s)	
RC203	989.44	62.55	962.98	67.16	-2.67%
RC204	819.84	59.31	807.92	63.69	-1.45%
RC205	1206.34	72.89	1185.98	66.90	-1.69%
RC206	1109.64	66.39	1096.09	65.45	-1.22%
RC207	1005.02	68.78	994.20	67.16	-1.08%
RC208	831.75	68.54	812.13	65.66	-2.36%
Average	1013.86	62.46	997.00	65.14	-1.80%

5.3. Results on BPRP Instances

Since BPRP is a new problem, we generate 10 small BPRP benchmark instances from classical VRP benchmark instances. The depot and the first 20 customers of the VRP instances are used to generate BPRP instances, along with their coordinates, and the customer demands. Time windows, vehicle capacity, and spoilage time are then added to the instances. We set the time window for all nodes to be [60, 660]. The time window represents the opening and closing time of the nodes. Vehicle capacity ranges from 50 to 140 to maintain the feasibility. The spoilage time is set to be 360s. CPLEX solver is used to solve these instances for comparison purpose. The maximum running time of CPLEX is 5 hours.

To evaluate the solution quality of the proposed SA_RS, this research compares results from both SA and SA_RS to the CPLEX results. As can be seen in Table 2, both SA and SA_RS find optimal solutions to 5 out to the 10 instances, and better solutions for the remaining instances. The comparative results reveal that the proposed algorithms can provide good BPRP solutions in a reasonable amount of time.

Table 2. Small Comparative Results on BPRP Instances.

Instance	C[a]	CPLEX		SA			SA_RS		
		TD[b]	CT(s)[c]	TD	CT(s)	Gap(%)	TD	CT(s)	Gap(%)
BPRP_cmt1	50	454.30	18000	454.30	21.84	0.00%	454.30	22.34	0.00%
BPRP_cmt2	60	418.89	18000	413.19	20.30	-1.36%	413.19	20.58	-1.36%
BPRP_cmt3	70	393.61	18000	389.12	20.41	-1.14%	389.12	21.03	-1.14%
BPRP_cmt4	80	368.81	18000	357.11	20.01	-3.17%	357.11	20.42	-3.17%
BPRP_cmt5	90	339.29	18000	339.29	19.45	0.00%	339.29	19.44	0.00%
BPRP_cmt6	100	329.12	18000	329.12	19.15	0.00%	329.12	20.05	0.00%
BPRP_cmt7	110	329.03	18000	325.13	19.26	-1.19%	325.13	19.82	-1.19%
BPRP_cmt8	120	319.23	18000	314.40	18.63	-1.51%	314.40	19.06	-1.51%
BPRP_cmt9	130	297.88	18000	297.88	17.90	0.00%	297.88	18.68	0.00%
BPRP_cmt10	140	293.28	18000	293.28	18.29	0.00%	293.28	18.26	0.00%
Average		354.34	18000	351.28	19.52	-0.84%	351.28	19.97	-0.84%

[a] Capacity.

[b] Total Distance.

[c] CPU Time.

6. Conclusions

This research introduces the BPRP for optimizing the routes of blood collection and proposes an SA_RS heuristic for the problem. The proposed algorithm is tested on a set of small BPRP instances generated by this study. Computational experiments reveal that SA_RS performs well on solving these BPRP instances. Testing the performance of SA_RS on large BPRP instances and real cases and studying the trade-offs between service level and logistics costs in BPRP are interesting areas to explore. This research assumes that the supply at each donation site is deterministic. Future research can consider stochastic blood supply to bring the problem closer to reality. Future research may also focus on the optimization of blood delivery to further enhance the efficiency of the blood supply chain. Lastly, future research may develop other metaheuristics for BPRP and then compare them to the proposed SA_RS.

References

[1] Lee W-C and Cheng B-W 2011 An intelligent system for improving performance of blood donation. *Journal of Quality* **18**(2) 173-185

[2] Yi J 2003 Vehicle routing with time windows and time-dependent rewards: A problem from the american red cross. *Manufacturing & Service Operations Management* **5**(1) 74-77

[3] Dumas M B and Rabinowitz M 1977 Policies for reducing blood wastage in hospital blood banks. *Management Science* **23**(10) 1124-1132

[4] Alfa A S, Heragu S S and Chen M 1991 A 3-opt based simulated annealing algorithm for vehicle routing problems. *Computers & Industrial Engineering* **21**(1-4) 635-639

[5] Van Breedam A 1995 Improvement heuristics for the vehicle routing problem based on simulated annealing. *European Journal of Operational Research* **86**(3) 480-490

[6] Kuo Y 2010 Using simulated annealing to minimize fuel consumption for the time-dependent vehicle routing problem. *Computers & Industrial Engineering* **59**(1) 157-165

[7] Yu V F, Lin S-W, Lee W and Ting C-J 2010 A simulated annealing heuristic for the capacitated location routing problem. *Computers & Industrial Engineering* **58**(2) 288-299

[8] Lin S-W, Yu V F and Lu C-C 2011 A simulated annealing heuristic for the truck and trailer routing problem with time windows. *Expert Systems with Applications* **38**(12) 15244-15252

[9] Xiao Y, Zhao Q, Kaku I and Xu Y 2012 Development of a fuel consumption optimization model for the capacitated vehicle routing problem. *Computers & Operations Research* **39**(7) 1419-1431

[10] Iswari T, Yu V F and Asih A M S 2016 Simulated annealing for the blood pickup routing problem. *International Journal of Information and Management Sciences* **27**(4) 317-327

[11] Lin S-W and Yu V F 2015 A simulated annealing heuristic for the multiconstraint team orienteering problem with multiple time windows. *Applied Soft Computing* **37** 632-642

[12] Lin S-W, Ying K-C, Lee Z-J and Hsi F-H. Applying simulated annealing approach for capacitated vehicle routing problems. 2006 IEEE International Conference on Systems, Man and Cybernetics, 2006, 639-644.

[13] Lin S W, Yu V F and Chou S Y 2009 Solving the truck and trailer routing problem based on a simulated annealing heuristic. *Computers & Operations Research* **36**(5) 1683-1692

[14] Lin S W and Yu V F 2012 A simulated annealing heuristic for the team orienteering problem with time windows. *European Journal of Operational Research* **217**(1) 94-107

[15] Beliën J and Forcé H 2012 Supply chain management of blood products: A literature review. *European Journal of Operational Research* **217**(1) 1-16

[16] Cumming P D, Kendall K E, Pegels C C, Seagle J P and Shubsda J F 1976 A collections planning model for regional blood suppliers: Description and validation. *Management Science* **22**(9) 962-971

[17] Doerner K F, Gronalt M, Hartl R F, Kiechle G and Reimann M 2008 Exact and heuristic algorithms for the vehicle routing problem with multiple interdependent time windows. *Computers & Operations Research* **35**(9) 3034-3048

[18] Rabbani M, Farshbaf-Geranmayeh A and Haghjoo N 2016 Vehicle routing problem with considering multi-middle depots for perishable food delivery. *Uncertain Supply Chain Management* **4**(3) 171-182

[19] Song B D and Ko Y D 2016 A vehicle routing problem of both refrigerated-and general-type vehicles for perishable food products delivery. *Journal of Food Engineering* **169** 61-71

[20] Rabbani M, Ramezankhani M-J, Farrokhi-Asl H and Farshbaf-Geranmayeh A 2015 Vehicle routing with time windows and customer selection for perishable goods. *International Journal of Supply and Operations Management* **2**(2) 700-719

[21] Osvald A and Stirn L Z 2008 A vehicle routing algorithm for the distribution of fresh vegetables and similar perishable food. *Journal of food engineering* **85**(2) 285-295

[22] Metropolis N, Rosenbluth A W, Rosenbluth M N, Teller A H and Teller E 1953 Equation of state calculations by fast computing machines. *The journal of chemical physics* **21**(6) 1087-1092

[23] Pop P C, Zelina I, Lupşe V, Sitar C P and Chira C 2011 Heuristic algorithms for solving the generalized vehicle routing problem. *International Journal of Computers Communications & Control* **6**(1) 158-165
[24] Coy S P, Golden B L, Runger G C and Wasil E A 2001 Using experimental design to find effective parameter settings for heuristics. *Journal of Heuristics* **7**(1) 77-97

Acknowledgments

This research was partially supported by the Ministry of Science and Technology of the Republic of China (Taiwan) under grants MOST 103-2221-E-011-062-MY3 and MOST 106-2410-H-011-002-MY3. This support is gratefully acknowledged.

Single stage queueing/manufacturing system model that involves emission variable

P S Murdapa[1,2*], I N Pujawan[2], P D Karningsih[2] and A H Nasution[3]
[1]Industrial Engineering Study Program, Faculty of Engineering, Widya Mandala Catholic University, Madiun, Indonesia
[2]Department of Industrial Engineering, Institut Teknologi Sepuluh Nopember, Kampus ITS Sukolilo-Surabaya 60111, Indonesia
[3]Department of Business Management, Institut Teknologi Sepuluh Nopember, Kampus ITS Sukolilo-Surabaya 60111, Indonesia

*setya.murdapa@gmail.com

Abstract. Queueing is commonly occured at every industry. The basic model of queueing theory gives a foundation for modeling a manufacturing system. Nowadays, carbon emission is an important and inevitable issue due to its huge impact to our environment. However, existing model of queuing applied for analysis of single stage manufacturing system has not taken Carbon emissions into consideration. If it is applied to manufacturing context, it may lead to improper decisisions. By taking into account of emission variables into queuing models, not only the model become more comprehensive but also it creates awareness on the issue to many parties that involves in the system. This paper discusses the single stage M/M/1 queueing model that involves emission variable. Hopefully it could be a starting point for the next more complex models. It has a main objective for determining how carbon emissions could fit into the basic queueing theory. It turned out that the involvement of emission variables into the model has modified the traditional model of a single stage queue to a calculation model of production lot quantity allowed per period.

1. Introduction

A queue system where there is only one service stage is called a single stage queue system. Entity arrival process occurs randomly. An entity must join the queue line and wait for some time if upon its arrival, all the servers are busy. Once there is one server available, one entity in the queue immediately gets the process or service provided by the server. The service takes a certain time, called the service time (or processing time). After completion of service the entity leaves the system (called the departure process). The departure process is generally also a random process.

A single queue modeling is an interesting and important study in technology and management [1] . It is an important foundation for modeling more commplex manufacturing systems [1] [3] [4] [5] . It is able to reflect the behavior of a single step or single work station manufacturing system [3] . The assessment of the single queue system will provide a fundamental, important, and useful knowledge for manufacturing system analysis. Discussions about a single queue system include the following cases: random characteristics of natural variables (arrival and departure process), number of servers (single server or multi-server), number of entity types (single class or multi-class), produce-to-order or produce-to-stock, queue capacity (finite or infinite), queuing priority, processor failures and many others [1] [1] [7] [3] .

Knowledge of a single queue system model has been very well established proven by the availability of many fundamental books about it. Some of them are Cohen [1] [8] [9] The discussion of queuing theory also appears in several chapters in books on introductory probability theory and operations research [10] [11] [12] [13] Several more books discuss queuing theories that lead to application in manufacturing systems [6][7] [5]

Of the several types of manufactured systems being modeled, Buzacott and Shanthikumar [7] uncovered models for single stage systems that operate to produce to order and produce to stock. The difference between produce to order and produce to stock lies in the availability of the warehouse after the finished workpiece is processed (finished product stock). In the case of produce to order there is no stock of finished products then it is assumed that the finished product is directly sent or received by the customer. While in produce to stock the product is stored in a warehouse before finally taken by customer. This system is the simplest manufacturing system where there is only one machine or process facility (and possibly a worker operating the machine or facility). Interarrival time and service time are important stochastic quantities. Various assumptions of the probability distribution on two quantities can be taken, for example exponential, phase-type, general or otherwise.

However, the authors argue that the modeling of this single queue system should be further developed to cover the global issues facing the world today, that is emissions. Indeed, emissions have become a critical issue. Carbon emissions arise because of the large use of fossil fuels for vehicles, power plants, and industry. Carbon dioxide, CO2, is the main constituent of greenhouse gases. CO2 levels in the atmosphere in 2016 has reached 406 ppm [16] According to Wang and Choi [17] the growth rate of Carbon content in the atmosphere is about 2.2% per year. At that rate, there will be considerable impacts, such as the occurrence of typhoons in various places, etc. [17] Furthermore, if the rate does not decrease, the greenhouse gases will rise up to 530 ppm by 2030.

Inclusion of emission variables into current system analysis models is important when global warming and climate change and its impacts are increasingly felt. In the lotsizing decision of production Wang and Choi [17] and Hua et al. [18] accommodated the amount of these emissions into EOQ-like formulas. In these papers the amount of emissions is expressed in the form of intensive quantities. Similarly, in this paper the inclusion of emission quantities will be made into the model of the single stage queue associated with a single stage manufacturing system.

Given that the issue of emissions has become critical, the single-queue system modeling should pay attention to this. So, some questions to be answered at the end of in this paper is formulated as follows:

a) How do we put emission variables into a single stage queuing/manufacturing system model?
b) How do inclussion of emission variable could give contribution on the analysis of production lotsize?

2. Problem description

This paper will focus the study on inclusion of emission variables into a single queuing systems model that is limited to: single class entity, infinite queue capacity, produce to order, and reliable server. Let the raw material for workparts is assumed to be available in abundance. The queue process occurs if the incoming order (demand) can not be immediately fulfilled because there are previous orders that have not been processed yet.

When an order arrives, a raw material is taken and marked. The marked raw material becomes the system entity. So the time between the occurrence of the order will be the entity's arrival to the system. For ease of analysis, in this paper the traditional $M/M/1$ queuing model is used where the inter-entity arrivals and service times are distributed exponentially, each having an average rate of λ and μ. The capacity of the waiting space is considered infinite (figure 1a) rather than finite (figure 1b).

(a) $M/M/1$ model (b) $M/M/1/N$ model

Figure 1. Single stage queueing system

In the system, one by one entity will be processed by the server. The average arrival rate, λ, is smaller than the average service rate, μ, so that the system would reach steady state. In such cases the workpiece will generally experience the waiting process before the time comes to be processed.

It is assumed that the queuing system operates using the electrical energy generated from the burning of fossil fuels. Entities are on a queue line facilitated by a powered material handling system, and processed in a CNC-based machining system. So, here we will pay attention to emissions that occur because the work ought to be waiting, and processed on the machine.

3. The model

3.1. The basic model

The formulation of the $M / M / 1$ model is widely available in the literature. The model is Markovian, so we can define the states of the system, n, as the number of entities in the system. Then a transition diagram can be made (figure 2).

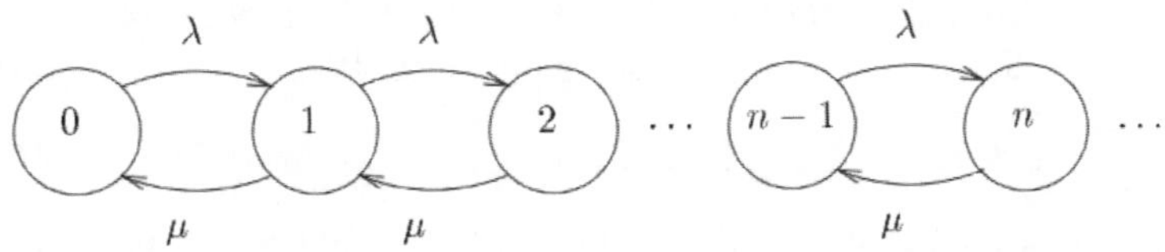

Figure 2. States transition diagram for $M/M/1$ model

If the system has reached steady state, then the following equation applies: $arrival\,rate = departure\,rate$, which can be imposed on each successive two state transitions. If P_n is defined as the probability that the system is at state-n (n = 0,1,2, ...), where $\lim_{n\to\infty} P\{N(t) = n\}$ then we can find the equations:

$$\lambda P_0 = \mu P_1$$

and

$$(\lambda+\mu)P_n = \lambda P_{n-1} + \mu P_{n+1} \text{ untuk } n = 1, 2, ...$$

so we get:

$$P_n = (1-\rho)\rho^n \text{ untuk } n = 0, 1, 2, ... \quad (1)$$

where

$$\rho = \lambda/\mu \quad (2)$$

Equation (1) is very useful for deriving important criterion formulas in the $M / M / 1$ queuing model such as residence time in the system, number of average entities in the system, etc. On the $M / M / 1$ system, the expected value of residence time in the system, $E\{\mathcal{T}\}$ is (see [11]):

$$E\{\mathcal{T}\} = \frac{1}{\mu - \lambda} \quad (3)$$

the average number of entities in the system, L:

$$L = \frac{\lambda}{\mu - \lambda} \quad (4)$$

and queuing time is calculated with:

$$E\{\mathcal{T}_w\} = E\{\mathcal{T}\} - \frac{1}{\mu}$$
$$= \frac{\lambda}{\mu(\mu-\lambda)}$$

Hence,

$$E_w = \frac{\epsilon_w \lambda}{\mu(\mu-\lambda)} \tag{5}$$

and

$$E_p = \frac{\epsilon_p}{\mu} \tag{6}$$

If the queuing number is called L_q then the value will be equal to the average entity in the system minus the average entity in the machine, that is:

$$L_q = L - \rho = \frac{\lambda}{\mu-\lambda} - \frac{\lambda}{\mu} = \frac{\lambda^2}{\mu(\mu-\lambda)} \tag{7}$$

Viewed as a single stage manufacturing system, the throughput is, $Q_0 = \lambda$. (at steady state).

3.2. Model with emission variable

The basic model would be extended to accomodate the emission variable. Defined, ϵ_w and ϵ_p, respectively, are the intensive quantities that show how much emission occurs (in kg Carbon/ minute/workpiece unit) because the workpiece has to wait and be processed (in this case $\epsilon_w \ll \epsilon_p$). Thus, the amount of emissions that occur when a workpart has to wait, E_w, is:

$$E_w = \epsilon_w \text{ x } average\ time\ of\ waiting \tag{8}$$

and to be processed, E_p, is:

$$E_p = \epsilon_p \text{ x } average\ time\ of\ processing. \tag{9}$$

Since the average processing rate is μ, then each entity unit undergoes processing for an average time of $1/\mu$. Meanwhile, to get the queue time we need to calculate the residence time of the entities in the system, $\mathcal{T}$.

The system will generate total emissions per period of $(L_q\ E_w + E_p)\ Th$, where L_q = the average number of queuing entities (entity units), and Th = throughput (in entity units per week). Th can be viewed as a production quantity or lotsize (Q) which can be produced for a certain period (e.g., week). At a time epoch, there would be L_q units of entities in waiting, and λ/μ being in process, averagely. Since the amount of emissions in that period is limited to no more than a certain limit (i.e., emission cap, $\mathcal{C}$), then:

$$[L_q\ E_w + (\lambda/\mu)E_p]Th \le \mathcal{C} \tag{10}$$

Substituting L_q, E_w, and E_p by equations (5) and (6) we get:

$$\left(\frac{\lambda^2}{\mu(\mu-\lambda)} \frac{\lambda\, \epsilon_w}{\mu(\mu-\lambda)} + \left(\frac{\lambda}{\mu}\right)\left(\frac{\epsilon_p}{\mu}\right)\right) Q \le \mathcal{C} \tag{11}$$

The time unit of Q has to be the same with the time unit of $\mathcal{C}$. For example, if the unit of $\mathcal{C}$ is kg of Carbon per week, then the unit of Q should be unit of entity per week. Equation (11) implies that once the value of $\mathcal{C}$ is reached the manufacturing system has to immediately re-establish the emission quota ownership for the next period. In order for Q to be larger (for example, because its own production capacity allows for that), the only way it can be done is that the system immediately provides a larger $\mathcal{C}$ through cap-and-trade mechanism [19] [20]

3.3. Numerical example

A hypothetical case is taken arbitrarily to give a calculation example. Suppose a single stage manufacturing system is modeled as a single $M/M/1$ queuing system. The arrival of the order follows the Poisson process with an average rate of 10 units/hour. The average rate of process in the machine is 15 units/hour. The amount of unit emissions per entity per minute is 0.01 kg Carbon. Whereas, the machine's processing of emissions per unit of entities per minute is 0.075 kg Carbon. It is known that the emission cap is 250 kg Carbon per week.

We will calculate the amount of production quantity per week allowed. It should be noted that the arrival and departure rates have to be expressed in units/mins by dividing them by 60. By entering the known values into equation (11) then calculating the value of Q, we get

$$\left[\frac{(10/60)^2}{(15/60)(15/60-10/60)}\frac{0.01\,(10/60)}{(15/60)(15/60-10/60)}+\frac{(10/60)}{(15/60)}\frac{0.075}{15/60}\right]Q \le 250$$

which gives quantity of production equal to Q = 815 units/week. The 250 kg carbon dioxide emissions per week does not allow the manufacturing system to produce over 815 units per week. Although in that week there was still some residual capacity, for example, the new quota could only be used for production for the following week. To increase the amount of production per week, the manufacturing system must raise the quota per week.

4. Experimental design

Based on the above numerical examples, it is interesting to examine the effect of emissions on the value of production lot, as well as the maximum production lot (regardless of emissions). Defined Q_E and Q_0 as the production lot if the emissions are noticed and not. Thus, a simple experimental design has been constructed as shown in figure 3.

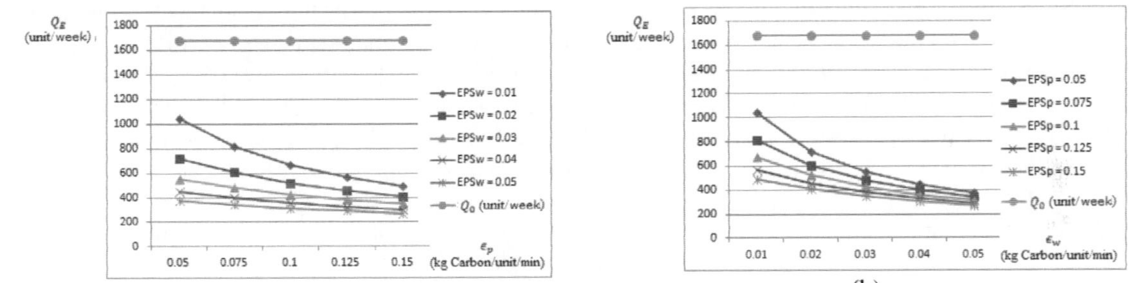

(a) (b)

Figure 3. Value of Q_E and Q_0 at at various value of ϵ_w and ϵ_p
($\lambda = 10\,unit/hour = 1680\,unit/week$; $\mu = 15\,unit/hour$; C = 250 kg of Carbon per week)

From figure 3 it is clear that emission regulation forces a manufacturing system to reduce its production lots. The greater the value of the emission intensity the greater the reduction of the lot of production.

5. Conclusion

Emission issues force a manufacturing system to account the amount of emissions it generates so as not to exceed its own quota. Including emission quantities into a simple performance model, in this case, a single stage $M/M/1$ queue stage is a good start to trigger the inclusion of these magnitudes into more complex system models.

In this paper, the author deliberately takes a simplest example of the system, since this topic has never been expressed in the literature. So far, the discussion of the effect of the emission quantity is done by mathematical optimization model, never with the performance model.

The amount of emissions can be infiltrated into the traditional $M/M/1$ queuing/manufacturing system model by using an intensive amount of kg of Carbon emissions/kg of activity/minute entities. The two emission driver activities seen are waiting and processing activities. A particular emission cap forces a manufacturing system has to limit the quantity of production that may be produced during a certain period (weeks). This production quantity might be called the emission-controlled production quantity.

6. Development opportunities

The real manufacturing systems are very complex cases. Thus, research with similar or similar ideas can be applied to more complex systems. For example in the case of flow line, transfer line, flexible manufacturing system, and others.

References

[1] Cohen J W 1982 *The single server queue.* Revised (Amsterdam: North-Holland Publishing Company)
[2] Altiok T 1989 *Performance Evaluation*, **9** (2) pp 93–102.
[3] Gurgur C and Altiok T 2004 *Annals of Operations Research* **125** (1–4) pp 95–116
[4] Buzacott J A and Shanthikumar JG 1992 *Queueing Systems* **12** (1–2) pp 135–213
[5] Gershwin S B 2003 *Analysis and modeling of manufacturing systems, International series in operations research & management science* (New York: Springer Science+Business Media)
[6] Altiok T 1997 *Performance Analysis of Manufacturing Systems* (New York: Springer Science+Business Media)
[7] Buzacott J A and Shanthikumar J G 1993 *Stochastics Models of Manufacturing Systems* (Englewood Cliffs, NJ: Prentice Hall)
[8] Benjaafar S, Li Y and Daskin M 2012 *IEEE Transactions on Automation Science and Engineering* **10** (1) pp 1–18
[9] Haviv M 2013 *Queues: A Course in Queuing Theory, International series in operations research & management science* (New York: Springer Science+Business Media)
[10] Winston W L 2003 *Operations Research: Applications and Algorithms* (Duxbury Press)
[11] Taha H A 2007 *Operations Research: An Introduction* Eighth Ed (Pearson Prentice Hall)
[12] Ross S M 2014 *Introduction to Probability Models* Eleventh Ed (Academic Press Elsevier Inc)
[13] Hillier F S and Lieberman G J 2001 *Introduction to Operations Research* Seventh Ed (McGraw-Hill)
[14] Viswanadham N and Narahari Y 1993 *Performance Modeling of Automated Manufacturing Systems* (Englewood Cliffs, New Jersey: Prentice Hall)
[15] Gershwin S B 1994 *Manufacturing System Engineering* (Englewood Cliffs New Jersey: Prentice-Hall Inc)
[16] U.S. Environmental Protection Agency 2016 *Climate change indicators in The United States,* Fourth Ed EPA 430-R-16-004 (Available at: www.epa.gov/climate-indicators)
[17] Wang X J and Choi S H 2016 *International Journal of Production Research* **7543** (2) pp 1–18
[18] Hua G, Cheng T C E and Wang S 2011 *International Journal of Production Economics* **132** (2) pp 178–185
[19] Liu B et al 2014 *International Journal of Production Research* **53** (2) pp 532–548
[20] Kleinrock L 1975 *Queueing Systems* (John Wiley & Sons)

Acknowledgement

The authors would like to thank the Kemenristek Dikti and LPDP who have made this research possible through the BUDI-DN scholarship program.

IOP Conf. Series: Materials Science and Engineering **337** (2018) 012009 doi:10.1088/1757-899X/337/1/012009

Cost analysis of an electricity supply chain using modification of price based dynamic economic dispatch in wheeling transaction scheme

Wahyuda[1,2*], Budi Santosa[1], and Ahmad Rusdiansyah[1***]**
[1]Department of Industrial Engineering, Institut Teknologi Sepuluh Nopember, Kampus ITS Sukolilo-Surabaya 60111, Indonesia
[2]Industrial Engineering Department, Universitas Mulawarman, Samarinda Indonesia

*wahyuda@gmail.com, **shima1907@yahoo.com, ***arusdian@ie.its.ac.id

Abstract. Deregulation of the electricity market requires coordination between parties to synchronize the optimization on the production side (power station) and the transport side (transmission). Electricity supply chain presented in this article is designed to facilitate the coordination between the parties. Generally, the production side is optimized with price based dynamic economic dispatch (PBDED) model, while the transmission side is optimized with Multi-echelon distribution model. Both sides optimization are done separately. This article proposes a joint model of PBDED and multi-echelon distribution for the combined optimization of production and transmission. This combined optimization is important because changes in electricity demand on the customer side will cause changes to the production side that automatically also alter the transmission path. The transmission will cause two cost components. First, the cost of losses. Second, the cost of using the transmission network (wheeling transaction). Costs due to losses are calculated based on ohmic losses, while the cost of using transmission lines using the MW - mile method. As a result, this method is able to provide best allocation analysis for electrical transactions, as well as emission levels in power generation and cost analysis. As for the calculation of transmission costs, the Reverse MW-mile method produces a cheaper cost than the Absolute MW-mile method

1. Introduction

The restructuring of the electric power market has created a new deregulation. The deregulation transforms the original monopoly power system with vertical integration into a competitive system. Vertical integration, generation, transmission, and distribution are managed by one party [1]. Whereas in deregulation, the system is separate and competitive.

Deregulation divides the electricity market into 3 types: pool system, bilateral contract and hybrid system. Pool system is the most widely applied market systems. In this system, the parties involved are divided into Gencos, Transcos, Discos and managed by an ISO (Independent System Operator). As a producer, Gencos will offer price to market, as well as Discos. Discos are a term for a group of customers. Then ISO will determine the winner of the transaction. The ISO will also arrange the scheduling of each Gencos operation that is tailored to the network load for the security of the system. Bilateral is an electric market between two parties without involving ISO. Two parties, namely Gencos and Discos conduct transactions bilateral. The result of the agreement is the amount of electricity transacted reported to ISO, then ISO will schedule shipments from producer to consumer tailored to network security [2]-[4].

This study discusses the allocation of electrical load on the bilateral contract. In this type of contract, the price agreement involves only two parties namely Gencos and Discos. Therefore, Gencos should be able to optimize the performance of plants owned so that it can produce low-cost electricity by minimizing fuel consumption. While Discos as the buyer will pay 2 types of costs, which is the cost of each kWh purchased and the cost of transmission. Transmission costs in the bilateral system are also known as wheeling transactions. There are 2 types of wheeling that is, own use and not own use. Wheeling for own use occurs when there are parties/customers who have their own power station in an area, and want to deliver electricity to its production facilities to other distant places. For that must use the network of the other party. Wheeling is not for own use when the party/customer wants to run electricity at his production facility, and for that he must buy electricity from other parties.

The type of wheeling used is not its own use. Deregulation enables all power companies to compete with each other to provide cheaper, quality power supplies. On the other hand, deregulation has also transformed the transmission system into a transportation service provider system. As a service function, the transmission should be able to provide transportation at a low cost, fair and transparent. Cheaper transportation is expected to reduce the total cost of electricity. Transparent transportation means that the fees paid by transportation service users can be clearly illustrated. Equitable transportation means that all transport service users are not subject to discriminatory treatment.

Cheaper power supplies can be achieved by minimizing fuel costs. A load allocation aimed at minimizing fuel costs using economic dispatch is present in [5]–**Error! Reference source not found.**. Economic dispatch is also called static economic dispatch because it does not consider changes in the demand side. Economic dispatch that considers changing demand is called dynamic economic dispatch as in [3], **Error! Reference source not found.**–**Error! Reference source not found.**. However, the two types of models are not suited to the deregulation of the electricity market due to differences in purpose structure and function. Therefore, the price based dynamic economic dispatch (PBDED) model emerges. This model fits into market deregulation. In this model as the goal is to maximize profit as in [20]. All of the above models result in the allocation of loads with minimal cost. By reducing fuel costs, it is expected to reduce the final price of electricity so that it is competitive. Once electricity is produced, electricity must be transmitted through the transmission network. To be able to produce competing products then the transmission should also be available at a cheap price.

Transmission cost will be cheap, fair and transparent when the transmission costs can be presented in detail [22]. This can be realized by dividing the transmission operation into several parts according to the conditions. The condition of the transmission operation is that in order to reach the customer, the electricity generated by the generator must pass through several substations first connected by a network. There are many substations and paths, so that electricity will have many combinations of lines before it reaches the customer. This combination of paths and substations raises a new problem of losses on the network.

The loss in the transmission network is the loss of electrical energy due to the nature of the conductor used or known as ohmic losses. This loss is influenced by the magnitude of current and resistance. The losses on the transmission line can be approximated by a model proposed by Bamigbola et al. **Error! Reference source not found.**. In the model, the greater the distance between the plant and the load, the greater the loss occurs. This network loss causes the plant to produce more. This is a compensation for network losses that occurs so that the amount of energy produced equal to the energy demanded. Thus, the optimization should not only be on the production side, but also on the transmission side to minimize the total cost.

Thus, the total cost of a load allocation actually consists of 3 components. First, the cost of fuel consumption due to the load. Second, the cost of fuel consumption due to transmission losses. Third, the cost of transmission. This research proposes the development of PBDED model called Multi-Echelon PBDED. This model is a combination of 2 types of models that are multi-echelon models as in [24] and PBDED models as in [20]. The model can answer the goal of profit maximization by minimizing fuel costs, minimizing transmission losses and minimizing transmission costs. The cost of transmission is determined based on the MW-mile method. Proposed models have been tried on systems with artificial data consisting of 3 generators, 3 substations, and 6 load centers. The result is a load allocation with detailed transmission costs and their impact on total costs and emissions.

2. PBDED model and multi-echelon distribution

The PBDED model as in[21] aims to maximize profit with the following objective functions:

$$maximize\ PF = RV - TC \tag{1}$$

In this case,

$$TC = \sum_{t=1}^{T}\sum_{i=1}^{N}\left[C_i\left(P_{(i,t)} + ST_t\right]\right. \tag{2}$$

$$RV = \sum_{t=1}^{T}\sum_{i=1}^{N}\left[\sigma_g(t)\ (P_{(i,t)}I_{(i,t)})\right] \tag{3}$$

C_i: production cost of unit i, $P_{(i,t)}$: Output of generator i at time t. $I_{(i,t)}$ commit or not commit at time t. ST_t start-up cost at time t. t: time of dispatch. i: index generator. N is the number of generating units. $\sigma_g(t)$ load forecasting at time t. $C_i\ (P_{(i,t)})$ generation cost of unit i

1. Demand Constraint

$$\sum P_{(i,t)}\, I_{(i,t)} \leq D_t \quad t = 1, \dots, Tm \tag{4}$$

2. Generator Constraint

$$P_{i,min} \leq P_{(i,t)}I_{(i,t)} \leq P_{i,max} \tag{5}$$

$$-DR_i \leq P_i - P_i^0 \leq UR_i \tag{6}$$

$$\max\{P_i^0 - DR_i, P_{i,min}\} \leq P_i \leq min\{P_i^0 + UR_i, P_{i,max}\} \tag{7}$$

3. Multi-echelon distribution model

Multi echelon distribution as in[24] as follows:

Minimize $$\sum_j f_j X_j + \sum_i\sum_j\sum_m\sum_k c_{ijm}^k\ Y_{ijm}^k \tag{8}$$

Constraints $$\sum_i\sum_m\sum_k Y_{ijm}^k \leq MX_j \tag{9}$$

$$\sum_j\sum_m Y_{ijm}^k \geq h_i^k \tag{10}$$

$$\sum_i\sum_j Y_{ijm}^k \leq S_m^k \tag{11}$$

$$Y_{ijm}^k \geq 0 \tag{12}$$

$$X_j = 0, 1 \tag{13}$$

h_i^k = Demand
f_j = fixed cost
c_{ijm}^k = production cost of unit I at plant m.
S_m^k = Plant capacity.

4. Proposed model using multi-echelon PBDED

The PBDED model has managed to maximize profit, but cannot present optimization of the transmission network, whereas the transmission network leads to the emergence of two types of costs. First, the costs which is caused by losses on the network. Second, the cost which is caused by the use of transmission (wheeling transaction). While in multi-echelon distribution model (formula 8 to 13), this model can present optimization on the distribution line, but it is less suitable for electricity. Therefore, to utilize the advantages of both models, this paper proposes a multi-echelon PBDED model. The objective function of this proposed model is the same as the formula (1) - (3) while the additional restrictions include:

$$F = \sum_{t=1}^{T}\sum_{i=1}^{I} F_{it}(P_{it}) \tag{16}$$

$$F_w = \sum_{t=1}^{T}\sum_{i=1}^{I} F_{it}(P_{it})_w \tag{17}$$

$$C_w = F_w - F + C_{ijt}^y + C_{jkt}^y \tag{18}$$

$$F_{it}(P_{it}) = a_i P_{it}^2 + b_i P_{it} + c_i \tag{19}$$

$$C_{ijt}^{y} = \frac{P_{ijt}^{y}}{P_{ijt}} z_{ij} \quad (20)$$

$$C_{jkt}^{y} = \frac{P_{jkt}^{y}}{P_{jkt}} z_{jk} \quad (21)$$

$$E_{it}(P_{it}) = \alpha_i P_{it}^2 + \beta_i P_{it} + \delta_i \quad (22)$$

The formula (16) is the cost of fuel with a regular load. The formula (17) is the cost of fuel when wheeling occurs. The formula (18) is the transmission cost. The formula (19) is a quadratic function of fuel costs. Formulas (20) and (21) MW-mile method calculations for transmission of 500 KV and 150 KV. While the formula (22) is a quadratic function of the resulting emissions.

5. Numerical experiment

Models are tried on systems with artificial data. The system consists of 3 generators, 3 substations and 6 loads, with the following details:

Table 1. Distance between Generators and Transmission Stations 500 kV and 150 kV (in km)

	GT1	*GT2*	*GT3*			*GD1*	*GD2*	*GD3*	*GD4*	*GD5*	*GT6*
P1	50	300	500		GT1	5	10	30	15	20	15
P2	200	50	600		GT2	10	5	20	10	20	10
P3	200	600	50		GT3	30	20	5	10	20	30

Table 2. Generator Data

	a	*b*	*c*	α	β	δ	*Max (MW)*	*Min (MW)*	*UR (MW)*	*DR (MW)*
P1	0,04	50	15	0.0400	0.09	0.10	8.000	500	1.000	1.000
P2	0,35	22	20	0.0200	0.09	0.90	5.000	500	1.000	1.000
P3	0,25	10	30	0.0100	0.01	0.05	8.000	500	1.000	1.000

Table 3. Load Data

Time (t)	*Load GD1*	*Load GD2*	*Load GD3*	*Load GD4*	*Load GD5*	*Load GD6*	*Scenario*
1	1,000	2,000	2,000	1,100	1,700	2,000	+ 200
2	1,100	1,500	1,700	1,500	2,000	1,600	+150
3	1,100	1,700	1,500	1,700	1,900	2,000	+100

6. Results

At base load, the total profit for 3 periods is Rp 31,836,139 (income Rp 69,650,728 Expenses Rp 37,814,589). Emissions 6,109,994 and losses of 15,153. With the utilization of the plant between 46 and 84%, except in the third period, the utilization of the plant reaches 100% for the 2nd plant. The low utilization will have a bad impact on the return on investment, but on the contrary, up to 100% utilization is at high risk due to electricity demand which fluctuates.

Furthermore, there are two scenarios used for this model. Scenario 1 is a condition when a request occurs on GD1 (retailer1). Scenario 2, is a condition when additional demand occurs on GD4 (Retailer 2). The results of both scenarios are shown in the following table:

Table 4. Profit, Emission and network loss for wheeling cases on GD1 and GD4

t	Scenario 1 (on GD1)				Scenario 2 (on GD4)			
	Revenue	*Expense*	*Emission*	*Losses*	*Revenue*	*Expense*	*Emission*	*Losses*
1	21,203,763	10,386,790	2,222,548	4,936	21,250,202	10,464,871	2,226,951	4,967
2	27,020,181	15,894,890	2,900,362	6,108	27,095,314	16,004,907	2,915,597	6,137
3	26,029,078	14,547,499	2,335,343	5,311	26,057,472	14,557,974	2,343,724	5,328

t	Scenario 1 (on GD1)				Scenario 2 (on GD4)			
	Revenue	*Expense*	*Emission*	*Losses*	*Revenue*	*Expense*	*Emission*	*Losses*
Total	74,253,022	40,829,179	7,458,252	16,355	74,402,988	41,027,752	7,486,272	16,432
Profit		33,423,843				33,375,236		

Table 5. The cost of transmission with Reverse and Absolute MW mile on GD1 and GD4

		Scenario 1	Scenario 2
Reverse MW mile	The difference between fuel costs	3,014,590	3,213,163
	wheeling 500 KV	33,202	33,664
	wheeling 150 KV	-14,733	-12,591
	Total Cost	3,033,059	3,234,236
Absolute MW mile	The difference between fuel costs	3,014,590	3,213,163
	wheeling 500 KV	62,870	63,067
	wheeling 150 KV	27,207	28,929
	Total Cost	3,104,667	3,305,159

With the same amount of load, both scenarios produce different profits. Profit on scenario 1 is bigger, that is Rp. 33,423,843. This profit is earned from income for 3 periods of Rp. 74,253,022. And fuel costs are Rp. 40,829,179. With total emissions of 7,458,252 and transmission losses is Rp. 16,355. Although scenario 1 generates a smaller total revenue than scenario 2, it produces a larger total profit. This is because scenario 1 can result in a more efficient allocation of loads with lower total cost (fuel costs) than scenario 2. This can also be seen from the resulting losses. Where losses in scenario 1 are smaller than scenario 2. Losses will increase generation fuel costs. Losses can be reduced by the selection of appropriate transmission routes through optimization. With regard to wheeling transaction, scenario 1 also provides a lower total transmission cost, which is Rp. 3,033,059 and Rp. 3,104,667 using Reverse MW-mile and Absolute MW-mile respectively.

The calculation of wheeling transaction using Reverse MW-mile method will result in lower total transmission cost, which is Rp. 33,202 on the transmission of 500 kV and -Rp. 14.733 on a 150 kV transmission. Negative signals on transmission costs mean a decrease in the amount of electricity that flows on the network. Negative transmission becomes revenue for transmission service users. Thus, additional demand (wheeling transaction) will be more advantageous if done on GD1 (retailer 1) than on GD4 (retailer 2).

7. Conclusions

The Multi Echelon Priced Based Dynamic Economic Dispatch method has successfully been built to maximize profit electrical allocation by presenting a more detailed and fair transmission cost calculation for the parties involved. Maximum profit can be obtained by two ways: first, maximizing revenue; and second, reducing fuel and transmission costs. This paper uses a reverse MW-mile and an absolute MW-mile method, based on 2 scenarios of wheeling transactions performed. The Reverse MW-mile method produced the cheapest transmission cost. In general, it can be concluded that transactions on GD1 (scenario 1) result in greater profit and lower transmission costs. Greater profits on scenario 1 are due to cheaper generation fuel costs and smaller losses.

References

[1] Wahyuda and Santosa B 2015 Dynamic Pricing in Electricity: Research Potential in Indonesia *Procedia Manuf.* **4** Iess pp 300–306

[2] Shivaie M and Ameli M T 2015 An environmental / techno-economic approach for bidding strategy in security-constrained electricity markets by a bi-level harmony search algorithm

Renew. Energy **83**

[3] Liao G 2012 Integrated Isolation Niche and Immune Genetic Algorithm for solving Bid-Based Dynamic Economic Dispatch *Electr Power Energy* **42** pp 264–275

[4] Shahidehpour M and Alomoush M 2001 *Restructured Electrical Power Systems* (New York: Marcel Dekker, Inc.)

[5] Niknam T, Mojarrad H D, Meymand H Z and Firouzi B B 2011 A new honey bee mating optimization algorithm for non-smooth economic dispatch *Energy* **36** (2) pp 896–908

[6] Singh N and Kumar Y 2015 Multiobjective Economic Load Dispatch Problem Solved by New PSO *Hindawi* **2015** pp 1–7

[7] Palanichamy C and Babu N S 2008 Analytical solution for combined economic and emissions dispatch *Electr. Power Syst. Res.***78** pp 1129–1137

[8] Kumar R, Sadu A, Kumar R and Panda S K 2012 A novel multi-objective directed bee colony optimization algorithm for multi-objective emission constrained economic power dispatch *Int. J. Electr. Power Energy Syst.* **43** (1) pp 1241–1250

[9] Kumar R, Sharma D and Sadu A 2011 A hybrid multi-agent based particle swarm optimization algorithm for economic power dispatch *Int. J. Electr. Power Energy Syst.***33** (1) pp 115–123

[10] Kumar S and Kumar S 2015 A Comparative Study of Solution of Economic Load Dispatch Problem in Power Systems in the Environmental Perspective *Procedia - Procedia Comput. Sci.* **48** Iccc pp 96–100

[11] Zhou J, Wang C, Li Y, Wang P, Li C, Lu P and Mo L 2017 A multi-objective multi-population ant colony optimization for economic emission dispatch considering power system security *Applied Mathematical Modelling* **31** (45) pp 684-704

[12] Mandal B, Roy P K and Mandal S 2014 Economic load dispatch using krill herd algorithm *International journal of electrical power & energy systems* **31** (57) pp 1-10

[13] Vahidinasab V and Jadid S 2010 Joint economic and emission dispatch in energy markets: a multiobjective mathematical programming approach *Energy* **35** (3) pp 1497-504.

[14] Zhao X, Wu L and Zhang S 2013 Joint environmental and economic power dispatch considering wind power integration: Empirical analysis from Liaoning Province of China *Renewable energy* **52** pp 260-5

[15] Pandit M, Chaudhary V, Dubey H M and Panigrahi B K 2015 Multi-period wind integrated optimal dispatch using series PSO-DE with time-varying Gaussian membership function based fuzzy selection *Int. J. Electr. Power Energy Syst* **73** pp 259-72

[16] Nwulu N I and Xia X 2015 Multi-objective dynamic economic emission dispatch of electric power generation integrated with game theory based demand response programs. *Energy Convers. Manag.* **89** pp 963-74

[17] El-sobky B and Abo-elnaga Y 2014 Multi-objective economic emission load dispatch problem with trust-region strategy *Electric Power Systems Research* **108** pp 254-9

[18] Zhong H, Xia Q, Chen Y and Kang C 2015 Energy-saving generation dispatch toward a sustainable electric power industry in China *Energy Policy* **83** pp 14-25.

[19] Arriagada E, Lopez E, López M, Blasco-Gimenez R, Roa C and Poloujadoff M 2015 A probabilistic economic dispatch model and methodology considering renewable energy, demand and generator uncertainties *Electric Power Systems Research* **121** pp 325-32

[20] Wu Z L, Ding J Y, Wu Q H, Jing Z X and Zhou X X 2016 Two-phase mixed integer programming for non-convex economic dispatch problem with spinning reserve constraints *Electric Power Systems Research* **140** pp 653-62

[21] Columbus C C and Simon S P 2013 Profit based unit commitment for GENCOs using parallel NACO in a distributed cluster *Swarm and Evolutionary Computation* **10** pp 41-58

[22] Wahyuda, Santosa B and Rusdiansyah A 2016 Load Allocation of Power Plant using Multi Echelon Economic Dispatch *Annual Conference on Industrial and System Engineering (ACISE)* pp 42–47
[23] Bamigbola O M, Ali M M and Oke M O 2014 Mathematical modeling of electric power flow and the minimization of power losses on transmission lines *Applied Mathematics and Computation* **241** pp 214-21
[24] Daskin M S 1995 *Network and Discrete Location. Model, Algorithm, and Applications* vol. 1 (Canada: John Wiley & Sons, Inc.)

IOP Conf. Series: Materials Science and Engineering **337** (2018) 012010 doi:10.1088/1757-899X/337/1/012010

Power plant allocation in East Kalimantan considering total cost and emissions

Muslimin[1*] and D S Utomo[2]**
[1] Electrical Engineering Department, Universitas Mulawarman, Samarinda- Indonesia
[2] Industrial Engineering Department, Universitas Mulawarman, Samarinda- Indonesia

*wahyuda@gmail.com, **musculi@gmail.com

Abstract. The fulfillment of electricity need in East Kalimantan is the responsibility of State Electricity Company/Perusahaan Listrik Negara (PLN). But PLN faces constraints in the lack of generating capacity it has. So the allocation of power loads in East Kalimantan has its own challenges. Additional power supplies from other parties are required. In this study, there are four scenarios tested to meet the electricity needs in East Kalimantan with the goal of minimizing costs and emissions. The first scenario is only by using PLN power plant. The second scenario is by combining PLN + Independent Power Producer (IPP) power plants. The third scenario is by using PLN + Rented power plants. The fourth scenario is by using PLN + Excess capacity generation. Numerical experiment using nonlinear programming is conducted with the help of the solver. The result shows that in the peak load condition, the best combination is scenario 2 (PLN + IPP). While at the lowest load condition, the cheapest scenario is PLN + IPP while the lowest emission is PLN + Rent.

1. Introduction

Related to power energy, two major issues in the world which have always been a topic of discussion is the decrease in fossil-fueled fuel and increased emissions. Reduced fuel sources will result in all human activities, whether production, construction, transportation, etc. Furthermore, uncontrolled emissions will trigger an increase in earth temperature and harmful gas for humans.

The main emissions from energy combustion are CO_2 and SO_2. It is estimated that 70% of total CO_2 emissions comes from coal combustion, while the rest comes from burning fossil fuels such as oil and natural gas. Furthermore, 90% of total SO_2 emissions come from coal combustion while the rest are oil burning [1].

Electrical energy has become a major need for human beings to support their daily activities. However, not all people have access to electricity. More than 1.6 billion people worldwide have no access to electricity and about 80% are in Asia and Africa especially in rural areas [2].

Electrical system consists of three subsystems namely: generating subsystem, transmission subsystem and distribution subsystem. Each subsystem has a specific problem. In the case of Indonesia, the main problem in the power plant system is the determination of the type and number of power plants that need to be involved in meeting the fluctuating electricity demand with multi power plants. Multi generators will be a problem when viewed from an emissions standpoint. Generators with low cost not necessarily produce low emissions as well [3]. The current problem of electricity generation is the determination of the optimal load allocation at the plant to obtain the cheapest cost with low emissions.

As an illustration is the case of Indonesia. In 2014, the use of fuel oil for electricity generation is 12.35% of total fuel used, and then coal fuel is 51.35% while gas is 23.41%. These data suggest that the choice of cheap fuels does not necessarily provide a single solution for the procurement of electrical energy to meet demand. The main reason is emissions; all countries including Indonesia are committed to lower emission levels in their respective countries.

2. Economic Dispatch Model

Initially the main objective of economic dispatch is to simply minimize the total cost [3]–[9]. Furthermore, economic dispatch is required to also consider environmental factors by including emissions [10]–[18]. The purpose of this electrical load allocation model is to minimize the operational costs of a plant while minimizing the impact on the environment through reducing emissions. Generally, between cost minimization and emission minimization are two contradictory things. Reduce costs easily by using the cheapest fuel such as coal. However, the move will result in an increase in emissions.

The model in this paper is based on 4 scenarios. First, the load allocation model using only power plant owned by State Electricity Company/Perusahaan Listrik Negara (PLN). Second, the load allocation model using PLN + Independent Power Producer (IPP) combined power plant. Third, the model of load allocation by using a combination of PLN + rented power plant. Fourth, the load allocation model using PLN + excess power plant.

3. Load Allocation Model with PLN Power Plant.

The general form of allocation of generation loads by economic dispatch model is as follows:

Minimize $\sum_{i=1}^{N} F_i\,(P_i) = a_i\,P_i^2 + b_iP_i + c_i$ (1)

Constraints

$\sum_{i=1}^{N} P_i = PD$ (2)

$Pmin_i \leq P_i \leq Pmax_i$ (3)

The above model aims to minimize fuel costs. The first limitation indicates that the amount of energy generated must be equal to the demand for electricity, this limitation is called the power balance constraints. The second limitation describes the limits of the generator. The power released by each generator has a maximum value depending on the type of plant used. The plant also has a minimum limit in order to operate with a stable

4. Load allocation model with PLN + IPP power plant

Minimize $\sum_{i=1}^{N} F_i\,(P_i) + \sum_{i=1}^{N} F_{iIP}\,(P_{iIP}) \;= a_i\,P_i^2 + b_iP_i + c_i + a_{iIP}\,P_{iIP}^2 + b_{iIP}P_{iIP} + c_{iIP}$ (4)

Constraints

$\sum_{i=1}^{N} P_i + \sum_{i=1}^{N} p_{iIP} = PD$ (5)

$P_{imin} \leq P_i \leq P_{imax}$ (6)

$P_{iIPmin} \leq P_{iIP} \leq P_{iIPmax}$ (7)

5. Load allocation model with PLN + Rent power plant

Minimize $\sum_{i=1}^{N} F_i\,(P_i) + \sum_{i=1}^{N} F_{iR}\,(P_{iR}) \;= a_i\,P_i^2 + b_iP_i + c_i + a_{iR}\,P_{iR}^2 + b_{iR}P_{iR} + c_{iR}$ (8)

Constraints

$\sum_{i=1}^{N} P_i + \sum_{i=1}^{N} p_{iR} = PD$ (9)

$P_{imin} \leq P_i \leq P_{imax}$ (10)

$P_{iRmin} \leq P_{iR} \leq P_{iRmax}$ (11)

6. Load allocation model with PLN + Excess power plant

Minimize $\sum_{i=1}^{N} F_i\,(P_i) + \sum_{i=1}^{N} F_{iE}\,(P_{iE}) \;= a_i\,P_i^2 + b_iP_i + c_i + a_{iE}\,P_{iE}^2 + b_{iE}P_{iE} + c_{iE}$ (12)

Constraints

$\sum_{i=1}^{N} P_i + \sum_{i=1}^{N} p_{iE} = PD$ (13)

$P_{imin} \leq P_i \leq P_{imax}$ (14)

$P_{iEmin} \leq P_{iE} \leq P_{iEmax}$ (15)

7. Results

The experiment was conducted using 4 scenarios and 2 conditions. The four scenarios used are: (1) using PLN's power plant, (2) using PLN + IPP power plant, (3) using PLN + Rental power plant, and (4) using PLN + Excess power plant. While the two conditions considered are peak load conditions and low load conditions.

Scenario 1 with peak load conditions yields the largest total cost of 1,159,917.51 and produce the largest emission of 173.7. While under low load conditions, this scenario yields a total cost of 209,487.94 and an emission of 93.9. Scenario 2 with peak load conditions resulted in the lowest total cost of 273,136.5 and the lowest emission of 142.0. For low load conditions, this scenario is also capable of producing the lowest cost of 118.133.1 but the resulting emissions of 95.5. Scenario 3 with peak load conditions resulted in total cost of 634,620.08 and emissions of 155.9. While at low load this scenario generates a total cost of 227,274.53 and emissions of 86.5. Scenario 4 with peak load conditions resulted in total cost of 753,529.92 and emissions of 166.7. While under low load conditions, this scenario generates a total cost of 332,655.37 with a total emission of 98.5 (table 1)

Table 1: Result of numerical experiment

Scenarios	Conditions	Total Cost	Emission
1	Peak	1,159,917.51	173.7
	Low	209,487.94	93.9
2	Peak	273,136.5	142.0
	Low	118,133.1	95.5
3	Peak	634,620.08	155.9
	Low	227,274.53	86.5
4	Peak	753,529.92	166.7
	Low	332,655.37	98.5

8. Discussion and Further Research

The maximum load on the plant system in East Kalimantan is 343.4 kW. While the minimum load of 194.86 kW. This load can be met from interconnected power plants. Based on the ownership of the plant, the power plant is divided into 4. First, the power plant owned by PLN. Secondly, the plant is owned by Independent Power Producer (IPP). Third, the plant is owned by another party but is leased by PLN. Fourth, the plant has the other party and the excess of its production can be bought by PLN.

Based on the ownership of the power plant, there are 4 scenarios to fulfill the electrical load. Scenario 1: using PLN's power plant. Scenario 2: using PLN and IPP power plants. Scenario 3: using PLN's power plant and rental generator. Scenario 4: Own generator + excess capacity. There are 2 conditions of electrical load that is peak load conditions and minimal load conditions. When the system is in peak load condition, the PLN + IPP scenario will produce two advantages at a minimum value, i.e., price and emissions. Conversely, scenario 1 in the form of use of PLN's power plant alone will result in the highest cost and emission.

While the system is in the lowest load condition, the lowest fuel cost is the combination of PLN + IPP generator, while the most expensive is the combined generator between PLN + Excess. While the lowest emission is when using a combination of power plants owned by PLN + Rent.

Based on the 4 scenarios above, it is found that the result of load allocation for peak load condition is not the same as low load condition. This raises two types of problems which lead to further research. The first one is static economic dispatch to dynamic economic dispatch. The above load allocation uses load allocation assuming that the electrical load does not change, so load allocation is made based on 1 type of load. The allocation of these loads does not match the real conditions. Seen from 4 plant scenarios for 2 types of time, they do not present the same results. Peak load time is more appropriate using the second scenario, which is a combination of PLN + IPP. While the low load time is more appropriate using the scenario, PLN + IPP while the lowest emissions from the scenario PLN + Rent.

Second issue is reliability. Reliability system is defined as the ability of the electric power system to supply additional electrical power caused by damage to one of the generators. In the Mahakam system, the largest power plant is P11 with a maximum power of 200 MW. Based on the total power owned by all power plants, then when there is disruption at the largest power plant, the power system will not be able to supply the required power, so from the point of view of reliability still need to be added new generator to do the allocation of the load.

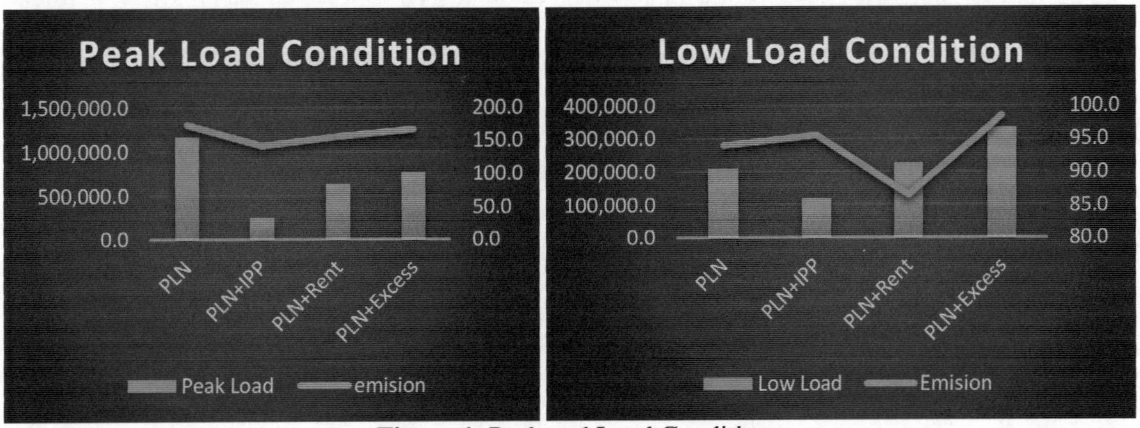

Figure 1. Peak and Load Condition

9. Conclusion

The allocation of load in East Kalimantan would be better if it combines PLN + IPP power plant. The combination of both propellers will result in an allocation of electrical loads with a minimum total cost and the smallest emissions for peak load conditions. As for the low load, the combination of PLN + IPP power plant will still provide the smallest total cost but the smallest emission is generated from the combined PLN + Rent power plant. All of the above scenarios assume that there is no disruption to the entire plant. Further research needs to be conducted to find proper allocation of the load if one of the generators is damaged.

Reference

[1] He Y, Liu Y, Wang J, Xia T and Zhao Y 2014 Low-carbon-oriented dynamic optimization of residential energy pricing in China *Energy* **66** pp 610–623

[2] Mohammed Y S, Mustafa M W, and Bashir N 2014 Hybrid renewable energy systems for off-grid electric power : Review of substantial issues *Renew. Sustain. Energy Rev* **35** pp 527–539

[3] Wahyuda and Santosa B 2015 Dynamic Pricing in Electricity: Research Potential in Indonesia *Procedia Manuf* **4** pp 300–306

[4] Niknam T, Mojarrad H D, Meymand H Z and Firouzi B B 2011 A new honey bee mating optimization algorithm for non-smooth economic dispatch *Energy* **36** pp 896–908

[5] Zhong H, Xia Q, Chen Y and Kang C 2015 Energy-saving generation dispatch toward a

sustainable electric power industry in China *Energy Policy* **83** pp 14–25

[6] Chen C -L, Chen Z -Y and Lee T -Y 2014 Multi-area economic generation and reserve dispatch considering large-scale integration of wind power *Int. J. Electr. Power Energy Syst* **55** pp 171–178

[7] Mandal B, Kumar P and Mandal S 2014 Economic load dispatch using krill herd algorithm," *Int. J. Electr. POWER ENERGY Syst.* **57** pp 1–10

[8] Meng A, Li J and Yin H 2016 An efficient crisscross optimization solution to large-scale non-convex economic load dispatch with multiple fuel types and valve-point effects *Energy* **113** pp 1147–1161

[9] Zou D, Li S, Wang G, Li Z and Ouyang H 2016 An improved differential evolution algorithm for the economic load dispatch problems with or without valve-point effects *Appl. Energy*, **181** pp 375–390

[10] Zhao X, Wu L and Zhang S 2013 Joint environmental and economic power dispatch considering wind power integration: Empirical analysis from Liaoning Province of China *Renew. Energy* **52** pp 260–265

[11] Singh N and Kumar Y 2015 Multiobjective Economic Load Dispatch Problem Solved by New PSO *Hindawi* vol. **2015** pp 1–7

[12] Pandit M, Chaudhary V, Dubey H M and Panigrahi B K 2015 Multi-period wind integrated optimal dispatch using series PSO-DE with time-varying Gaussian membership function based fuzzy selection *Int. J. Electr. Power Energy Syst.* **73** pp 259–272.

[13] Palanichamy C and Babu N S 2008 Analytical solution for combined economic and emissions dispatch *Electr. Power Syst. Res.* **78** pp 1129–1137

[14] Kumar R, Sadu A, Kumar R and Panda S K 2012 A novel multi-objective directed bee colony optimization algorithm for multi-objective emission constrained economic power dispatch *Int. J. Electr. Power Energy Syst.* **43** pp 1241–1250

[15] Nwulu N I and Xia X 2015 Multi-objective dynamic economic emission dispatch of electric power generation integrated with game theory based demand response programs *Energy Convers. Manag.* **89** pp 963–974

[16] Kumar R, Sharma D and Sadu A 2011 A hybrid multi-agent based particle swarm optimization algorithm for economic power dispatch *Int. J. Electr. Power Energy Syst.* **33** pp 115–123

[17] Ghasemi M, Aghaei J, Akbari E, Ghavidel S and Li L 2016 A differential evolution particle swarm optimizer for various types of multi-area economic dispatch problems *Energy* **107** pp 182–195

[18] Zhou J, Wang C, Li Y, Wang P, Li C, Lu P and Mo L 2017 A multi-objective multi-population ant colony optimization for economic emission dispatch considering power system security **45** pp 684–704

Shift scheduling model considering workload and worker's preference for security department

A Herawati, D R Yuniartha[1*], I L I Purnama[1] and LT Dewi[1]
Department of Industrial Engineering, Universitas Atma Jaya Yogyakarta, Yogyakarta, Indonesia

* dena@mail.uajy.ac.id

Abstract. Security department operates for 24 hours and applies shift scheduling to organize its workers as well as in hotel industry. This research has been conducted to develop shift scheduling model considering the workers physical workload using rating of perceived exertion (RPE) Borg's Scale and workers' preference to accommodate schedule flexibility. The mathematic model is developed in integer linear programming and results optimal solution for simple problem. Resulting shift schedule of the developed model has equally distribution shift allocation among workers to balance the physical workload and give flexibility for workers in working hours arrangement.

1. Introduction

Personnel scheduling is a complex problem since it has many factors to be considered, not only satisfying workforce requirement, management rules, and government regulation, but also the human factor itself as the main role. Biologically people work at morning to afternoon, but nowadays increasing productivity drives many sectors operate for more than 8 hours a day and also in weekend day, mainly in public service such as health center, security service, transportation service, hotels, and retail. Many researchers have found that shift scheduling could cause fatigue and influences workers' performance, such as in [1][2][3][4][5]. So that the shift arrangement to the workers become one of the important factor considered in managing workers.

Research in [6], [7], and [8] have observed 3 departments of 20 hotels in Yogyakarta, Indonesia, i.e. Security Department, Front Office Department, and Housekeeping Department. There have been identified weaknesses of applied shift schedule for all observed departments [6]. Therewith, [7] has measured workers' physical and psychosocial workload from the same observed hotels. Since that the 3 observed departments have different job description then the worker's physical workload has measured using rating of perceived exertion (RPE) Borg's Scale, conversion of measured worker's heart rate in performing their job; and using Copenhagen Psychosocial Questionnaire (COPSOQ) for the psychosocial workload to assess worker's subjective perception of individual interaction with their job and work environment. Some of observed hotels for all the departments in [7] have indicated workers dissatisfaction, shown by moderate level of workers' psychosocial workload. This indication may be caused by weaknesses of the applied shift schedule, i.e. long working hours in a day, short between-shift break time, and consecutive night shift [6]; in spite of it has found that there is no direct relation between applied shift schedule and workers' workload of the observed hotels [8].

This research focus on developing shift scheduling model for Hotel Security Department, to eliminate the shift schedule weaknesses in [6] and consider the workers' workload. Particularly, security departments operate for 24 hours every day and definitely apply shift system for managing its workers. The workers have to be in good physical condition moreover in night shift. There is strong relationship between workers preference, attitudes, and retention indicators toward shift schedule, especially for working in evening or night shift and weekend day [9]. Furthermore, [9] has found that the key of reducing negative workers reactions to working outside standard hours and day is managing better work schedule.

The workload balance has been considered in personnel scheduling in many ways; [10] has develop integer programming model for shift scheduling of a hotel's engineering department. The shift scheduling model in [10] has balanced the workers' workload by limits the working hours in a week and allocates adequate rest hours between shift allocations. Shift scheduling models considering working hours to balance worker's workload have been studied in [10][12][13][14]. The shift scheduling model could manage the workload balance among workers by considering the shift allocation or the off-day as discussed in [15][16][16][18][19]. The allocation of shift and off-day to balance workers' workload balance could be arranged in cyclic scheduling's [10][15][20]. Besides balancing the workload, researches in personnel scheduling also focus on maintaining work-life balance to accommodate workers' social and family life. Shift scheduling in [12][14][20][21] has considered workers' preference to develop schedule flexibility accommodating work-life balance. The developed shift scheduling model in this research would balance the workers' physical workload in rating of perceived exertion (RPE) Borg's Scale to maintain workers' physical performance. The psychosocial workload relates with workers' social and family time which is in evening or night and weekend day. So that the developed shift scheduling model would consider schedule flexibility applying worker's preference of not assigned in certain shift in certain day. The management rule considered in the model would base on shift scheduling parameters in [6], referred to [20] for the similar parameters and [19]for balancing the physical workload.

2. Shift Scheduling Model

The developed shift scheduling model in this research would eliminate schedule weaknesses founded in [6] by developing shift allocation constraint and consecutive night shift constrain. The physical workload balancing would manage the shift allocation as in [19] and consecutive workdays arrangement in workstretch. The main objective of the developed shift scheduling model is worker's preference of assigning in certain shift in certain day fulfillment. So that the objective function of the mathematic model is minimizing violation of workers' preference fulfillment.

$$\sum_{i=1}^{N}\sum_{j=1}^{3}\sum_{k=1}^{P} R_{ijk} Q_{ijk} \tag{1}$$

Parameters and variables in the model are:

i : index for worker, $i = 1, 2, \ldots, N$
j : index for shift, $j = 1, 2, 3$
k : index for day, $k = 1, 2, \ldots, P$
m_j : minimum number of workers in shift j
n_j : maximum number of workers in shift j
A : physical workload in RPE scale for shift 1
B : physical workload in RPE scale for shift 2
C : physical workload in RPE scale for shift 3
S : physical workload deviation among workers
Y_{ijk} : binary value for worker's preference, 1 if worker i ask for assigned in shift j of day k
R_{ijk} : weight of worker's preference of assigned in shift j of day k, i.e. {1, 2, 3, 4}
X_{ijk} : binary value of worker assignment, 1 if worker i assigned in shift j of day k

Constraints of the mathematic model are:

A worker would be assigned only one shift in one day or had off-day

$$\sum_{j=1}^{3} X_{ijk} \leq 1 \qquad \forall i, \forall k \tag{2}$$

A worker who have be assigned in shift 3, would not be assigned in shift 1 on the next day

$$X_{i3k} + X_{i1(k+1)} = 1 \qquad \forall i, \forall k \tag{3}$$

A worker who have be assigned in shift 2, would not be assigned in shift 1 on the next day

$$X_{i2k} + X_{i1(k+1)} = 1 \qquad \forall i, \forall k \tag{4}$$

A worker would be assigned in shift 3 for maximum 2 consecutive days

$$X_{i3k} + X_{i3(k+1)} + X_{i3(k+2)} = 1 \qquad \forall i, \forall k \tag{5}$$

Maximum and minimum number of workers in a shift

$$\sum_{i=1}^{N} X_{ijk} \leq n_j \qquad \forall j, \forall k \tag{6}$$

$$\sum_{i=1}^{N} X_{ijk} \geq m_j \qquad \forall j, \forall k \tag{7}$$

Balancing the physical workload

$$\sum_{i=1}^{N}\sum_{k=1}^{P} AX_{i1k} + \sum_{i=1}^{N}\sum_{k=1}^{P} BX_{i2k} + \sum_{i=1}^{N}\sum_{k=1}^{P} CX_{i3k} = L \tag{8}$$

$$\frac{L}{N} = S \tag{9}$$

$$\sum_{k=1}^{P} AX_{i1k} + \sum_{k=1}^{P} BX_{i2k} + \sum_{k=1}^{P} CX_{i3k} \leq L + S \qquad \forall i \tag{10}$$

$$\sum_{k=1}^{P} AX_{i1k} + \sum_{k=1}^{P} BX_{i2k} + \sum_{k=1}^{P} CX_{i3k} \geq L - S \qquad \forall i \tag{11}$$

Worker's preference fulfillment

$$\left| X_{ijk} - Y_{ijk} \right| = Q_{ijk} \tag{12}$$

3. Model Evaluation and Discussion

The developed shift scheduling model has been evaluated using actual data investigated in [6] and physical workload data in [7]. Table 2 shows the example of scheduling result for security department which has 4 workers and other scheduling parameters in Table 1.

Table 1. Scheduling parameter for security department with 4 workers.

Paramater	Shift 1	Shift 2	Shift 3
Number of minimum workers	1	1	1
Number of maximum workers	1	1	2
RPE scale of physical workload	10.7	10.3	10.3
Schedule period		30 days	
Physical workload deviation among workers		1	

Table 2. Schedule Result for security department with 4.

Day	Worker			
	A	B	C	D
1	2	1		3
2	2	3	1	3

Day	Worker			
	A	B	C	D
3	2	3	1	
4	3		2	1
5	3	2	3	1
6		2	3	1
7	1	2		3
8	2	3	1	3
9	2	3	1	
10	3		1	2
11	3	1	3	2
12		1	3	2
13	1	2		3
14	1	3	2	3
15	1	3	2	
16	3		2	1
17	3	1	3	2
18		1	3	2
19	2	1		3
20	2	3	1	3
21	2	3	1	
22	3		1	2
23	3	1	3	2
24		1	3	2
25	1	2		3
26	1	3	2	3
27	1	3	2	
28	3		2	1
29	3	3	2	1
30		3	2	1

The shift scheduling model has resulted better schedule in shift allocation arrangement and consecutive night shift. Resulted schedule has eliminated long working hours by assigned a worker only one shift in a day in constraint (2), short break time between shift by applying forward rotation in constraint (3) and (4), and long consecutive night shift by constraint (5). To get the forward rotation, the resulted schedule will assign off-day after 2 consecutive night shifts.

The resulted schedule has balanced the workers' physical workload so that the shift evenly allocated among workers. Physical workload in RPE scale for the workers in the resulted schedule of Table 2 are 260.3, 260.7, 260.7, and 260.3, respectively for worker A, B, C, and D. The schedule flexibility will assign the worker's preference for certain shift in certain day that also fulfills the workforce requirement in each shift but restricted off-day assignment. For the case in Table 2, worker A, B, C, and D request for assigneing in shift 2 in day 2 as tight worker's preference. To fulfill the minimum number of workers in all shifts and forward rotation, the model only assign worker 2 in shift 2 of day 2, and other workers in shift 1 and 3.

Model evaluation has been conducted to identify the model behavior by modifying the RPE scale of physical workload and number of workers. For the same value of physical workload for all shift would resulted equal number of shift allocation among workers for each shift. When one of the shifts has physical workload value far different among others, than the model would evenly allocate the highest physical workload shift among workers. As the consequences, the other shifts allocation would not be evenly distributed to fulfill the minimum number of workers in a shift and forward rotation.

Increasing number of workers as logically would resulted more balance physical workload as well as number of each shift allocation among workers; however it would need more computation time.

4. Conclusion and Future Research

The developed shift scheduling model has resulted optimal solution for simple problem but it would time consuming for more number of workers. Resulted schedule has assigned workers in forward rotation shift allocation to give adequate break time between shifts. The model has equally distributed shift allocation among workers to balance the physical workload. As security department operates for 24 hours, the worker's preference of shift assignment in certain day has provided flexibility for workers to arrange their work hours. The shift scheduling model considering workload for housekeeping department has been already conducted in [22] and for front office department has been considered.

References

[1] Chiang F F T, Birtch T A and Kwan H K 2010 The moderating roles of job control and work-life balance practices on employee stress in the hotel and catering industry *International Journal of Hospitality Management* **29** 25–32

[2] Puttonen S, Härmä M and Win C H 2010 Shift work and cardiovascular disease – pathways from circadian stress to morbidity *Scandinavian Journal of Work, Environment & Health* **36** 96-108

[3] Lee G, Magnini V P and Kim B C P 2011 Employee satisfaction with schedule flexibility: Psychological antecedents and consequences within the workplace *International Journal of Hospitality Management* **30** 22–30

[4] Eldevik M F, Flo E, Moen B E, Pallesen S and Bjorvatn B 2013 Insomnia, excessive sleepiness, excessive fatigue, anxiety, depression and shift work disorder in nurses having less than 11 hours in-between shifts *PloS ONE* **8** (8)

[5] Louly M A O 2013 A goal programming model for staff scheduling at a telecommunications center *Journal of Mathematical Modelling and Algorithms* **12** 167–178

[6] Purnama I L I and Yuniartha D R 2014 Shift-scheduling characteristic identification of non-star hotels industry in Yogyakarta Indonesia *Proceeding of The 15th Asia Pacific Industrial Engineering and Management Systems Conference* 1442-1448.

[7] Dewi L T, Yuniartha D R and Purnama I L I 2014 Psychosocial and physical work load hotel's shift worker in Yogyakarta Indonesia *Proceeding of The 15th Asia Pacific Industrial Engineering and Management Systems Conference* 367-372

[8] Yuniartha D R, Dewi L T and Purnama I L I 2015 Work load identification of hotel's shift scheduling in Yogyakarta Indonesia *Proceeding of The 16th Asia Pacific Industrial Engineering and Management Systems Conference* 929-935

[9] Wittmer J L S, Shepard A K and Martin J E 2015 Schedule preferences, congruence and employee outcomes in unionized shift workers *American Journal of Business* **30** 192-110

[10] Kassa B A and Tizazu A E 2013 Personnel scheduling using an integer programming model - an application at Avanti Blue-Nile Hotels *SpringerPlus 2013* 2 333

[11] Rocha M, Oliveira J F and Carravilla M A 2014 A constructive heuristic for staff scheduling in the glass industry *Ann Oper Res* 217 463–478

[12] Han A F, Li E C 2014 A constraint programming-based approach to the crew scheduling problem of the Taipei mass rapid transit system *Ann Oper Res* 223 173–193

[13] Todovic D, Makajic-Nikolic D, Kostic-Stankovic M and Martic M 2015 Police officer scheduling using goal programming *Policing: An International Journal of Police Strategies & Management* **38** (2) 295-313

[14] van Veldhoven S, Post G, van der Veen E and Curtois T 2016 An assessment of a days off decomposition approach to personnel shift scheduling *Ann Oper Res* 239 207–223

[15] Rocha M, Oliveira J F and Carravilla M A 2013 Cyclic staff scheduling: optimization models for some real-life problems *J Sched* **16** 231–242
[16] Smalley H K and Keskinocak P 2016 Automated medical resident rotation and shift scheduling to ensure quality resident education and patient care *Health Care Manag Sci* **19** 66–88
[17] Dahmen S and Rekik M 2015 Solving multi-activity multi-day shift scheduling problems with a hybrid heuristic *J Sched* **18** 207–223
[18] van der Veen E, Hans E W, Post G F and Veltman B 2015 Shift rostering using decomposition: assign weekend shifts first *J Sched* **18** 29–43
[19] Dewi D S and Septiana T 2015 Workforce scheduling considering physical and mental workload: a case study of domestic freight forwarding *Proc. of 3rd Industrial Engineering and Service Science* 531–538.
[20] Maenhout B and Vanhoucke M 2017 A resource type analysis of the integrated project scheduling and personnel staffing problem *Ann Oper Res* 252 407–433
[21] Labidi M, Mrad M, Gharbi A and Louly M A 2014 Scheduling IT Staff at a Bank: A Mathematical Programming Approach *The Scientific World Journal* 2014
[22] Jafari H and Salmasi N 2015 Maximizing the nurses' preferences in nurse scheduling problem: mathematical modeling and a meta-heuristic algorithm *J Ind Eng Int* **11** 439–458
[23] Jockvom, Yuniartha D R, Purnama I L I and Dewi L T 2016 Shift scheduling model considering workload for housekeeping department *Proc. of The 17th Asia Pacific Industrial Engineering and Management Systems Conf.*

Acknowledgments
This work is supported by General Directorate of Higher Education, Ministry of Research Technology and Higher Education, Republic of Indonesia under Research Grant on Competitive Grant Scheme No. DIPA-042.06-0.1.401516/2016.

Split delivery vehicle routing problem with time windows: a case study

E Latiffianti[1*], N Siswanto[1], R A Firmandani[1]
Department of Industrial Engineering, Institut Teknologi Sepuluh Nopember, Kampus ITS Sukolilo-Surabaya 60111, Indonesia

*effi_l@ie.its.ac.id

Abstract. This paper aims to implement an extension of VRP so called split delivery vehicle routing problem (SDVRP) with time windows in a case study involving pickups and deliveries of workers from several points of origin and several destinations. Each origin represents a bus stop and the destination represents either site or office location. An integer linear programming of the SDVRP problem is presented. The solution was generated using three stages of defining the starting points, assigning busses, and solving the SDVRP with time windows using an exact method. Although the overall computational time was relatively lengthy, the results indicated that the produced solution was better than the existing routing and scheduling that the firm used. The produced solution was also capable of reducing fuel cost by 9% that was obtained from shorter total distance travelled by the shuttle buses.

1. Introduction

A vehicle routing problem is a combinatorial integer programming problem that search for an optimal set of routes for a set of vehicles that move from a depot to one or more destinations to make a delivery to a set of given customers. Typical objective functions may include minimizing costs, fulfilling demands, maximizing profits, and maximizing utilities. The classical model that was first introduced in 1959 [1] has been extended to many different types such as: vehicle routing problem with time windows (VRPTW), vehicle routing problem with deliveries and pickups (VRPDP), split delivery vehicle routing problem (SDVRP), and capacitated vehicle routing problem (CVRP). More VRP extensions and solving methods were studied for both theoretical and practical purposes. As more manufacturers and distributors realized that transportation expense is one significant factor contributes to the cost of goods sold, the application of computer-generated solutions increased significantly. The transportation cost is especially more significant for a developed country like Indonesia where logistics sector like road and freight transport and distribution services underperformed [2][3]. Although VRP were mostly applied in product distribution, some were also applicable in other cases such as school bus [4][5], worker shuttle bus, other public transport [6], and marine border security patrol.

A study has revealed that computer-generated solution may reduce transportation costs by 5% to 20% [7]. With the advantages a VRP possibly offers, this paper aims to implement an extension of VRP for a case study in a coal-mining firm. In the case study, the firm provided shuttle buses for the workers that picked them up from several locations and drop them off at several destinations. However, the current approach the firm used has made the cost of transporting workers soaring. The transportation cost included fuel cost, maintenance cost, and driver cost. Thus, a better routing and scheduling solution were required. The application of VRP, specifically SDVRP, in this case study was expected to produce a better routing and scheduling decisions for worker shuttle buses. The SDVRP is one of VRP extensions that drawn numerous attentions. It is applicable for many problems where customers or points can be visited more than once. It was introduced by Dror and Trudeau [8][9] and became widely studied later [10][11][12][13].

This paper presents the implementation of SDVRP and an exact method to generate a solution for the selected case study. The focus of the paper is on the construction of the problem formula and its implementation in the case study. No further analysis has been made in the performance of method used or comparison with any other methods. The remainder of the paper is organized as follows. The next section provides a description of the case study to be solved. It is followed by a mathematical formulation of the SDVRP, the computational procedure, results and discussion, and then concluding remarks.

2. Problem Description

The case study takes place in a coal-mining firm. The firm business process includes exploration, production, marketing, and distributing the products worldwide. The first two processes involve the use of heavy vehicles. Thus, to minimize the risk of accident in the work place and to provide a safer working environment, the firm decided to establish an access restriction for several offices and sites. This includes restricting any private vehicles use by the workers, except for the very important persons. As a consequence, the firm should provide official vehicles for the management and shuttle buses for all workers. When this study was performed, the firm was employing 49 shuttle buses to transport 1726 workers from 7 pickup points to 12 destinations – sites or offices – with different working hours. The shuttle bus schedule planning was done manually by considering only the capacity of the bus and the start of working hours. The shuttle buses were available in different capacity sizes; the small ones can take up to 29 persons, while the larger buses varied in sizes, ranged from up to 42 persons to 47 persons. More detailed information about the number of available buses and the capacity are described in Table 1. Working schedule at destination points, workers pickups and delivery points, as well as the number of workers to be picked up at each point are shown in Table 2.

Table 1. Capacity and number of busses available.

Max capacity	29	42	43	44	45	46	47
Number of buses	10	2	3	3	21	5	5

Table 2. Schedule (time windows) and numbers of workers at each points.

Destination	Work	Number of workers waiting at pick up points						
(sites/offices)	starts at	1	2	3	4	5	6	7
1	7.00 am	3	0	1	8	2	38	2
2	7.00 am	18	7	0	6	1	36	0
3	7.00 am	42	0	23	0	0	0	25
4	6.00 am	7	14	3	5	5	9	3
5	6.00 am	7	9	1	9	2	21	3
6	6.00 am	34	35	16	92	16	73	8
7	7.00 am	56	49	56	11	5	44	16
8	7.00 am	42	56	25	31	11	21	24
9	6.00 am	31	27	10	26	17	30	8
10	7.00 am	39	16	25	0	4	12	5
11	6.30 am	54	24	63	0	11	0	34
12	6.00 am	26	34	15	36	12	43	11

3. Solving Procedure

While trying to solve the problem using exact method, we completely understand the consequence that it may take very long time to find the solution. Hence, we set up initial conditions that enable the exact method to work faster. The solving procedure was divided into three steps. The first step involved a set covering method to find the best possible points from where buses shall depart. This allowed the buses to start from the selected points instead of any points in the set. These selected points were then called as starting points. The second step is to assign each bus to a specific starting point that was selected in the first step. This made the SDVRP to be simpler to solve. And finally the last step aimed to find the routes that minimize the total distance travelled as formulated in Section 3. Formulation of set covering and bus assignment are described as follows.

3.1. Set Covering

The set covering aims to find a minimum number of starting points that covers all points on the given set. In this case, let L be the set of points with a total number of 7. The formulation can be described as follows.

Indexed Sets:
$i = \{1, 2, \ldots, 7\}$; point index
$j = \{1, 2, \ldots, 7\}$; point index

Parameters:
n: the number of points
C_{ij}: the value is 1 if point i is can covers point j

Decision Variables:
x_i: the value is one if point i is chosen as starting point

Objective:

$$\text{Minimize} \quad Z = \sum_{i \in L} x_i \tag{1}$$

Constraints:

$$\sum_{j \in L} C_{ij} x_j \geq 1 \qquad \forall i \in L \tag{2}$$

$$\sum_{i \in L} x_i \leq 3 \tag{3}$$

$$x_i \in \{1, 0\} \qquad \forall i \in N \tag{4}$$

The objective is represented by (1), which is to minimize the total number of starting point. Constraint (2) ensures the selected points can cover all other points in the set. Constraint (3) was added because the firm would like to limit the number of selected starting points up to 3. Constraint (4) sets the decision variables to have binary value. The input value for constraint (4) is given in Table 3. A value of 1 is given for a point (in row) that may cover another point (in column). This value is given based on two considerations; the first one is that they are close in distance, and the second is that they are in the same direction given the last stops are the drop-off points.

Table 3. Set covering input matrix.

	1	2	3	4	5	6	7
1	1	0	0	0	0	0	0
2	1	1	0	0	0	0	0
3	1	1	1	0	0	0	0
4	1	1	1	1	0	0	0
5	1	1	1	1	1	1	1
6	0	0	0	0	0	1	1
7	0	0	0	0	0	0	1

3.2. Bus Assignment

For the bus assignment, let P be the set of starting points and M is a set of vehicles. The formulation can be explained as follows.

Indexed Sets:

$i = \{1, 2, \ldots, p\}$; starting point index
$k = \{1, 2, \ldots, m\}$; vehicle index

Parameters:

p: the number of stopping points
m: the number of vehicles
Q^k: the capacity of vehicle k
d_i: the number of workers at stopping point i
A^k: the available number of vehicle k

Decision Variables:

x_i^k: the number of vehicles allocated in starting point i

Objective:

$$\text{Minimize} \quad Z = \sum_{k \in M} \sum_{i \in P} x_i^k \tag{5}$$

Constraints:

$$\sum_{k \in M} x_i^k Q^k \geq d_i \qquad \forall i \in P \tag{6}$$

$$\sum_{i \in P} x_i^k \leq A^k \qquad \forall k \in M \tag{7}$$

$$x_i^k \in \text{Integer} \qquad \forall k \in M; i \in P \tag{8}$$

The objective of the bus assignment is to minimize the total vehicles assigned as represented by (5). Constraint (6) guarantees that the total capacity at each point is greater than the number of workers to be transported. Constraint (7) ensures that the number of vehicle allocated does not exceed the availability and constraint (8) sets the value of decision variables to be integer.

3.3. The SDVRP Model with Time Windows

The SDVRP formulation can be defined as follows. Let S be the set of destination points, while $N = S \cup \{p\}$ is a set of all possible stopping points for all vehicles including the starting point (p), and M is a set of vehicles. The variables in the SDVRP formulation include the following.

Indexed Sets:

$i = \{1, 2, \ldots, n\}$; stopping point index

$j = \{1, 2, ..., n\}$; stopping point index
$k = \{1, 2, ..., m\}$; vehicle index

Parameters:
n: the number of stopping points
m: the number of vehicles
Q^k: the capacity of vehicle k
J_{ij}: the distance from stopping point i to stopping point j
W_{ij}: the travel time from stopping point i to stopping point j
d_i: the number of workers at stopping point i
e_i: opening time of location i
l_i: the schedule at which the work starts of location i
r: distance travelled for each unit of fuel consumption (km/litre)

Decision Variables:
$x_{i,j}^k$: 1 if vehicle k travels from stopping point i to stopping point j
y_i^k : the proportion of workers that are delivered to stopping point i by vehicle k
v_i^k : the number of workers that are delivered to stopping point i by vehicle k
t_i^k : departure time of vehicle k at stopping point i
u_i^k : the remaining number of workers in vehicle k after delivery at stopping point i

Objective:

$$\text{Minimize} \quad Z = \sum_{i \in N} \sum_{j \in N} \sum_{k \in M} \frac{1}{r} J_{ij} x_{ij}^k \tag{9}$$

Constraints:

$$\sum_{i \in N} x_{ip}^k = \sum_{j \in N} x_{pj}^k \qquad \forall p \in N; k \in M \tag{10}$$

$$\sum_{k \in M} y_{ik}^k = 1 \qquad \forall i \in S \tag{11}$$

$$\sum_{k \in M} v_i^k = d_i \qquad \forall i \in S \tag{12}$$

$$v_i^k \geq x_{ij}^k \qquad \forall i \in S; k \in M \tag{13}$$

$$\sum_{i \in S} d_i y_i^k \leq Q^k \qquad \forall k \in M \tag{14}$$

$$\sum_{j \in S} x_{ij}^k \geq y_i^k \qquad \forall i \in S; k \in M \tag{15}$$

$$d_i y_i^k \geq v_i^k \qquad \forall i \in S; k \in M \tag{16}$$

$$x_{ij}^k + x_{ji}^k \leq 1 \qquad \forall i \in N; j \in N; i \neq j; k \in M \tag{17}$$

$$u_i^k - u_j^k - Q^k x_{ii}^k \geq v_j^k - Q^k \qquad \forall i \in S; j \in S; i \neq j; k \in M \tag{18}$$

$$d_i \leq u_i^k \leq Q^k \qquad \forall i \in S; k \in M \tag{19}$$

$$x_{ijk} \leq \sum_{\substack{p \in N \\ p \neq i}} x_{jpk} \qquad \forall i \in S; j \in S; i \neq j; k \in M \tag{20}$$

$$t_i^k + Wi_j - t_j^k + 2M \; x_{ij}^k \leq 2M \qquad \forall i \in S; j \in S; i \neq j; k \in M \tag{21}$$

$$e_i \leq t_i^k \leq l_i \qquad \forall i \in N, k \in M \qquad (22)$$

$$x_{ijv} \in \{0,1\} \qquad \forall i \in N; j \in N; k \in M \qquad (23)$$

$$v_{ik} \in Integer \qquad \forall i \in N; k \in M \qquad (24)$$

The objective function is represented by (9), which is to minimize the total distance travelled. Constraint (10) ensures that each route continues and finally ends up at the last destination. Constraints (11) and (12) ensure all workers with destination *i* have been served. Constraint (13) guarantees that if location *i* is visited, then at least one worker is served. Constraint (14) makes sure that the number of workers in a vehicle does not exceed the capacity of the vehicle. Constraint (15) guarantees that if location *i* is visited by vehicle *k*, then the same vehicle leaves the location. Constraint (16) arranges the workers to be delivered to node *i* by vehicle *k*. Constraints (17) to (19) ensure that sub tours are eliminated, while constraint (20) ensures the fractional cycle is eliminated. Time windows are represented by constraints (21) and (22). Finally, constraints (23) and (24) make sure that variable are binary or integer based as intended.

Input data for the SDVRP are given as follows. Table 4 shows the distance between two points within the stopping point set. Other data used in this case study include the distance between each starting point found in the set covering and the destinations as shown in Table 5 and the travel time required between a starting point and destinations as shown in Table 6.

Table 4. Distance matrix between stopping points in kilometer (km).

	1	**2**	**3**	**4**	**5**	**6**	**7**
1	-	1.6	3.9	4	7.1	8.1	12.3
2	1.6	-	2.2	2.4	5.4	6.4	10.7
3	3.9	2.2	-	0.12	3.2	4.2	8.4
4	4	2.4	0.12	-	3	4	8.3
5	7.1	5.4	3.2	3	-	2.2	6.5
6	8.1	6.4	4.2	4	2.2	-	4.3
7	12.3	10.7	8.4	8.3	6.5	4.3	-

Table 5. Distance matrix between starting points and destinations in kilometer (Km).

		Starting points		Destinations (offices/sites)											
		1	2	1	2	3	4	5	6	7	8	9	10	11	12
Starting points	1	-	-	10.4	21.5	10.5	10	11.2	9.4	15.8	20.9	10.4	9.8	15.8	9.4
	2	-	-	6.5	21	10	9.5	4.7	4.4	6.4	17	5.5	4.9	11.9	3.2
Destinations (offices/sites)	1	10.4	6.5	-	15.1	4.1	3.7	1.8	2.1	8.5	10.5	0.65	2.4	5.4	6
	2	21.5	21	15.1	-	12.4	12.1	15.9	16.2	22.5	25.6	15.3	16.5	20.5	20.1
	3	10.5	10	4.1	12.4	-	0.65	4.8	9.2	11.6	14.6	4.1	5.5	9.1	9.2
	4	10	9.5	3.7	12.1	0.65	-	4.5	4.8	11.2	14.2	3.7	5.1	9.1	8.7
	5	11.2	4.7	1.8	15.9	4.8	4.5	-	0.23	6.7	12.3	0.85	0.6	7.2	4.2
	6	9.4	4.4	2.1	16.2	9.2	4.8	0.23	-	6.5	12.1	1.2	0.4	7	4.1
	7	15.8	6.4	8.5	22.5	11.6	11.2	6.7	6.5	-	19	7.4	6.1	14	3.2
	8	20.9	17	10.5	25.6	14.6	14.2	12.3	12.1	19	-	11.2	13	5.1	16.5
	9	10.4	5.5	0.65	15.3	4.1	3.7	0.85	1.2	7.4	11.2	-	1.4	6.1	5.1
	10	9.8	4.9	2.4	16.5	5.5	5.1	0.6	0.4	6.1	13	1.4	-	7.8	3.7
	11	15.8	11.9	5.4	20.5	9.1	9.1	7.2	7	14	5.1	6.1	7.8	-	11.4
	12	9.4	3.2	6	20.1	9.2	8.7	4.2	4.1	3.2	16.5	5.1	3.7	11.4	-

Table 6. Travel time between starting points and destinations in minutes.

		Starting points		Destinations (offices/sites)											
		1	2	1	2	3	4	5	6	7	8	9	10	11	12
Starting points	1	-	-	24	38	24	23	24	24	34	37	23	25	32	30
	2	-	-	12	31	18	16	9	9	11	25	10	10	25	6
Destinations (offices/sites)	1	24	12	-	21	9	8	3	4	17	15	1	4	10	10
	2	38	31	21	-	18	17	22	22	33	38	20	23	30	29
	3	24	18	9	18	-	2	9	10	19	20	8	11	12	16
	4	23	16	8	17	2	-	7	7	18	22	7	8	17	14
	5	24	9	3	22	9	7	-	1	11	20	1	1	13	7
	6	24	9	4	22	10	7	1	-	11	19	2	1	12	7
	7	34	11	17	33	19	18	11	11	-	45	12	10	35	6
	8	37	25	15	38	20	22	20	19	45	-	17	21	8	27
	9	23	10	1	20	8	7	1	2	12	17	-	3	13	9
	10	25	10	4	23	11	8	1	1	10	21	3	-	13	6
	11	32	25	10	30	12	17	13	12	35	8	13	13	-	21
	12	30	6	10	29	16	14	7	7	6	27	9	6	21	-

4. Results and Discussion

Performing the three stages procedure as aforementioned above, we started by defining two starting points where the busses should depart based on the result from set covering. At the second stage, we found that the total number of buses required in this case study was 25 at the first departure point and 12 for the second point. At the last stage, the SDVRP produces 23 routes in total, 12 routes for the first departure point and 11 routes for the second departure point. All buses were successfully scheduled within the given time windows.

The computational time required to solve the problem using the procedure was relatively long. Although the set covering and bus assignment were solved in relatively no time, the SDVRP in this case was solved in about 2.5 to 11 hours depending on the number of iterations that ranged between 10 million to 50 million. The SDVRP was solved for 2 different starting points. The first one did not improve after 20 million iterations, while for the second starting point we found the best result in 40 million iterations.

Based on the solution, several cost improvements were found. With the smaller number of buses to operate, the firm can save operational cost that included driver cost, bus maintenance cost, and fuel cost. With only fuel cost data available, we also found that this solution can save about 9% of the fuel cost in comparison with the existing routing and scheduling. In the case where the buses are not yet available, the use of computational solution would also benefit the company with less investment cost for buses. When more complete data is available, a study will be able provide important information on total cost reduction instead of fuel cost only.

5. Conclusion

The solution to the problem from the case study was found through 3 stages of: 1) defining the starting point and clustering using set covering, 2) assigning bus from each starting point, and 3) determining the route by solving the SDVRP with time windows using an exact method. With 7 pickup points and 12 destinations, this combinatorial problem required a significant duration to find the global optimum solution when an exact method is used. This is not a problem when the VRP does not need to be solved quickly and it does not need to be run many times (i.e. when the problem is multiple), otherwise the use of metaheuristics is deemed more practical. Based on our solution, we found that all the pickups and deliveries can be performed using 37 buses out of the available 49 buses that previously used. Different schedules for each bus were applied. Implementation of these schedules and routes allowed all shuttle buses to operate without violating the time windows and speed limit policy set by the firm. Additionally, this solution is capable of decreasing the fuel cost by 9% of the given initial cost. Other potential savings include reduction in investment cost (of buses) and operational cost that includes drivers and maintenance cost.

6. Acknowledgement

The work presented in this paper was funded by Institut Teknologi Sepuluh Nopember under the funding scheme of *Penelitian Pemula – Dana Lokal ITS Tahun 2017*.

7. References

[1] G. B. Dantzig and J. H. Ramser, 1959. The Truck Dispatching Problem, *Management Science,* Vol. **6**, No. 1, pp. 80-91.

[2] OECD, 2015. *OECD Economic Surveys Indonesia March 2015 Overview,* [Online] Available at: https://www.oecd.org/eco/surveys/Overview-Indonesia-2015.pdf

[3] World Economic Forum, 2016. *The Global Competitiveness Report 2016-2017,* available online at <www.weforum.org/gcr>

[4] L.Y.O. Li and Z. Fu, 2002. The school bus routing problem: a case study, *Journal of the Operational Research Society,* Vol. **53**, No. 5, pp. 552-558

[5] J. Park and B.I. Kim, 2010. The school bus routing problem: A review, *European Journal of Operational Research,* Vol. **202**, No. 2, pp. 311-319

[6] P. N. Kechagiopoulos and G. N. Beligiannis, 2014. Solving the Urban Transit Routing Problem using a particle swarm optimization based algorithm, *Applied Soft Computing,* Vol. **21**, pp. 654-676.

[7] P. Toth and D. Vigo, 2002. The vehicle routing problem, *Society for Industrial and Applied Mathematics,* Philadelphia.

[8] M. Dror and P. Trudeau, Savings by Split Delivery Routing, 1989. *Transportation Science,* Vol. **23**, No. 2, pp. 141-145.

[9] M. Dror and P. Trudeau, 1990. Split Delivery Routing, *Naval Research Logistics,* Vol. **37**, pp. 383-402.

[10] M. Dror, G. Laporte, and P. Trudeau, 1994. Vehicle Routing with Split Deliveries, *Discrete Applied Mathematics,* Vol. **50**, No. 3, pp. 239-254.

[11] C. Archetti and M.G. Speranza, 2006. An Overview on the Split Delivery Vehicle Routing Problem, *Operations Research Proceedings,* Part IV, pp. 123-127.

[12] D.J. Gulczynki, B. Golden and E. Wasil, 2008. Recent Developments in Modeling and Solving the Split Delivery Vehicle Routing Problem, *Tutorials in Operations Research,* INFORM, pp. 170-180.

[13] J.H. Wilck IV and T.M. Cavalier, 2012. A Construction Heuristic for the Split Delivery Vehicle Routing Problem, *American Journal of Operations Research,* Vol. **2**, pp. 153-162

IOP Conf. Series: Materials Science and Engineering **337** (2018) 012013 doi:10.1088/1757-899X/337/1/012013

Optimization of municipal waste collection scheduling and routing using vehicle assignment problem (case study of Surabaya city waste collection)

M N Ramdhani[1*], I Baihaqi[2], and N Siswanto[1]
[1]Department of Industrial Engineering, Institut Teknologi Sepuluh Nopember, Kampus ITS Sukolilo-Surabaya 60111, Indonesia
[2]Department of Business Management, Institut Teknologi Sepuluh Nopember, Kampus ITS Sukolilo-Surabaya 60111, Indonesia

*mukh.nasirramdhani@gmail.com

Abstract. Waste collection and disposal become a major problem for many metropolitan cities. Growing population, limited vehicles, and increased road traffic make the waste transportation become more complex. Waste collection involves some key considerations, such as vehicle assignment, vehicle routes, and vehicle scheduling. In the scheduling process, each vehicle has a scheduled departure that serve each route. Therefore, vehicle's assignments should consider the time required to finish one assigment on that route. The objective of this study is to minimize the number of vehicles needed to serve all routes by developing a mathematical model which uses assignment problem approach. The first step is to generated possible routes from the existing routes, followed by vehicle assignments for those certain routes. The result of the model shows fewer vehicles required to perform waste collection asa well as the the number of journeys that the vehicle to collect the waste to the landfill. The comparison of existing conditions with the model result indicates that the latter's has better condition than the existing condition because each vehicle with certain route has an equal workload, all the result's model has the maximum of two journeys for each route.

1. Introduction

Population growth in a country can affect the rate of waste production. This waste problem must be handled properly, especially in urban areas. Municipal solid waste (MSW) is a multidisciplinary activity that includes several activities such as source separation, storage, collection, transfer and transport, processing and recovery and also disposal [1]. The complexity of this MSW problem will increase the importance of planning or policy-makers to find an effective and sustainable way for the collection and disposal of MSW [2] [3].

A municipal waste transfer and transportation activity covers the transportation process, determining vehicles route and scheduling vehicles departure process. This problem is specific for Stationary Container System (SCS) where the vehicle is traveling to several nodes and pick up the garbages until the container is full. The vehicles route is determined in the beginning of process by optimizing the fastest travel time on the sequence of visits combination at each point of garbage collection. From the selected route then the vehicle move to serve to the other routes. This distribution problem involves some key considerations, from number of fleet and vehicle's routes including their scheduling [4]. Waste transport system can be modeled as a variant of the vehicle routing problem (VRP) in the presence of the compound and facilities between intermediate facility [5].

The research of [6] aims to collect garbage using waste collection vehicle routing problem with time windows (WCVRPTW) model to minimize the cost route by considering all waste must be transported to the disposal site within each customer time window. In this planning process, each vehicle has a departure schedule to serve each route which is determined by the start opening hour of the depot and the cycle time for each route. Route Sequencing on each route is intended to minimize the number of vehicles needed to serve all existed routes. [7] designs the daily truck schedule with each truck has a fixed route and empty cargo in several recycling operations. [2] founds that a route scheduling garbage collection is influenced by several factors, including the number of stops, the number of accumulated trash, road access, workers' schedules, and capacity of the container.

Surabaya is a the largest population city in East Java . This will affect the level of waste in Surabaya which triggers problems related to garbage. Surabaya provides several places and its vehicles for garbage collection on any residential areas to final processing. Waste volume is significantly increased while the activity of transporting waste transfer and transport are not yet optimally applied, so that city officials need to conduct operational strategy to be able to optimize the use of the vehicle in the depot to all routes of transporting waste. Surabaya has 37 compactor trucks with a capacity of 40 garbage containers to be operated in 60 locations in transporting garbage collection points and one landfill. Each location has a deterministic garbage collection request.

In this research, transportation and scheduling problems in the municipal waste is modeled by developing a mathematical model in the form of Integer Linear Programming (ILP) with considering the cycle time for all existing routes. The objective of this research is to minimize the number of vehicles used to serve meeting the demand for garbage collection, to schedule departure time of the vehicles, and to determine the sequence route for vehicles used on each existing route.

2. Literature review

2.1 Municipal solid waste

Municipal Solid Waste is universally recognized and used to calculate the expenses of solid waste management [8]. Thus, planners or policymakers try to find an efficient and sustainable way for collection and disposal of municipal solid waste [3]. To reduce waste collection and transportation cost, the city government is motivated to develop new strategies for collection and transportation of waste, especially in downtown [1].

2.2 Assignment problem

Generally, assignment problem is one of the fundamental combinatorial optimization problems in optimization which tries to match all supplies and demands equal to one [9]. Thus, the assignment problem is characterized by how to assign each point of supply for each point of demand. Assignment Problem becomes more complicated if the incidence varies in duration, and each event must occupy only one task for the entire duration of the assignment [10]. One of the implementation of this method is to determine the class into classroom by proposing a simple heuristic procedure [11].

2.3 Route sequencing

Route Sequencing determines the vehicle assignment on a route considering the time takes for a vehicle on one route. [12] develops a model that vehicle routes is designed from existing route and scheduling methods. It is assumed that the existing service is assigned to a particular vehicle. If the service has a short duration of time, the use of vehicle is not optimal, based on existing time horizon. It is intended that a route starts after the first service is completed, the vehicle then is available to be assigned to the second route, and so on until specified time horizon is completed. The example of route sequencing in vehicle routing problem is shown in figure 1. Therefore, the number of vehicle needed could be determined by sequencing these placements. Sequencing the route with this method has an objective to minimize the number of vehicle needed to serve all existed routes. Although these procedures are manually sorted, the computer program can be used to integrate routing and vehicle scheduling to provide a plan for the entire vehicle.

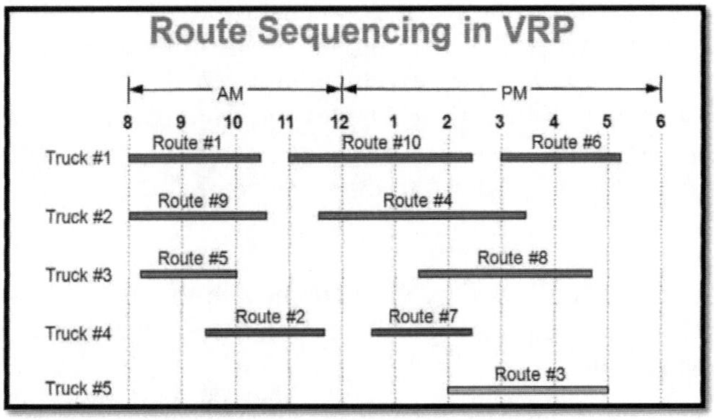

Figure 1. Route sequencing.

3. Model of Vehicle Assignment Problem

This study develops a model based integer linear programming. The model starts from existing condition which a route consists of multi trips. The route sequencing is calculated based on the calculation of operational travel time (cycle time), composed ofthe vehicle's travel time and service time at each temporary disposal site and the number of demand transported by truck compactor. The objective of this research is to minimize the number of vehicles used to serve all temporary disposal site and to schedule the vehicles departure times for each destination route.

The formulation used for this study can be defined as (i) for vehicle destination route and (j) for vehicle departure time per hours. This parameters and decision variables of this model are as follow:

Parameters:

Di = Number of demand on route i.

TS_{ij} = Time slot for every destination route i in departure time j

n = Minimal number of cycle from departure vehicles every hour

m_i = Maximum number of cycle on destination route i

CT_i = Cycle time for every destination route i

Decision variables:

Y = Number of vehicles used

Rit_i = Number of cycle for every destination route i.

X_{ij} = The cycle for destination route i in departure time j.The mathematical model of this vehicle assignment problem is as follow:

$$Minimize\ :\ Y \tag{1}$$

Subject to:

$$\sum_{j=1}^{J} TS_{ij} \times X_{ij} = Rit_i \qquad \forall i \in I \tag{2}$$

$$\sum_{i=1}^{I} Rit_i \times CT_i \le Y \times J \tag{3}$$

$$\sum_{j=1}^{J} TS_{ij} \times X_{ij} \ge D_i \qquad \forall i \in I \tag{4}$$

$$\sum_{i=1}^{I} TS_{ij} \times X_{ij} \le n \qquad \forall j \in J \tag{5}$$

$$TS_{ij} \times X_{ij} \leq m_i \qquad \forall i \in I, \forall j \in J \tag{6}$$
$$\sum_{i=1}^{I} \sum_{j=1}^{J} TS_{ij} \times X_{ij} \leq n \times J \tag{7}$$
$$X_{ij} = Integer \tag{8}$$
$$Y = Integer \tag{9}$$

The objective function (1) is to minimize the number of vehicles to meet the waste transportation demands. Constraint (2) is to find the total cycles for each destination route. Constraint (3) is to ensure that number of vehicles are enough to serve the system. Moreover, constraint (4) ensures that all destination routes must meet the demand. Constraint (5) ensures that every vehicle in one hour on the transport cyce must in the minimum number of vehicle dispatched. Constraint (6) explains that every destination route and every departure time should be less or equal than the maximum number of cycle on destination route. Constraint (7) ensures that transport cylce in one day should not exceed the total cycle. Constraint (8) and (9) enforce decision variables are integer.

4. Discussion

This existing routes defined by Surabaya City Council become the input in calculating cycle time in every route to determine the route sequencing. The calculation of cycle time for each vehicle on the existing routes can be seen in table 1.

Based on the data of table 1, the number of vehicle that can be used during waste collection process is calculated. The result shows that the minimum vehicle that need to be used are 19 compactors of which has 6 cycle. Maximum number of cycle in every route is 3 cycles.

Based on the existing route, a new route is generated by considering the truck compactor's capacity and the number of containers in temporary disposal site. If A new destination is created when a container cannot be transported by the existing vehicle, it must be served by another vehicle by considering travel time between locations and by using with the nearest neighbour algorithm. The travel time is calculated based on the in real time from Google Maps. The cycle time of a new route can be seen in table 2. The new route shows that the optimal minimum number of vehicle is 13 truck compactors of which has 5 cycles for all routes and the maximum number of cycles in one route is two.

Table 1. The cycle time for each existing routes.

Routes	Ritase	Cycle Time (Hours)	Routes	Ritase	Cycle Time (Hours)
Route 1	1	3.07	Route 17	1	3.68
Route 2	2	5.97	Route 18	1	3.79
Route 3	1	3.20	Route 19	1	3.12
Route 4	2	6.40	Route 20	1	3.25
Route 5	1	3.28	Route 21	1	2.97
Route 6	1	3.23	Route 22	1	3.45
Route 7	1	3.69	Route 23	1	3.05
Route 8	1	2.96	Route 24	1	2.51
Route 9	1	3.45	Route 25	1	3.10
Route 10	1	3.19	Route 26	2	5.49
Route 11	1	3.75	Route 27	2	5.14
Route 12	1	3.18	Route 28	2	5.11
Route 13	1	3.30	Route 29	1	3.09
Route 14	1	3.43	Route 30	1	3.81
Route 15	1	3.20	Route 31	1	3.40
Route 16	1	2.97			

Table 2. The cycle time on the new routes.

Routes	Ritase	Cycle Time (Hours)	Routes	Ritase	Cycle Time (Hours)
Route 1	1	3.07	Route 19	1	2.51
Route 2	1	2.63	Route 20	1	3.12
Route 3	1	3.17	Route 21	1	3.25
Route 4	1	3.02	Route 22	1	2.81
Route 5	0	2.00	Route 23	2	4.97
Route 6	2	5.21	Route 24	1	3.24
Route 7	3	8.93	Route 25	1	3.71
Route 8	2	5.26	Route 26	1	2.51
Route 9	2	5.43	Route 27	2	5.41
Route 10	1	2.72	Route 28	1	2.50
Route 11	1	2.65	Route 29	2	5.62
Route 12	1	3.52	Route 30	1	3.77
Route 13	1	3.19	Route 31	0	1.77
Route 14	2	6.33	Route 32	1	2.53
Route 15	1	3.18	Route 33	1	3.07
Route 16	2	5.39	Route 34	1	2.63
Route 17	2	5.56	Route 35	2	4.41
Route 18	2	4.91			

Comparing the departure schedules between the existing data and the new route cane be seen in table 3 and table 4 as seen in below:

Table 3. Vehicle Assignment Problem and Departure Schedule vehicles on existing routes.

Vehicles	Departure Scheduled Time Routes		Vehicles	Departure Scheduled Time Routes		Vehicles	Departure Scheduled Time Routes	
Vehicle 1	5 Route 27	10 Route 22	Vehicle 8	7 Route 16	11 Route 3	Vehicle 14	9 Route 30	
Vehicle 2	5 Route 29	10 Route 24	Vehicle 9	7 Route 17	11 Route 4	Vehicle 15	9 Route 32	11 Route 13
Vehicle 3	5 Route 35	10 Route 25	Vehicle 10	8 Route 6		Vehicle 16	9 Route 33	11 Route 15
Vehicle 4	6 Route 7		Vehicle 11	8 Route 8		Vehicle 17	9 Route 34	12 Route 2
Vehicle 5	6 Route 18	10 Route 26	Vehicle 12	8 Route 9		Vehicle 18	10 Route 20	12 Route 10
Vehicle 6	6 Route 23	11 Route 1	Vehicle 13	9 Route 28	11 Route 12	Vehicle 19	10 Route 21	12 Route 11
Vehicle 7	7 Route 14	12 Route 19						

Table 4. Vehicle Assignment Problem and Departure Schedule Vehicles on a new routes.

Vehicles	Departure Scheduled Time Routes			Vehicles	Departure Scheduled Time Routes			Vehicles	Departure Scheduled Time Routes	
Vehicle 1	5	9	11	Vehicle 6	7	8	11	Vehicle 10	7	10

	Route 26	Route 14	Route 11		Route 24	Route 22	Route 9		Route 31	Route 6
Vehicle 2	5 Route 27	9 Route 15	12 Route 8	Vehicle 7	7 Route 25	9 Route 17	11 Route 10	Vehicle 11	8 Route 19	10 Route 7
Vehicle 3	5 Route 28	9 Route 16	11 Route 23	Vehicle 8	7 Route 29	9 Route 18		Vehicle 12	8 Route 20	10 Route 12
Vehicle 4	6 Route 2	11 Route 1		Vehicle 9	7 Route 30	10 Route 5		Vehicle 13	8 Route 21	10 Route 13
Vehicle 5	6 Route 4	11 Route 3								

Both tables show the vehicle assignment problems on every route of which table 3 shows on each existing route while table 4 shows on each new route. Each vehicle can be assigned to transport on all routes due to homogeneous compactor vehicles. The differences in the number of route served by one vehicle are influenced by the cycle time on each route. If the cycle time on one route reaches the deadline of the specified departure time, then the vehicle can serve one or two routes more. This activity will forms route sequencing.

Having the model for vehicle assignment problems, we perform a sensitivity analysis. The sensitivity analysis is done by changing the demand in destination route's parameter (D_i). The changes of demand value will affect the number of truck compactor used which consequently limit the number of vehicles depart from depot for each hour (n), the maximum number of cycle performed by vehicle on each destination route (m) and the total cycle time (CT_i) for each condition. The sensitivity result of changes in demand can be seen in table 5:

Table 5. Sensitivity Analysis Results of changed parameter D_i.

Total *Demand* (D_i)	Number of route	Objective function	Number of Ritase Vehicle Depart in every hour (n)	Maximum ritase (m)	Total Cycle Time (CT) per day
1297 (*Existing*)	35 Route	19 Vehicle	6	3	133.98
1297 (New Route)	31 Route	13 Vehicle	5	2	113.25
1427 (Increase 10%)	32 Route	17 Vehicle	5	2	124.51
1556 (Increase 20%)	32 Route	20 Vehicle	6	3	136.59
1816 (Increase 40%)	30 Route	28 Vehicle	9	4	154.77

The increase demand in temporary disposal site will be directly proportional to the number of truck compactor needed, maximum number of cycle transported by vehicle and the total cycle time required to transport the demand to all destinations per day. As seen in table 5 above, the changes in demand has a significant influence to another parameter.

5. Conclusions

Based on the results using the model of vehicle assignment problems on generating a new route, it is obtained that the minimum number of vehicle required is 13 out of 37 truck compactors owned by Surabaya City Council. Also, the minimum number of cycle vehicle for each departure time is 5 cycle and maximum number of cycle is two for each destination route. Total number of truck compactor and its departure schedules are used for daily operations of transporting waste to all temporary disposal site served by the Surabaya City Council. Changes in demand affects the number of vehicle needed from existing destinations. The larger demand, the larger number of vehicle to transport the waste. Also, the larger the demand, the more frequency the truck serving in temporary disposal site. The increasing demand in temporary disposal site also affects the cycle time on each of destination route. The cycle time for this destination route is influenced by time to transport in each temporary disposal site and number of demand that must be transported by truck compactor on each destination route. In other words, the increasing demand will be directly proportional to the number of vehicle used, the number cycle of vehicle in every departed hour, and cycle time needed for a vehicle to serve every destination route by truck compactor.

For future research, the model of vehicle assignment problem can be developed to a heterogeneous vehicle with different capacity. Also, the model can be developed to area of decision support system, in which the system will be integrated with Google Maps to generate solution directly when the rerouting of vehicle used is necessary.

6. References

[1] Das S and Bhattacharyya B K 2015 Optimization of municipal solid waste collection and transportation routes *Waste Manag.* **43** pp 9–18

[2] Huang S H and Lin P C 2015 Vehicle routing-scheduling for municipal waste collection system under the “Keep Trash off the Ground” policy *Omega (United Kingdom)* **55** pp 24–37

[3] Xue W, Cao K and Li W 2015 Municipal solid waste collection optimization in Singapore *Appl. Geogr.* **62** pp182–90

[4] Bodin L, Golden B, Assad A and Ball M 1981 *The state of the art in the routing and scheduling of vehicles and crews*

[5] Fitria L, Susanty S and Suprayogi S 2009 Penentuan Rute Truk Pengumpulan dan Pengangkutan Sampah di Bandung *J. Tek. Ind.* **11** pp --51

[6] Buhrkal K, Larsen A and Ropke S 2012 The Waste Collection Vehicle Routing Problem with Time Windows in a City Logistics Context *Proc. - Soc. Behav. Sci.* **39** pp 241–54

[7] Li J Q, Borenstein D and Mirchandani P B 2008 Truck scheduling for solid waste collection in the City of Porto Alegre, Brazil *Omega* **36** pp 1133–49

[8] Jacobsen R, Buysse J and Gellynck X 2013 Cost comparison between private and public collection of residual household waste: Multiple case studies in the Flemish region of Belgium *Waste Manag.* **33** pp 3–11

[9] Kaur P, Sharma A, Verma V and Dahiya K 2016 A priority based assignment problem *Appl. Math. Model.* **0** pp 1–12

[10] Phillips A E, Waterer H, Ehrgott M and Ryan D M 2015 Integer programming methods for large-scale practical classroom assignment problems *Comput. Oper. Res.* **53** p 12

[11] Glassey C R and Mizrach M 1986 A decision support system for assigning classes to rooms *Interfaces (Providence).* **16** pp 92–100

[12] Ballou R H 2004 *Business logistics/supply chain management: planning, organizing, and controlling the supply chain* (New Jersey : Person Education)

Implementation of system dynamic simulation method to optimize profit in supply chain network of vegetable product

I P Tama[1*], Z Akbar[1], and A Eunike[1]
[1]Department of Industrial Engineering, Universitas Brawijaya, Malang, Indonesia

*kangdith@ub.ac.id

Abstract. Vegetables are categorized as a perishable product, which is a product with short lifespan thus requires proper handling and planning to reduce losses caused by the short lifespan. In order to reduce the losses, coordination among the players in the supply chain is required. On the other hand, the decision in the supply chain of vegetables and other farming products in the traditional market of developing country is independent among the players. This research is conducted by using System Dynamic Simulation method to develop model and scenario by coordinating the supply quantity amongst players in the supply chain. The scenarios are developed based on newsboy inventory model. This study aims to compare scenarios combining tiers involved in coordination program. The result shows that coordination in supply chain increases total supply chain profit, although there will always be players who experienced decrements in profit. The scenario of coordination among the farmer, the distributor, and the wholesaler resulted in the highest increase in total supply chain profit compared to other coordination scenarios, with an increased value of 10.49%.

1. Introduction

Vegetables are categorized as a perishable product, which has short lifespan thus requires proper handling and planning to reduce losses caused by the short lifespan. Hence, these products should be delivered to the hands of consumers quickly. Malang is a potential area in Indonesia for the development of food crops including mustard greens. In 2013, Malang contributed 1.93% (12,241 tons) of national mustard greens production. Mustard greens have a very short lifespan and require many processes to reach the hands of consumers. There are many players involved in the distribution of mustard vegetable products ranging from farmers to markets.

This research is conducted on mustard greens in traditional market, which the players in the supply chain are still independent in doing the business processes. However, there are some obstacles in developing this potential such as highly dynamic environment, especially in the traditional market. Therefore, several programs of agricultural development are necessary including a systems approach to solve the problems. These concerns motivate this paper to study the coordination model of the agriculture supply chain in traditional market system of agriculture product. Supply chain is a concept of integrating network facility among supplier, manufacturer, and warehouse in distributing materials and products [1]. However, coordination of supply chain is particularly a challenge for the conventional supply chain. The coordination usually leads to higher costs in the short term, while the revenues are uncertain [2]. In long term, integral cost may become lower, because the cost factors in the supply chain are better known, while risk and profit sharing stimulate the quality improvement and customer satisfaction.

To analyse the behavior of supply chain toward the coordination design, the active players need to be informed about the long-term impact [3]. Hence, to illustrate the long-term dynamic complexity between variables in supply chain, system dynamic approach is chosen to build the desired model. In addition, several other advantages of system dynamic are more informative in presenting the forecasted data, reflecting more accurate in short-and medium-term behavior, and more in depth of scenario development [4]. Therefore, system dynamic can develop a policy related to the profit in supply chain effectively [5] [6].

Fresh vegetable products are categorized as seasonal products because fresh vegetables are at risk of expiration if not sold during their sales season. Therefore, the inventory policy of the seasonal product is to order the product with a quantity that meets the demand of the selling season [6]. This character becomes a reference in designing the scenarios for order quantity coordination between the players. For products with seasonal demand, the order quantity is calculated by determining a balance between overstock cost and understock cost. Seasonal products have a high risk if they are not sold during the sales season.

2. Research method

In this research, data collection is done in accordance with the facts that occur in the supply chain of mustard green, which is further interpreted based on theory and literature associated with the system. Furthermore, the analysis is done by system dynamic simulation approach in determining the best scenario policy to optimize the profit in the supply chain. The steps are as follows.

2.1. Problem identification

Observing the mustard greens supply chain to acquire some information related to the business processes in order to analyze the factors affecting the total supply chain profit based on the observation and literature study. This research is conducted to optimize total profit in the supply chain. This research uses primary data and secondary data. Primary data is acquired by interviewing the players in supply chain about the information of selling price, production cost, market demand, harvest quantity and time of harvest. While the secondary data consists of distribution flow, inflation rate, and weather condition.

2.2. Modeling simulation

Modeling simulation begins with illustrating the relationship between variables in a causal loop diagram. The causal relation among the variables is entered according to the observation and literature study on the previous stage. The next step is to build simulation model in a stock and flow diagram based on the causal loop diagram. The stock and flow diagram is including mathematical equations for each variable according to the statistical data acquired in the previous stage. Furthermore, verification and validation are conducted to compare the model structure and behavior with the actual system. The modeling is done using Vensim simulation software.

After the model is declared valid [8], the model is used to simulate the scenarios of coordination that consist of order quantity combination. The purpose of scenario in this research is to optimize profit on the supply chain. The results from the simulation will be analyzed to learn how the policy affects the changes in profit.

3. Result and discussion

3.1. Identification of supply chain players

The supply chain players are identified from seed supplier to final consumers. However, in this study, the observed players are farmers, central market, and the main market, because this study only covers the local market that is supplied only by the farmers of Tumpang District. Table 1 shows the players in supply chain and the description.

Table 1. Players in the supply chain.

Player	Description
Farmer	Farmers of Tumpang district
Distributor	Traders in Kedung Boto central market
Wholesaler	Traders in Gadang main market

3.2. Identification of variables

The identification of variables is performed through interviews, observation, and literature. The identification is divided based on the players in supply chain. Table 2 shows the variables in the supply chain.

Table 2. Variable in the supply chain.

Player	No	Variable	Type	Unit
Farmer	1	Inflation Rate Effect	Auxiliary	-
	2	Farmer Incremental Price	Rate	IDR/(Kg*Month)
	3	Base Price	Stock	IDR/Kg
	4	Farmer Product Price	Auxiliary	IDR/Kg
	5	Quantity Multiplier	Auxiliary	-
	6	Supply Demand Ratio	Auxiliary	-
	7	Farmer Sales	Auxiliary	Kg/Month
	8	Output quantity	Auxiliary	Kg/Month
	9	Desired Production	Auxiliary	Kg/Month
	10	Productivity	Constant	
	11	Farming area	Constant	M2
	12	Season Multiplier	Auxiliary	-
	13	Incremental Cost	Rate	IDR/(Kg*Month)
	14	Base Cost	Stock	IDR/Kg
	15	Production Cost	Auxiliary	IDR/Kg
	16	Farmer Revenue	Rate	IDR/Month
	17	Farmer Production Expense	Rate	IDR/Month
	18	Farmer Profit	Stock	IDR
Distributor	19	Distributor Price Adjustment	Constant	-
	20	Distributor Product Price	Auxiliary	IDR/Kg
	21	Sales to wholesaler	Auxiliary	Kg/Month
	22	Other Wholesaler Demand	Constant	Kg/Month
	23	Distributor Demand	Auxiliary	Kg/Month
	24	Distributor Sales	Auxiliary	Kg/Month
	25	Distributor Labor Cost	Constant	IDR/Month
	26	Distributor Transportation Cost	Constant	IDR/Month
	27	Distributor Revenue	Rate	IDR/Month
	28	Distributor Expense	Rate	IDR/Month
	29	Distributor Profit	Stock	IDR
	30	Distributor Overstock	Auxiliary	Kg/Month
	31	Distributor Shortage	Auxiliary	Kg/Month
Wholesaler	32	Wholesaler Incremental Price	Auxiliary	-
	33	Wholesaler Product Price	Stock	Kg/Month
	34	Wholesaler Demand	Auxiliary	Kg/Month
	35	Wholesaler Sales	Auxiliary	IDR/Month
	36	Overstock Product Price	Auxiliary	IDR/Kg
	37	Wholesaler Revenue	Rate	IDR/Month
	38	Wholesaler Expense	Rate	IDR/Month
	39	Wholesaler Profit	Stock	IDR
	40	Wholesaler Overstock	Auxiliary	Kg/Month
	41	Wholesaler Shortage	Auxiliary	Kg/Month

3.3. Causal loop Diagram

Causal loop diagram shows the cause and effect relationship between the 41 identified variables. The relationship between variables is described with arrows. Each arrow is assigned a polarity, either positive (+) or negative (-) depending on the interaction between the variables. The blue arrow indicates positive relationship (+), whereas the red arrow indicates the negative relationship (-). The mark (II) on the arrow indicates a delay in the relationship between the variables. Figure 2 shows the causal loop diagram.

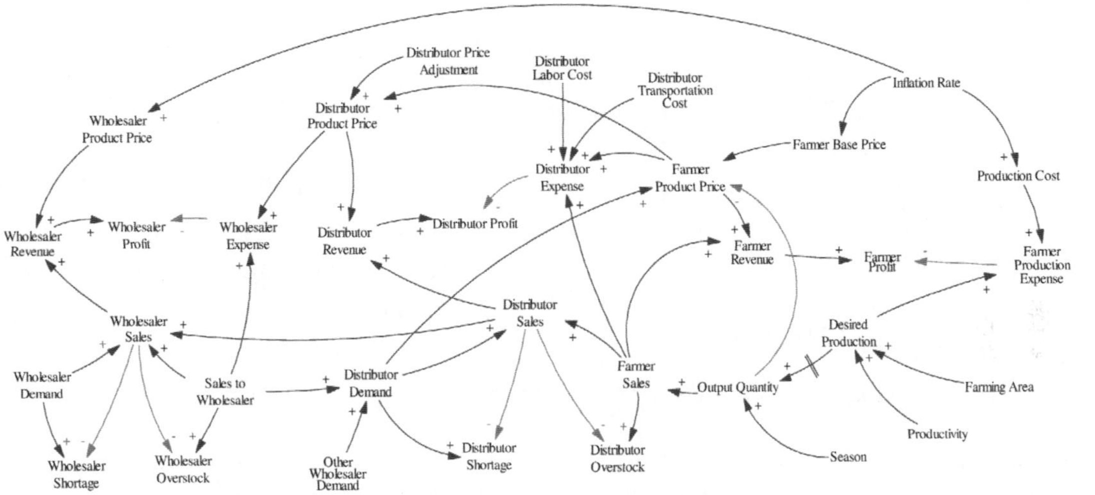

Figure 1. Causal loop diagram.

The causal loop diagram shows that the profit acquired by players in the supply chain are interrelated. Each profit is linked to a variety of the identified variables. Therefore, coordination between players could affect the total profit gained in supply chain.

3.4. Stock and flow diagram

Stock and flow diagram is divided into three sub models based on players in supply chain. The sub models include farmer profit sub model, distributor profit sub model, and wholesaler profit sub model. Figure 2 displays farmer profit sub model. This sub model showed that farmer profit is affected by season, farming area, productivity, production cost, and product price.

3.5. Scenario development

Scenario is developed by changing the parameters that affect the model. Based on the observation, players in the supply chain still interact independently in a two-party transaction, rather than transacting with a consideration of the entire players in the supply chain. Thus, it is necessary to develop a coordination scenario, which is expected to increase profit for the supply chain.

This research proposes newsboy inventory model as the coordination scenario for the supply chain. Newsboy inventory model determines order quantity based on optimal service level (CSL^*) derived from a balance between overstock (c_o) cost and understock costs (c_u) for seasonal products such as fresh vegetables [3]. The optimal service level can be calculated as follows:

$$CSL^* = \frac{c_u}{c_u + c_o} \tag{1}$$

Therefore, the optimal order quantity (O^*) is as follows:

$$O^* = \text{NORMINV}\,(CSL^*, \mu, \sigma) \tag{2}$$

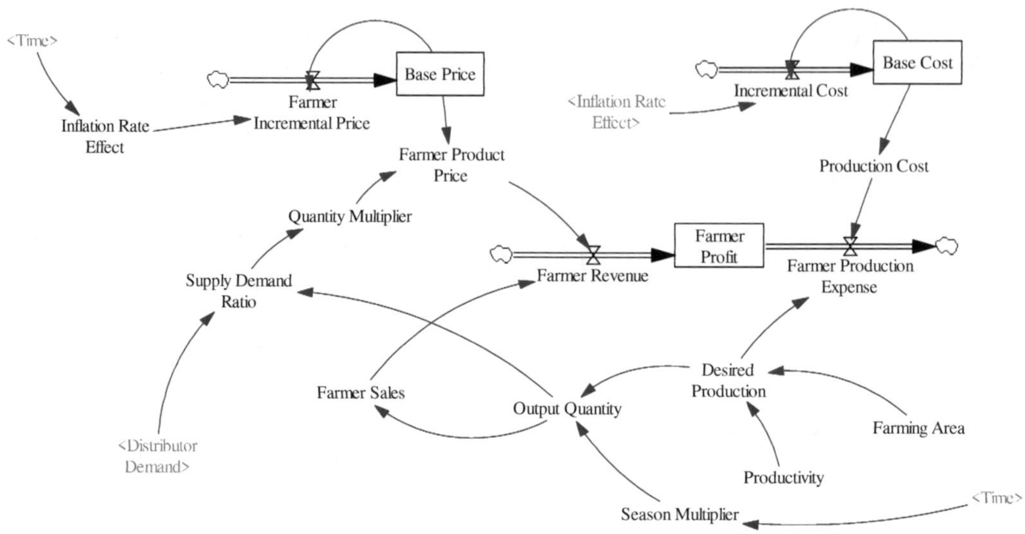

Figure 2. Stock and flow diagram of farmer profit sub model.

Three coordination scenarios are developed in this research including coordination between farmer and distributor (scenario 1), coordination between distributor and wholesaler (scenario 2) and coordination between farmer, distributor, and wholesaler (scenario 3).

3.5.1. Scenario 1. The purpose of scenario 1 is to determine the order quantity amount between farmer and distributor in the dry season, since the amount of supply from the farmer in rainy season is less than the optimal quantity due to the lack of harvest quantity. According to the calculation, the optimal order value is 12076 Kg / Month. In stock and flow diagram, the value of the supply delivered from farmers to the distributors is the value of the output quantity variables. Variable output quantity value cannot be directly changed as the value is influenced by desired production variables and season multiplier variables. Hence, another calculation is necessary to determine the value of desired production variable.

3.5.2. Scenario 2. The purpose of scenario 2 is to determine the order quantity amount between distributor and wholesaler. According to the calculation, the optimal order value is 3500 Kg / Month. In stock and flow diagram, the value of the supply delivered from distributor to the wholesaler can be directly changed.

3.5.3. Scenario 3. Scenario 3 is a combination of scenario 1 and scenario 2. Scenario 1 determines the amount of supply delivered from farmer to distributor, while scenario 2 determine the amount of supply delivered from distributor to wholesaler. However, an adjustment need to be done since the amount of supply from farmer to distributor is influenced by the amount of supply from distributor to wholesaler. According to the calculation the optimal order value from farmer to distributor is 12576 Kg/Month and the optimal order value from distributor to wholesaler is 3500 Kg/Month.

3.6. Result analysis
Scenario results are presented in Table 3 which shows the value of final income for each player in the supply chain and the total supply chain profit for the existing conditions and each scenario. The blue values show the best performance for each supply chain player. Whereas the red value show the worst

performance for each supply chain player. Total supply chain profit (Total SC) is obtained by summing all profits on each supply chain player.

The three scenarios require coordination between players in determining the amount of supply for each player. Table 3 shown that scenario 3 produces the highest profit for supply chain despite there is a decrease in profit for wholesaler.

Table 3. Final income for each scenario.

Condition		Supply chain player			
		Farmer	Distributor	Wholesaler	Total SC
Existing	IDR	1.185.000.000	1.809.000.000	1.514.000.000	4.508.000.000
	%	0%	0%	0%	0%
Scenario 1	IDR	1.156.000.000	2.138.000.000	1.460.000.000	4.754.000.000
	%	-2,45%	18,19%	-3,57%	5,46%
Scenario 2	IDR	1.327.000.000	2.271.000.000	1.344.000.000	4.942.000.000
	%	11,98%	25,54%	-11,23%	9,63%
Scenario 3	IDR	1.246.000.000	2.391.000.000	1.344.000.000	4.981.000.000
	%	5,15%	32,17%	-11,23%	10.49%

4. Conclusion

The profit that is acquired by each player in the supply chain can be affected not only by the players itself, but also by many other variables. Hence, by managing the order quantity, they could increase the total supply chain profit, although there will always be a player that experienced a decrement in profit. From the result of the simulation, the scenario of coordination between the farmer, the distributor, and wholesaler resulted in the highest increase of total supply chain profit compared to other coordination scenarios with an increased value of 10.49%. This research proposes newsboy inventory model as a policy in the developed scenario. Further research could develop more scenarios, such as risk sharing and revenue sharing in supply chain to obtain better results.

5. References

[1] Daniel J G and Fengqi Y 2015 *Computers and Chemical Engineering* **81** 153-170.
[2] P M (Nel) W, Harry B, Jacques H T, Jack G A J van der V and Jacqueline M B 2011 *Advanced Engineering Informatics* **25** 65-76.
[3] John D S 2000 *System Thinking and Modeling for a Complex World* (USA: McGraw-Hill) p 22
[4] James M L 2000 *System Dynamics Review* **16** 3-25.
[5] Ishardita P T, Agustina E, Rahmi Y, Sugiono and Yoganda R P 2017 *JEEST* **4** 90-102.
[6] Rafi R P L and Maghsoud A 2016 *Expert Systems With Applications* **51** 231-244.
[7] Nyoman P and Mahendrawati E R 2010 *Supply chain Management* vol 2. (Surabaya: Guna Widya)
[8] Yaman B 1996 *System Dynamic Review* **12** 183-210.
[9] Sunil C and Peter M 2013 *Supply chain Management Strарegy, Planning, and Operation* vol 5 (New Jersey: Pearson) pp 358-376

IOP Conf. Series: Materials Science and Engineering **337** (2018) 012015 doi:10.1088/1757-899X/337/1/012015

A review of agent-based modeling approach in the supply chain collaboration context

N I Arvitrida[1*]
[1] Department of Industrial Engineering, Institut Teknologi Sepuluh Nopember, Kampus ITS Sukolilo-Surabaya 60111, Indonesia

*arvietrida@gmail.com

Abstract. Collaboration is considered as the key aspect of supply chain management (SCM) success. This issue has been addressed by many studies in recent years, but there are still few research employs agent-based modeling (ABM) approach to study business partnerships in SCM. This paper reviews the use of ABM in modeling collaboration in supply chains and inform the scope of ABM application in the existing literature. The review reveals that ABM can be an effective tool to address various aspects in supply chain relationships, but its applications in SCM studies are still limited. Moreover, where ABM is applied in the SCM context, most of the studies focus on software architecture rather than analyzing the supply chain issues. This paper also provides insights to SCM researchers about the opportunity uses of ABM in studying complexity in supply chain collaboration.

1. Introduction

Collaboration between companies along a supply chain is believed as the core driver to achieve a long-term business success. In supply chain management (SCM) concept, a success collaboration practice is characterized by the integration of all operations from the upstream channel (the suppliers of raw materials) to the downstream side (end consumers) of the supply chain. Nevertheless, this ideal collaboration is hard to achieve. Most enterprises claim that integrating supply chain planning and operations activities is a complex issue although the alignment is only at a corporate level[1]. The challenge becomes more complicated when the integration involves other organization in the supply chain.

Agent-based modeling (ABM) is one of the simulation approaches that is increasingly employed to study behavior complexity in strategic partnerships, such as conducted by Robertson and Caldart [1] in strategic management research and Axelrod [2] in social science. Even though it has been considered as an innovative approach to study business relationships, the implementation of ABM in SCM context is still less popular than the other research field, particularly in strategic management.

This paper aims to explore the application of ABM in supply chain collaboration with a view to identifying the nature of their use. The nature of use is defined in terms of the aspect of collaboration, which considers supply chain functions. The contribution provided by this work is twofold. First, it documents the scope of supply chain collaboration issues that have been considered in the current literature. Secondly, the review provides an insight about potential uses of ABM in the future, in the context of studying partnership complexity in supply chains.

The remaining sections of this paper is structured as follows. In the next section, the literature review of supply chain collaboration and ABM work in SCM are presented. Then it is followed by a description of the review approach adopted, the findings and discussion of the review. Finally, Section 4 concludes the paper and presents suggestions for further studies.

2. Review on collaboration studies

Here is the brief literature review that provides the existing collaboration studies in both SCM and ABM field. The first subsection describes how ABM approach is appropriate in studying collaboration in supply chains, and the second subsection lists the current ABM work in supply chain collaboration.

2.1 Supply chain collaboration

Much research has discussed supply chain collaboration problems by applying analytical approach. Several examples of these studies are as follows. Cachon and Lariviere [3] study revenue sharing contracts by incorporating the newsvendor model for the collaboration approach. Dimitriou, Robinson, and Kotiadis [4] investigate the performance of newsvendor model under the bounded rational decision. Wu and Chen [5] perform a laboratory experiment to investigate rationally bounded behavior in newsvendor settings under several theoretical supply chain contracts. However, these studies only focus on inventory and logistics policy. Moreover, they tend to model collaboration issue in the form of supply chain contract, which is only appropriate to implement in the downstream level of the supply chain rather than for supply chain partnerships in general.

Other SCM studies apply empirical approach to view collaboration issues in a more comprehensive scope than inventory and logistics decisions. Holweg, Disney, Holmström, and Småros [6] consider the geographical dispersion of customers and supplier plants, demand pattern of products, and product characteristics as the key factors that drive supply chain collaboration. Cao and Zhang [7] study the nature of supply chain collaboration to explore its impact on company's performance. They took U.S. manufacturing enterprises as case studies. Purwaningrum and Evers [8] also find that culture has a significant effect on the mechanism of supply chain collaboration. However, none of these works explores how these complexities emerge from interactions between firms during their collaboration.

Studying complexities is hard to perform by using both analytical and empirical approach. Analytical models are mostly available for operational level problems where data can be obtained and follows a certain statistical distribution. Nevertheless, analytical solutions only work in a circumstance with a low complexity level. Meanwhile, empirical approach enables observation of many factors. However, exploring the impacts of sensitive factors in strategic partnerships through what-if analysis is difficult to conduct in the real world. It potentially causes problems of research ethics and consumes a great amount of time. Thus, simulation can be the best approach to study complexities that cannot be explained by both analytical and empirical approaches, particularly when the research aims to explore the impact of individual behavior on the supply chains.

2.2 ABM studies in SCM

ABM employs a bottom-up modeling, which starts with defining the individual agent. Then, the agents interact with each other and create a resulting emergent behavior at system level. ABM is a non-aggregated method that allows system perspective analysis to the emergent results [9].

The earliest and most popular ABM simulation in SCM is the beer game [9] although it is more popular to be modelled in system dynamics approach. The beer game simulates the increase of demand volatility as it moves further up a supply chain, which is known as the bullwhip or whiplash effect. This effect emerges because each company inside the supply chain is a rationally bounded entity and does not coordinate with each other in their decision-making process.

SCM research that employs ABM in analyzing the collaboration issue is still limited. Several studies consider supply chain collaboration as firm's integration. For instance, Xue, Li, Shen and Wang [10] employ ABM to address collaboration issue in construction supply chain, but they concentrate on the information flow and negotiation. Zhang, Anosike, Lim, and Akanle [11] present an ABM as an approach for e-manufacturing to provide flexibility, robustness, and adaptability to the rapid changes. Zhu [12] also models supply chain collaboration, but it does not consider the collaboration as integration between firms; the study focuses on investigating the impact of information sharing in a single *two-echelon* supply chain. Chen, Ong, Tan, Zhang, and Li [13] conduct a literature review on the use of

ABM in supply chain risk management (SCRM). They consider that SCRM is a result of collaboration success in a supply chain.

Other studies consider supply chain collaboration only in the scope of inventory decision, such as Dimitriou, Robinson, and Kotiadis [4] and Robinson, Dimitriou, and Kotiadis [15]. The study examines the effect of bounded rational decisions in a classical inventory model for perishable products (the Newsvendor inventory model) by combining ABM and multiple linear regressions.

Trust between collaborating firms has also modelled and investigated by using ABM, but not many studies can be found in this topic so far. Only Mohamed, Omar, and Wei [15] examine this issue in SCM context through an empirical approach in Malaysian industries.

Meanwhile, other works focus on modeling and analyzing collaboration issue in the downstream level of supply chain, such as conducted by Caridi, Cigolini, and De Marco [16]. They review the literature on ABM applications in managing supply chain processes, particularly in collaborative planning, forecasting, and replenishment (CPFR). In SCM, CPFR involves procedures and guidelines for sharing sales and forecast information between buyer and seller.

With regards to all ABM studies in supply chain collaboration, they concentrate mostly on software architecture than investigating the problem. These studies tend to employ ABM as a part of intelligent system in decision making rather than exploring the firm behavior. The following are several examples of these studies described in brief. Swaminathan, Smith, and Sadeh [17] utilize ABM as a multi-agent approach to develop a supply chain modeling framework. It addresses supply chain configuration, coordination, and contracts issues, which deal with inventory decisions. Julka, Srinivasan, and Karimi [18] propose an ABM framework for developing a decision support system prototype to integrate supply chain processes in a refinery supply chain. However, the goal of the system is optimizing a firm's performance, not the supply chain's performance. Jiao, You, and Kumar [19] applies an ABM system to develop a framework of collaborative negotiation in a supply chain. The framework incorporates supply chain network and inventory decisions. Zarandi, Pourakbar, and Turksen [20] employ ABM to reduce the bullwhip effect by coordinating all entities along the supply chain to minimize the total costs. Cheng [21] proposes an agent-based supply chain collaboration model that studies production and logistics processes at enterprise-level. The model comprises a single two-stage supply chain, which involves a manufacturer and a supplier. Kwon, Im, and Lee [22] propose an agent-based web approach to support supply chain collaboration in e-business. It models a three-stage supply chain that consists of suppliers, manufacturers, and retailers. The framework focuses on inventory decisions and allows flexibility in coping with partnerships changes. Santos, Coutinho, Cretan, and Jardim-Goncalves [23] develop a prototype of an agent-based framework for negotiation. The system is intended to support a supply chain collaboration network by improving the interoperability in the single supply chain. Hsieh and Lin [24] proposes ABM model with multi-agent system (a distributed agent-based modeling) to manage collaborative workflows and schedule the activities within a firm.

Besides supply chain collaboration, ABM has been widely applied in many SCM issues. Most of them focus on simulation software development instead of adopting ABM to analyze the supply chain problem. This is because they are conducted under the research area of computer science, not operational research and management science or SCM. For example, Barbuceanu, Teigen, and Fox [25] model a supply chain system that focuses on the information architectures, and Shen and Norrie [26] survey the application of agent distributed computing in supporting the mechanism of manufacturing systems.

3. The research approach, findings and discussion

The findings in this paper is based on a review of journal articles that mainly apply ABM in supply chain collaboration. The literature review performed follows two phases: identification of the journal articles, and categorizing the papers by the collaboration issues addressed. SCM involves a wide variety of topics, and this causes difficulties in categorizing the nature of the papers. A readily classification of supply chain collaboration issues was not available in the existing literature. Moreover, each paper uses different terms, such as production planning is used in place of inventory management. Therefore, the papers are classified based on author's own interpretation and judgment on the main information provided in each article.

A summary of previous research on competition and collaboration in SCM is presented in Table 1. The table outlines the scope of application for each research, which are classified into 12 issues: supply chain planning, bullwhip effect, network/supply chain configuration, scheduling, trust, inventory, product development, logistics, supply chain risks, information sharing, supply chain financial aspect, and negotiation. This categorization represents the scope of supply chain problems that is popularly discussed in SCM literature. According to the area of applications, it can be seen that most of these research defines collaboration aspect in supply chains based on the supply chain operations.

Table 1. Previous ABM work in supply chain collaboration.

Author(s)	Scope											
	Plan	Bull	Netw	Sche	Trus	Inve	Pro	Log	Risk	Info	Fina	Neg
[10]		✓				✓						
[11]												✓
[12]											✓	
[13]										✓		
[14]									✓	✓		
[5,15]						✓						
[16]					✓							
[17]	✓	✓				✓						
[18]			✓			✓						
[19]						✓		✓				
[20]			✓			✓						
[21]		✓										
[22]							✓	✓				
[23]						✓						
[24]			✓									
[25]				✓								

Note:

Plan	:	Supply chain planning	Pro	:	Product development
Bull	:	Bullwhip effect	Log	:	Logistics
Netw	:	Network/supply chain configuration	Risk	:	Supply chain risks
Sche	:	Scheduling	Info	:	Information sharing
Trus	:	Trust	Fina	:	Supply chain financial aspect
Inve	:	Inventory	Neg	:	Negotiation

From the literature review, it can be interpreted that ABM approach has been well implemented in SCM studies. However, as shown in Table 1, it is obvious that the number of ABM work in supply chain collaboration is still limited. All of these studies observe a particular single supply chain with a limited attempt to explore the effectiveness of the collaboration approach under supply chain competition. Moreover, the perspective used in the literature still tends to side with one company rather than improving and maintaining the whole supply chain performance for the long-term. Furthermore, most ABM research in SCM is conducted through computer science research, so the works tend to focus on software architecture rather than analyzing a problem of the proposed topic. This indicates that analyzing collaboration dynamics by using an ABM approach can be a potential SCM research topic in this field. In other words, there is many opportunities to apply ABM in collaboration in supply chains.

Furthermore, ABM has a unique feature to model and observe a problem. ABM employs a bottom-up approach that enables researchers to understand an emergent behavior at macro level by investigating the behavior at the micro level of individual agent. This approach is appropriate to model a phenomenon that is difficult to explain empirically and analytically.

4. Conclusion

The findings of this work provide evidence on the extent use of ABM approach in SCM, particularly in collaboration issues. The nature of use is defined based on the popular topics on supply chain collaboration found in this review, which are classified into 12 issues: supply chain planning, bullwhip effect, network/supply chain configuration, scheduling, trust, inventory, product development, logistics, supply chain risks, information sharing, supply chain financial aspect, and negotiation. The review also offers new ideas and insights about the application of ABM in studying partnerships in supply chains as well as the potential uses of ABM in collaboration issues. A review of ABM approach in supply chain collaboration has never been conducted in the existing literature both in SCM and ABM studies, so this paper is useful to motivate SCM researchers to benefit ABM to enrich the current body of knowledge, in terms of exploring complexity in supply chain partnerships.

This paper employs a literature-based approach to classifying the collaboration aspects modelled using ABM, and it does not consider other factors or criteria in business collaboration. Future work will be performed by taking a more specific scope of supply chain environment, such as partnerships under a competitive market.

5. References

[1] Fawcett S E and Magnan G M 2002 The rhetoric and reality of supply chain integration *International Journal of Physical Distribution and Logistics Management* **32**(5) pp 339–361

[2] Robertson D A and Caldart A A 2009 *The dynamics of strategy: Mastering strategic landscapes of the firm* (New York: Oxford University Press)

[3] Axelrod R 1997 *The complexity of cooperation: Agent-based models of competition and collaboration* (New Jersey: Princeton University Press)

[4] Cachon G P and Lariviere M A 2005 Supply chain coordination with revenue-sharing contracts: Strengths and limitations *Management Science* **51**(1) pp 30–44

[5] Dimitriou S Robinson S and Kotiadis K 2009 The impact of human decision makers' individualities on the wholesale price contract's efficiency: Simulating the newsvendor problem *Proceedings of the 2009 Winter Simulation Conference (WSC)* pp 2353–2364

[6] Wu D Y and Chen K Y 2014 Supply chain contract design: Impact of bounded rationality and individual heterogeneity *Production and Operations Management* **23**(2) pp 253–268

[7] Holweg M Disney S Holmström J and Småros J 2005 Supply chain collaboration: Making sense of the strategy Continuum *European Management Journal* **23**(2) pp 170–181

[8] Cao M and Zhang Q 2011 Supply chain collaboration: Impact on collaborative advantage and firm performance *Journal of Operations Management* **29**(3) pp 163–180

[9] Purwaningrum F Evers H D and Yaniasih 2012 Knowledge flow in the academia-industry collaboration or supply chain linkage? Case study of the automotive industries in the Jababeka cluster *Procedia - Social and Behavioral Sciences* **52** pp 62–71

[10] North M J and Macal C M 2007 *Managing business complexity: Discovering strategic solutions with agent-based modeling and simulation* (New York: Oxford University Press)

[11] Xue X Li X Shen Q and Wang Y 2005 An agent-based framework for supply chain coordination in construction *Automation in Construction* **14**(3) pp 413–430

[12] Zhang D Z Anosike A I Lim M K and Akanle O M 2006 An agent-based approach for e-manufacturing and supply chain integration *Computers and Industrial Engineering* **51**(2) pp 343–360

[13] Zhu X 2008 *Agent based modeling for supply chain management: Examining the impact of information sharing* Kent State University

[14] Chen X Ong Y S Tan P S Zhang N S and Li Z 2013 Agent-based modeling and simulation for supply chain risk management - A survey of the state-of-the-art *International Conference on Systems, Man, and Cybernetics* IEEE pp 1294–1299

[15] Robinson S Dimitriou S Kotiadis K 2016 Addressing the sample size problem in behavioural operational research: simulating the newsvendor problem *J Oper Res Soc* Palgrave Macmillan UK

[16] Mohamed M I Omar S S and Wei L C 2015 Trustworthiness in buyer-supplier relation on supply chain collaboration among SMEs *International Symposium on Technology Management and Emerging Technologies* Langkawi, Kedah, Malaysia pp 454–458

[17] Caridi M Cigolini R and De Marco D 2005 Improving supply-chain collaboration by linking intelligent agents to CPFR *International Journal of Production Research* **43**(20) pp 4191–4218

[18] Swaminathan J M Smith S F and Sadeh N M 1998 Modeling supply chain dynamics: A multiagent approach *Decision Sciences* **29**(3) pp 607–632

[19] Julka N Srinivasan R and Karimi I 2002 Agent-based supply chain management—1: framework *Computers and Chemical Engineering* **26**(12) pp 1755–1769

[20] Jiao J You X and Kumar A 2006 An agent-based framework for collaborative negotiation in the global manufacturing supply chain network *Robotics and Computer-Integrated Manufacturing* **22**(3) pp 239–255

[21] Zarandi M H F Pourakbar M and Turksen I B 2008 A fuzzy agent-based model for reduction of bullwhip effect in supply chain systems *Expert Systems with Applications* **34**(3) pp 1680–1691

[22] Cheng R 2011 A study on multi-agent based supply chain collaboration operation model facing time of delivery *Proceedings - 2011 8th IEEE International Conference on e-Business Engineering ICEBE 2011* pp 102–104

[23] Kwon O Im G P and Lee K C 2011 An agent-based web service approach for supply chain collaboration *Scientia Iranica* **18**(6) pp 1545–1552

[24] Santos T Coutinho C Cretan A and Jardim-Goncalves R 2013 Agents and rules for the negotiation of interoperability solutions *Proceedings of the IEEE International Technology Management Conference and 19th ICE Conference (ICE 2013) The Hague, Netherlands*

[25] Hsieh F S and Lin J B 2014 A multiagent approach for managing collaborative workflows in supply chains *Proceedings of the 2014 IEEE 18th International Conference on Computer Supported Cooperative Work in Design CSCWD 2014* pp 71–76

[26] Barbuceanu M Teigen R and Fox M S 1997 Agent based design and simulation of supply chain systems *Proceedings of IEEE 6th Workshop on Enabling Technologies: Infrastructure for Collaborative Enterprises* Los Alamitos, California: The Institute of Electrical and Electronics Engineers, Inc. pp 36–41

[27] Shen W and Norrie D H 1999 Agent-based systems for Intelligent manufacturing: A state-of-the-art survey *Knowledge and Information Systems* **1** pp 129–156

Modeling of shallot supply decisions: the case of Indonesia

N F Prabawati[1*], I N Pujawan[1], E Widodo[1***]**
[1] Department of Industrial Engineering, Institut Teknologi Sepuluh Nopember, Kampus ITS Sukolilo-Surabaya 60111, Indonesia

*nuriliafitrip@gmail.com, **pujawan@gmail.com, ***erwin_widodo@yahoo.com

Abstract. To optimize supply chain role, the players of supply chain need to integrate its function. One of the general problems in supply chain was the unbalanced quantity of sales and quantity of supply. This paper focused on modelling a simple method to manage the gap between the demand and the supply. The gap might cause an overstock or a loss. This paper propose a buffer quantity in order to handle the gap by using import decision. The case study was about shallot supply - demand in Indonesia. In this study we model the supply decisions of shallot in Indonesia. While the demand was quite stable over time, the supply was heavily affected by the yield from the farms. The shortage could result in the government importing shallot from other countries. Hence, the government also needed to have a proper buffering mechanism in order to ensure the supply was sufficient and the price was quite stable. The initial model of this research was built by stochastic parameters and the extended model to gain pricing mechanism was built by Shapley value principal with modification. The primary variables were supply quantity, demand quantity, buffer and purchased quantity (stock needed), actual consumption, and price for three players. The validation proved that the result of price at each player presented a significant difference. Therefore, the model could be applied to decide the stock quantity needed and to keep the price stable at each player especially at the end player which would influence the market price.

1. Introduction

Supply chain is a sequential activity, therefore a decision may cause a whole change among supply chain players. One of the general problems is the unbalanced quantity demand and supply. The aims of this research are to propose a model for buffer quantity in order to gain a particular stock quantity and to extend the initial model into pricing mechanism. This research took a study of shallot import in Indonesia so that the model would help the decision maker to control the supply and price by controlling purchased quantity. The goal of the initial model was modeling the stock needed by using a buffer stock obtained from import. Furthermore, the price mechanism was built to control the market price by cutting the supply chain structure. Therefore, this research used three players.

The study was about shallot import problem because it influenced the inflation in Indonesia. The reason was unstable price because of supply chain structure which involving many players and uncertain supply because of season and field condition. The complexity of supply chain increased with import-export activities, not only focused on the balance of sales and purchased quantity, but also widespread to social welfare. Bloechl [1] showed that trading benefits the two cooperating nations. Nevertheless, there were still research opportunities that measure social welfare in the supply chain model of a commodity. Similar earlier research was the study by Crucini and Davis [2] which arranged the distribution of capital on short and long run import demand elasticity. While in this study, the focus of research was on the distribution of price allocation. This research used Shapley value modification which is game theory approach to control the price at each player in supply chain.

The research on game theory in the supply chain has been done previously by Zhang [3]. Choi and Messinger [4] examined the role of fairness in competitive supply chain relationships. The research used supply chain model among supplier-retailer with some replication which produced model where each supply chain player tended to choose almost the same margin. In addition, Lambo and Wambo [5] suggest on how uniformly shared expenses (solidarity expenses) were shared. The study looked at how the funds transfer between players using Nowak and Radzik model.

Sodhi and Tang [6] in their research provided a model of supply chain in which the follower was a supplier or distributor. The study aimed to make models that can be used as the basis for further research relating to "n-middleman". Research was motivated by the company's efforts to employ the weak (the poor). The model gave a reference to what the optimum price should be given by retail when there were as many n-middlemen and how much optimum price allocation would be obtained by upstream parties (farmers, fishermen, or other micro-entrepreneurs). The other related researches were Xu et al. [7], Wahyudin et al. [8], Macho-stadler et al. [9], Xiaofeng and Aiqing [10], Anglano et al. [11], Tan et al. [12], and Khmelnitskaya et al. [13].

This study had similarity with the research of Wahyudin et al. [8] which studied about shallot pricing policy. Shallot supply had often been a problem in Indonesia. While demand could be considered quite stable over time, the supply from domestic farming was very much affected by various factors, including the weathers. In some season, the yield was very good, in others it could be as low as 50% of the normal yield. The government often had to import from other countries. The fluctuated supply also results in fluctuated price. This paper was an attempt to model the supply decision of shallot where import was decided when domestic supply was deemed insufficient. We model the decision as a *Monte Carlo* simulation.

2. Decision Models

2.1. Research Model

This research was performed by a sequential model. Figure 1 describes that supply, consumption (demand), and import activities were processed on a quarterly basis. It was assumed that the short lead time controls the import decision so that the supply lead time was shorter than the planning period. Calculations in one period did not affect the calculation of other periods. However, it might occur a period when production (supply) quantity could be very high and thus terms would affect the excess of inventory or shortage. The model used normal distribution for the demand pattern, the formulation follows Chopra & Meindl [14].

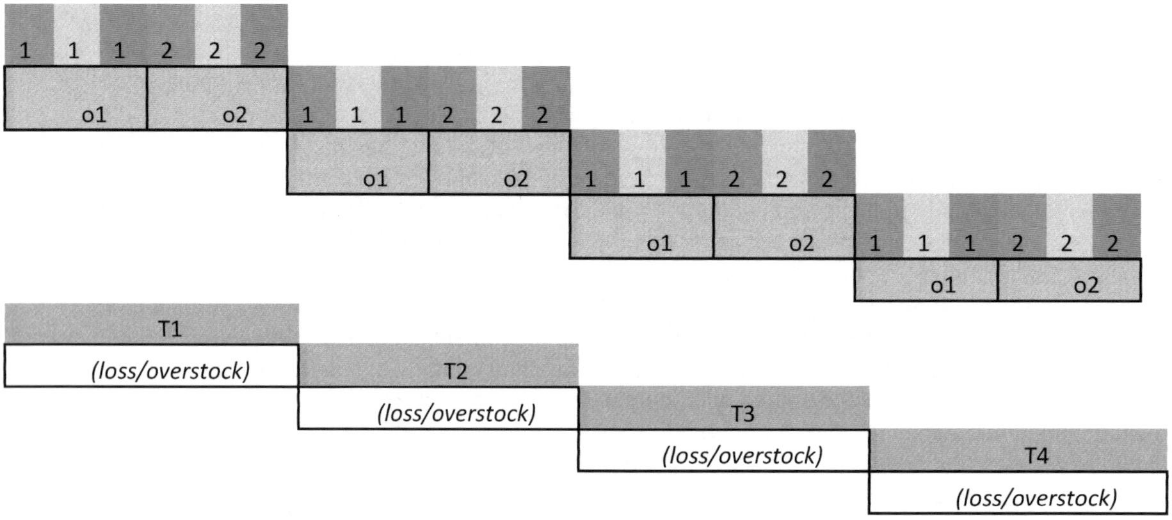

Figure 1. Construct of Simulation.

There were variables used in this research, the variables notation are:

q_a supply prediction
q_b consumption (demand) prediction
q_m actual supply
q_n actual consumption
I_i preliminary import decision
O_i overstock signal
r_I Cumulative Distribution Function of I_i
r_O Cumulative Distribution Function of O_i
q_y preliminary stock decision
B_i buffer quantity
Q_i buffer and purchased quantity (stock needed)
r_Q Cumulative Distribution Function of q_y

In addition, the other variables notation are $p_{retailer}, p_{wholesaler}, p_{supplier}$

After defining the variables, here were the steps of simulation

Step 1 Basic variables were generated by stochastic parameters

$$\boldsymbol{q_a, q_b, q_m, q_n}$$

Step 2 Observing preliminary import decision

$$I_i = q_b - q_a \quad (1)$$

Step 3 Calculating overstock signal

$$O_i = q_m + I_i - q_n \quad (2)$$

Step 4 Collecting report from *Monte Carlo* simulation,

$$\mu_I, \sigma_I^2, \mu_O, \sigma_O^2 \quad (3)$$

Step 5 Calculating CDF of I_i and O_i

$$r_I = F_I(I_i, \mu_I, \sigma_I^2) \text{ and } r_O = F_O(O_i, \mu_O, \sigma_O^2) \quad (4)$$

Step 6 Calculating preliminary stock decision

$$q_y = q_a - q_m - q_n \text{or} \quad (5)$$

$$q_y = q_a - O_i - I_i \text{ or} \quad (6)$$

$$q_y = q_a - \left(F_O^{-1}(r_O, \mu_O, \sigma_O^2) - F_I^{-1}(r_I, \mu_I, \sigma_I^2)\right) \quad (7)$$

Step 7 Collecting report from preliminary stock decision about recalculating of overstock/loss (update) obtained from preliminary stock decision. The report would be used to build the buffer parameters.

Step 8 Calculating buffer

$$B_i = q_y - q_b - F_B^{-1}\left(F_y\left(q_y, \mu_y, \sigma_y^2\right), \mu_B, \sigma_B^2\right) \quad (8)$$

where μ_B was set based minimum, average, or maximum value of buffer parameters

Step 9 Calculating buffer and purchased quantity (stock needed)

$$Q_i = q_y + B \quad (9)$$

$$Q_i = 2q_y - q_b - F_B^{-1}\left(F_y\left(q_y, \mu_y, \sigma_y^2\right), \mu_B, \sigma_B^2\right) \quad (10)$$

Step 10 Calculating CDF of buffer and purchased quantity

$$r_Q = F_Q(Q_i, \mu_n, \sigma_n^2) \quad (11)$$

Step 11 Calculating the price for each player

$$p_{retailer} = F_{retail}^{-1}(1 - r_Q, \mu_{pretail}, \sigma_{pretail}^2) \quad (12)$$

$$p_{wholesaler} = F_{wholesale}^{-1}(1 - r_Q, \mu_{pwholesale}, \sigma_{pwholesale}^2) \quad (13)$$

$$p_{supplier} = p_{retailer} - \frac{p_{retailer} - p_{wholesaler}}{1 - (r_Q/\beta)} \quad (14)$$

where β was the denominator to decrease r_Q degree

Firstly, the value of the variables was generated using *Monte Carlo* simulation for 10000 periods to obtain the prediction of production quantity (supply) and the prediction of consumption quantity (demand), the gap between those two variables was used as the preliminary of import decision. This model assumed that the import decisions had been made before actual production took place. Then, those three variables were compared with the actual supply and the actual demand using stochastic parameters. All of the variables were reported and analyzed. Figure 2 shows this preliminary model briefly. The second step, the model was evaluated. Adopting the model of normal distribution, overstock/loss which had been obtained from the first model would be calculated to evaluate the import decision. The number of import decision which had been evaluated known as buffer. Figure 3 shows the second step of this initial model.

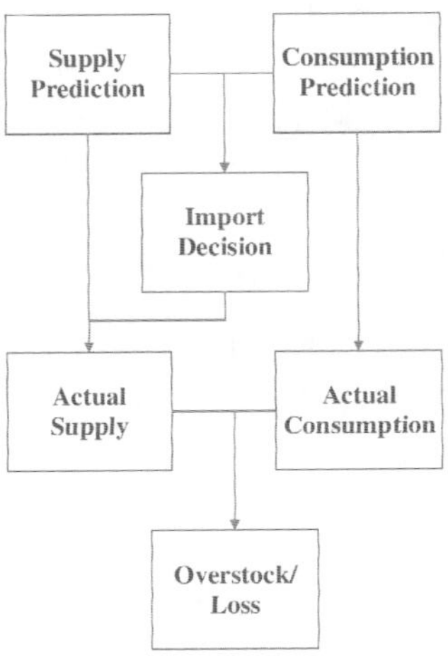

Figure 2 Preliminary Model.

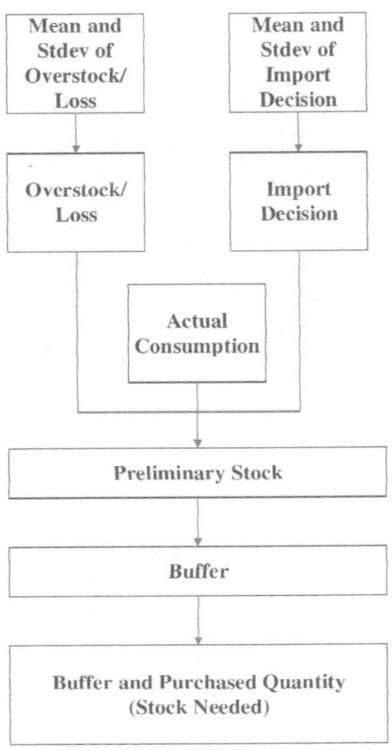

Figure 3 Buffer and Stock Model.

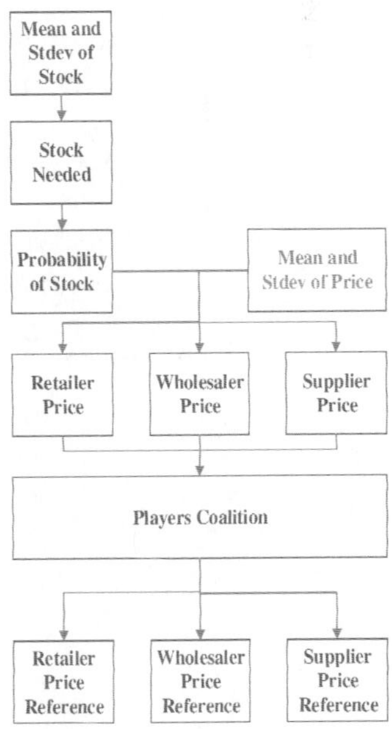

Figure 4 Price Allocation Model.

After completing the initial model, the size of buffers would be added to the preliminary stock decision. Those variables produce a new value known as a stock needed. The inverse of the number of stock needed would be calculated to build a price for each player. The price was generated by considering means of the actual price. Figure 4 shows the step to get the price at each player.

Furthermore, Shapley value methodology was adopted to build price allocation for each player. However, the coalition was built from the additional price at each player and the grand coalition was the total price at each player added by the average of price at each player. Therefore, the Shapley core was 0 and the grand coalition core for three players was the average of price at each player. It would shift the price cheaper than the initial price which had be obtained from the initial model. However, to calculate the ratio, this case should subtract the grand coalition core by the player's Shapley value before dividing by the grand coalition core.

All of those steps were experimented. The outcome would be discussed on the results and discussion section. The validation of the model proved that there were significant differences with the secondary data. Therefore, the model could be applied to decide the buffer quantity and to keep the price under control at each player.

3. Results and Discussion

3.1. Simulation Results

This research used the data which were collected by the government agencies who had responsibility to analyze the production in the field and the market price of shallot commodity. The report of initial model in Table 1 shows that the average and the standard deviation of overstock/loss vary substantially from one quarter to the other.

Table 1. Overstock/Loss Type in Preliminary and Correction Model.

		Average (Ton)	St.dev (Ton)
Overstock/Loss (Preliminary)	Quarter I	692	29109.810
	Quarter II	102	31897.761
	Quarter III	78262	37774.465
	Quarter IV	20182	8761.421
Overstock/Loss correction	Quarter I	-148	8291.334
	Quarter II	-88	11335.989
	Quarter III	-368	22365.497
	Quarter IV	-282	19220.397

IOP Conf. Series: Materials Science and Engineering **337** (2018) 012016 doi:10.1088/1757-899X/337/1/012016

Table 2. Buffer Parameters $\mu_\delta \sigma_\delta^2$.

Period	Decision	Buffer based on Minimum Value	Buffer based on Average Value	Buffer based on Maximum Value	St.dev
Quarter I	Stock	167778	272778	372888	29110
	Overstock/Loss Percentage	-10.95%	-0.15%	10.94%	3.10%
Quarter II	Stock	216915	333424	455080	31898
	Overstock/Loss Percentage	-13.34%	-0.15%	12.14%	3.46%
Quarter III	Stock	13045	329356	511360	43982
	Overstock/Loss Percentage	38.82%	0.62%	20.65%	7.09%
Quarter IV	Stock	241063	307516	375290	19254
	Overstock/Loss Percentage	27.02%	0.49%	17.88%	6.36%

From the model of overstock/loss correction, model of buffer and purchased quantity were made (stock needed). The quantity of buffer and purchased led the management to decide how large buffers were needed and how many quantities should be purchased in order to provide the ideal stock for the shallot market. Buffer and purchased quantity tended to give a positive number for the final overstock/loss. The decision was exactly influenced by the parameters of buffer quantity. Table 2 shows the parameters which were influencing the decision of buffer quantity.

The decision of stock needed was extended into the model of probability. That form was used to help calculate the price for each player in this shallot supply chain model. Pricing model showed the price decreasing in the case of quantity increasing. Those price were formed in each quarter. The purpose was to control the stability of price within the year by dividing the period whenever the quantity seemed not as normal as usual. Figure 5 shows the pricing model for quarter I as an example. All of the models in the other quarter were built using the same parameters by changing the quantity which depended on each quarter.

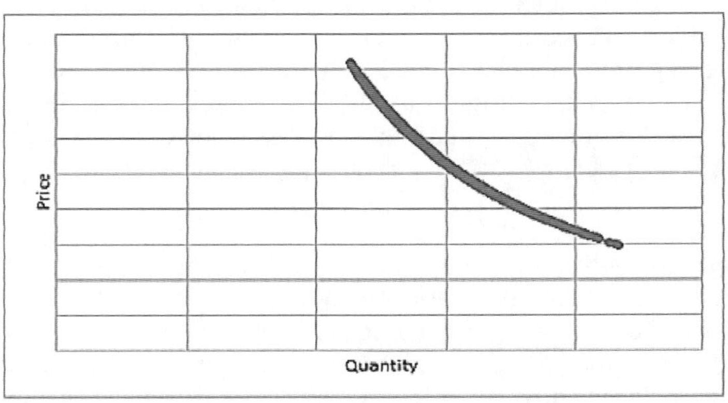

Figure 5 Pricing Model (10000 events)

Figure 6 shows the form of pricing model in another view. It also shows the relation between price and quantity. Furthermore, the price would be shifted in order to decrease the price as much as it could be, to obtain the new value of price which could be considered as the standard of market price. This final model would be explained in the next section.

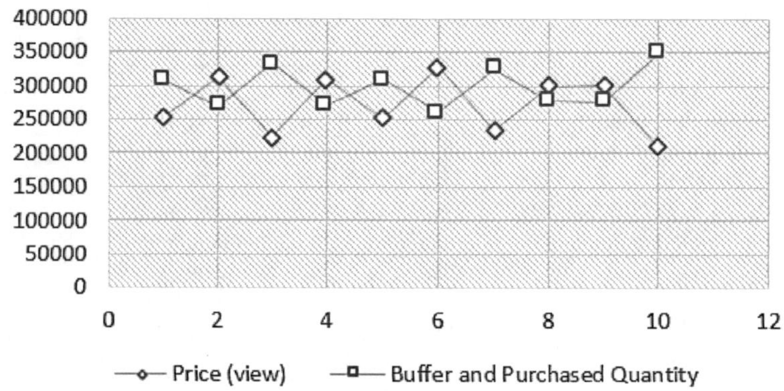

Figure 6 The Relationship between Price and Stock Needed

3.2. The Market Price Model

This model was built based on Shapley value principal, however the grand coalition was built by the average of each player price. The core of the other coalitions but grand coalition were 0. This model gave a satisfied result which can be seen in Table 3 that each player tended to have the same price within the year. Table 3 shows the result of the average price of shallot market price built by this model.

Table 3. Market Price.

Price	Quarter I (Rp/kg)	Quarter II (Rp/kg)	Quarter III (Rp/kg)	Quarter IV (Rp/kg)
Retailer	23207.46	23205.97	23511.48	23277.62
Wholesaler	17334.7	17333.25	1763.28	17404.1
Supplier	15069.15	15054.5	15413.69	14979.27

However, the modification of this model gave a little difference when calculating the ratio of Shapley value with the grand coalition but still had the same principal. When using the average price of each player based on the price obtained from the initial model, the difference between the value of grand coalition and the Shapley value of a player divided by the value of grand coalition would give a ratio that if all ratio were summarized, the result was 100%. Therefore, the principal of Shapley value was still confirmed by this modification. The validation model also showed there were significant differences between the price in finding and the secondary data.

4. Conclusions

The research findings showed that the average and standard deviation of overstock/loss based on import decision could be shifted especially for Quarter III and Quarter IV. During the Quarter III, the supply prediction was high and there was no decision of import, however the actual supply may fall and not as good as prediction, therefore the average was 78262 (indicate overstock) but the standard deviation also high 37774.465. It meant the range of overstock/loss in Quarter III had a high difference. By correcting the model, the average of overstock in Quarter III could be reduced to become -368 with the standard deviation of 22365.497. Buffer parameters made another correction so that the loss occurrence might happen not too often. Furthermore, pricing model which had been built based on the stock decision provided report that the average price within the year was quite stable, no matter the fluctuation of the supply. Price at the *retailer* approximately was above Rp 23000,-/kg in each Quarter, price at the *wholesaler* was above Rp 17000,-/kg, and price at the shallot farmer (supplier) was Rp 15000,-/kg.

5. References

[1] Bloechl F 2013 *Sugar and Rice Trade between the United States and Cuba Trade Potential and Welfare Analysis* (University of Arkansas)

[2] Crucini M J and Davis J S 2016 Distribution capital and the short- and long-run import demand elasticity *J. Int. Econ.* **100** pp 203–219

[3] Zhang X 2014 Game Theoretical Approach in Supply Chain Management

[4] Choi S and Messinger P R 2016 The role of fairness in competitive supply chain relationships: An experimental study *Eur. J. Oper. Res.* pp 1–16

[5] Lambo L D and Wambo P 2015 The Fairness of Solidarity Bills under the Solidarity Value of Nowak and Radzik *Game Theory*

[6] Sodhi M S and Tang C S 2014 Supply-Chain Research Opportunities with the Poor as Suppliers or Distributors in Developing Countries *Production and Operations Management* **23**(9) pp 1483–1494

[7] Xu G, Dai H, Hou D, and Sun H 2016 A-potential function and a non-cooperative foundation for the Solidarity value *Oper. Res. Lett.* **44** pp 86–91

[8] Wahyudin M, Maksum M, and Yuliando H 2015 The Shallot Pricing in the View of Import Restriction and Price Reference *Ital. Oral Surg.* **3** pp 132–136

[9] Macho-stadler I, Pérez-castrillo D, and Wettstein D 2007 Sharing the surplus : An extension of the Shapley value for environments with externalities *Econ. Theory* **135** pp 339–356

[10] Xiaofeng X and Aiqing R 2012 Physics Procedia The Construction of Shapley Value in Cooperative Game and its Application on Enterprise Alliance *Phys. Procedia* **24** pp 1377–1384

[11] Anglano C, Guazzone M, and Sereno M 2014 Maximizing profit in green cellular networks through collaborative games *Comput. Networks* **75** pp 260–275

[12] Tan R, Andiappan V, Kin Y, Ng R T L, and Ng D K S 2015 Chemical Engineering Research and Design An optimization-based cooperative game approach for systematic allocation of costs and benefits in interplant process integration *Chem. Eng. Res. Des.* **106** pp 43–58

[13] Khmelnitskaya A, Selçuk Ö, and Talman D 2016 The Shapley value for directed graph games *Oper. Res. Lett.* **44** pp 143–147

[14] Chopra S and Meindl P 2007 *Supply Chain Management : Strategy, Planning, and Operation*, 3rd ed. Upper Saddle River, USA: Pearson Prentice Hall

The Effect of Supply Disruptions on Customer Service Levels: a Case for Delivering Fertilizer Products using Maritime Transportation

N Siswanto[1*], U Kurniawati[1], S E Wiratno[1], A Rusdiansyah[1]
[1]Department of Industrial Engineering, Institut Teknologi Sepuluh Nopember, Kampus ITS Sukolilo-Surabaya 60111, Indonesia

*siswanto@ie.its.ac.id

Abstract. Delivering a product to customers can have a series of activities. It starts with the production of the product and then transporting it to the customers. However, uncontrollable and undesirable chance of disruption can occur during the delivery either at the production facility/supply side or in the process of transporting the product. Many researches has been conducting in the process of delivering the product. However not many considers these disruptions, although the disruptions has negative impacts on company such as reduce the profit, produce unbalanced inventory, and affect its reputation. This research will focus on the effect of supply disruption on customer service levels in the maritime transportation problem in order to maintain inventory level both in the supply and destination warehouses during predetermined planning horizon. The system considered consists of one loading port and two discharge ports for distributing one product. By using discrete event simulation, the result showed that supply disruption affects unbalanced inventory in the destination warehouses so that it will also influence company's service level. The results show that there is a significant decreasing both in delivery service level, about 14%, and production service level, about 15% when the disruption occurs. A scenario to increase production rate is simulated to improved the service level.

1. Introduction

Delivering a product to customers can have a series of activities. It starts with the production of the product and then transporting it to the customers. When distributing the product, the company needs to manage their transportation efficiently in order to minimize the cost because this cost has the biggest proportion which almost reaches 66% of the total distribution cost [1]. In this case, this efficiency of distribution process may increase the company's profit.

A company may distribute its product by using either land, air or maritime transportation. In the case of distributing fertilizer products in Indonesia, the producer uses both land and maritime transportation. The company uses trucks wherever the customer warehouse are in the same island with the factories. On the other hand, the company manages ships to deliver the products wherever the warehouses are in the different islands.

This research focus on distributing one particular fertilizer products by using maritime transportation. The product needs to be loaded from supply port to two discharge ports as seen in Figure 1. Three heteregenous (in term of capacity) time charter ships move around to serve these three ports to maintain the warehouses at those ports at minimum and maximum levels during the planning horizon. To select which demand port needs to be selected for the next destination after loading the product at supply port, the company is assumed to use the minimum coverage day criteria. Coverage day is defined as the number of days that a warehouse can cover the demand of a product from its storage before it runs out [2][3]. By using this definition, the smallest number will show that the warehouse at discharge port is very critical and needs to be visited in order not to stock out. Thus, a port which has a warehouse with smallest number coverage days will be chosen for the next destination of a ship from supply port. After arriving at demand port, a ship will unload the products into warehouse. The ship, then, sails back to the supply port for the next assignment. The ship will travel back and forth during predefined planning horizon so that the problem can be categorised as inventory routing problem (IRP). According to Al-Khayal and Hwang [4], inventory routing problem is concerned with the status inventory at each demand node and route selection in order to to find the least total cost of inventory and distribution. This problem integrates two concepts: inventory control and vehicle routing problem, where the route selection and inventory policy are simultaneously determined. The IRP that used ships as transportation mode is called as maritime inventory routing problem (mIRP) or ship inventory routing problem (sIRP).

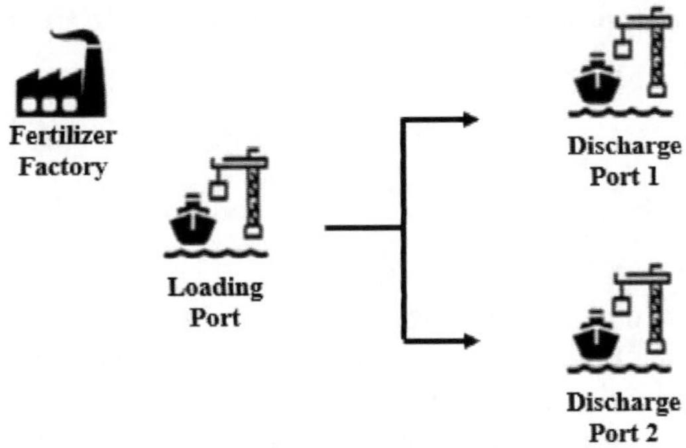

Figure 1. Distribution process of fertilizer products using ships

During distribution process, it may occur not only uncontrollable and undesirable events but also schedule events called as disruptions [5]. Disruption is defined as everything that may disrupt the system [6]. This also can happen in the maritime trtansportation. Brouer et al. [7] states that a journey of a ship to a port may about 70% - 80% not arrive on time. The delay may be caused by weather, port regulation, congestion either for shift in or shift out, and ship's machine or facility breakdown. Disruption may also occur at the production facility such as either schedule maintenance or unschedule factory breakdown. These breakdowns are influenced by the age of factory which has been operated for more than twenty years and also are caused by the late arrival of raw materials. These facility disruptions which called also as supply disruptions can affect distribution line. This supply disruption becomes the focus of this paper. The transportation disruption will be discussed in other our paper.

Table 1. Total downtime days for one particular production facility in year 2015

Downtime	Actual (days)
Scheduled downtime	61.2
Unscheduled downtime	19.28

Tabel 1 shows that the production facility had a total of scheduled downtime 61.2 days and unschedule downtime 19.28 days in 2015. However, these disruptions did not occur in once time, but several times. The time between downtime was distributed exponential with mean 27 days, while the length of downtime had the value minimum one hour, average 17 hours and maximum 15.75 days. These supply disruptions may decrease the quantity of product distributed and may delay the departure of a ship if the level of warehouse is less than the compartment capacity of the ship. This delay called as demurrage may decrease the utility of the ship as well as increase the ship's operational cost. If the ship is departed with the quantity loaded less than the ship's capacity, the shipment cost per weight of product will also increase.

Disruption needs to be taken attention seriously because of the severe affects to the company, such as increase cost, unmet demand, customer loss, unbalance inventory, and decrease company's reputation [5]. Because of those previously stated, this paper try to minimize the effect of disruption on customer satisfaction.

To model this disruption system, we build a discrete event simulation. This method is used because it can handle complex system [8]. The complexity can be seen from the variablity and interdependency of the elements, such as ships, warehouse, production facility, and others. The variability elements are the time between disruptions and the length of disruptions, the product loading time, the sailing time, congestion time before shift in and shift out to/from port, unconstant demand and other factors. Maritime transportation is shown to have a high degree of uncertainty [9]. The interdependency in this system can be seen from the interaction of activities, such as the length of supply disruption will affect sailing time and congestion time. To solve this system, there are several scenarios will be generated and tested to find the best recovery scenario in order to increase the preformance of the system.

2. Model Development

In this section, the steps for developing the model are described and explained as follow:

2.1 The system elements

System elements include entity, activities, resource and control of the system. The entities of the system are the ships which distribute the product and the product itself. The observed activity is maritime line distribution system as seen in the Figure 2 below. Resources of this system are the facilities at port, such as berth and cranes. The system is controlled by some rules or directions, such as ship's assignment to a discharge port with the least coverage days, port operational hours, and loading/unloading procedures.

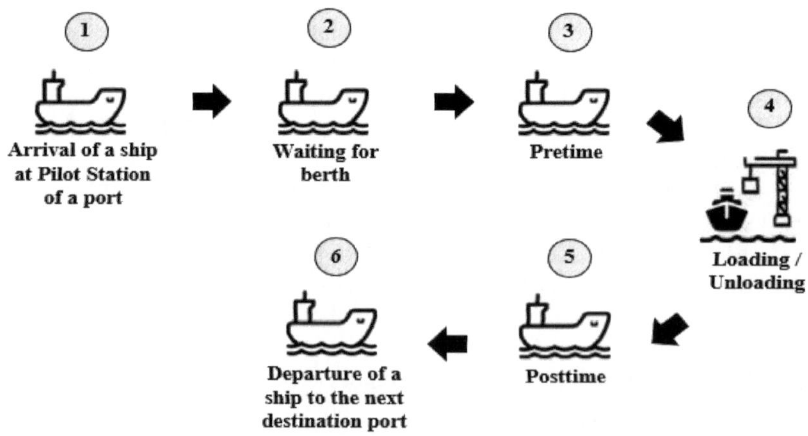

Figure 2. Maritime distribution activities

2.2 System variables

In simulation, the variables consist of decision variables, respond variables and status variables. In this maritime transportation system, the decision variables are the number of ships and mitigation schema used. The respond variables are inventory levels at the warehouses both in supply and discharge ports and total cost. Finally, the status variables are the status of the ships: idle or busy, loaded or unloaded; the status of the ports: empty or occupied. Tabel 2 shows the summary of these system variables.

Table 2. System variables for maritime transportation

Decision Variables	Variabel Respon	State Variabel
The number of ships	Delivery lead time	Ship status
Disruption mitigation scheme	Service level	Port available status
	Total cost	

2.3 Key performance indicator

Key Performance Indicator is a quantitative indicator which is used to measure the performance of the system. The criteria for this system is service level for fulfilling the demand of the customers which has formulation as below:

$$\textit{Service Level} = \frac{\text{The number of delivery}}{\text{total of target delivery}} \quad (1)$$

2.4 Data collection

In this subsection, the data as the input of the simulation model is collected. The first step is to identify and explore data which is needed. In simulation, there are three kind of data: structural, operational and numerical. The structural data is the data to indicate the structure of the system, such as delivery line, loading port location, and discharge port location. The second, operational data is the data to demonstrate the way of system works, such as delivery route. Lastly, the numerical data is the quantitative data in the system. Table 3 shows the data needs to be collected for the system as below.

Table 3. Types of data for the maritime transportation

	Structural Data	
Ship:	**Loading Port:**	**Discharge Port:**
Type of ships	Loading port location	Discharge port location
The number of ship	Type of loading port	Type of discharge port
	Operational Data	
Ship:	**Loading Port:**	**Discharge Port:**
Schedulling	Time Windows	Time Windows
	Numerical Data	
Ship:	**Loading Port:**	**Discharge Port:**
Ship's capacity	Warehouse capacity	Warehouse capacity
Fuel consumption	Production rate (normal and recovery condition)	Demand rate
Operational cost	Loading rate	Unloading rate
	Shift in time, shift out time	Shift in time, shift out time
	Distance	
Charter cost	Breakdown Data / Supply facility disruption	Port congestion data

2.5 Data processing

In this step, the data that collected from the previous step is processed which is called fitting distribution. The objective of this process is to have what kind of distributions of the data and their parameters. The processed data are production rates either at normal or recovery conditions, demand rates at destination warehouses, loading and unloading rates at every ports, pre-loading and post-loading times both at loading port and discharge ports, time between facility breakdown and its repair duration.

2.6 Building simulation

There are two steps to build the simulation model. The first is to make the conceptual model which the logic flow diagram is used. There are five conceptual submodels developed: the arrival of facility downtime submodels, the update of facility inventory level submodel, the selection of discharge port destination submodel, the shipment of product submodel, and lastly, the sail of ships to the loading port submodel. Based on these submodels, simulation model is developed by entering the input data that has been fitted of distribution.

2.7 Generating scenarios

When the existing model has been verified and validated, scenario needs to be generated based on the results of decision variables. There are two scenarios are generated. The first scenario is without increasing the production rate, and the second one is increasing the production rate by 106% after occuring facility downtime.

3. Result and Discussion

In this section, the simulation experiment is conducted which can be divided into three conditions: no supply disruption model, supply disruption model with normal production rate, and finally supply disruption model with recovery production rate. All of these models used three ships with the shipment rule full ship load. The detail of these three models are described below.

3.1 No disruption Model

The objective of building this model is to verify and validate the model developed. The result of this simulation model shows that warehouse inventory levels of this model can be seen in Figure 3 as below. The delivery service level by using equation (1) has a value 98%. It means that the delivery performance is about 2% under target. On the other hand, the production service level has a value 100% which means that the production facility has achieved its target. Without any disruption, the average round trip days of a ship to deliver a product to the demand port 1 is 21.47 days, while to demand port 2 is 17.3 days.

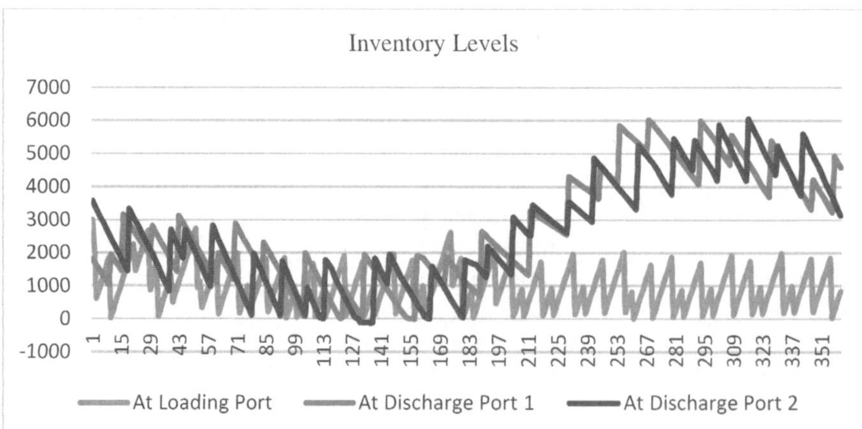

Figure 3. Inventory levels at three warehouses for no disruption model

From Figure 3, the inventory levels at discharge port 1 and discharge port 2 are relatively stable. There exist stockout at discharge port in about day 130, however this stockout quantity is only 2% from the delivery target. After day 180, the Inventory levels at discharge ports are increasing for about six months due to low demands (because it is not in the plant season). The inventory levels reach about 6000 tons in about day 270.

3.2 Supply disruption model with normal production rate

For this model, there exist facility disruptions called as supply disruptions in the form of factory breakdowns. These breakdowns are influenced by two factors: the age of factory which has been operated for more than twenty years and the late arrival of raw materials. When the disruption occur, factory cannot produce the product so there is loss of production. After the disruption occur, the production is back to the normal condition. Based on the experiment of this simulation model, the delivery service level is 84.5%, below about 14% from the model of without any disruption as in the first model. The production service level decrease to 84,93% from 100% if without any disruptions. It means that supply disruptions will affect the loss of production by about 15% from the target. With supply disruption, the average round trip days of a ship to deliver a product to the demand port 1 is 25,8 days, longer than the first model which is 21.47 days, while to demand port 2 is 19.95 days, longer about 3 days from the first model. 17.3 days. Figure 4 depicts the warehouse levels of the case supply disruption without recovery scenario as below.

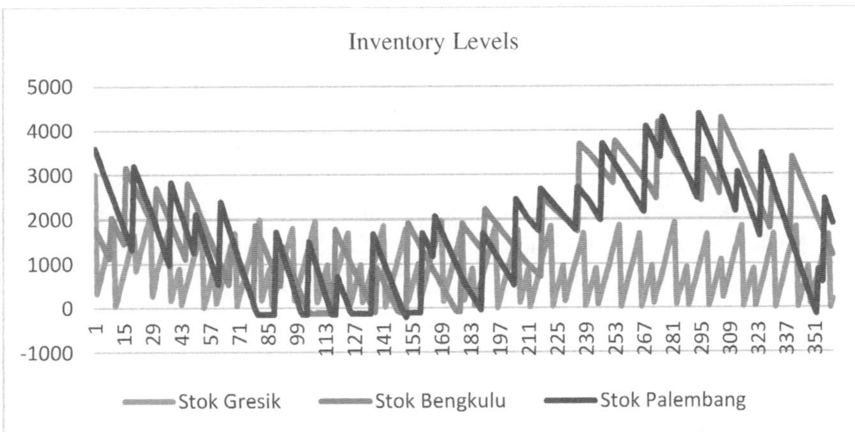

Figure 4. Inventory levels at three warehouses for supply disruption model with normal production rate

From Figure 4, the inventory levels at discharge port 1 and discharge port 2 seems not stable at the beginning period. There are plenty of stockout condition at discharge port 1 and dicharge port 2. This stockout condition is higher than the model without any disruptions. The Inventory levels at discharge ports for the six month and so on is relatively high due to the dry season and not in plant season. Compared to model without any disruption, the inventory level reach 4000 tons when entered the nineth and tenth months, below 2000 tons compared to the first model.

3.3 Model with supply disruption combine with recovery rate

This model also consiederd facility disruptions. When the disruption occur, factory cannot produce the product so there is loss of production. However, there exists recovery option to increase the production rate by 106% whenever after breaking down occurred. This recovery rate has been done to compensate the loss of production. Based on the experiment of this model, the delivery service level can be increased 7% to 91.5%. The production service level also increase to 92,3%, above also 7% compared without recovery scenario. With additional production rate, the average round trip days of a ship to the demand port 1 is 24,2 days, shorter than without recovery with 1,5 days, while to demand port 2 is 19.03 days, shorter about 1 days from the scenario without recovery. Figure 5 depicts the warehouse levels of the case supply disruption with recovery scenario as below.

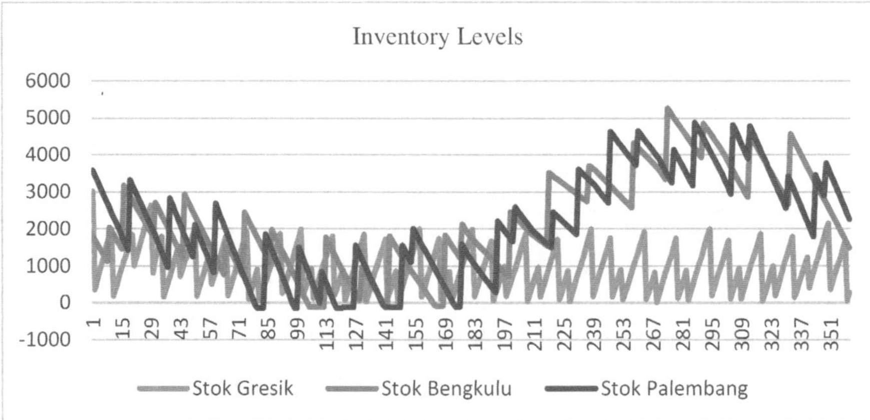

Figure 5. Inventory levels at three warehouses for supply disruption model combined with recovery production rate

From Figure 5, there are plenty of stockout condition at discharge port 1 and dicharge port 2. This stockout condition is higher than model without any disruption, but lower than model with supply disruption without recovery rate. When entered the nineth and tenth months, the inventory level reach 5000 tons, below 1000 tons compared to the first model, and above 1000 tons compared to model with supply disruption without recovery rate.

4. Conclusion

The disruption may occur in the maritime transportation for distributing a product. The disruptions may be in the form either the supply disruption when the breakdown of the facilities occur or the transportation disruption as the congestion to wait a berth at a port. Based on the experiments in the previous section, it can be concluded that disruptions will affect both delivery or production service level. When the supply disruption occurs, the delivery service level decreases about 14% and the production also decreases 15%. If the system has an option to increase production rate after breaking down, the delivery and production service levels can be increased both to 7% compared to without recovery scenario. The inventory levels in destination warehouses are also lower when the disruption occur.

The calculation for round trip days also gives the same result as the ones for production and delivery service levels. An interesting question is that whether the transportation disruptions will give the same consequences or worst than the supply disruption. This will be our research agenda.

5. Reference

[1] Ghiani G , Laporte G and Musmanno R 2004 *Introduction to Logistic System Planning and Control* California : John Wiley and Sons

[2] Siswanto N , Essam D , Sarker R (2011a) *Solving the ship inventory routing and scheduling problem with undedicated compartments Computers and Industrial Engineering* 61 289-299

[3] Siswanto N , Essam D and Sarker R (2011b) *Multi-heuristics based Genetic Algorithm for solving maritime inventory routing problem Proc of the 2011 IEEE Industrial Engineering and Engineering Management Conference* Singapore116-120

[4] Hwang S J and Al Khayyal F 2007 *Inventory Constrained Maritime Routing and Scheduling for Multi Commodity Liquid Bulk European Journal of Operational Research* pp 106-130

[5] Paul S K , Sarker R and Essam , D 2016 *A Reactive Mitigation for Managing Supply Disruption in a Three Tier Supply Chain Journal of Intelligent Manufacturing*

[6] Hishamuddin , Sarker and Essam, D 2012 *A Disruption Recovery Model for a Single Stage Production Inventory System European Journal of Operational Research* pp 464-473

[7] Brouer B D , J Dirksen, et al (2013) *"The Vessel Schedule Recovery Problem (VSRP)–A MIP model for handling disruptions in liner shipping " European Journal of Operational Research 224(2)*: 362-374
[8] Kelton W D , Sadowski R P , and Swets N B *Simulation with Arena*
[9] Christiansen M , Agra A , Delgado A and Hvattum L M 2015 *A Maritime Inventory Routing Problem with Stochastic Sailing and Port Times Computers and Operations Research,* Issue 61 pp 18-20

Acknowledgments
Authors wishing to acknowledge the funding of this research is from Ministry of Research Technology and Higher Education, Republic of Indonesia through International Research Collaboration Scheme.

IOP Conf. Series: Materials Science and Engineering **337** (2018) 012018 doi:10.1088/1757-899X/337/1/012018

A Joint Replenishment Inventory Model with Lost Sales

N L Devy[1], T J Ai[1*] and R D Astanti[1]
[1]Department of Industrial Engineering, Universitas Atma Jaya Yogyakarta, Indonesia

*jinai@staff.uajy.ac.id

Abstract. This paper deals with two items joint replenishment inventory problem, in which the demand of each items are constant and deterministic. Inventory replenishment of items is conducted periodically every T time intervals. Among of these replenishments, joint replenishment of both items is possible. It is defined that item i is replenished every Z_iT time intervals. Replenishment of items are instantaneous. All of shortages are considered as lost sales. The maximum allowance for lost sales of item i is S_i. Mathematical model is formulated in order to determining the basic time cycle T, replenishment multiplier Z_i, and maximum lost sales S_i in order to minimize the total cost per unit time. A solution methodology is proposed for solve the model and a numerical example is provided for demonstrating the effectiveness of the proposed methodology.

1. Introduction

In the research on multi-item inventory policy decision problems, there are many possible ways to be conducted in order to obtain cost efficiency. One of the possible way is conducting joint replenishment of items. Some researchers in the past have been proposed a problem called the joint replenishment problem (JRP) in order to address multi item inventory system, in which several items may be jointly replenished from the same supplier [1][2][3][4].

After the basic joint replenishment problem proposed [5], some researchers have been working on the methodology to solve this problem [6][7][8]. While, other researchers have been working on the variants of this problem, such as stochastic demand [9][10], quantity discount [11], discrete time replenishment [12], and continuous unit cost decrease [13].

The condition of shortage, i.e. whenever demand is exist while the inventory level is zero, is common in the inventory system. Especially in the discussion on stochastic demand, shortage is common to be included in the model. However, to the best of authors' knowledge in the JRP model, all shortage is usually treated as backorder. It is well known in the practice that some customers are not willing to wait whenever no stock in the inventory. In other words, the shortage is not backorder but lost sales. Therefore, it is necessary to develop JRP model that treated shortage as lost sales.

The remainder of this paper is organized as follows: Problem and mathematical formulation is defined in section 2. After that, solution methodology of the proposed mathematical model is proposed in section 3. Section 4 presents some numerical examples. Finally, some the conclusion of this study is presented with some suggestions for further research in this research area.

2. Problem and Mathematical Formulation

The problem can be formulated as follow. An n items inventory problem is considered here, where the demand of each items are constant and deterministic. The demand rate of item i is D_i units per unit time. Replenishment of item i is instantaneous with size of Q_i units. A joint replenishment is conducted periodically every T time intervals. However, all items may not be included in each replenishment. Item i is only included every Z_iT time intervals. In other word, Z_iT is the cycle time of item i. A replenishment incurs of major ordering cost \$$K$ and minor ordering cost \$$k_i$ for every item i included in the replenishment. The inventory holding cost of item i is \$$h_i$ per unit per unit time. All of shortages are considered as lost sales. The maximum allowance for lost sales of item i is S_i and the lost sales cost of item i is \$$p_i$ per unit. The decision to be taken in this situation is determining the basic time cycle T, replenishment multiplier Z_i, and lost sales quantity S_i in order to minimize total cost per unit time.

Inventory level of this problem situation with 2 items, Z_1=1, and Z_2=2 is illustrated in Figure 1. The total lost sales of item i is S_i can be calculated as t_i D_i. Therefore, the optimization problem can be converted into minimizing total cost C as the function of T, Z_i, and t_i.

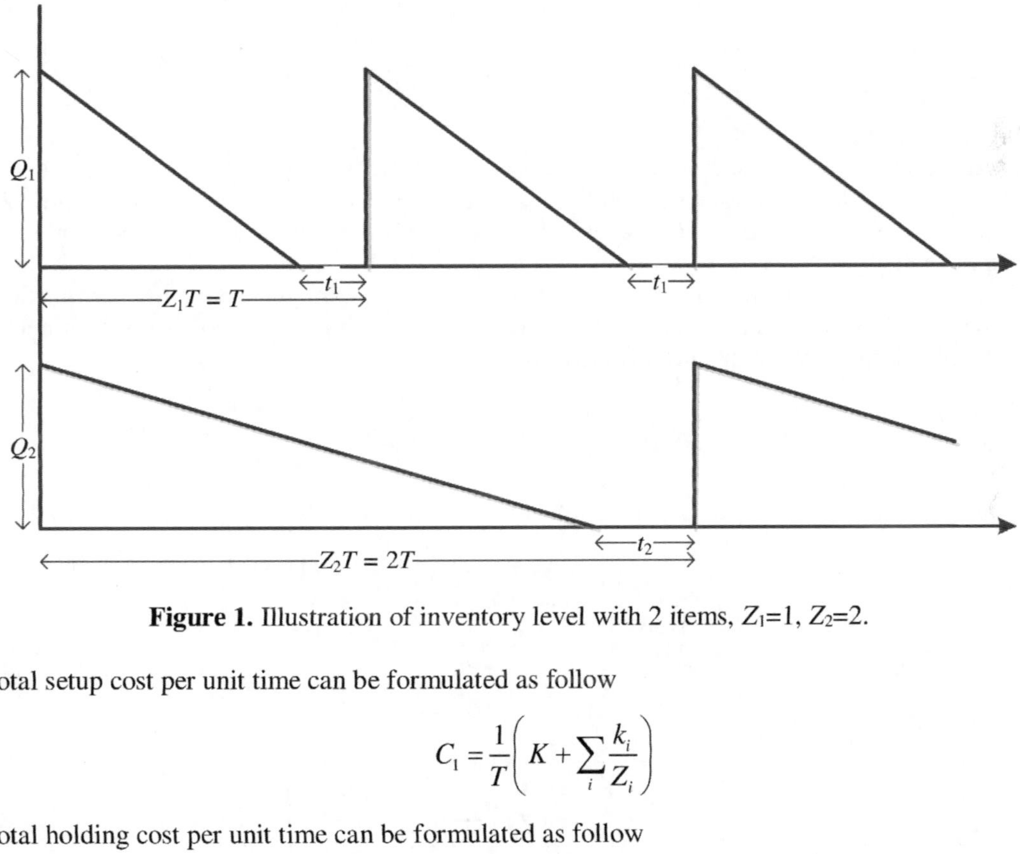

Figure 1. Illustration of inventory level with 2 items, Z_1=1, Z_2=2.

The total setup cost per unit time can be formulated as follow

$$C_1 = \frac{1}{T}\left(K + \sum_i \frac{k_i}{Z_i}\right) \tag{1}$$

The total holding cost per unit time can be formulated as follow

$$C_2 = \sum_i \frac{D_i\left(Z_iT - t_i\right)^2 h_i}{2Z_iT} \tag{2}$$

The total lost sales cost per unit time can be formulated as follow

$$C_3 = \sum_i \frac{D_i t_i p_i}{Z_iT} \tag{3}$$

Therefore, the total cost can be summarized as follow

$$C = C_1 + C_2 + C_3$$
$$= \frac{1}{T}\left(K + \sum_i \frac{k_i}{Z_i}\right) + \sum_i \frac{D_i\left(Z_i T - t_i\right)^2 h_i}{2Z_i T} + \sum_i \frac{D_i t_i p_i}{Z_i T} \quad (4)$$

After some algebra, Equation (4) can be written as

$$C = \frac{1}{T}\left(K + \sum_i \frac{k_i + D_i p_i t_i}{Z_i}\right) + \sum_i \left(\tfrac{1}{2} D_i h_i Z_i T - D_i h_i t_i + \frac{D_i h_i t_i^2}{2Z_i T}\right) \quad (5)$$

It is also seen from Figure (1) that following constraints are necessary:

$$T \geq 0 \quad (6)$$

$$0 \leq t_i \leq TZ_i, \quad \forall i \quad (7)$$

$$Z_i \in \text{integer}, \quad \forall i \quad (8)$$

3. Solution Procedure

It is noted that the objective function in Equation (5) consists of two types of variables, which are real variables, i.e. T and t_i, and integer variables, i.e. Z_i. Therefore, multi variables classical optimization technique cannot be directly applied here. However, some of the classical optimization technique can be partially applied.

Let first consider the situation of fixed value of t_i and Z_i. In this situation, the objective function can be considered as function of single variable T. Therefore, necessary condition to obtain minimum value of C is the first derivative of C with respect to T is equal to zero.

$$\frac{d}{dT}C = -\frac{1}{T^2}\left(K + \sum_i \frac{k_i + D_i p_i t_i}{Z_i}\right) + \sum_i \left(\tfrac{1}{2} D_i h_i Z_i - \frac{D_i h_i t_i^2}{2Z_i T^2}\right) = 0 \quad (9)$$

After some algebra, the expression of optimal T can be obtained as follow

$$\left(T^*\right)^2 = \frac{2\left(K + \sum_i \frac{k_i}{Z_i} + \sum_i \frac{D_i h_i t_i^2}{2Z_i} + \sum_i \frac{D_i p_i t_i}{Z_i}\right)}{\sum_i \left(D_i h_i Z_i\right)} \quad (10)$$

Therefore,

$$T^* = \sqrt{\frac{2\left(K + \sum_i \frac{k_i}{Z_i} + \sum_i \frac{D_i h_i t_i^2}{2Z_i} + \sum_i \frac{D_i p_i t_i}{Z_i}\right)}{\sum_i \left(D_i h_i Z_i\right)}} \quad (11)$$

Second, let consider the situation of fixed value of T and Z_i. In this situation, the objective function can be considered as function of multi variable t_i. Therefore, necessary condition to obtain minimum value of C is the first derivative of C with respect to t_i is equal to zero.

$$\frac{\partial}{\partial t_i}C = \frac{1}{T}\left(\frac{D_i p_i}{Z_i}\right) - D_i h_i + \frac{D_i h_i t_i}{Z_i T} = 0 \quad (12)$$

After some algebra, the expression of optimal T can be obtained as follow

$$t_i^* = Z_i T - \frac{p_i}{h_i} \tag{13}$$

It is noted that due to constraint (7), which is $t_i \geq 0$, following expression is also required

$$T \geq \frac{p_i}{Z_i h_i} \tag{14}$$

Next, let consider the situation of fixed value of T and t_i. In this situation, the objective function can be considered as function of multi variable Z_i, in which all variable Z_i are integer. To find an optimal value of Z_j, let consider the situation whenever the other values of Z_i $(i \neq j)$ are fixed. Since Z_j is integer, following condition is required to minimize C.

$$C\left(Z_1, Z_2, \ldots, Z_j - 1, \ldots, Z_n\right) \leq C\left(Z_1, Z_2, \ldots, Z_j, \ldots, Z_n\right) \leq C\left(Z_1, Z_2, \ldots, Z_j + 1, \ldots, Z_n\right) \tag{15}$$

Substituting equation (5) into condition above and after some algebra, following condition is obtained for optimal value of Z_i

$$Z_i^*\left(Z_i^* - 1\right) \leq \frac{2}{T^2}\left(\frac{t_i^2}{2} + \frac{1}{h_i}\left(\frac{k_i}{D_i} + t_i p_i\right)\right) \leq Z_i^*\left(Z_i^* + 1\right) \tag{16}$$

Based on some optimality conditions describe above, following iterative procedure is proposed in order to find the optimal values of T, t_i, and Z_i.

Step 1. Set all $t_i = 0$ and all $Z_i = 1$. Set iteration index $j = 1$.

Step 2. Calculate T using equation 11, with considering constraints in equation 14. In particular, for any item i, if current value of T is smaller than $p_i / Z_i h_i$ then update $T = p_i / Z_i h_i$

Step 3. Update the values of t_i using equation 13

Step 4. Find the optimal values of Z_i using conditions in equation 16

Step 5. If updated values of T, t_i, and Z_i are similar with previous iteration, then stop. Otherwise, return to Step 2.

4. Numerical Example

A numerical example is considered here in order to show the application of the proposed solution procedure. Let consider the joint replenishment problem with 2 items. Demand of item 1, D_1, is 900 unit/year and demand of item 2, D_2, is 600 unit/year. Holding cost of item 1 and 2 are \$4/unit/year and \$2.5/unit/year, respectively. Major setup cost is \$100/order, while minor setup cost of item 1 and 2 are \$10/order and \$7/order, respectively. The lost sales cost of item 1 and 2 are \$2/unit and \$1/unit, respectively.

Following proposed procedure described in previous section, the iteration steps of this example are presented in the Table 1. It is obtained that the optimal values of decision variables are $T = 0.5$, $t_1 = 0$, $t_2 = 0.1$, $Z_1 = 1$, $Z_1 = 1$ with total cost $C = \$1494$/year.

5. Conclusion

This paper is successfully presented a mathematical model for joint replenishment inventory system with lost sales. The methodology for solving the model is also proposed. For the future work, the extension of this model and solution methodology is expected in the area of stochastic demand. Other possibility is to extend this model to partial backlog case, that is also common in the inventory practice.

Table 1. Iteration Steps to Find T, t_1, t_2, Z_1, and Z_2.

Iteration	T	t_1	t_2	Z_1	Z_2	C
1	1.4096	1.0000	1.0000	1	1	2089.12
2	1.3451	0.9096	1.0096	1	1	2071.12
3	1.2774	0.8451	0.9451	1	1	2054.53
4	1.2059	0.7774	0.8774	1	1	2035.16
5	1.1298	0.7058	0.8058	1	1	2012.10
6	1.0482	0.6298	0.7298	1	1	1984.03
7	0.9598	0.5482	0.6482	1	1	1948.81
8	0.8623	0.4597	0.5597	1	1	1902.76
9	0.7523	0.3622	0.4622	1	1	1838.85
10	0.6231	0.2522	0.3522	1	1	1741.25
11	0.5000	0.1230	0.2230	1	1	1571.26
12	0.5000	0.0000	0.1000	1	1	1494.00
13	0.5000	0.0000	0.1000	1	1	1494.00

References

[1] Khouja M and Goyal S 2008 *European Journal of Operational Research* **186** 1–16
[2] Moon IK, Goyal SK and Cha BC 2008 *Int. Journal of Systems Science* **39** 629–37
[3] Tsai CY, Tsai CY and Huang PW 2009 *Int. Journal of Production Economics* **117** 30–41
[4] Wang L, He J, Wu D and Zeng YR 2012 *International Journal of Production Economics* **135** 190–8
[5] Shu FT 1971 *Management Science* **17** B-406–10
[6] Kaspi M and Rosenblatt MJ 1991 *Int. Journal of Production Research* **29** 107–14
[7] Porras E and Dekker R 2006 *European Journal of Operational Research* **174** 1595–615
[8] Olsen AL 2005 *Computers and Industrial Engineering* **48** 223–35
[9] Johansen SG and Melchiors P 2003 *Journal of the Operational Research Society* **54** 283–90
[10] Atkins DR and Iyogun O 1988 Management Science **34** 791–6
[11] Cha BC and Moon IK 2005 *OR Spectrum* **27** 569–81
[12] Klein CM and Ventura JA 1995 *Journal of the Operational Research Society* **46** 643–57
[13] Khouja M, Park S and Saydam C 2005 *Int. Journal of Production Research* **43** 311–26

Acknowledgment

This work is partially supported by Ministry of Research, Technology, and Higher Education, Republic Indonesia under International Research Collaboration and Scientific Publication Research Grant No. 045/HB-LIT/IV/2017.

IOP Conf. Series: Materials Science and Engineering **337** (2018) 012019 doi:10.1088/1757-899X/337/1/012019

Model of refrigerated display-space allocation for multi agro-perishable products considering markdown policy

D Satiti[1*], A Rusdiansyah [1]**
[1]Agrifood Supply Chain (AfSC) Research Group, Logistics and Supply Chain Management (LSCM), Department of Industrial Engineering, Institut Teknologi Sepuluh Nopember, Kampus ITS Sukolilo-Surabaya 60111, Indonesia

[*]satiti.dyah92@gmail.com, [**]arusdian@ie.its.ac.id

Abstract. Problems that need more attention in the agri-food supply chain are loss and waste as consequences from improper quality control and excessive inventories. The use of cold storage is still being one of favourite technologies in controlling product quality by majority of retailers. We considerate the temperature of cold storage in determining the inventory and pricing strategies based on identified product quality. This study aims to minimize the agri-food waste, utility of cold storage facilities and maximize retailer's profit through determining the refrigerated display-space allocation and markdown policy based on identified food shelf life. The proposed model evaluated with several different scenarios to find out the right strategy.

1. Introduction

In recent years, most of modern retailers not only selling processed food products, but also the fresh ones, like fruits and vegetables. Fruits and vegetables fall into agricultural product, and have perishable character. It means that their qualities are decreasing over time as a result of their natural metabolism. If the product is still not being sold at the end of its self life, either fruits or vegetables should be discarded because they cannot be consumed and caused food borne disease. Therefore they require an appropriate inventory management [1] and cold storage to control their quality [2][3]. The loss due to expired products is quite high, reaching 15% [4]. According to Herbon et al. [5], the right pricing strategy can increase sales number and reduce waste number of product. One of pricing strategy that suitable for agricultural product is markdown policy.

Product characteristics affect complexity level of food supply chain management [6]. The increasing of public awareness according to food health, safety, and quality, as well as the diversity of demand and prices, and limited lifetime of the product add the complexity of fresh food supply chain management [7]. As discussed before, there is a need of an appropriate inventory management to handle excessive inventory, especially in fresh food supply chain problem. There were much research on inventory management for perishable products [8][9][10]. One of them developed an EOQ model for food product by considering stochastic demand using a probability approach [11]. But those researches were not considering capacity of the cold storage facility yet. In this study, a model is proposed that can be used as a basis for managerial decision in operational activities (determining ordering number, display-space allocation, discount value, and discount time). The aim is to minimizing the number of waste and maximizing retailer's profit.

2. Literature review

Perishable products can be divided into three terms based on obsolescence reason according to Jia and Hu [12]. First, because of losing value such as air tickets, cinema tickets, hotel rooms, and newspapers; second, due to technological developments such as electronic goods: cell phones, laptops, television and so forth; third because of physical changes as a result of deterioration product quality caused by natural processes (metabolic) and the environment influence (temperature, humidity, and moisture content) such as fruits, vegetables, meat, fish and other food products. Out of them, the third one become a concern for some scientists in the last decade [13][14][15][16]

Some scientist have developed inventory model for perishable products [8][9][11][17]. However, on those studies did not consider capacity of storage facility and quality deterioration for perishable food products yet. Capacity of storage facility was take into consideration in inventory model by Bai and Kendall [18], and Piramuthu and Zhou [19]. Bai and Kendall [18] solved space-allocation and inventory management problem for various fresh food products by considering its freshness through non-linier optimization. There were several lack of model which built by Bai and Kendall [18]. First, they assumed that fresh product display on non-refrigerated facility (room temperature). In facts, some fresh products need cold storage to maintain quality. It means, their models would be appropriate for products that didn't need cold storage. Second, the value of decay rate was already known without numerical calculation. As a result, that assumption made the model was less able to capture the quality declining process in more specific way. Third, the discount value did not considering the identified quality of product yet. It would made the price that already set was too cheap for retailer or too expensive for customer according to the identified quality. Therefore, it needs a solution to accommodate both of them, which the discounts is adjusted to the remaining quality of products through a more specific approach.

The given discount value can influence inventory level. This is related to the change of sales level as the effect of differentiation price. Herbon et al. [5] mentioned that the different pricing strategy could increase sales number of perishable products. That is why, it is really important to include discount value as consideration in inventory decision. Markdown based on quality had been discussed previously [20]. In the study they conducted a pricing strategy approach based on identification of product lifetime. The aims of study were reducing the amount of wasted food products and maximizing retailer's profit. Product lifetime was identified by evaluating deterioration rate of product based on standard temperature and length of storage time before eventually sold or wasted. Zanoni and Zavanella [21] developed the study of Rong et al. [22] which's using coefficient of performance ratio as a consideration in determined storage cost. The study also explained that the precise standard temperature determination could reduce energy costs. Based on that study, storage cost for products which's using a refrigerated facility can be determined based on reference and ideal storage temperature.

From those explanations, it is really important to determine the right storage temperature for refrigerated display facility that still using by retailer to storage (display) fresh product. In other hand, set temperature can influence product lifetime and decay rate which have impact on the right pricing strategy during the selling periods. Storage temperature, product quality and pricing strategy can be used as consideration in determine the order number and space allocation for each product in accordance with the refrigerated display capacity. Therefore, based on the studies of Wang and Li [20] Zanoni and Zavanella [21], and Bai and Kendall [18] we develop refrigerated display-space allocation model by considering markdown policy for multi agro-perishable products multi temperature with aim to maximize the retailer's profit.

3. Problem definition and notation

This study develops refrigerated display-space allocation and inventory management model considering markdown policy according to quality deterioration over times until its end lifetime. The system studied consist a retailer that procure and selling the fresh perishable food product like fruit and vegetable (item $= 1,2,3,..k$) which have different attribute: size, shape, and ideal storage temperature. The model development on this study conducted based on several model that already built by Wang and Li [20], Zanoni and Zavanella [21], and Bai and Kendall [18]. On proposed model, there are two markdown policies based on the frequency ($n = 1,2$). First, single price markdown ($n = 1$) and, second, double price markdown ($n = 2$). The formulation of purposing model for each item's profit (F_i) per hour at selling period T_i can be seen on this equation:

$$F_i = T_i^{-1}[p_i ED_{0i} + \sum_{n=1}^{l} p_{ni} ED_{ni} - (Q_i - ED_{0i})C_p - \dots - (Q_i - \sum_{n=1}^{l} ED_{n-1i})C_p - Q_i\left(UC_i + C_o + \tfrac{1}{2}HC\rho_{T_R}T_i\right) - A_i AC_i S_i - OC_i] \quad (1)$$

with p_i is normal selling price at the beginning of selling period, ED_0 expected demand with normal selling price, $p_{ni}$is discount price, ED_{ni} expected demand with discount price, Q_i ordering number, $A_i$displaying number (allocation number), S_i Space that needed by a unit product, and some costs that involve such as ordering cost (OC_i), unit cost (UC_i), operational cost (C_o), holding cost based on ratio of COP ($HC\rho_{T_R}$), and additional cost for used a unit space (AC_i).

Then, the objective function of this proposed model becomes:

$$\text{Max} \sum_{i=1}^{l} F_i \quad (2)$$

Subject to

$$\sum_{i=1}^{k} A_i S_i \leq W \quad (3)$$

$$L_i \leq A_i \leq U_i \quad (4)$$

$$\sum_{n=0}^{l} ED_{ni} \leq Q_i \quad (5)$$

$$Q_i \geq 0 \qquad i = 1,2,3,\dots,k \quad (6)$$

Constraint (3) ensures the used space is not more than display-space capacity (W). Constraint (4) ensures that displayed product numbers are not less than the lower limit and exceed the upper limit. Constraints (5) and (6) ensure that the total expected demand does not exceed procured number, in other words all customer requests can be fulfilled and procured number is more than or equal to 0.

In this study, some of the procured product will automatically fill display facility, and rest of them will be storage as back inventory in the facility which's has been set at same storage temperature. It assumed that display facility will always in full state, in other words, as soon as the number of displayed product is decrease it will be filled by back inventory product as much as the decreasing number. Demand of this system is continue-deterministic and influenced by price and identified quality of product over times. Therefore, it needs a right markdown strategy that appropriate to identified quality of product which afterwards this decision is used as consideration in procuring product (number of order). In other words, hopefully the order number can fulfil costumer's demand in certain price and quality until the end of product lifetime. Markdown frequency which maximizes retailer's profit becomes one of variables that must be found.

There are several steps on numerical calculation to get the maximum profit, as follows:

1. Determine the reference storage temperature T_R.
2. Calculate the COP ratio value between reference storage temperature and ideal storage temperature for each product.

$$T_{Ei} = \frac{T_{Ei_L} + T_{Ei_H}}{2} \quad (7)$$

$$COPcooling = \frac{T_{cold}}{T_{hot} - T_{cold}} \quad (8)$$

$$\rho_{T_R} = \frac{COP_{T_{Ei}}}{COP_{T_R}} \quad (9)$$

with T_{Ei} ideal storage temperature of product i, T_{Ei_L} minimum temperature recommendation for product i, T_{Ei_H} maximum temperature recommendation for product i, T_R storage temperature on *retail* (K), T_{cold} temperature that used for cold storage, T_{hot} is 293.15 K (environment temperature).

3. Calculate the quality deterioration rate (σ), quality level ($q_{(t)}$) and the remaining life of product due to storage temperature differences.

$$\sigma = k_A e^{-\left[\frac{E_A}{RT_R(t_h)}\right]} \quad (10)$$

$$q(t) = q_0 - \sum_{h=1}^{m} \sigma_h t_h \quad (11)$$

With k_A is quality deterioration constants, E_A activation reaction energy which control quality *loss* [J/unit], R gas constants [J/unit °K], q_0 initial quality of product (it assumed 1), and t_h storage duration $(0, T)$ [hour].

4. Determine the discount value for each product based on markdown startin time and quality deterioration rate.
5. Find the number that must be ordered and displayed, as well as retailer's profit based on used set parameters.

4. A numerical simulation and discussion

In this section, the proposed model approach is applied in an illustrative case study. In order to capture model performance numerical simulation is performed with generated data which resembles the real conditions on the retailer. Parameters in proposed model are simulated with various values to investigate their impacts on model performance. By using the proposed model from equation (1) until equation (11) we get the result that represent in table 1. It shows that uniform discount in single price markdown ($n = 1$) with storage temperature 275,15 K gives the biggest profit 7,30495 £/hour. Meanwhile, quality discounts in double price markdown ($n = 2$) with storage temperature 281,15 K gives the smallest profit 0,37876 £/hour. The use of higher set temperature (T_R) than its ideal temperature for some product increasing its deterioration rate and fastening the decaying process. This condition will influence visual performance of product which can decreasing consumer's interest. It also affects consumer's willingness to pay. That is why, the use of higher temperature will decreasing total profit per hour.

Table 1. Profit's profile for each discount type.

Discount Type	Discount Frequency	Tm1	Tm2	TR	Total profit (£/hour)
Uniform	n=1	50%	-	275,15	7,30495
		50%	-	278,15	6,92976
		50%	-	281,15	1,91413
	n=2	50%	75%	275,15	6,17360
		50%	75%	278,15	5,94875
		50%	75%	281,15	2,27950
Quality Based	n=1	50%	-	275,15	6,04061
		50%	-	278,15	5,84738
		50%	-	281,15	1,95256
	n=2	50%	75%	275,15	2,95605
		50%	75%	278,15	2,82077
		50%	75%	281,15	0,37876

Table 2. Order quantity (Q_i*) for each product dependent on markdown policy.

item	S(n=1)	K(n=1)	S(n=2)	K(n=2)
1	61	78	83	102
2	45	58	62	76
3	27	27	36	42
4	49	59	66	80
5	25	27	33	39
6	54	83	74	94
7	90	122	123	152
8	73	98	99	123
9	92	122	125	155
10	64	92	87	110

S : uniform discount
K : quality based discount

The order quantities are different from one discount strategy to others. On single price markdown, the discount types give different order quantities. This is related to given discount value where uniform discount give same value which is 40% of the initial sales price, while quality based discount give different value for each product depend on remaining product lifetime, initial product quality and quality deterioration rate. The gap of given discount moves the number of expected demand and impacted on order quantity. It also applies to double price markdown. Because of order quantity more than the upper limit of displayed product, it makes displayed product number is same with the upper limit for each product, 13 unit per item. This is reported in table 2.

Quality based discount gives smaller value than uniform discount, either single or double price markdown policy. Profit per hour for each unit product is decreasing by the increasing of markdown frequency. Even so, for item 3 profit per hour is negative on double price markdown with quality based discount strategy (figure 1). It means that for some product, double price markdown could not be appropriate because it makes the consumer lose their confidence in the product quality itself.

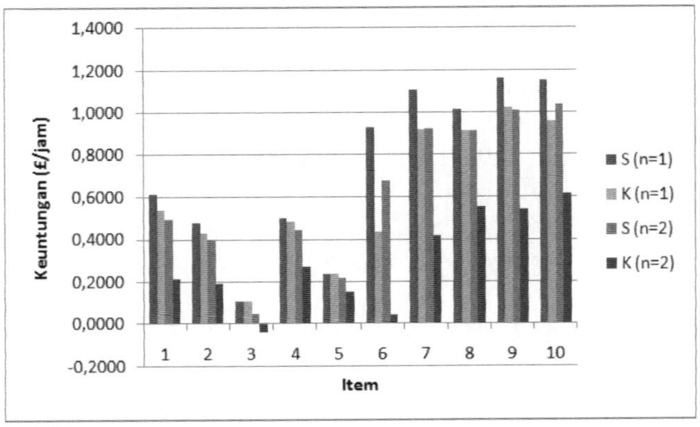

Figure 1. Profit for each product depend on markdown policy.

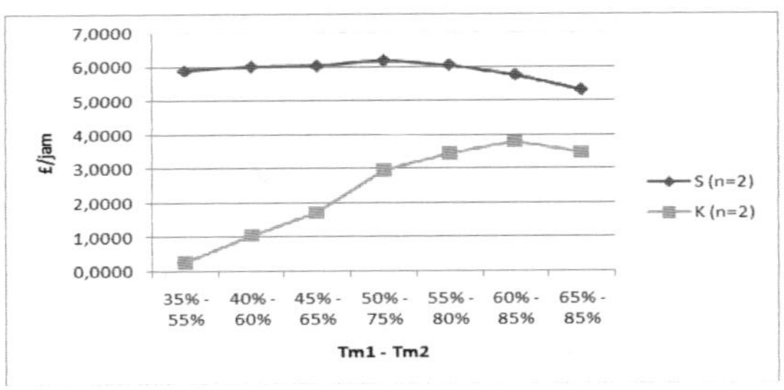

Figure 2 Single price markdown's total profit ($T_R = 275{,}15\ K$).

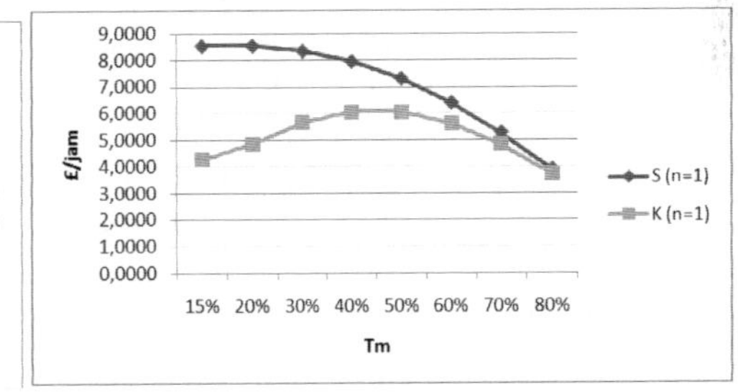

Figure 3. Double price markdown's total profit ($T_R = 275{,}15\ K$).

Profit fluctuation pattern for single and double price markdown can be seen on figure 2 and figure 3. Figure 3 shows markdown starting time which gives biggest profit per hour is 20% from remaining product lifetime with uniform discount strategy. After 20% point, total profit per hour is decreasing continuously with the increasing of T_m value. Figure 4 shows uniform discount strategy's profit reach the highest point at T_{m1}= 50% and T_{m2}= 75%, then decreasing after the markdown starting time move to T_{m1}= 55% dan T_{m2}= 60%. Meanwhile, on quality-based discount strategy, retailer's profit increasing continuously until T_{m1}= 60% dan T_{m2}= 85% point, then decreasing after it moves to T_{m1}= 65% dan T_{m2}= 85%.

5. Conclusion and future research

This paper attempts to develop allocation space and inventory model by considering deterioration quality of various fresh food products multi temperature to improve the performance of perishable food supply chain. This approach aims to avoid waste at the end life of perishable product. Based on result of proposed model, decision maker can make the right decision to solve operational problem (procurement number, displayed number, markdown policy, and markdown time). As the consequences, we can minimize number of waste and maximize profit. From the numerical simulation known that low storage temperature gives greater profit than the high ones; and single price markdown with markdown starting time delayed until certain point gives optimal display-space allocation and order quantity, it also give maximum profit.

The limitation of the research is, for simplification, that a linier deterministic demand is assumed, and the upper and lower displayed number is assumed. It will be useful to investigate further research problems using stochastic, and each product have different priority displaying level.

6. References

[1] Beshara S, El-kilany K S and Galal N M 2012 Simulation of Agri-Food Supply Chains *Int. J. Mech. Aerospace, Ind. Mechatron. Manuf. Eng.* **6** pp 899–904

[2] Song L, Chen H, Gao H, Fang X, Mu H, Yuan Y, Yang Q and Jiang Y 2013 Combined modified atmosphere packaging and low temperature storage delay lignification and improve the defense response of minimally processed water bamboo shoot. *Chem. Cent. J.* **7** p 147

[3] Xiao X, He Q, Fu Z, Xu M and Zhang X 2016 Applying CS and WSN methods for improving efficiency of frozen and chilled aquatic products monitoring system in cold chain logistics *Food Control* **60** pp 656–66

[4] Ferguson M and Ketzenberg M E 2006 Information Sharing to Improve Retail Product Freshness of Perishables Prod. *Oper. Manag.* **15** pp 57–73

[5] Herbon A, Levner E and Cheng T C E 2014 Perishable inventory management with dynamic pricing using time-temperature indicators linked to automatic detecting devices *Int. J. Prod. Econ.* **147** pp 605–13

[6] Agustina D, Lee C K M and Piplani R 2014 Vehicle scheduling and routing at a cross docking center for food supply chains *Int. J. Prod. Econ.* **152** pp 29–41

[7] Balaji M and Arshinder K 2016 Modeling the causes of food wastage in Indian perishable food supply chain *Resour. Conserv. Recycl.* **114** pp 153–67

[8] Dobson G, Pinker E J and Yildiz O 2016 An EOQ model for perishable goods with age-dependent demand rate *Eur. J. Oper. Res.* **0** pp 1–5

[9] Duan Y, Li G, Tien J M and Huo J 2012 Inventory models for perishable items with inventory level dependent demand rate *Appl. Math. Model.* **36** pp 5015–28

[10] Dye C-Y and Hsieh T-P 2011 Deterministic ordering policy with price- and stock-dependent demand under fluctuating cost and limited capacity *Expert Syst. Appl.* **38** pp 14976–83

[11] Muriana C 2016 An EOQ model for perishable products with fixed shelf life under stochastic demand conditions *Eur. J. Oper. Res.* ***255*** pp 388–96

[12] Jia J and Hu Q 2011 Dynamic ordering and pricing for a perishable goods supply chain *Comput. Ind. Eng.* **60** pp 302–9

[13] Huang J-Y and Yao M-J 2006 A new algorithm for optimally determining lot-sizing policies for a deteriorating item in an integrated production-inventory system *Comput. Math. with Appl.* **51** pp 83–104

[14] Chung K-J and Huang T-S 2007 The optimal retailer's ordering policies for deteriorating items with limited storage capacity under trade credit financing *Int. J. Prod. Econ.* **106** pp 127–45

[15] Osvald A and Stirn L Z 2008 A vehicle routing algorithm for the distribution of fresh vegetables and similar perishable food *J. Food Eng.* **85** pp 285–95

[16] Nguyen C, Dessouky M and Toriello A 2014 Consolidation strategies for the delivery of perishable products *Transp. Res. Part E Logist. Transp. Rev.* ***69*** pp 108–21

[17] Taleizadeh A A, Mohammadi B, Cárdenas-Barrón L E and Samimi H 2013 An EOQ model for perishable product with special sale and shortage *Int. J. Prod. Econ.* **145** pp 318–38

[18] Bai R and Kendall G 2008 A Model for Fresh Produce Shelf-Space Allocation and Inventory Management with Freshness-Condition-Dependent Demand INFORMS *J. Comput.* **20** pp 78–85

[19] Piramuthu S and Zhou W 2013 RFID and perishable inventory management with shelf-space and freshness dependent demand *Int. J. Prod. Econ.* **144** pp 635–40

[20] Wang X and Li D 2012 A dynamic product quality evaluation based pricing model for perishable food supply chains *Omega* **40** pp 906–17

[21] Zanoni S and Zavanella L 2012 Int . J . Production Economics Chilled or frozen ? Decision

strategies for sustainable food supply chains *Int. J. Prod. Econ.* **140** pp 731–6
[22] Rong A, Akkerman R and Grunow M 2011 An optimization approach for managing fresh food quality throughout the supply chain *Int. J. Prod. Econ.* **131** pp 421–9

A modeling of dynamic storage assignment for order picking in beverage warehousing with Drive-in Rack system

M Z Hadi[1*], T Djatna[1,2], Sugiarto[1]
[1]Department of Agro-Industrial Technology, Bogor Agricultural University, Bogor, Indonesia
[2]Department of Computer Science, Bogor Agricultural University, Bogor, Indonesia

*zaky_hadi@apps.ipb.ac.id

Abstract. This paper develops a dynamic storage assignment model to solve storage assignment problem (SAP) for beverages order picking in a drive-in rack warehousing system to determine the appropriate storage location and space for each beverage products dynamically so that the performance of the system can be improved. This study constructs a graph model to represent drive-in rack storage position then combine association rules mining, class-based storage policies and an arrangement rule algorithm to determine an appropriate storage location and arrangement of the product according to dynamic orders from customers. The performance of the proposed model is measured as rule adjacency accuracy, travel distance (for picking process) and probability a product become expiry using Last Come First Serve (LCFS) queue approach. Finally, the proposed model is implemented through computer simulation and compare the performance for different storage assignment methods as well. The result indicates that the proposed model outperforms other storage assignment methods.

1. Introduction

Storage assignment is an order picking organization and operational policies attempts to find an effective method of locating products in a warehouse to improve the operational of order picking performance [1]. De Koster et al. [2] says a storage assignment method is a set of rules which can be used to assign products into warehouse storage locations. It influences almost warehouse key performance indicators such as order picking time and cost [1].

Recent trends in the order picking operational or tactical level have resulted in order picking storage assignment design and management becoming more important and complex. One of the challenge and complexity in order picking storage assignment is storing beverage products in warehouse with drive-in rack system under dynamic orders from customer. So, beverage warehouse with drive-in rack need a dynamic storage assignment method to face the dynamic and complexity challenge in order to increase warehouse responsiveness to customer order.

Drive-in rack is a pallet rack type that has column, row and depth. Typically, Drive-In racking requires operators to drive into the racking system from one side and remove pallets from the same entry point. A simple way to identify a Drive-In Rack Systems is its LIFO queue. Product is moved in a Last-In, First-Out queue way because of the single entry and exit point. Many food and beverage industries use this rack type because of low installation cost and extremely efficient storage density. But, the problem is this rack type can increase total product expiry if the management does not set the order picking strategy well.

Dynamic storage assignment is a storage assignment in warehouse operational level which allows decision maker re-arrangement, re-warehousing, re-assignment products in storage location follows dynamic order from customer requests. According to De Koster et al. [2] the storage assignment is a set of rules which can be used to assign products to storage locations. According to previous passage, we can say that the dynamic order picking is a set of dynamic rules in a put away process which can be used to re-arrangement, re-warehousing, re-assignment product in storage location to satisfy dynamic customer request. This warehouse operational type become necessary since researcher found that the storage assignment policy can reduce travel distance, reduce picking time and congestion.

Storage assignment method has been extensively investigated. Most of researches is focused on method development with steady state assumption and no product type consideration. There are many storage assignment researches with specific warehouse case but drive-in rack has not yet led to attention now by researcher. De Koster et al.[2] says the dynamic order picking is an important future research caused by the rapid development of internet and technology to fulfil customer request more responsive. Hence, we can conclude that this research in dynamic order picking for beverage products with drive-in rack is important research.

According to the problems, complexities and challenges, the purpose of this paper is to develop a dynamic storage assignment model to solve storage assignment problem (SAP) for beverages order picking in a drive-in rack warehousing system to determine the appropriate storage location and space for each beverage products dynamically so that the performance of the system can be improved. The storage assignment model is developed under beverage limited shelf life characteristic, dynamic orders condition, product adjacency, Cube per Order Index (COI), order frequency, and arrangement rule for drive-in representation. The performance of the algorithm is validated and compared with random storage assignment approach (the most common approach for storage assignment process) by calculating rule accuracy (for product adjacency), expected total distance from storage position to loading area and probability a product become expiry using Last Come First Serve (LCFS) and First Come First Serve (FCFS) queue combination.

The rest of the paper is organized as follows. In the next section, we present methods framework to solve the problem. It is started by problem statement of dynamic storage assignment in beverage warehouse with drive-in rack system, then present research methodology and performance calculation. In section 3, we present a real study case to test our proposed model. In Section 4, we present computational simulation experiments. This is followed by performance comparison with random assignment approach and by the conclusion.

2. Problem Statement and Research Methodology

2.1. Problem Statement

Drive-in rack is three-dimension rack that has column, row and depth. The warehouse with this rack type has aisles, product source, loading area, storage area, picker and palletized beverage product (see figure 1). The design of adaptive order picking system for beverage warehousing with drive in rack system in this research will use the following assumptions.

- The warehouse receives product from production department
- The beverage packed into secondary packaging and stacked on pallet
- One pallet or one pallet position only storage one product type
- Order rate equal or higher than product arrival rate
- Put away and picking process follow Poisson distribution
- The warehouse only storage products with same/identic behavior

The problem of this study can be outlined in the follow manner. Orders and pallet products come to the warehouse, and picked by several pickers with limited capacity. Pallets from production area are picked and placed into specific storage area or pallet position according to storage assignment and zoning rules.

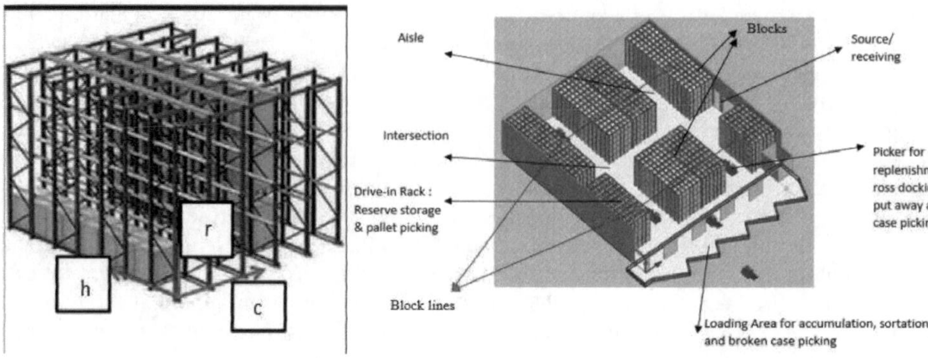

Figure 1. Drive-in rack construction and warehouse structure with drive-in rack

Because orders from customers come dynamically, the rule for storage assignment process will change follows dynamic orders from customers. So, we can reduce the total of travel distance for a specific order fulfillment case from storage position to loading area by determining the storage location dynamically according to its order frequency, shelf life and order volume. According to the problem statement the notation used in this research is presented in table 1.

Table 1. Notation used in this research

Notation	Description	Notation	Description
X	Antecedent of the rule	*irule*	Index rule generated by association rule (1,2,3,…., *nrule*)
Y	Consequent of the rule		
h	Depth index (1,2,3,….,H)	y_{irule}	Data set generated by association rule
c	Column index (1,2,3,….,C)	$\hat{y}_{irule}$	Data set generated from real order from customer
r	Row index (1,2,3,….,R)		
i	index product (1,2,3,…I)	*E(D)*	Expected total distance for specific order
p^{expiry}	probability product will expiry	$d_{i\to l}$	Distance from specific product storage location to loading area l
p^{pick}	Probability product will picked by operator	n	Total product in storage position (queue)
		γQ	Expected rate of expired product

2.2. Research Methodology

This research frameworks are adopted, combined and redesign from Pan et al. [1], De Koster et al.[2], and Chuang et al. [3] for storage assignment and order picking concepts. As we mention before that storage assignment is set of rules, so the model is designed by set of stages to find the dynamic storage assignment rule. Three stages are used to model dynamic storage assignment with the output is set of rules that represent storage assignment process in warehouse with drive-in rack.

2.2.1. Stage 1: Finding product adjacency rule.

Association rule is used in this research to find product adjacency. Association rule is an expression $X \rightarrow Y$, where X and Y are item sets [4]. X is the antecedent of the rule and Y the consequent. Different rule interestingness measures try to quantify the dependence between the consequent and antecedent. Support and confidence is used in this association rule method as rule assessment to find product adjacency of the storage assignment rule. The support $sup(X \rightarrow Y)$ of the rule is defined as the number of transactions that contain both X and Y (1), that is

$$sup(X \rightarrow Y) = sup(XY) = |t(XY)| \quad (1)$$

The confidence$con(X \rightarrow Y)$ of a rule is the conditional probability that a transaction contains the consequent Y given that it contains the antecedent X (2):

$$conf(X \rightarrow Y) = P(Y \mid X) = P(XY)/P(X) \quad (2)$$

2.2.2. Stage 2: Design class based storage rule.
The idea of class based storage is to group products into classes in such a way that the fastest moving class [2]. Each class is then assigned to a dedicated area of the warehouse. Classes are determined by some measure of demand frequency of the products, such as Cube per Order Index (COI) or pick volume. Fast moving items are generally called A-items. The next fastest moving category of products is called B-items, and so on. Product class is used in this research according to volume, frequency, shelf life and COI. COI is Average required cubic storage footage per average numbers of orders [5][6]. The higher the frequency, volume and short shelf life then the closer to the loading area and the small COI value the closer to the loading area

2.2.3. Stage 3: Design of Arrangement Rule.
The design of arrangement rule is conducted to ensure that each assignment of product follow rule that must fulfil all row first in a rack column and depth. This rule is important to avoid an empty rack between 2 assigned pallet rack positions. This research designs an arrangement rule for drive-in storage rule. The rule is modified from stack algorithm from Cormen [7]. The illustration of the rule is described as follow.

- Search empty row r in a depth h head in each column c where h,r,c is started from 1
- If row r in a depth h head in each column c is empty then assign a specific product with product adjacency and class based storage consideration
- If not, and $r \neq R$ go to step 1 with $r+1$
- If $r=R$ and $h \neq H$ then $h+1$
- If $h=H$ and $c \neq C$ then $c+1$
- If $c=C$ then warehouse status is over capacity

2.2.4. Rule Combination.
After we determine rule for product adjacency, class based storage, and arrangement rule, we propose rule combination algorithm from these rules to determine an appropriate storage location and arrangement of the product according to dynamic orders from customers. The algorithm is described as follow

- Assign product into a block according to generated rule from class based storage in a block line
- Check the adjacency. If the adjacency is appropriate then assign product according to arrangement rule. The assignment process is finish
- Else if assign in the product to another block according to adjacency rule but in the same block line then assign product according to arrangement rule
- Back to step 2

2.3. Performance Calculation

The performance calculation for storage assignment rule is considered by rule accuracy, probability expiry product and total picking distance. The formulation of rule accuracy is summation of total accurate rule compared with current data (3). Total picking distance *E(D)* is total minimum distance of travelling path (3). The probability there are *n* expiry product (4) is adopted from LCFS concept [8].

$$RuleAccuracy = \frac{1}{nrule}\sum_{irule=1}^{nrule} I(y_{irule} \rightarrow \hat{y}_{irule}),\; E(D) = \sum_{i=1}^{n} \min(d_{i\rightarrow l}) \tag{3}$$

$$P_n^{\text{expiry}} = 1 - P_n^{pick} \text{ where } P_n^{pick} = \frac{\gamma Q}{\lambda} \tag{4}$$

The model is compared with random policy storage approach. Random policy is most common storage assignment approach for manual or automated warehouse operation. Random policy is the simplest one among various storage policies and is applied widely in practice and often used as a benchmark to evaluate the performance of other policies in many research [2].

3. A Case Study in a Beverages Warehousing with Drive-in Rack System

This study case derived from x company produces 5 beverage products (Tea based, Milk Based, Jelly based packed in secondary packaging and stacking on pallet). They have warehouse with 3000 racking pallet position with drive-in rack, 4 forklifts and 5 hand pallets, source product from production department, and the construction of warehouse.

4. Result and Discussion

The rule combination algorithm is integrated with computer Graphical User Interface (GUI) to create computer-based model for simulation purpose. The GUI is developed by using java programming. Figure 2 shows part of the computer-based model from rule.

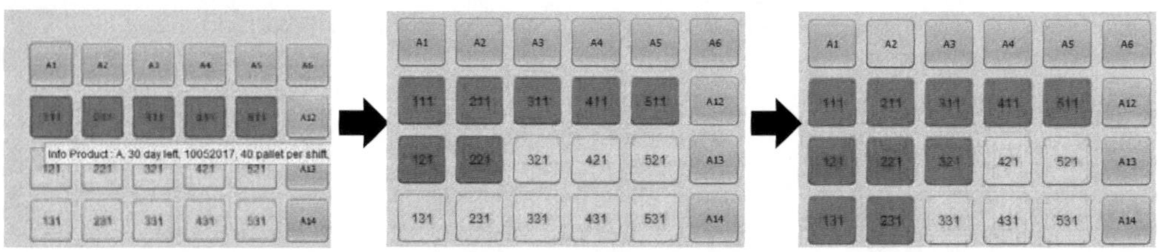

Figure 2. Part of the computer based model for assignment in drive-in rack

The grey square represents aisle lanes, the red square represents full storage position and green square represent the storage position already assigned but not full yet. Each storage position in the GUI combined with graph model and stack algorithm as drive-in rack representation. The GUI represents a block rack which is combined with association rule, class based storage, and arrangement algorithm.

This research implements the storage assignment model for study case in the x company. For 5 products with specific volume, frequency, shelf life, and COI, 5 top rules result are

- [product2] → [product5, product3]: confidence: 0.5857142857142857
- [product2, product1] → [product5, product3]: confidence: 0.5714285714285714
- [product1] → [product5, product3]: confidence: 0.5384615384615384
- [product1] → [product2, product5]: confidence: 0.5384615384615384
- [product3] → [product2, product5]: confidence: 0.5189873417721519

e.g. first rule mean product 2 must assigned near product 5 and 3 and so on. This research compares the rule for 20 instance current data and got the accuracy of rule is 93%.

Table 2. The study case dynamic storage assignment result from proposed model

Product (before and After Data Change)				Frequency Order (Pallet/Shift)	COI	Block
Before	Expiry	After	Expiry			
2	6 months	5	3 months	23	1.09	A
1	6 months	3	6 months	18	1.39	B
5	3 months	1	6 months	16	1.56	C
3	6 months	2	6 months	14	1.79	D
4	1.5 years	4	1.5 years	5	5	E

This accuracy means there are 93% matching rule compared with current data instance. For class based storage we got the position of each product as presented in table 2. From table 2 For this study case we set scenario by switch product's order pattern. The proposed model can determine an appropriate storage location of the product according to dynamic orders from customers (data pattern change). According to table 2 the block changed when the data changed. For example, product 2 must assign in block A because has lowest value of COI and higher order frequency, and then the product 2 assign in block D because the frequency and COI changed.

This paper use order case 8 pallet products 4, 6 pallet products 3 are total distance picking 0.8 km and total distance put away 0.98 km comparing with random storage 1.4 and 1.8 km. From the study case with case "there are probability 10 product will expire", this research found the probability of expiry product from the proposed model is 0.01 comparing with random storage assignment with the probability is 0.3. So, we conclude that our proposed storage assignment model outperforms random storage assignment approach.

Table 3. Comparing the result with random storage assignment approach

Model	Before data pattern change		Before data pattern change	
	Picking distance	Put away distance	Picking distance	Put away distance
Proposed Model	0.8	0.98	0.89	1.5
Random Assignment	1.4	1.9	1.2	1.8

5. Conclusion and Future Research

This research focused on dynamic storage assignment with the main task is generate rule for storage assignment process in beverage warehousing with drive-in rack system. The proposed model was compared with random storage assignment approach as the benchmark and this research found the proposed model is outperforms the benchmark. For further research, this research suggests considering other warehouse operational process i.e. cross docking, considering motion and handling aspect and it impact to order picking, considering shipping process, because the available of truck can affect warehouse operational, build a system that integrating this model with RFID technology and other warehouse technology, other type of warehouse i.e. Drive-thru, forward-reserve warehouse etc., and considering other type of product.

References

[1] Pan J C H, Shih P H, Wu M H, and Lin J H 2015 A storage assignment heuristic method based on genetic algorithm for a pick-and-pass warehousing system *Computers & Industrial* Engineering **81** 1-13

[2] De Koster R, Le-Duc T, and Roodbergen K J 2007 Design and control of warehouse order picking: A literature review. *European Journal of Operational Research* **182**(2) 481-501

[3] Chuang Y F, Lee H T, and Lai Y C 2012 Item-associated cluster assignment model on storage allocation problems *Computers & industrial engineering* **63**(4) pp 1171-1177
[4] Zaki M J and Meira W 2014 *Data mining and analysis: fundamental concepts and algorithms* (Cambridge University Press)
[5] Tsige M T 2013 *Improving order-picking efficiency via storage assignments strategies* (Master's thesis University of Twente)
[6] Kofler M, Beham A, Wagner S, Affenzeller M, and Achleitner W 2011 Re-warehousing vs. healing: Strategies for warehouse storage location assignment In Logistics and Industrial Informatics (LINDI) *2011 3rd IEEE Int. Symp. on* pp 77-82
[7] Cormen T H 2009 *Introduction to algorithms* (MIT press)
[8] Jouini O 2012 Analysis of a last come first served queueing system with customer abandonment *Computers & Operations Research* **39**(12) 3040-3045

IOP Conf. Series: Materials Science and Engineering **337** (2018) 012021 doi:10.1088/1757-899X/337/1/012021

Model of delivery consolidation of critical spare part : case study of an oil and gas company

D Hartanto[1*], A Agustinita[1]
[1]Department of Industrial Engineering, Institut Teknologi Sepuluh Nopember, Kampus ITS Sukolilo-Surabaya 60111, Indonesia

*dodyhartanto@ie.its.ac.id

Abstract. The availability of spare parts in oil and gas industry is very important to prevent the occurrence of very high opportunity cost, that is the loss caused by exploitation equipment which must stop because of unavailability of the spare part. This is done by providing safety stock with a very high service level that leads to high inventory costs. If the company wants to lower inventory costs, the choices are not to lower the service level but to lower the ordering cost. One of the components of ordering cost is the delivery cost. Exploitation facilities are usually located in remote areas so that the cost of delivery is high. In addition, many spare parts are supplied by the same supplier. Therefore, there is an opportunity to lower the cost of delivery of spare parts by consolidation. In this paper,mixed integer linear programming (MILP) model is developed to plan the procurement of spare parts so that inventory costs which include holding and ordering cost for spare parts can be minimized. The model has been verified and validated. Using this model the company can lower inventory costs of the spare part by 32%.

1. Introduction

Oil and gas exploitation facilities are usually located in remote areas[1]. This causes the high cost of delivery the various needs for production, including the needs of spare parts. Spare part requirement is determined by maintenance schedule and historical data of spare part requirement[2][3]. So far, the company is planning the procurement of spare parts with lot for lot method that is procured the spare part with the amount and time as needed. This method will minimize holding cost but ordering cost will be high[4]. Improvement can be done by using other lot sizing methods such as economics order quantity (EOQ), periodic order quantity(POQ), least unit cost, wagner whitin and others. These methods try to find a procurement schedule that can minimize inventory costs that include both holding cost and ordering cost[5][6][7][8].

These lot sizing methods are used to minimize the cost of certain types of inventory spare parts. The method has not considered any potential savings by consolidating delivery from several spare parts supplied by the same supplier. The company manages thousands of spare parts. Many of these types of spare parts are supplied by the same supplier. Therefore, a model is needed to assist the company in determining the spare part procurement plan that considers the potential savings by combining the delivery of multiple spare parts supplied by the same supplier.

Delivery consolidation can lower the cost of delivery[9]. If the delivery of some type of spare part is combined in one delivery then the means transportation used is most likely to deliver spare parts in nearly full or full capacity. Cost reductions can be significant because oil and gas exploitation facilities are often located in remote areas.

Consolidated delivery can also lead to higher holding costs. Delivery consolidation often causes the spare part to be delivered early so that the spare part must be stored for a longer time. So the decrease in delivery costs due to consolidation can lead to an increase in holding cost. Therefore, the consolidation of delivery of spare parts must be done carefully so that minimum total cost can be obtained. In this paper, a mixed integer linear programming model is developed to determine procurement plans that can minimize total inventory costs. The objective function is to minimize the holding cost and ordering cost which also includes the delivery cost.

2. Model Development

Objective function in integer linear programming model is minimize the total cost of delivery, ordering cost and holding cost. The formulation can be described as follows.

Parameters:

D_{it} : requirement of spare part i in period t
H_i : holding cost per month of spare part i
S : Delivery cost per supplier
C : Ordering cost
SS_{it} : safety stock for each spare part i in period t
I_{it} : inventory spare part i for period t
Q_t : total group of spare parts ordered in period t

Decision variables :

Xi_t : the number of spare parts i that is ordered in period t
Z_{it} : 1 if there is delivery of spare part i in period t, 0 otherwise
Y_{1t}: 1 if there is more than one spare part which is scheduled to be delivered in period t,0 if it does not place an order
Y_{2t}: 1 if there is only one spare part that is scheduled to be delivered in period t,0 if it does not place an order
Y_{3t} : 1 if there is no delivery, 0 if there is a delivery
I_{it}: inventory spare part i in period t

Ordering cost is calculated by multiplying ordering cost and ordering frequency. Holding cost can be calculated by multiplying the average inventory in a year and holding cost. While the delivery cost is calculated by considering three conditions that are delivery without consolidation, consolidated delivery and no delivery. Consolidated delivery requires a smaller cost because the conveyance can deliver at full or near full capacity. Delivery costs can be calculated by multiplying the delivery cost and the frequency of delivery.

Objective function:

Minimize:

$$Z = \sum_{t=1}^{T} \left(\left(C + \frac{S}{Q_t}\right)Y1_t + (C + S)Y2_t\right) + \sum_{i=1}^{I}\sum_{t=1}^{T} H_i\, I_{it} \qquad (1)$$

Constrains :

$$I_{it} = \sum_{i=1}^{I} \sum_{t=1}^{T} X_{it} + \sum_{k=1}^{t} I_{i(t-1)} - \sum_{t=1}^{T} D_{it} \qquad \forall i; \forall t \qquad (2)$$

$$X_{it} \leq M\, Z_{it} \qquad \forall i; \forall t \qquad (3)$$

$$2 - \sum_{t=1}^{T} Z_{it} \leq M * (1 - Y1_t) \qquad \forall i; \forall t \qquad (4)$$

$$1 - \sum_{t=1}^{T} Z_{it} \leq M * (1 - Y2_t) \qquad \forall i; \forall t \qquad (5)$$

$$\sum_{t=1}^{T} Z_{it} - 1 \leq M * (1 - Y2_t) \qquad \forall i; \forall t \qquad (6)$$

$$\sum_{t=1}^{T} Z_{it} \leq M * (1 - Y3_t) \qquad \forall i; \forall t \qquad (7)$$

$$Y1_t + Y2_t + Y3_t = 1 \qquad \forall t \qquad (8)$$

$$\sum_{t=1}^{T} I_{it} \geq \sum_{t=1}^{T} D_{it} + \sum_{i=1}^{I}\sum_{t=1}^{T} SS_{it} \qquad \forall i;\forall t \qquad (9)$$

$$Z_t\ , Y1_{it}, Y2_{it}, Y3_{it} \in \{0,1\} \qquad \forall i;\forall t \qquad (10)$$

Constraint (2) ensures that the amount of inventory during a period is the sum of inventory in the previous period and the number of units received during that period reduced by demand in that period. Constraint(3) is required to determine the value of the binary variable Z_{it} based on the value of the X_{it} decision variable. If there is a spare part delivery schedule in period t, then Z_{it} should have a value of 1. This is ensured by using big M on the constraint. Constraint (4), (5), (6) and (7) are constraints needed to determine the value of variables Y_1, Y_2 and Y_3. The decision variables Y_1, Y_2, and Y_3 are binary variables. Y_1 will have a value of 1 if consolidating delivery for some spare part type in one delivery. Y_2 will have a value of 1 if sending only for one type of spare part. Whereas Y_3 will have a value of 1 if there is no delivery in period t. Constraint (8) is required to make sure that Y_1, Y_2 and Y_3 cannot have a value of 1 simultaneously. The value of the variables Y_1, Y_2 and Y_3 will be multiplied by the ordering cost. Ordering cost is a fixed cost whose amount is determined by the frequency of delivery and does not depend directly on the quantity of the spare part delivered. Constraint(9) is necessary to ensure that spare part requirements in t periods must always be met. This is done by ensuring that the amount of inventory in the period t is always equal or greater than the requirement in that period plus safety stock. Constraint (10) are an integer decision variable.

3. Results and Discussion

The spare part procurement process in the company involves 4 divisions: operations, warehouse, procurement and inventory. At the beginning of the year, the operation division makes maintenance plans and spare parts needed for maintenance activities both preventive maintenance and corrective maintenance. Recapitulation of spare part requirement is informed to the inventory division, then the inventory division checks whether the spare part inventory is sufficient to meet the requirement of maintenance program of the operation division. If the supply is still sufficient to meet the requirement within one year then there is not procurement but if the supply is not sufficient then the order is made. Procurement planning which includes the number of spare part and when the delivery should be done for each spare part. If by chance there are spare parts supplied by the same supplier, scheduled to be delivered at the same time and can be consolidated in the same mode of transportation then the spare parts will be delivered simultaneously. Delivery consolidation can reduce delivery costs. The decrease in delivery costs is due to mode of transportation used (truck, train, ship, plane) can deliver spare part in full or almost full capacity so that the delivery cost per unit will be lower.

The developed model is intended to minimize the total cost. Lowering delivery costs and also ordering costs by consolidating delivery can lead to higher holding costs. Conversely, the decrease in holding cost leads to higher delivery costs as well as ordering costs. Holding cost is a cost that includes warehouse cost, material handling, worker, obsolence, and cost of capital. The cost of leasing the warehouse, material handling and workers is estimated by dividing the warehouse costs by the number of spare parts stored in a year. Spare parts usually become dead stock if after five years the spare part is not used. Therefore, obsolence cost each year is obtained by dividing the spare part value by five. The main component of holding cost, cost of capital, is estimated using the prevailing loan interest in the market. The loan interest rate in the market is 10.25%.

Ordering cost is the cost required to procure spare parts. Ordering cost is not directly dependent on the quantity of spare parts delivered in a single delivery, but depends on the frequency of delivery. Ordering cost includes administrative costs, loading, unloading, quality check and delivery costs. Most of spare part suppliers are located in Surabaya and Jakarta. Distance from Surabaya to the company's warehouse is about 500 km and Jakarta to the company warehouse is about 2000 km. In Table 1 can be seen distance of supplier to company warehouse.

Table 1. Distance and cost of delivery from each Supplier to company's warehouse.

Supplier	City	Distance (Km)	Delivery Cost (IDR/delivery)
1	Surabaya	566	4.015.960
2	Surabaya	546	3.974.760
3	Jakarta Utara	2084	9.216.040
4	Jakarta Utara	2090	9.228.400
5	Surabaya	550	3.983.000
6	Jakarta Selatan	2068	9.183.080
7	Surabaya	548	3.978.880
8	Jakarta	2072	9.191.320
9	Surabaya	564	4.011.840

Therefore, the cost of delivery is large cost. There is a need to reduce the cost of delivery. Delivery costs can be lowered by consolidating delivery of some spare part. Delivery costs are estimated by taking into account the costs of drivers, fuel, tolls, levies, and truck rentals. Delivery costs of nine suppliers can be seen in table 1. In table 1 also can be seen that the cost of delivery is a considerable cost. Decrease in delivery costs by consolidation can be calculated by dividing shipping cost in one shipment by truck capacity ratio with average shipping quantity.

Grouping spare parts is needed to determine the spare parts that can be shipped in one delivery. There are three factors to be considered in the grouping. These three factors are the size, shape and the suppliers that supply the spare parts. Spare part grouping can be seen in table 2. KT is group of spare part of big dimension so that its delivery cannot be combined with other spare part. SH and JT are spare part groups in the form of sheets and pipes so that the delivery also cannot be combined with other spare parts. While group EA, CT and SET is a group of spare parts in the delivery can be combined with other spare parts because the spare part has a similar dimension.

Table 2. Classification of spare parts.

Function of the spare part	Group of Spare Part	UOM	Group
pare Part for Pipe Fits	08	EA	1
		JT	3
Spare Part for Blower	11	EA	2
Spare Part for Valve	12	EA	1
Spare Part for Cartepillar Engine	13	EA	5
		KT	6
Spare Part for Genset	14	EA	1
Spare Part for Compressor	15	EA	2
Spare Part for Water Pump	16	EA	1
Electrical Component	18	EA	1
Spare Part for Control Valve	19	EA	1
Spare Part for Fire Safety	22	EA	1
		SET	1
Spare Part for Material Handling	40	EA	1

The developed model is run for each supplier and for each group of spare parts that can be delivered simultaneously. Output of the model shows that the model has been able to produce a spare part procurement plan that minimizes costs including delivery costs, holding cost and ordering cost. For example, output of the model for supplier 1 can be seen in table 5.

Table 3. Output model for supplier 1.

Spare Part	Month											
	1	2	3	4	5	6	7	8	9	10	11	12
11-950-0220-A	0	0	0	0	0	0	30	0	0	0	1	0
11-950-0237-A	0	0	0	0	0	0	0	0	0	0	4	6
11-950-0253-A	0	0	0	0	0	0	30	0	0	0	12	0
11-950-0258-A	0	0	0	0	0	0	0	0	0	0	1	0
11-950-0581-A	0	0	0	0	0	0	22	0	0	0	51	27
11-950-0709-A	0	0	0	0	0	0	2	0	0	0	1	1

In table 4, it can be seen that there are four types of spare parts are sent simultaneously that is spare part 11-950-0220-A, 11-950-0253-A, 11-950-0581-A and 11-950-0709-A. In the 11th month, there is delivery for all types of spare parts and in the 12th month, there are delivery of three types of spare parts that are 11-950-0237-A, 11-950-0581-A, and 11-950-0709- A. Delivery schedule will be different if lot sizing is done for every spare part and if by chance there are spare part that is scheduled to be delivered simultaneously it will be consolidated delivery. Procurement planning using this model can provide savings of 32%. The amount of inventory cost savings for each spare part supplied by each supplier can be seen in Figure 1.

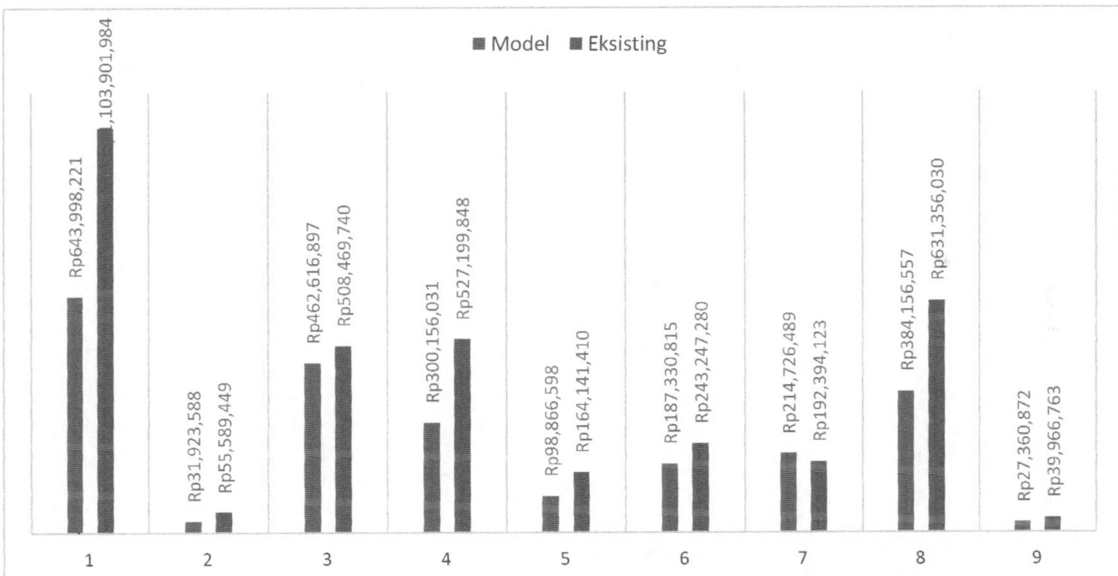

Figure 1. Total inventory cost of spare part supplied by each supplier.

4. Conclusion

Oil and gas companies are often located in remote areas so that the cost of delivery spare parts and other materials requires high costs. The cost of spare part delivery can be reduced by arranging the delivery schedule of the spare part that considers the possibility of consolidating the delivery. Consolidated delivery can be made on spare parts supplied by the same supplier and have characteristics that allow to be shipped in one delivery. In determining the delivery schedule must also consider the holding cost because the delivery schedule will have a direct impact on the holding cost. In this research has been developed integer linear programming model to determine the delivery schedule of spare parts.

The objective function of integer linear programming is minimization of delivery cost and inventory cost which includes holding cost and ordering cost. The model has been validated and the use of the model to plan the procurement of spare parts can reduce the total cost of spare part inventory by 32%. In the future research, the integer linear programming model can be further developed by considering some factors such as discount, warehouse capacity, multimode transportation, supplier capacity and budget for each period.

5. References

[1] James G Speight 2011 *Handbook of Industrial Hydrocarbon Processes* (Oxford: Elsevier) p 86

[2] Mohammad A A Fawzan et al. 2000 *Maintenance, Modeling and Optimization* Mohamed Ben-Daya, Salih O. Duffuaa, Abdul Raouf (New York : Springer Science + Business) p 83

[3] Gopalakrishnan, A K Banerji 2013 *Maintenance and Spare Parts Management* (New Delhi : PHI Learning Limited) p 349

[4] Steven Nahmias, Tava Lennon Olsen 2015 *Production and Operations Analysis* (Illinois: waveland Press Inc.) p 216

[5] L A Wolsey, 2002. *Solving Multi-Item Lot-Sizing Problem with an MIP Solver Using Classification and Reformulation* Management Science

[6] David A Collier 1980 A comparison of MRP lot sizing methods considering capacity change costs J. Operations Management Vol 1 Issue 1 Pp 23-29

[7] M Mohammadi, Masine M Tap 2012 A Mixed Integer Programming Model Formulation for Solving the Lot-Sizing Problem International Journal of Computer Science Pp 28-36

[8] Woarawichai, Kullpattaranirun and Rungreunganun 2011 *Applying Genetic Algorithms for Inventory Lot-Sizing Problem with Supplier Selection under Storage Space* International Journal of Computer Science Issues Vol. 8 Issue 2

[9] Robert L Mccarthy, Kenneth W Schafermeyer, Kimberly S Plake 2012 *Introduction to Health Care Delivery* (London : Jones Bartlett Learning)

IOP Conf. Series: Materials Science and Engineering **337** (2018) 012022 doi:10.1088/1757-899X/337/1/012022

Innovation in healthcare services – creating a Combined Contingency Theory and Ecosystems Approach

Per Engelseth[1] and Duangpun Kritchanchai[2*]
[1]Department of Logistics, Molde University College, Specialized University in Logistics, Molde, Norway
[2]Department of Industrial Engineering, Faculty of Engineering, Mahidol University, Nakhonpathom, Thailand

*duangpun.skr@mahidol.ac.th

Abstract. The purpose of this conceptual paper is to develop an analytical framework used for process development in healthcare services. Healthcare services imply a form of operations management demanding an adapted research approach. This study therefore highlights first in the introduction challenges of healthcare services as a reasoning of this study. It is a type of service that has high societal and therefore ethical concern, but at the same time needs to be carried out efficiently to economise service production resource use. Combined business and ethics concerns need to be balanced in this service supply system. In the literature review that is the bulk of this paper, first, particularities of the service industry processes are considered. This is followed by considering literature on contingency theory to consider the nature of the supply chain context of the healthcare service processes highlighting interdependencies and appropriate technology use. This developed view is then expanded to consider an ecosystems approach to encompass the environment expanding analyses to considering in balanced manner features of business, society and nature. A research model for directing both further researches on the healthcare service industry an innovation of such services in practice is introduced.

1. Introduction

This conceptual paper aims to develop a research model that can provide guidance in the development of healthcare services as well as foundation for further research within this industrial sector. This choice of research topic is founded on research of healthcare services that points to how this form of operations management (OM), is poorly developed and thus inefficient [1][2]. Hospitals face obstacles, related to e.g. inventory management and transportation costs. A study by Butt and Run [3] revealed that the information management infrastructure of hospitals in Malaysia is poorly developed, showing how patients in Malaysian hospitals rarely receive the right service at the right time. In healthcare services of Singapore, one of leading countries in medical tourism, Leng [4] revealed that the government to a greater degree has prioritised healthcare services compared with its neighbouring countries. Still, hospitals in Singapore continue to cope with weak operational performance statistics. Patients in Singapore exhibit a long treatment cycle time despite expressing satisfaction with waiting between 15-30 minutes [5]. Empirical findings also suggest that in Thailand hospitals also suffer from inefficient healthcare service system development. This is especially related to inventory management [6]. Through a case study, Kritchanchai [7] revealed that in Thailand medicine product standardization in the information systems used by healthcare services is weak impeding supply chain integration. The agents in the studied Thai healthcare supply chains tend to use different codes for the same drugs obstructing a seamless flow of information and thus affecting information exchange efficiency within Thai hospitals.

How to improve healthcare services? Parker [8] states that: "Measuring performance is something that all organizations do". Measuring operations performance in healthcare service organizations enables them to evaluate, control, budget, motivate, promote, celebrate, learn, and improve these operations [9]. Performance measurement has been applied used in various industries and in the public sector [10][11][12]. Introducing an effective performance measurement system in healthcare entails that quality shortfalls can be detected on a daily and continuous basis more supporting efficiency in quality shortfalls detection and thereby also efficient process improvement [13]. The ultimate objective of performance measurement in healthcare services is to increase service performance [14]. Healthcare process effectiveness is associated with quality operations practices that provide the appropriate service for its recipients. Both effectiveness and efficiency objectives are vital in healthcare and these aims are interdependent. The further development of a healthcare information system (HIS) is therefore a priority in many countries [15].

To strengthen national HISs, a global partnership, the Health Metrics Network (HMN), was established in 2005. Its goal is to increase the availability, accessibility, quality and use of healthcare information in healthcare service organizations so support professional decision making at both a national and global level. This implies considering healthcare services also as an inter-organizational phenomenon. HMN is predominately applied to assess healthcare service performance at the national level. Research-based literature on operational performance measurement for healthcare services as type of industry is still limited. This study accordingly focuses on developing healthcare services as an intra-organisational process integration problem. The supply chain is this considered as the immediate context of these service processes and national and global healthcare control mechanism as environmental characteristics. Healthcare service *processes* represent the unit of analysis in this approach. A process focus implies focus on resource use and transformation; the dynamics. "Context" implies direct interaction between hospitals and other actors in a supply chain system. "Environment" implies forces not necessarily interacting, but still impacting on healthcare services in a more one-way manner. Figure 1 below expresses this approach:

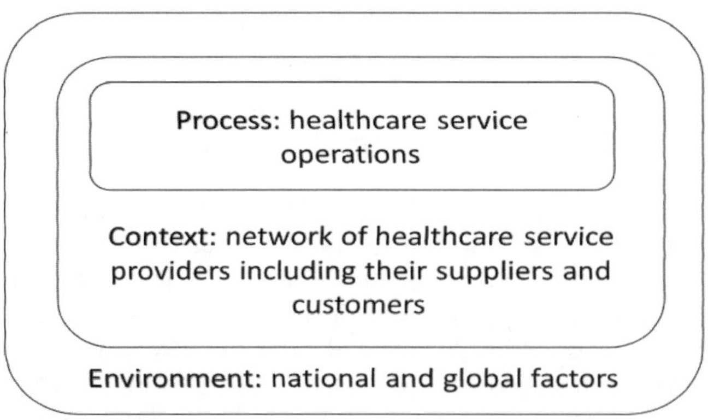

Figure 1. The healthcare service process as embedded unit of analysis

To improve healthcare services an approach is needed to elaborate on industrial particularities. Being a service, research needs to be founded on research that elaborates on service industry particularities. To enable this, we start by considering features of the service industry in general and strengthen this approach focusing on particularities by analysing interdependencies. Within contingency theory Thompson [16] later, Stabell and Fjeldstad [17], have developed an approach focusing on the role of power relations in the supply chain that enhances service industry particularities enabling process development schemes better adapted to service. Furthermore, the embedded nature of healthcare in society call for ecosystems thinking [18] to integrate a wider array of environmental concerns into studying healthcare services. Literature concerning these two lines of research are first discussed and then integrated to create a research model concerning process development in healthcare services.

2. Services particularities

This section concerns the process layer embedded in its industrial particularities. Particularities reflect features of process in their immediate manageable context and wider less manageable environment (see fig. 1). Healthcare is a services type of industry. The physical distribution of goods represents the dominant industrial interest within operations management. Only recently has there been development to consider services likewise as goods supply from an operations management perspective. To include services in supply chain management (SCM), Ellram et al. [19] integrate the concept of "service" in their SCM definition: "Supply chain management is the management of information, processes, capacity, service performance and funds from the earliest supplier to the ultimate customer". They also present a service supply chain model, where they identify six managerial processes; 1) capacity management, 2) demand management, 3) customer relationship management, 4) supplier relationship management, 5) service delivery management, and 6) cash flow management. Services are commonly classified as intangible, heterogenic, inseparable, and perishable [20]. However, this static classification provides, according to Spring and Araujo [20], provides limited value when analysing services.

Fundamental to this critical view of using a static classification of services in academia is that service production demands a different form of organising of the resource structure and processes within this structure [21][22]. The "people" resource and human interaction is fundamental to services [23][24], and is highlighted in the more recent service-dominant logic that highlights the importance of customer value in supply [25]. Given the importance of interaction, service supply chains are therefore bidirectional in nature [26], as opposed to physical distribution, where outbound flows are the dominant mode of producing value. Sampson and Froehle [26] state in their effort to conceptualise service process that: "With service processes, the customer provides significant inputs into the production process". Sampson and Froehle [26] also point out that from a supplier's perspective, they may experience that the quality of customer inputs in the service process interactions can vary.

Producing a service cannot start before the customer provides the supplier with some form of resource input. According to Sampson and Froehle [26], three types of customer inputs can be found in services. They are (1) the customer person, (2) physical resources such as customer belongings, tools and other tangible objects, and (3) information. These resources are pooled and used in combination to produce a service. Another distinct feature of service supply is that these supply chains can be characterised as having a hub configuration. They are, according to Sampson and Froehle [26], also short since more than two levels are rarely found. Regarding the service process itself, Sampson and Froehle [26] site empirical evidence of quality issues predominant in the service supply chains including (1) random arrivals, (2) inconsistent specification, and (3) varying input quality that influence service processes as capacity and demand management and quality management. From a Lean perspective Bicheno and Holweg [27] point to typical forms of waste (muda) found in services are represented by (1) delay, (2) duplication, (3) unnecessary movement, (4) unclear communication, (5) incorrect inventory, (6) poor customer service, and (7) transaction and production errors. To support quality service provision tools such as reservation systems, price incentives and promotion of off-peak demand, and customer self-service are may be used. In addition, capacity management involves a mix of resources such as people, tools and goods. Following up service quality is often organized to be carried out through a specialized customer service department that needs to network both internally with the company's different own functions as well as other service providers and customers [28].

Important sub-conclusions regarding service process particularities are (1) that regarding *structure*, services are carried out by combining various interlinked resources in an integrated network structure, (2) regarding the *dynamics*, that manging these services demands interaction, and (3) regarding *learning and development*, that experiences form the services industry indicate that service quality is poor much due to the complex nature of services process coupled with weak understanding of manging services as emergent phenomena.

3. Contingency theory

This section concerns the context of healthcare processes, the supply chain structure. This focus on context is in line with contingency theory that is founded on the presumption that processes are environmentally contingent [16]. This section elaborates on this immediate context, in management literature commonly termed as a "supply chain" or alternatively as a "distribution channel". The first encompasses focus on resource transformation through processes (goods and service flow) while the latter encompasses focus on the transaction flow. A key feature of the supply chain is that it is a network consisting of multiple interconnected actors [29]. When approaching how to understand the nature of healthcare process contingency in relation to the supply chain context, analysis is dependent on revealing and the analyst thus understanding the nature of this context.

One of the key features of any network is the strength of coupling between the network entities. Following Weick [30], network interactions take place through business relationships where this coupling varies on a continuum ranging from weak to strong. The nature of coupling impacts importantly on the loyalty of actors in the network, which then again may be viewed as expression of degree of power and trust; the "network atmosphere" [31]. One of the fundamental reasons for networking is associated with interdependencies [16]. Since resources are scarce and investments have made production resources specialized to a single firm, these actors need to interact to produce. Emerson [32] points accordingly to that power is fundamental characteristic of this interaction. In a network, relationships can be characterised as degrees of being imbalanced or balanced.

Change in power imbalance is associated with corporate strategy change. In supply chains power is associated with efficient and effective resource control; vital in securing production. Following Pfeffer and Salancik [33], this form of control is based on a mix of resource ownership, access, use and ability to make the rules regarding resource use. Power is often associated with coercion. Based on the writings of the process-focused sociologist Elias, Stacey [34] argues how power both enables and constrains in production processes in industry. According to Pfeffer and Salancik [33] and Leonardi [35], interdependencies can be managed, be increased, reduced, or the dominant interdependency in a dyadic relationship changed. Managing interdependency of the network context is, following contingency theory, the core feature of strategic corporate management.

Integrating at a strategic level to coordinate resource use involves, following Thompson [16], taking into consideration whether interdependency is mainly *pooled*, *sequential* or *reciprocal*. Interdependency describes fundamentally how economies of complementarity are reaped through interaction in a business relationship. Thompson [36] states that pp.101-102), "...human action emerges from interaction of (1) the individual, who brings aspirations, standards, and knowledge or beliefs about causation; and (2) the situation, which presents opportunities and constraints". In line with Parsons [37], the management of technical production activities is embedded in an institutional layer. This is important since, decision-making is embedded in discourse, a developed business culture. Perceptions of interdependencies accordingly impact on how processes are managed.

Interdependencies are impacted by uncertainty which is defined by Burns and Stalker [38] as "...the ignorance of the person who is confronted with a choice about the future in general, and in particular about the outcomes of which may follow any of his possible lines of action". Interaction helps soothe uncertainty through exchange mechanisms; fundamentally involving information sharing [33]. *Mutual adjustment* is typical of reciprocal interdependency using exchange mechanisms founded on intensive technology to coordinate production processes. This is a costly form of organisation in developed economies since it is predominately manual. Alternatively, services may be increasingly pooled using mediating technology. This implies increasing standardization in the network. Since services are predominately characterized by reciprocal or pooled interdependence [17], automating service processes entail increasing strategically pooled interdependence by reducing reciprocal interdependence in individual or sets of business relationships. Managing interdependency provides a pathway to increased service process efficiency.

4. Ecosystems theory

Considering ecosystems involves accounting for the environmental level presented in figure 1. One of the fundamental characteristics of Thompson's [16] interdependency theory is that it is associated with systems theory. Ecosystems also represent system thinking, meaning function and interconnectedness within defined boundaries are characteristic of such a system. Thompson's [16] discussion, is however, limited to use of closed as well as open systems, and for his purpose, seeking understanding on what determines how and when organizations act. More precisely, he states that "...we will conceive of organizations as open systems, hence indeterminate and faced with uncertainty, but at the same time as subject to criteria of rationality and hence needing determinateness and certainty" [39]. Management understanding the nature of interdependencies is associated with increasing rationality in decision-making in the network. However, interactions in supply chain networks account for only a part of these influences. Following Leonardi [35], interdependency change is viewed as subject to incremental and iterative adaptations; a process view that implies interaction both within the supply chain and a wider social and natural environment. While business systems tend to conceptually be governed by management, ecosystems places weight ion how both nature and society together interplay making the system more self-governed; more out of reach to the manger.

Expanding supply chain management to regard it as an ecosystem involves taking account of not only interaction in the supply chain network to manage production flows, but to expand management discourse to encompass also societal and nature concerns. A direct impact of this expansion is not only widening the scope of systemic description and investigation, but also expanding the researched time frame. Since ecosystems are associated with sustainability, this means that the time frame of analytical scrutiny is expanded to considering interests of future generations. Furthermore, an ecosystem will be form the perspective of the manger be perceived as uncertain and inherently complex. Ecology has its own logic of organising that may be different from that of mangers, e.g. in a hospital.

Finally, some key concepts related ecosystems thinking. Ecology was termed by Haeckel in 1866 as the science of relations between organism and the surrounding outer world [40]. "Ecosystems" indicate accordingly considering nature, society and business as integrated from a system's perspective. Systems thinking finds its roots in the natural sciences, based on observations of how biological organisms function. As Capra and Luisi [41] state based on recent studies in many fields of natural sciences that "...nature does not show us any isolated building blocks, but rather appears as a complex web of relationships between the various parts of a unified whole". Systems are found in nature regardless of the glasses the researcher wears. In sum, an ecosystem understanding of healthcare processes implies using systems thinking encompassing economic, societal and nature concerns; an expansion of systems border that entails increased complexity.

5. Concluding with a research model

Ecosystems are considered as objective entities. A fundamental view is that they do function this way as micro-purposeful interaction that also can be reflected upon from a holistic perspective. This is an understanding that can change how we understand and thereby use SCM principles focusing on integration to better collaborate and coordinate processes as a managerial imperative. SCM thus should not be limited to organizational, economic and technological considerations. Nature and society envelopes production and is a part of it. Developing SCM as an ecosystem represents an integrated view that considers the economy as an environmentally contingent network and thus also faces a wider range of challenges than normally conceived of when using this managerial philosophy. Therefore, in this study we choose to remain simple in our modelling effort and leave many questions and considerations unwrapped. Engelseth [42] has proposed that causality in the influence of society and nature on management is filtered through economic considerations. This view is followed in figure 2:

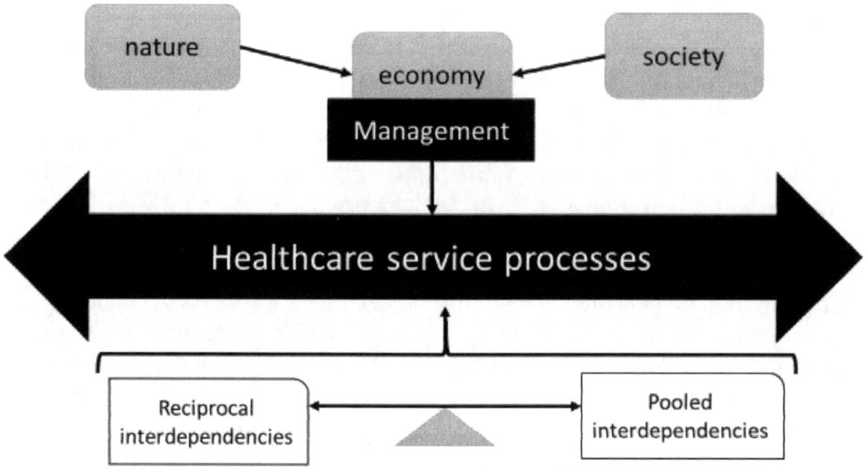

Figure 2. An ecosystem view of interdependency in healthcare service processes

Figure 2 characterises healthcare service processes as involving varying degrees of reciprocal and pooled interdependency; they are usually more or less imbalanced. This implies that sequential interdependency, dominant in manufacturing, is reduced merely to a role of describing healthcare service processes as flows. Timing is accordingly understood as strategically managing reciprocal and pooled interdependency found in healthcare service processes. Management, associated with operational as well as strategic level decision-making, is primarily influenced by economic concerns. These concerns are, following the developed ecosystems view (figure 2), considered as impacting on, and thereby intertwined with economic considerations. This view is substantiated by a view that managerial cognition is human perception and therefore limited. Management's inherent prime concern is survival in the volatile market place; normally considered a purely economic concern. Adding societal and nature concerns to this commonplace view implies therefor heightening of ethical as well as long-term thinking on the behalf of management. This is rooted in a continuous learning process. It also implies sustainability as a key factor when designing and implementing change in healthcare services. This view is high in the level of abstraction and needs therefore to be further substantiated with empirical evidence to increase the detail in the model, lowering abstraction, and moving towards the operational level. This needs to be done, due to the complexity of the model, through two potentially parallel research stages. First, what are the factors considered as "society" and "nature" considered as impacting on healthcare service? Also, studies may study the detailed nature of pooled and reciprocal interdependencies in health care; evaluate more closely the role of sequential interdependency, to create more detailed roadmaps for healthcare service development; creating support for automating healthcare service processes.

This developed analytical framework will also be administered to existing data from in-depth site visit in twenty hospitals. This research focuses are on the resource transformation through processes (goods and service flow) and the transaction flow. Six dimensions are included. These are logistics information management; forecasting and inventory management, purchasing processes, warehouse management, transportation management and organisation function for logistics. Initially it has been revealed through this research that the hospitals with good inventory visibility have a good performance in safety stock level and purchasing processes. Data standards in hospitals significantly influence the information flow within the hospitals and logistics information system. Furthermore, the hospitals which have an organisation structure for logistics function obviously has a good performance in inventory and transportation management.

References

[1] Haszlinna M N and Potter A 2009 *Supply Chain Management: An International Journal*, **14** p 234
[2] Hall R 2014 *Patient flow* AMC **10** p 12
[3] Butt M M and Run E C 2010 *International Journal of Health Care Quality Assurance* **23** p 658
[4] Leng C H 2010 *Global Social Policy* **10** p 336
[5] Lim P C and Tang N K H 2000 *International Journal of Health Care Quality Assurance* **13** p 290
[6] Kritchanchai D 2012 *Operations and Supply Chain Management: An International Journal* **5** p 103
[7] Kritchanchai D and Suwandechochai R 2010 *International Journal of Services, Economics and Management*, **2** p 211
[8] Parker C 2000 *Work Study* **49** p 63
[9] Behn R D 2003 *Public administration review* **63** p 586
[10] Schmitz J and Platts K W 2004 *International Journal of Production Economics* **89** p 231
[11] Kald M and Nilsson F 2000 *European Management Journal* **18** p 113
[12] Palmer A J 1993 *Public Money & Management* **13** p 31
[13] Purbey S, Mukherjee K and Bhar C 2007 *International Journal of Productivity and Performance Management* **56** p 241
[14] Baker G R, Brooks N, Anderson G, Brown A, McKillop I, Murray M and Pink G 1998 *Hospital Quarterly* **2** p 22
[15] Haux R 2006 *International Journal of Medical Informatics* **75** p 268
[16] Thompson J D 1967 *Organizations in Action* (New York: McGraw Hill)
[17] Stabell C B and Fjeldstad Ø D 1998 *Strategic Management Journal* **19** p 413
[18] Capra F and Luisi P L 2014 *The Systems View of Life. A Unifying Vision* (Cambridge UK: Cambridge University Press)
[19] Ellram L M, Tate W L and Billington C 2004 Journal of Supply Chain Management **40** p 17
[20] Spring M and Araujo L 2009 *International Journal of Operations & Production Management* **29** p 444
[21] Chase R & Garvin D 1989 *Harvard Business Review* **67** p 61
[22] Oliva R and Kallenberg R 2003 *International Journal of Service Industry Management* **14** p 160
[23] Grönroos C 1990 *Service management and marketing: managing the moments of truth in service competition* (Lexington, MA - Lexington Books)
[24] Normann R 2001 *Reframing business: When the map changes the landscape* (West Sussex UK: Johan Wiley & Sons Ltd)
[25] Lusch R F and Vargo S 2014 *Service-Dominant Logic, Premises, Perspectives, Possibilities* (Cambridge UK: Cambridge University Press)
[26] Sampson S E and Froehle C M 2006 *Production and Operations Management* **15** p 329
[27] Bicheno J and Holweg M 2009 *The Lean Toolbox: The Essential Guide to Lean Transformation* (Buckingham UK - PICSIE Books)
[28] Engelseth P, Wagner A and Farrukh A 2014, Developing Lean customer service in a Lean manufacturing firm *Proceedings of the 21st international EurOMA conference*
[29] Christopher M 2016 *Logistics and Supply Chain Management (5 ed.)* (London - Financial Times Press)
[30] Weick K E 1976 *Administrative Science Quarterly* **21** p 1

[31] Gadde L E, Håkansson H and Persson G 2010 *Supply Network Strategies* (Chichester UK: John Wiley & Sons)
[32] Emerson R 1962 *American Sociological Review* **27** p 31
[33] Pfeffer J and Salancik G R 1978 *The External Control of Organizations* (Stanford CA: Stanford Business Books)
[34] Stacey R D 2003 *Complexity and Group Processes* (Milton Park UK: Routledge)
[35] Leonardi P M 2013 Nicolini, D, Langley, A & Tsoukas, H *How Matter Matters. Objects, Artefacts, and Materiality in Organization Studies* (Oxford UK: Oxford University Press) 142
[36] Thompson J D 1967 *Organizations in Action* pp 101-102 (New York: McGraw Hill)
[37] Parsons T 1960 *Structure and processes in modern societies* (New York: The Free Press of Glencoe)
[38] Burns T and Stalker G M 1961 *The Management of Innovation* (London: Tavistock Institute) p 112
[39] Thompson J D 1967 *Organizations in Action* p 10 (New York: McGraw Hill)
[40] Haeckel E 1866 *Generelle Morphologie der Organismen* (Berlin: Reimer)
[41] Capra F and Luisi P L *The Systems View of Life. A Unifying Vision* p 68 (Cambridge UK: Cambridge University Press)
[42] Engelseth P and Sandvik M 2017 *Proceedings in Food System Dynamics 2017* forthcoming

IOP Conf. Series: Materials Science and Engineering **337** (2018) 012023 doi:10.1088/1757-899X/337/1/012023

Using DEMATEL approach to develop relationships of performance indicators on sustainable service only supply chain performance measurement

EB Leksono [1, 2*], Suparno [2] and I Vanany [2]
[1] Department of Industrial Engineering, Universitas Muhammadiyah Gresik, Gresik 61121. Indonesia
[2] Department of Industrial Engineering, Institut Teknologi Sepuluh Nopember, Kampus ITS Sukolilo-Surabaya 60111, Indonesia

*eko_budileksono@umg.ac.id

Abstract. Service only supply chain (SOSC) concept is service supply chain (SSC) implementation on pure services. The globalization and stakeholder pressure makes operation of SSC should give the attention to the environment effect, community, economic and intangibility assets. SOSC performance measurement (SOSCPM) may be developed for measuring of performance for sustainability aspects and intangibility assets to meet customer satisfaction. This article discusses sustainable SOSCPM based on balanced scorecard (BSC), include sustainability aspects, intangibility and relations between perspectives and indicators. From literature review, it is found 34 performance indicators that must be confirm to expert and SC actors by survey. From survey validation using weighted average and level of consensus, it is found 29 valid indicators for processed by DEMATEL. From DEMATEL, it is found 26 indicators can be used on sustainable SOSCPM. Furthermore, innovation and growth perspective most influence to other, and customer perspective most important. Intangibility indicators incorporated on innovation and growth perspective very related with human resources. Finally, relations between perspectives and indicator used to design of BSC strategy maps.

1. Introduction

SSC implementation has become a trend because service sector has significant contribution to *gross domestic product* (GDP) **Error! Reference source not found.**-[3] and affect to the world economic [4]. SSC has two categories, service only supply chains (SOSC) and product service supply chains (PSSC) [5]. SOSC is system which output orientation is pure services, for example: healthcare, hospitality, consultant, bank, etc. PSSC is system which output orientation are service and physical product, for example: restaurant, mass customization, etc. In 2013, services sector contribution to Indonesia GDP is 39, 87% [6] which most contributions from pure services, so pure services existence can be developed in Indonesia. The supply chain and performance measurement can used to develop of the pure service (service only).

The globalization and stakeholder pressure makes operation of SSC should more attention to sustainability issue which consists of economic, social and environment simultaneously [7]. Sustainabilty orientation on SC called sustainable SC [8]. The sustainable SC implementation on services can minimize of negative operation effect to environment and social and maximize of profit

[9]. Main characteristic of services sector are intangibility assets that related with human resources [10]-[12], so SSC implementation must be attend to intangibility characteristic [2].

Sustainable SC needs performance measurement to success evaluation of: competitiveness increase [13], maximize of profit, minimize of environment negative effect and beyond of stakeholder expectation [14]. Limitations of sustainable SSCPM frameworks are less of interaction between SC actors on performance indicators determine [15], less on relationship between indicators based on sustainability aspect [7],[13], less on social aspect and focal company centralize [13],[16]. Besides, there aren't models of sustainable SSCPM that focus to the intangibility characteristics. From the limitations, so sustainable SOSCPM need new model which all actors can participate on performance indicators define and describe relationship between indicators. Furthermore, the new model has attended to intangibility characteristic and sustainability aspects. Finally, the decision-making trial and evaluation laboratory (DEMATEL) can be used to develop sustainable SOSCPM based on BSC.

2. Literature review

2.1. The balanced scorecard (BSC) on SSCPM

The BSC is performance measurement model that describe of relationship between perspectives and or indicators on BSC strategy map as business strategy [17]. The BSC is dynamic and innovative method that can be developed and collaborated with other method [18]. The BSC can be used simultaneously by several organizations collaborate [19].

The determination of the performance perspective is starting point on the BSC because the perspectives explain view point of organization. There are 5 perspectives use on services sector: Financial, Customers, Operational, Information and Innovation & Growth [20].

2.2. The decision-making trial and evaluation laboratory (DEMATEL)

DEMATEL is one method on multiple-attribute decision making (MADM) as tool to help of decision making [21]. DEMATEL can be used to view direct or indirect relationship between variable or attribute using matrix to get all of causal relationship [22]. Beside, DEMATEL can describe of important level and influence level of an attribute or a variable on system [22].

2.3. Performance indicators of sustainable SOSC

Performance indicators on sustainable SOSCPM taken from literature review with topics: performance measurement, service SC, service SCPM and sustainable SSC practices show on table 1.

Table 1. Performance indicators on sustainable SOSCPM.

Economic aspect		Environmental aspect		Social aspect	
Indicators	References	Indicators	References	Indicators	References
Operational costs	[23], [24]	Conservation	[25], [26]	CSR	[14], [27]
ROI	[15], [24], [28]	Green service	[14]	Customer satisfaction	[14], [15], [28]
Profit	[4], [14], [29]	Energy consumption	[26], [30], [31]	Stakeholder satisfaction	[25], [32]-[36]
Total Revenue	[15], [24], [37]	Water consumption	[30], [31]	Local suppliers	[9], [31]
Efficiency	[9], [26], [38]-[40]	CO_2 emission	[14], [26], [39]	Local human resources	[25], [30], [41]
Quality of service	[4], [24], [26], [42], [43]	Waste treatment	[26], [30], [31]	Health & safety	[9], [26], [27], [39]
ROA	[4], [14], [24], [26], [42]	Green material	[36]	Organization behaviour	[38], [39]
Delivery Time	[24], [28]	Reuse & recycle	[9], [36], [39]	Information sharing	[44]-[46]
Flexibility	[4], [15], [24], [28]	Green collaboration	**Error! Reference source not found.**, [36], [46]	Regulations & laws	[9], [26], [31], [33], [38], [39]
Level of inventory	[15], [28]	Integrated of IT	[23], [45], [47]		
Cost of TIC	[14], [28]				
Capacity of TIC	[28], [46], **Error! Reference source not found.**				
Human resources	[14], [31], [38]				

Qualification & competency	[11], [12], [49]
HRD	[9], [25], [31], [33], [36], [41]

3. Methodology

Model development used survey with in depth interview. In this study we asked three scholars and six managers from suppliers and provider of service only business. Stage on model development:

a. Validation of survey result of indicators with weighted average (WA) and level of consensus (LC). Indicator cut off : WA $\geq$ 4.2 & LC $\geq$ 0.5 (economic) and WA $\geq$ 4.5 & LC $\geq$ 0.7 (environmental and social)[50], and then classification based on BSC perspectives
b. Survey to level of influence between perspective and indicator.
c. Design of BSC strategy map based on DEMATEL steps: build matrix of direct relation, normalization, total relationship, calculate important level and influence level and make a relationship diagram with significant matrix [22].

4. Result and discussion

4.1. Validation and classification of indicators

From literature review, the indicators were validated with survey. Then, survey result processed with WA and LC. It is found five indicators that not valid, namely: ROA, cost of TIC, CO2 emission, green material and local supplier. Finally, there are 29 valid indicators that classified into performance perspectives on table 2.

Table 2. Indicators classification based on perspective.

Financial	Customer	Operational	Information	Innovation & Growth
Operational costs (X1)	Quality of service (X6)	Flexibility (X11)	Capacity of TIC (X20)	HR (X23)
ROI (X2)	Delivery Time (X7)	Level of inventory (X12)	Integrated of IT (X21)	Qualification & competency (X24)
Profit (X3)	CSR (X8)	Conservation (X13)	Information sharing (X22)	HRD (X25)
Total Revenue (X4)	Customer satisfaction (X9)	Green service (X14)		Organization behaviour (X26)
Efficiency (X5)	Stakeholder satisfaction (X10)	Energy consumption (X15)		Local HR (X27)
		Water consumption (X16)		Health & safety (X28)
		Waste treatment (X17)		Regulations & laws (X29)
		Reuse & recycle (X18)		
		Green collaboration (X19)		

4.2. Relation between perspectives and indicators

The DEMATEL steps used to design of relationship. Table 3 shows of influence level (D – R) and important level (D + R) of perspectives and indicators. Innovation and growth perspective is most influence and customer perspective is most important. The indicators of ROI (X2), water consumption (X16) and local human resource did not important and significant relations with other, so the BSC strategy map can develop without their indicators.

Table 3. Total relation matrix of perspectives and indicators.

Perspectives and Indicators		**D**	**R**	**D+R**	**D-R**
Financial		**3.27**	**3.756**	**7.026**	**-0.485**
Operational costs	X1	2.444	3.782	6.227	-1.33
ROI	X2	1.155	1.686	2.842	-0.53
Profit	X3	2.39	3.69	6.08	-1.3
Total revenue	X4	2.677	3.57	6.25	-0.89
Efficiency	X5	3.386	3.63	7.02	-0.24
Customer		**3.715**	**4.183**	**7.898**	**-0.468**
Quality of service	X6	2.694	3.378	6.072	-0.68
Delivery Time	X7	2.697	2.73	5.428	-0.03
CSR	X8	2.765	2.038	4.8	0.727
Customer satisfaction	X9	2.854	2.84	5.696	0.013
Stakeholder satisfaction	X10	3.485	3.78	7.266	-0.29
Operational		**3.518**	**3.924**	**7.442**	**-0.405**
Flexibility	X11	3.283	2.997	6.28	0.285
Level of inventory	X12	2.233	2.044	4.276	0.189
Conservation	X13	2.06	2.554	4.614	-0.49
Green service	X14	3.493	3.2	6.693	0.292
Energy consumption	X15	2.29	2.385	4.675	-0.09
Water consumption	X16	1.867	2.432	4.299	-0.56
Waste treatment	X17	2.83	2.5	5.331	0.33
Reuse & recycle	X18	2.955	2.71	5.666	0.244
Green collaboration	X19	3.5	3.135	6.634	0.364
Information		**3.181**	**3.194**	**6.375**	**-0.013**
Capacity of TIC system	X20	2.339	2.15	4.49	0.188
Integrated of inf. system	X21	3.048	2.664	5.712	0.384
Information sharing	X22	3.46	2.946	6.4	0.515
Innovation & Growth		**4.421**	**3.05**	**7.471**	**1.371**
HR	X23	3.17	2.8	5.978	0.37
Qualification & competency	X24	3.622	2.89	6.513	0.731
HRD	X25	3.667	3.056	6.724	0.611
Organization behaviour	X26	3.617	3.383	7.0	0.234
Local HR	X27	2.13	1.74	3.87	0.394
Health & safety	X28	2.954	2.66	5.615	0.292
Regulations & laws	X29	3.072	2.762	5.834	0.31

From the total relation matrix, we build significance matrix. The significance matrix describe of relationship between perspective and indicator and then will be input for BSC strategy map design. Figure 1 show of the BSC strategy map with twenty six indicators on five perspectives.

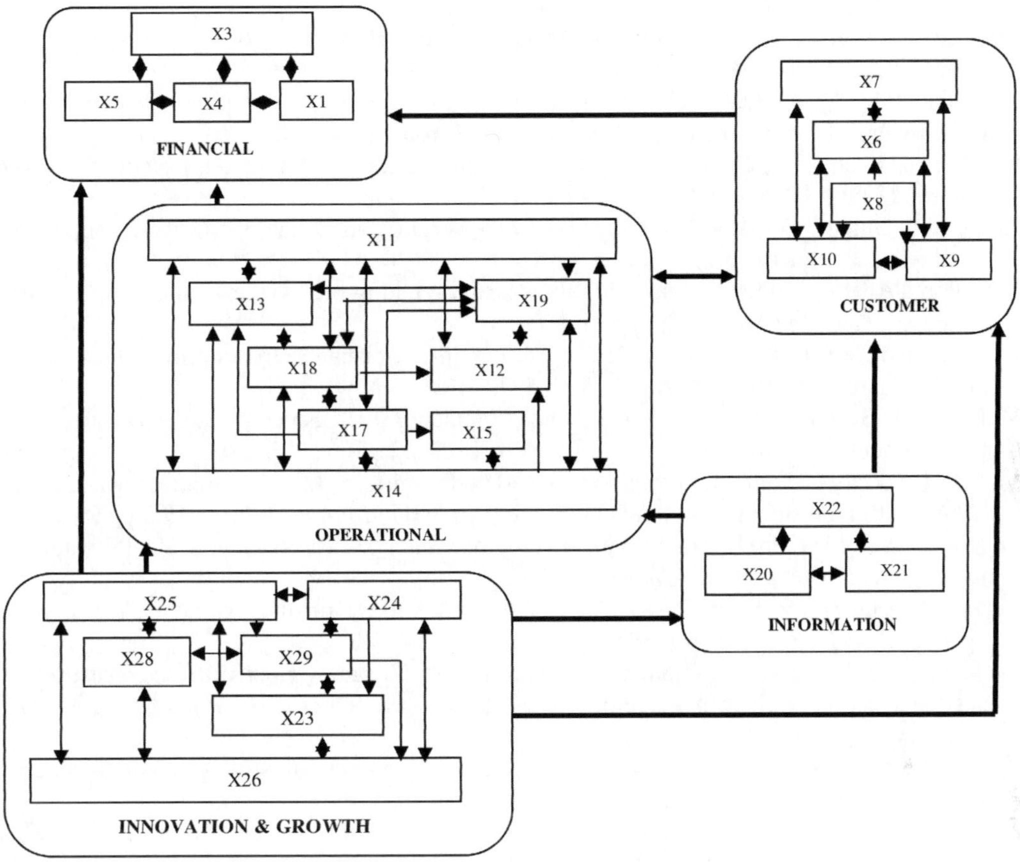

Figure 1. The BSC strategy map.

5. Conclusions

The sustainable SOSCPM can be developing based BSC and DEMATEL with five perspectives and twenty six indicators that oriented to sustainability dimensions and intangibility assets. We found two phenomenon's on the sustainable SOSC performance measurement: first, an intangibility indicator on innovation and growth perspective is most influence to other indicator, and second, existence of customer perspective is most important. From the phenomenon, the human resource and customer must be more attention by services business. Further research need calculate weight of indicator base on relationship between indicators, then simulate by system dynamic to predict of performance in the future. Furthermore, the model can be implemented on several service only with similar of intangibility as baseline for improving of sustainable SOSCPM.

6. References

[1] Baltacioglu T, Ada E, Kaplan MD, Yurt And O and Cem Kaplan Y 2007 A new framework for service supply chains *Serv. Ind. J* **27**(2) pp 105-24

[2] Giannakis M 2011 Conceptualizing and managing service supply chains *Serv. Ind. J* **31:**11 pp 1809–1823

[3] Drzymalski J 2012 *Supply Chain Frameworks for the Service Industry : A Review of the Literature*

vol. **1** December pp 31–42
[4] Cho D W, Lee Y H, Ahn S H and Hwang M K 2012 A framework for measuring the performance of service supply chain management *Comput. Ind. Eng.* **62**:3 pp 801–818
[5] Wang Y, Wallace S W, Shen B and Choi T 2015 Service supply chain management : A review of operational models *Eur. J. Oper. Res.* **247:**3 pp 685–698
[6] BPS 2014 Pertumbuhan Ekonomi Indonesia Triwulan II-2014 in *Berita Resmi Statistik* no. 16/02/Th. XVII pp 1–9
[7] Hassini E, Surti C and Searcy C 2012 A literature review and a case study of sustainable supply chains with a focus on metrics *Int. J. Prod. Econ.***140**:1 pp 69–82
[8] Linton J, Klassen R and Jayaraman V 2007 Sustainable supply chains: An introduction *J. Oper. Manag.***25**:6 pp 1075–1082
[9] Xu X and Gursoy D 2014 A conceptual framework of sustainable hospitality supply chain management *J. Hosp. Mark. Manag.* pp 1–31
[10] Marr B and Adams C 2006 The balanced scorecard and intangible assets : similar ideas , unaligned concepts *Meas. Bus. Excell.***8**:3 pp 18–27
[11] Zigan K and Zeglat D 2010 Intangible resources in performance measurement systems of the hotel industry *Facilities* **28** no. 13/14 pp 597–610
[12] Melnyk S A, Bititci U, Platts K, Tobias J and Andersen B 2014 Is performance measurement and management fit for the future? *Manag. Account. Res.***25**:2 pp 173–186
[13] Varsei M, Soosay C, Fahimnia B and Sarkis J 2014 Framing sustainability performance of supply chains with multidimensional indicators *Supply Chain Manag. An Int. J.***19**:3 pp 242–257
[14] Tajbakhsh A and Hassini E 2014 A data envelopment analysis approach to evaluate sustainability in supply chain networks *J. Clean. Prod.* Aug.
[15] Zhang X, Song H and Huang G Q 2009 Tourism supply chain management: A new research agenda *Tour. Manag.* **30**:3 pp 345–358
[16] Beske-janssen P, Johnson M P and Schaltegger S 2015 20 years of performance measurement in sustainable supply chain management – what has been achieved ? *Supply Chain Manag. An Int. J.* **20**:6 pp 664–680
[17] Kaplan R S and Norton D P 1996 Linking the Balanced Scorecard to strategy *Calif. Manage. Rev.* **39**:1 pp 53–79
[18] Kaplan R S and Norton D P 2006 Response to S. Voelpel et al., 'The tyranny of the Balanced Scorecard in the innovation economy,' Journal of Intellectual Capital, Vol. **7** No. 1, 2006, pp. 43-60 *J. Intellect. Cap.* **7**:3 pp 421–428
[19] Kaplan R S, Norton D P and Rugelsjoen B 2010 Managing alliances with the Balanced Scorecard *Harv. Bus. Rev.* no. February 2010 pp 114–121
[20] Leksono E B, Suparno and Vanany I 2016 Development of conceptual framework for sustainable supply chain performance measurement in service industries in *8th Widyatama Int. Seminar on Sustainability* pp 142–148
[21] Gölcük I and Baykasoglu A 2016 An analysis of DEMATEL approaches for criteria interaction handling within ANP *Expert Syst. Appl.***46** pp 346–366
[22] Chen F, Hsu T and Tzeng G 2011 A balanced scorecard approach to establish a performance evaluation and relationship model for hot spring hotels based on a hybrid MCDM model combining DEMATEL and ANP *Int. J. Hosp. Manag.* **30**:4 pp 908–932
[23] Wei Y, Hu Q and Xu C 2013 Ordering, pricing and allocation in a service supply chain *Int. J. Prod. Econ.***144**:2 pp 590–598
[24] Wu I L, Chuang C H and Hsu C H 2014 Information sharing and collaborative behaviors in enabling supply chain performance: A social exchange perspective *Int. J. Prod. Econ.* **148** pp 122–132
[25] Sigala M 2008 A supply chain management approach for investigating the role of tour operators on sustainable tourism: the case of TUI *J. Clean. Prod.***16**:15 pp 1589–1599
[26] Büyüközkan G and Cifci G 2013 An integrated QFD framework with multiple formatted and

incomplete preferences : A sustainable supply chain application *Appl. Soft Comput.* **13** pp 3931–3941

[27] Hussain M, Khan M and Al-aomar R 2016 A framework for supply chain sustainability in service industry with Confirmatory Factor Analysis *Renew. Sustain. Energy Rev.* **55** pp1301–1312

[28] Bhagwat R and Sharma M K 2009 Performance measurement of supply chain management using the analytical hierarchy process *Prod. Plan. Control* **18**: 8 pp 666–680

[29] Bhagwat R and Sharma M K 2007 Performance measurement of supply chain management : A balanced scorecard approach *Comput. Ind. Eng.* **53** pp 43–62

[30] Matos S and Hall J 2007 Integrating sustainable development in the supply chain: The case of life cycle assessment in oil and gas and agricultural biotechnology *J. Oper. Manag.* **25**:6 pp 1083–1102

[31] Schwartz K, Tapper R and Font X 2008 A sustainable supply chain management framework for tour operators *J. Sustain. Tour* **16**: 3 pp 298–314

[32] Wicks A M, St Clair L and Kidney C S 2007 Competing values in healthcare : Balancing the (Un) Balanced Scorecard *J. Healthc. Manag.* **52**: 5 pp 309–324

[33] Keating B, Quazi A, Kriz A and Coltman T 2008 In pursuit of a sustainable supply chain: insights from Westpac Banking Corporation *Supply Chain Manag. An Int. J.* **13**: 3 pp 175–179

[34] Smith B K, Nachtmann H and Pohl E A 2011 Quality measurement in the healthcare supply chain *Qual. Manag. J.* **18**:4 pp 50–60

[35] de Vries J and Huijsman R 2011 Supply chain management in health services: an overview *Supply Chain Manag. An Int. J.* **16**:3 pp 159–165

[36] Chithambaranathan P, Subramanian N and Gunasekaran A 2015 Service supply chain environmental performance evaluation using grey based hybrid MCDM approach *Int. J. Prod. Econ.* no. 2006 pp 1–14

[37] Fu N, Flood P C, Bosak J, Morris T and O'Regan P 2013 Exploring the performance effect of HPWS on professional service supply chain management *Supply Chain Manag. An Int. J.* **18**:3 pp 292–307

[38] Adriana B 2009 Environmental supply chain management in tourism: The case of large tour operators *J. Clean. Prod.* **17**:16 pp 1385–1392

[39] Gopalakrishnan K, Yusuf Y Y, Musa A, Abubakar T and H M Ambursa 2012 Sustainable supply chain management: A case study of British Aerospace (BAe) Systems *Int. J. Prod. Econ.* **140**: 1 pp 193–203

[40] Brindley C and Oxborrow L 2014 Aligning the sustainable supply chain to green marketing needs: A case study *Ind. Mark. Manag.* **43**:1 pp 45–55

[41] Font X, Tapper R, Schwartz K and Kornilaki M 2008 Sustainable supply chain management in tourism *Bus. Strateg. Environ.* **17** pp 260–271

[42] Elgazzar S H, Tipi N S, Hubbard N J and Leach D Z 2012 Linking supply chain processes performance to a company's financial strategic objectives *Eur. J. Oper. Res.* **223**:1 pp 276–289

[43] Lega F, Marsilio M and Villa S 2013 An evaluation framework for measuring supply chain performance in the public healthcare sector : evidence from the Italian NHS *Prod. Plan. Control* **24**:10–11 pp 931–947

[44] Chen D Q, Preston D S and Xia W 2013 Enhancing hospital supply chain performance: A relational view and empirical test *J. Oper. Manag.* **31**:6 pp 391–408

[45] Narayana S A, Kumar R and Vrat P 2014 Managerial research on the pharmaceutical supply chain – A critical review and some insights for future directions *J. Purch. Supply Manag.* **20**:1 pp 18–40

[46] Lee H K and Fernando Y 2015 The antecedents and outcomes of the medical tourism supply chain *Tour. Manag.* **46** pp 148–157

[47] Parker J and DeLay D 2008 *The future of the healthcare supply chain* Healthc. Financ. Manag **62**: 4 pp 66–69

[48] Wu T, Wu Y C, Chen Y J and Goh M 2014 Aligning supply chain strategy with corporate environmental strategy: A contingency approach *International Journal of Production Economics* **147** pp 220-9

[49] Hohenstein N O, Feisel E and Hartmann E 2014 Human resource management issues in supply chain management research *Int. J. Phys. Distrib. Logist. Manag.* **44**: 6 pp 434–463

[50] Feil A A, De Quevedo D M and Schreiber D 2015 Selection and identification of the indicators for quickly measuring sustainability in micro and small furniture industries *Sustain. Prod. Consum* **3** pp 34–44

Analysis of brand personality to involve event involvement and loyalty: A case study of Jakarta Fashion Week 2017

A H Nasution[1*] and Y A Rachmawan[1]
[1]Department of Business Management, Institut Teknologi Sepuluh Nopember, Kampus ITS, Sukolilo-Surabaya 60111, Indonesia

*armanhakim.nasution@gmail.com

Abstract. Fashion trend in the world changed extremely fast. Fashion has become the one of people's lifestyle in the world. Fashion week events in several areas can be a measurement of fahion trend nowadays. There was a fashion week event in Indonesia called Jakarta Fashion Week (JFW) aims to show fashion trend to people who want to improve their fashion style. People will join some events if the event has involvement to them, hence they will come to that event again and again. Annually and continuously event is really important to create loyalty among people who are involved in it, in order to increase positive development towards the organizer in organizing the next event. Saving a huge amount from the marketing budget, and creating a higher quality event. This study aims to know the effect of 5 brand personality dimension to event involvement and loyalty in Jakarta Fashion Week (JFW). This study use quantitative confirmative method with Structural Equation Model (SEM) analysis technique. The sample of this study is 150 respondents who became a participant of Jakarta Fashion Week 2017. Result show that there was significant effect of 5 brand personality dimension to 3 dimension of event involvement and loyalty. Meanwhile, there was one dimension of event involvement called personal self-expression that has not effect to loyalty.

1. Introduction

Fashion development in the world is very flexible and very fast changing. Fashion in passing is a dress, but more than that fashion is the way we reflect ourselves [1]. And fashion are symbols of the soul, and fashion cannot be separated from the development of human history and life [2]. To find out the current fashion trend benchmarks can be seen from the Fashion Week event, in the Fashion Week exhibition the latest collection featured from all the designers who follow the event is a benchmark of future fashion developments. In the development of the fashion industry in Indonesia, fashion is a sub-sector of the creative industry, in 2015 the creative industry sector has contributed Rp. 641.8 trillion to GDP, and fashion subsector in Indonesia accounted for 28.29% or equivalent to Rp. 181.5 trillion. Jakarta Fashion Week has been around since 2008. Besides being the biggest fashion show in Indonesia, Jakarta Fashion Week is the biggest fashion show in Southeast Asia. The objective of this research is to know how the brand personality of Jakarta Fashion Week and its influence on event involvement of the show, and how event involvement can influence the loyalty of people People involved in Jakarta Fashion Week

2. Literature Review

2.1. Brand Personality

Brand personality is "the set of human characteristic associated with a brand" which means a brand is a manifestation of the characteristics of a human who form an emotional relationship and describes the consumer expressing himself [7], and brand personality is something to measure or describe A personality of a brand itself, and defines that brand personality is "a set of human personality applicable to the brand, and relevant to the brand" [8]. The Power of Brand states that brand personality is the stage of development in a brand, which means the brand reflects the personality [9]. Basically the function of the brand is not just a description of a product, but the brand is a person who reflects its use. So, author [7] suggested that consumers perceive brands as having five basic personality traits, its sincerity, excitement, competence, sophistication, and ruggedness.

2.2. Event Involvement

Event involvement is a consumer's perception of an event or activity that is cognitive and affective [13]. Event involvement is an interesting, values, and needs someone in an object or event [14]. In previous research on recreation and organizing an event, it is known that in an activity that demands active audience participation, there are variables that support positive attitudes and enthusiastic personality traits in an event, such as commitment to an event, customer loyalty, psychological interrelation - motional and frequency of participation in the event [15]. Event involvement on recreational activities and hobby or leisure involvement according to past research [16] there are three main dimensions, namely attraction, self-expression, and centrality to lifestyle, its involve attraction, personal self expression, andpersonality centrality.

2.3. Loyalty

Author [17] in the Customer Loyalty book, states that the loyalty of an individual to a thing is can be shown by doing repetitive activities, recommending a product or service to others, showing the advantages of products or services owned from competitors to others . While Loyalty is divided into 2 main perspective in measuring a loyalty into behavior and attitude [18]. These two main perspectives have the same tendency to loyalty as reflected in behaviors and attitudes such as [15] suggest that the behavior and attitudes of a person who is loyal to leisure activities will give them the same intention or Even bigger for the same event next year, and tells about the satisfaction earned from the event through a positive worth-of-mouth for what the individual has experienced.

3. Methodology

3.1. Type of Data

Primary data is in the form of demographic characteristic data of respondents taken by survey method which represents sample from study population, this research uses purposive sampling, the sample size of each population is based on the division of interests of the people involved in Jakarta Fashion Week 2017.

3.2. Data Collection

This research was conducted by using questionnaire survey method that was distributed directly and offline. The population of this study is all the people involved in Jakarta Fashion Week 2017. The method used by purposive sampling, where the sample size is determined on the distribution of the interests of the people involved in the show, as many as 150 data obtained in Jakarta Fashion Week 2017 show on the 24th -28 October 2016.

3.3. Method and Variable

A structural model, in which it was hypothesized that: the five brand personality facets will influence the development of event involvement; and the three involvement dimensions will influence the development of event loyalty (behavioural intentions and W-O-M communications). The structural model is presented in Figure 1.

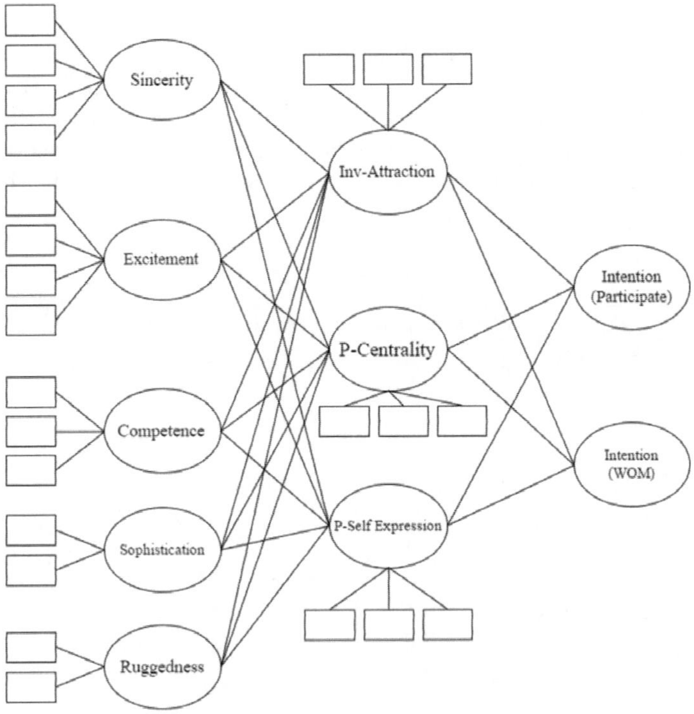

Figure 1.Structural Model.

A deductive methodological approach was adopted in the current study. A deductive approach is used when the goal of the study is to test causal relationships among variables, usually with the form of hypotheses and design a research strategy to test a theory [19].

4. Analysis and Review

4.1. Descriptive Analysis

Descriptive analysis serves to determine the distribution of respondents based on demography profile [10]. The following is the result of demographic analysis of respondents show that most people involved in Jakarta Fashion Week show are West Jakarta residents with 32 people, followed by Central Jakarta and South Jakarta with 29 people, North Jakarta 28 people and East Jakarta with 14 people From outside Jakarta 18 people. While most ages are present in Jakarta Fashion Week is the age of 26-35 years, and at least is at the age of 46-55 years, the last education of the people involved in the Jakarta Fashion Week show is known to most are undergraduate with a percentage of 53% or a number of 80 people, and the fewest are post graduate with a percentage of 105 or 15 people. The type of work of the people involved in Jakarta Fashion Week 2017 is at most private employees with a frequency of 49 people or a percentage of 33%, while the least is the TNI / POLRI with a percentage of 1% or a number of 2 persons.

The amount of interest of a respondent in the Jakarta Fashion Week 2017 event or can be said as the distribution of respondents who took part in the Jakarta Fahion Week 2017 is as follows, the audience is a number of 67 people or (45%) of the total respondents, then 28 people or (19%) are the committee, 17 respondents or (11%) are models, 12 respondents or 8% are designers, 9 respondents or (6%) are reporters / writers, 7 respondents or (5%) are Buyers, and 10 respondents or (7%) are photographer.

4.2. Confirmatory factor analysis

Confirmatory Factor Analysis (CFA) to test Validity and Cronbach's alpha to test reliability. Validity test is done to see if the measuring instrument used can represent precisely the variables studied. Data will be considered valid if the load factor value has exceeded the minimum limit of above 0.50 (Hair et al., 2014). Test Reliability is done to see the consistency of answers and useful to determine whether the instrument, in this case the questionnaire can be used more than once. The data will be considered reliable if the value of Cronbach's alpha has exceeded the minimum limit value of above 0.70, in this study the researchers used the minimum limit of 0.70 because with the minimal limit is more reliable and more able to provide support for internal consistency [12].

Table 1. CFA & Reliabilitas for Brand Personality.

Factor	Indicator	Estimate	Cronbach Alfa
Excitement	EX1	0.89	
	EX2	0.90	0.943
	EX3	0.90	
	EX4	0.89	
Sincerity	SC1	0.90	
	SC2	0.91	0.945
	SC3	0.91	
	SC4	0.88	
Copetence	CT1	0.92	
	CT2	0.81	0.907
	CT3	0.91	
Ruggedness	RU1	0.85	0.898
	RU2	0.96	
Sophistication	SH1	0.92	0.882
	SH2	0.86	

Table 2. CFA & Reliabilitas for Event Involvement.

Factor	Indicator	Estimate	Cronbach Alfa
Involve	IA1	0.92	
Attraction	IA2	0.88	0.927
	IA3	0.90	
Self	SE1	0.90	
Expression	SE2	0.91	0.927
	SE3	0.88	
Personal	PC1	0.87	
Centrality	PC2	0.89	0.914
	PC3	0.88	

Table 3. CFA & Reliablitas for Loyalty.

Factor	Indicator	Estimate	Cronbach Alfa
Loyalty	P	0.71	0.859
	WOM	0.70	

It is known based on the results of estimate and Cronbach alpha values that have been processed that all constructs in brand personality have been valid and reliable.

IOP Conf. Series: Materials Science and Engineering **337** (2018) 012024 doi:10.1088/1757-899X/337/1/012024

4.3. SEM Analysis

The fit model test is called the goodness of fit test to test the hypothesis whether the model is fit or not, the goodness of fit test is done to find out the feasibility of the structural model by comparing the goodness of fit and cut off tables. The variables found in this research are 2 independent variables, namely brand personality and event involvement variables, consisting of 5 dimensions on the brand personality variable, namely excitement (EX), sincerity (SC), competence (CT), ruggedness (RU), and Sophistication (SH) and 3 dimensions in event involvement variables, namely personal centrality (PC), personal self expression (SE) and involve attraction (IA).

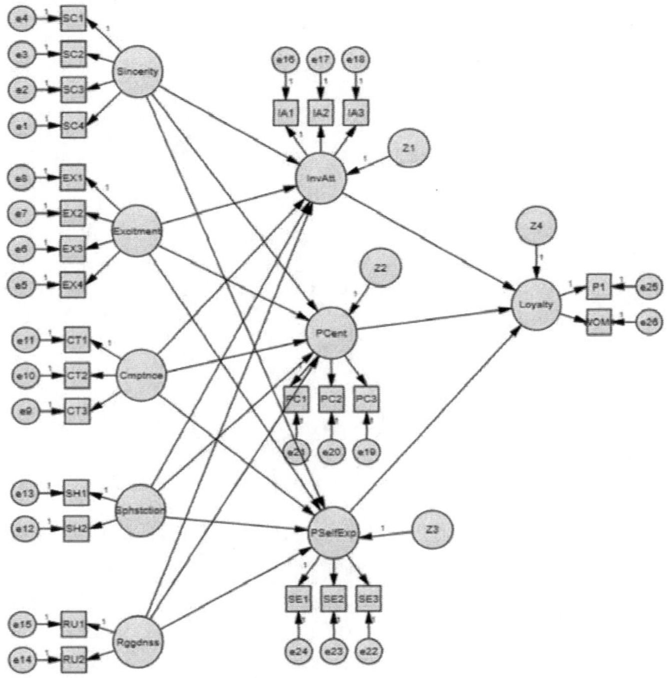

Figure 2. SEM.

In the goodness-of-fit test that has been done in this research is to see that the value of Chi - Square, P - Value, GFI, AGFI, CFI, RMSEA, CMIN / DF, TLI, and CFI must qualify. Model fit test in this research can be seen from table 4 that from the goodness of fit test results can be seen that the model has been fit and has passed the feasibility test. According to [20] if there have been two criteria of goodness-of-fit that meet the criteria then the model is said to be good or has been feasible. Then, in table 4, it can be seen that the results obtained are 2 criteria that decide that the model is not appropriate and does not meet the requirements of GFI criteria with cut off value of 0.876 and AGFI with cut off value of 0.846 and the rest decided that the model is appropriate. So it can be concluded, the model diagram that has been tested above has been suitable to be observed, but because there are 2 of 7 that decide not appropriate, so it needs to be modified in order to get a model that all criteria meet the comparative requirements.

IOP Conf. Series: Materials Science and Engineering **337** (2018) 012024 doi:10.1088/1757-899X/337/1/012024

Table 4. Goodness of Fit.

Criteria	Cut off	Comparison	Decision
Chi – Square	256.822	As small as possible	Marginal
P – Value	0.847	≥ 0.05	Accepted
GFI	0.876	≥ 0.9	Not good enough
AGFI	0.846	≥ 0.9	Not good enough
RMSE	0.000	≤ 0.08	Accepted
CMIN/DF	0.914	≤ 2	Accepted
TLI	2.077	≥ 0.9	Accepted
CFI	1	≥ 0.9	Accepted

4.4. Modification (SEM)

Modification of the model is done by connecting the error value of model measurement in one construct based on the biggest MI value, SEM modification will decrease the chi-square value and increase the GFI and AGFI value on the goodness of fit, after 3 SEM modifications to the data, The end result of this modification.

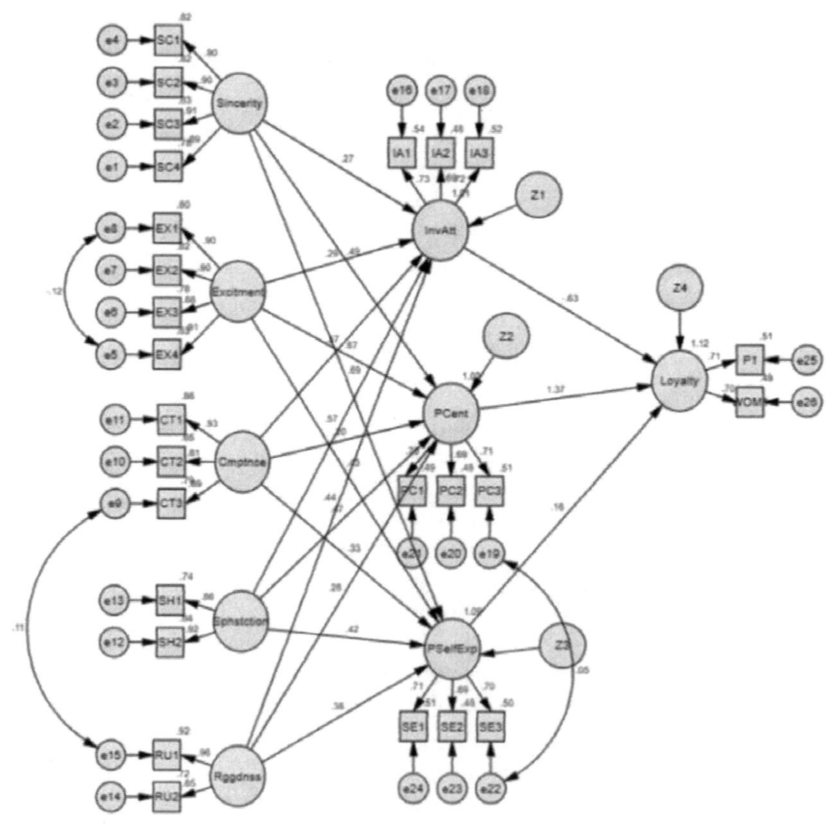

Figure 3. SEM after Modification.

And the value on the goodness of fit, turns into

Table 5. Goodness of Fit after SEM Modification.

Criteria	Cut off	Comparison	Decision
Chi – Square	235.245	As small as possible	Marginal
P – Value	0.970	≥ 0.05	Accepted
GFI	0.881	≥ 0.9	Good enough
AGFI	0.850	≥ 0.9	Good enough
RMSE	0.000	≤ 0.08	Accepted
CMIN/DF	0.859	≤ 2	Accepted
TLI	2.760	≥ 0.9	Accepted
CFI	1	≥ 0.9	Accepted

4.5 Hypothesis
After the model has been obtained, then performed hypothesis testing of each latent variable. Where here, want to know whether there is a relationship between brand personality with event involvement, and the relationship between event involvements with loyalty. Hypothesis testing is done by using p-value obtained from SPSS AMOS software

In table 6 below is a hypothesis testing table or also can be called by weight regression analysis. Hypothesis testing performed on each - each variable. Where if the value of p - value is more than alpha value 0.05, then the variable has no effect. In the table above can be seen that 18 variables significantly influence because the value of p - value is less than alpha value 0.05. While that does not have an effect is one variable, namely the variable self-expression to loyalty, because the value of p - value more than alpha value 0.05.

Table 6. Hypothesis.

Variabel		Variabel	Estimate	S.E.	C.R.	P – Value
InvAttraction	←	Sincerity	0.131	0.027	4.921	0.000
PCentrality	←	Sincerity	0.234	0.029	8.09	0.000
PSelfExpresion	←	Sincerity	0.323	0.032	10.017	0.000
InvAttraction	←	Excitement	0.141	0.027	5.167	0.000
PCentrality	←	Excitement	0.325	0.035	9.282	0.000
PSelfExpresion	←	Excitement	0.202	0.027	7.486	0.000
InvAttraction	←	Competence	0.284	0.033	8.726	0.000
PCentrality	←	Competence	0.098	0.024	4.025	0.000
PSelfExpresion	←	Competence	0.157	0.026	6.119	0.000
InvAttraction	←	Sophistication	0.296	0.035	8.355	0.000
PCentrality	←	Sophistication	0.242	0.03	7.946	0.000
PSelfExpresion	←	Sophistication	0.211	0.029	7.149	0.000
PSelfExpresion	←	Ruggedness	0.177	0.026	6.765	0.000
PCentrality	←	Ruggedness	0.134	0.023	5.949	0.000
InvAttraction	←	Ruggedness	0.208	0.029	7.1	0.000
Loyalty	←	InvAttraction	-0.674	0.191	-3.518	0.000
Loyalty	←	PCentrality	1.472	0.268	5.498	0.000
Loyalty	←	PSelfExpresion	0.179	0.237	0.756	0.45

5. Conclusion
Based on the result of structural equation model (SEM) analysis on the data of respondents who become participants in Jakarta Fashion Week 2017, it can be concluded that there are significant positive

influence between 5 model of personality brand that is Personal Sincerity, Personal Excitement, Personal Competence, Personal Ruggedness and Sophistication to 3 model Event involvement and loyalty. While one of the model of event involvement that is personal self-expression did not have a significant positive effect on loyalty model that is participate and WOM. Based on this it is known that the brand personality factor is really a very supportive event involvement of an event, because the brand personality of an event shows the identity of an event, the identity of the event is very influential on the people involved in the event activities. Based on 5 dimensions of brand personality that is Sincerity, Excitement, Competence, Ruggedness and Sophistication, Jakarta Fashion Week performance has been in accordance with the people involved and participate in it. Brand personality is very influential on the event involvement, because a good brand personality will create a high involvement event, the desire to participate and engagement in the performance event. It creates event involvement in the form of personal self-expression, involve attraction, and personal centrality high.

6.References

[1] Ibrahim I S 2011 Budaya Populer Sebagai Komunikasi (Yogyakarta: Jalasutra)
[2] Chaney D 2008 Lifestyle (Yogyakarta: Jalasutra)
[3] Kementrian Perindustrian Republik Indonesia 2015Sektor Fesyen Topang Ekspor
[4] Pramita D 2015 Kecantikan
[5] Keller K L 2008 Strategic Brand Management : Building, Measuring and Managing Brand Equity (New Jersey: Pearson-Prentice Hall)
[6] Eisend M and Stokburger-Sauer N E 2013 Brand personality: a meta-analytic review of antecedents and destinations *Marketing Letters* **24** (3) pp 205-216
[7] Aaker J L 1997 Dimensions of brand personality *Journal of Marketing Research* **34** (3) pp 347-356
[8] Azoulay A and Kapferer J 2003 Do brand personality scales really measure brand? *The Journal (of Brand Management* **11** (2) pp 143-155
[9] Rangkuti F 2004 The Power of Brand (Jakarta: PT. Gramedia Pustaka Utama)
[10] Ratna N K 2004 Teori, Metode, dan Teknik Penelitian Sastra (Yogyakarta: Pustaka Pelajar)
[11] Hair J, Black W C, Babin B J and Anderson R E 2014 Multivariate Data Anlysis 7th Edition (New Jersey: Pearson Prentice Hall)
[12] Maholtra N K and Birks D F 2012 Marketing research: An applied approach (UK: FT. Press)
[13] Robinson P, Wale and Dickson G 2010 Event Management (United Kingdom: CAB International)
[14] Zaichkowsky J L 1985 Conceptualizing Involvement *Journal of Advetising* **15** pp 4-34
[15] Alexandris K, Kouthouris C and Funk D 2008 The impact of constraints on motivation, constraints, involvement and attitudinal loyalty among recreational skiers *European Sport Management Quarterly* **8** 3 p. 247-265
[16] McIntyre N and Pigram J 1992 Recreation specialization reexamined: The case of vehicle-based campers *Leisure Research* **14** pp 3-15
[17] Griffin J 2005 Customer Loyalty (Jakarta: Erlangga)
[18] Tjiptono F 2002 Strategi Pemasaran (Yogyakarta: ANDI)
[19] Saunders M, Tornhill A and Lewis P 2003 Research Methods for Business student (London: Prentice Hall)
[20] Solimun 2006 Structural Equation Model (SEM): Aplikasi software Amos, Malang, Fakultas MIPA Pro-gram Pascasarjana Universitas Brawijaya.

IOP Conf. Series: Materials Science and Engineering **337** (2018) 012025 doi:10.1088/1757-899X/337/1/012025

Trade-offs mathematical modelling of 3DCE in new product development: real three dimensions and directions for development

M A Ilhami[1,2*], Subagyo[1], and N A Masruroh[1]
[1]Department of Mechanical and Industrial Engineering, Universitas Gadjah Mada
[2]Department of Industrial Engineering, Universitas Sultan Ageng Tirtayasa

*adha@untirta.ac.id

Abstract. In the last two decades, coordinating product, process, and supply chain has become the main focus in recent years as a growing body of research, which mathematical modelling is leading technique used in the early phase design. In this paper, we aim to conduct a comprehensive literature review of published paper and propose directions for future research, especially in mathematical modelling. Our findings exhibit fact that evidently there are only a few papers coordinate "real three dimensions". The other papers, in fact, show simply two dimensions. Finally, some suggestions are proposed such as paying more attention to "real three dimensions" research-based and more focus on minimizing time to market, product life cycle consideration, and product rollover.

1. Introduction

Three Dimensional Concurrent Engineering (3DCE) is a powerful model of new product development (NPD) supported by concurrent engineering, in which the traditional focus on an appropriate match between product and process is augmented by an additional consideration of supply chain configuration [1]. Decision making process is a key factor in product development that is made in every gate in Cooper's stage gate product development. The duration for making a decision is not always fast as needed, it takes weeks or even months to make decision. This long duration problem is caused by unspecified trade-offs that should be solved and considered for the decision making. Many researchers describe trade-offs in mathematical modelling and solve the trade-offs using common techniques in operation research. So, the decision making process can be made quicker as the trade-offs had been formulated and solved.

Literature review on 3DCE has been conducted several times, such as Ellram et al. [1], Forza et al.[2], Castellano and Dolado [3], Gan and Grunow [4], and Shahbazi et al.[5]. In Shahbazi et al. [5] review, mathematical modelling was the most common technique used that attracted attention most researchers to coordinate the three dimensions.

In distinct to above reviews, this paper goals are first to analyze on recent literature specifically on mathematical modelling trade-offs in 3DCE to support the "real three dimension" decision making. Second to identify the gaps in the existing literature and suggest directions for further research. Term "real three dimension" means that trade-offs are formulated considering product, process, and supply chain dimensions, since many mathematical models were formulated in two dimensions only.

This paper is organized as follows. Section 2 describe the background of the study. Section 3 is presented to define the method of collecting relevant papers. Section 4 provides highlight the mathematical model of selected relevant papers. Finally, in Section 5, the gaps and directions for further research are discussed.

2. Background

The concept of 3DCE as defined by Fine [6] is considered during product design phase. 3DCE means considering the requirements of manufacturing design and supply chain design in the product design phase. Surely, this consideration complicates the design problem that requires more complex multi objectives model with a large number of constraints, variables, parameters, and trade-offs. However, by considering all requirements of product design, process design, and supply chain design, we have a great chance of giving responsiveness, agility, variety, quality, and price to win the market. This also strengthened by Fine [6] statement that "when firms do not explicitly acknowledge and manage supply chain design and engineering as a concurrent activity to product and process design and engineering, they often encounter problems late in product development, or with manufacturing launch, logistical support, quality control, and production costs…".

Researches that modelled trade-offs with product, process, and supply chain as decision variables are conducted to help decision makers (designer) decide product, process, and supply chain configuration simultaneously (the real three dimensions). Simultaneously means all decisions are done together in the product design phase, so that manufacturing and supply chain configuration can be prepared directly after the decisions are made. Researches on mathematical modelling in 3DCE area are increasing in the last 15 years. There are only seven papers found during 2000 - 2005, and eight papers found during 2006 - 2010. However, during 2011 - 2015, there is a significant increase where there are 16 papers. This fact led to the questions which are become the objectives of our study that are "what is the most sub topics or focus of the researches?" and "what directions needed for researchers to develop 3DCE trade-offs modelling?"

3. Methodology

Selecting papers is the first step to review the literature. Journals in the areas of operation management, supply chain, product development, production economics etc. are collected using several databases such as Google Scholar, Scopus, Science Direct, Taylor & Francis, and Emerald. References of the paper are reviewed to add possible additional papers once the paper are identified.
We use several key words for searching the references such as "3DCE", "Three Dimensional Concurrent Engineering", "integrated product process supply chain alignment", "coordination product process and supply chain", "matching product process and supply chain", "integrated production and distribution planning", and "concurrent engineering". The papers with mathematical model are selected, we organize them into several categories based on sub topics or focus of the researches, and then the papers are analyzed.

4. Result and Discussion

After analyzing and organizing the papers, 35 papers are selected, the selected papers were presented in categories and can be seen in table 1, which classifies the papers based on dimensions. Fact that researches in trade-offs mathematical modelling are still scarce in 3DCE environment can be seen from the number of papers that is only five papers. We call these five papers as "real three dimensions" as they include all three dimension as decision variables and comprehensively coordinated all the three dimensions. The models presented in the other papers are limited to two-dimension model-based research, which are Product-Process, Product-Supply Chain, and Process-Supply Chain consideration.

IOP Conf. Series: Materials Science and Engineering **337** (2018) 012025 doi:10.1088/1757-899X/337/1/012025

Table 1. Papers selected based on dimensions.

Dimensions	Authors	Num. of Papers
Product- Process (A-2D)	Xu and Liang [7], Wu and O'Grady [8]	2
Process- Supply Chain (B-2D)	Cunha et al. [9], Graves and Willems [10], Kuang et al. [11], Shahrokhi et al.[12], Park et al. [13]	5
Product- Supply Chain (C-2D)	Amini and Li [14], Chiu and Okudan [15], Chiu and Okudan [16], Deng et al.[17], ElMaraghy and Mahmoudi [18], Ernst and Kamrad [19], Famuyiwa and Monplaisr [20], Gaber and Abdelsalam [21], Jiang et al. [22], Nepal et al.[23], Baud-Lavigne et al. [24], Khalaf et al. [25], Khalaf et al. [26], Lamothe et al. [27], Langenberg et al. [28], Metta and Badurdeen [29], Moon et al. [30], Rezapour et al. [31], Syam and Bhatnagar [32], Ulku and Schmidt [33], Wang et al. [34], Zhang et al. [35], and Zhang et al. [36].	23
Product-Process-Supply Chain (3D)	Fine et al. [37], Huang et al. [38], Jafarian and Bashiri [39], Shidpuor et al. [40] [41]	5

Table 2. Papers group based on sub topics.

Sub Topics	Authors	Dim	Num. of Papers
1. Modularity	Xu and Liang [7]	A-2D	9
	Cunha et al. [9]	B-2D	
	Famuyiwa and Monplaisr [20], Baud_Lavigne et al. [24], Khalaf et al. [25], Khalaf et al. [26], Zhang et al. [35], Ernst and Kamrad [19]	C-2D	
	Fine et al. [37]	3D	
2. Inventory and Supply Chain Consideration	Graves & Willems [10], Kuang et al. [11], Shahrokhi et al. [12], Park et al. [13]	B-2D	20
	Chiu and Okudan [15], Chiu and Okudan [16], ElMaraghy and Mahmoudi [18], Gaber and Abdelsalam [21], Jiang et al. [22], Nepal et al. [23], Lamothe et al. [27], Langenberg et al. [28], Rezapour et al. [31], Syam and Bhatnagar [32], Ulku and Schmidt [33], Wang et al. [34], Zhang et al. [36]	C-2D	
	Huang et al. [38], Shidpour et al. [40] [41]	3D	
3. Time to market	Deng et al. [17]	C-2D	2
	Jafarian and Bashiri [39]	3D	
4. Product life cycle	Metta and Badurdeen [29]	C-2D	1
5. Product rollover	Moon et al. [30]	C-2D	1
6. Impact of dynamic demand	Amini and Li [14]	C-2D	1
7. Design for assembly	Wu and O'Grady [8]	A-2D	1

Product-Supply Chain model is the highest number with 23 papers, this is a significant indication that inventory management in the supply chain is strong as topic of interest. Meanwhile, Product-Process or Process-Supply Chain model is far less then Product-Supply Chain model with two and five papers respectively.

After grouping, the papers based on dimensions, the papers then grouped based on their focus of interest. The focus grouping result can be seen in table 2, it shows that at least there are 5 sub topics of trade-offs mathematical modelling, which are: (1) inventory allocation or supply chain perspective; (2) product modularity; (3) product development time or minimizing time to market; (4) product life cycle consideration; and (5) product substitution or rollover. From table 2 we can see that sub topic 1 and 2 have more number of research. However, for sub topic 3, 4, and 5, there are only a few. Based on this fact, we highly recommend future research must be conducted in the area of sub topic 3, 4, and 5. Minimizing time to market guarantees the new product to be in the market as fast as possible which increase product delivery, meanwhile considering product life cycle and product rollover maintain company competitiveness by ensuring the arriving of the new product at the right time.

Another remarkable point is a very limited 3D mathematical model literature in all sub topics. As can be seen in table 2, product life cycle and product rollover has no 3D model at all. Therefore, research on mathematical modelling for decision making in new product development are need to be accelerated in all sub topics of new product development, especially product life cycle consideration and product rollover.

5. Conclusion

The "real three dimensions" mathematical modelling research are still scarce which is shown by the number of papers that is only 5 papers. This number is far less than two-dimension model-based research, especially Product-Supply Chain model-based research. From this point of view, we recommend more research on 3D mathematical modelling that emphasis all three dimensions.

Regarding the limitations of literature found in all sub topics, it is necessary to recommend 3DCE mathematical modelling research in the area time to market, product life cycle, and product rollover for the future research, as those sub topics are very limited. It is prominent to answer such questions of how to consider product life cycle in the model, or how to put time to market as an objective function or constraint in the model, and how to consider rollover product as constraint or variable in the model.

References

[1] Ellram L M, Tate W L, and Carter C R 2007 Product-process-supply chain: an integrative approach to three-dimensional concurrent engineering. *Int J Phys Distrib Logist Manag* **37** pp 305–330

[2] Forza C, Salvador F, and Rungtusanatham M 2005 Coordinating product design, process design, and supply chain design decisions: Part B. Coordinating approaches, tradeoffs, and future research directions. *J Oper Manag* **23** pp 319–324

[3] Castellano E, and Dolado J 2010 Product-processes-supply chain structures alignment for mass customization scenarios. A literature review. In: *4th International conference on Industrial Engineering and Industrial Management* San Sebastian pp 1805–1818

[4] Gan T S, and Grunow M 2013 Concurrent product supply chain design: A conceptual framework & literature review. *Procedia CIRP* **7** pp 91–96

[5] Shahbazi M, Soofi J B, Firouzabadi A K 2014 Coordination of Product , Process and Supply Chain Design ; Analysis of Recent Literature and Directions for Development. *Int J Sci Res Knowl* **2** pp 38–47

[6] Fine C H 1998 *Clockspeed: Winning industry control in the age of temporary advantage*. New York: Basic Books

[7] Xu Z, and Liang M 2005 Concurrent Optimization of Product Module Selection and Assembly Line Configuration: A Multi-Objective Approach. *J Manuf Sci Eng* **127** 875.

[8] Wu T, O'Grady P 1999 A Concurrent Engineering Approach to Design for Assembly. *Concurr*

Eng **7** pp 231–243
[9] Da Cunha C, Agard B, and Kusiak A 2007 Design for cost: Module-based mass customization. *IEEE Trans Autom Sci Eng* **4** pp 350–359
[10] Graves S C and Willems S P 2005 Optimizing the Supply Chain Configuration for New Products. *Manage Sci* **51** pp 1165–1180
[11] Kuang H, Jack Hu S, and Ko J 2016 Concurrent Design of Assembly Plans and Supply Chain Configurations Using AND/OR Graphs and Dynamic Programing. *J Manuf Sci Eng* **138**
[12] Shahrokhi M, Bernard A, Shidpour H. A hybrid method to select best process and suppliers, in the concurrent engineering environment. In: *IFAC Proceedings Volumes (IFAC-PapersOnline)* IFAC pp 6402–6406
[13] Park Y B **2005** An integrated approach for production and distribution planning in supply chain management. *Int J Prod Res* **43** pp 1205–1224
[14] Amini M, and Li H 2011 Supply chain configuration for diffusion of new products: An integrated optimization approach. *Omega* **39** pp 313–322
[15] Chiu M C, and Okudan G 2010 An Integrative Methodology for Product and Supply Chain Design Decisions at the Product Design Stage. In: *Proceedings of the 2010 Industrial Engineering Research Conference*. Epub ahead of print
[16] Chiu M C, and Okudan G 2011 An Integrative Methodology for Product and Supply Chain Design Decisions at the Product Design Stage. *J Mech* **133**
[17] Deng S, Aydin R, and Kwong C K 2014 Integrated product line design and supplier selection: A multi-objective optimization paradigm. *Comput Ind Eng* **70** pp 150–158
[18] ElMaraghy H A and Mahmoudi N 2009 Concurrent design of product modules structure and global supply chain configurations. *Int J Comput Integr Manuf* **22** pp 483–493
[19] Ernst R and Kamrad B 2000 Evaluation of supply chain structures through modularization and postponement. *Eur J Oper Res* **124** pp 495–510
[20] Famuyiwa O and Monplaisir L 2007 An Integrated Framework Matching Product Architecture with Supply Chain Design Policies. *MI* pp 113–122
[21] Gaber Y H and Abdelsalam H M 2015 A multi-objective optimization algorithm for the integrated product line selection and supply chain configuration problem with quality considerations. In: *10th IEEE Int. Conf. on Service Operations and Logistics, and Informatics, SOLI 2015 - In conjunction with ICT4ALL 2015* pp 165–170
[22] Jiang Z, Xuanyuan S, and Li L 2011 Inventory-shortage driven optimisation for product configuration variation. *Int J Prod Res* **49** pp 1045–1060
[23] Nepal B, Monplaisir L, and Famuyiwa O 2012 Matching product architecture with supply chain design. *Eur J Oper Res* **216** pp 312–315
[24] Baud-Lavigne B, Agard B, and Penz B 2012 Mutual impacts of product standardization and supply chain design. *Int J Prod Econ* **135** pp 50–60
[25] El R, Khalaf H, Agard B 2009 Product and supply chain design using a taboo search To cite this version : Product and supply chain design using a taboo search. In: *International Conference on Industrial Engineering and Systems Management*. Montreal
[26] Radwan EHK, Agard B, and Penz B 2011 Simultaneous design of a product family and its related supply chain using a Tabu Search algorithm. *Int J Prod Res* **49** pp 5637–5656
[27] Lamothe J, Hadj-Hamou K, and Aldanondo M 2006 An optimization model for selecting a product family and designing its supply chain. *Eur J Oper Res* **169** pp 1030–1047
[28] Langenberg KU, Seifert RW, and Tancrez J-S 2012 Aligning supply chain portfolios with product portfolios. *Int J Prod Econ* **135** pp 500–513
[29] Metta H and Badurdeen F 2013 Integrating sustainable product and supply chain design: Modeling issues and challenges. *IEEE Trans Eng Manag* **60 pp** 438–446
[30] Moon I, Park KS, and Hao J 2016 Joint decisions on product line selection, purchasing, and pricing. *Eur J Oper Res.*
[31] Rezapour S, Hassani A, and Farahani R Z 2015 Concurrent design of product family and supply

chain network considering quality and price. *Transp Res* **81** pp 18–35

[32] Syam SS and Bhatnagar A 2015 A decision support model for determining the level of product variety with marketing and supply chain considerations. *J Retail Consum Serv* **25** pp 12–21

[33] Ulku S and Schmidt G M 2011 Matching product architecture and supply chain configuration. *Prod Oper Manag* **20** pp 16–31

[34] Wang G, Huang S H, and Dismukes J P 2004 Product-driven supply chain selection using integrated multi-criteria decision-making methodology. *Int J Prod Econ* **91 pp** 1–15

[35] Zhang X, Huang G Q, and Rungtusanatham M J 2008 Simultaneous configuration of platform products and manufacturing supply chains. *Int J Prod Res* **46** pp 6137–6162

[36] Zhang X, Huang G Q, and Humphreys P K 2010 Simultaneous configuration of platform products and manufacturing supply chains: comparative investigation into impacts of different supply chain coordination schemes. *Prod Plan Control* **21** pp 609–627

[37] Fine CH, Golany B, and Naseraldin H 2005 Modeling tradeoffs in three-dimensional concurrent engineering: A goal programming approach. *J Oper Manag* **23** pp 389–403

[38] Huang G Q, Zhang X Y, and Liang L 2005 Towards integrated optimal configuration of platform products, manufacturing processes, and supply chains. *J Oper Manag* **23** pp 267–290

[39] Jafarian M and Bashiri M 2014 Supply chain dynamic configuration as a result of new product development. *Appl Math Model* **38** pp 1133–1146

[40] Shidpour H, Shahrokhi M, and Bernard A 2013 A multi-objective programming approach, integrated into the TOPSIS method, in order to optimize product design; In three-dimensional concurrent engineering. *Comput Ind Eng* **64** pp 875–885

[41] Shidpour H, Bernard A, and Shahrokhi M 2013 A group decision-making method based on intuitionistic fuzzy set in the three dimensional concurrent engineering environment: A multi-objective programming approach. *Procedia CIRP* **7** pp 533–538

IOP Conf. Series: Materials Science and Engineering **337** (2018) 012026 doi:10.1088/1757-899X/337/1/012026

An impact analysis of forecasting methods and forecasting parameters on bullwhip effect

R Y H Silitonga[1*] and N Jelly[1]
[1]Department of Industrial Engineering, Harapan Bangsa Institute of Technology, Bandung, Indonesia

*roland@ithb.ac.id

Abstract. Bullwhip effect is an increase of variance of demand fluctuation from downstream to upstream of supply chain. Forecasting methods and forecasting parameters were recognized as some factors that affect bullwhip phenomena. To study these factors, we can develop simulations. There are several ways to simulate bullwhip effect in previous studies, such as mathematical equation modelling, information control modelling, computer program, and many more. In this study a spreadsheet program named Bullwhip Explorer was used to simulate bullwhip effect. Several scenarios were developed to show the change in bullwhip effect ratio because of the difference in forecasting methods and forecasting parameters. Forecasting methods used were mean demand, moving average, exponential smoothing, demand signalling, and minimum expected mean squared error. Forecasting parameters were moving average period, smoothing parameter, signalling factor, and safety stock factor. It showed that decreasing moving average period, increasing smoothing parameter, increasing signalling factor can create bigger bullwhip effect ratio. Meanwhile, safety stock factor had no impact to bullwhip effect.

1. Introduction

In this globalization era, supply chain growth become faster and competition among supply chains become tighter. Supply chain that can survive is the ones who have good quality in serving and fulfilling customer needs. One of the ways to increase customer service quality is managing supply chain efficiently and effectively. Supply chain system needs collaboration with many parties such as factory, warehouse, distributor, wholesaler, retailer, and customer [1]. Company management has to pay attention about supply chain arrangement because there are many parties and processes involved. There is also uncertainty in demand and fulfilment rate. The most important supply chain processes are ordering and delivering goods [1]. Excess inventory and backlog can happen if supply chain parties concerned only about their own profit and lack of coordination in ordering goods.

From Figure 1, it can be seen that orders quantity become more fluctuated from downstream to upstream of supply chain. It means that in the upstream of supply chain, the value of orders variance amplification become greater. Retailer, wholesaler, distributor, and even manufacturer purposely order more than necessary to their own upstream of supply chain. They do that to anticipate demand fluctuation from downstream of supply chain. This condition is known as bullwhip effect.

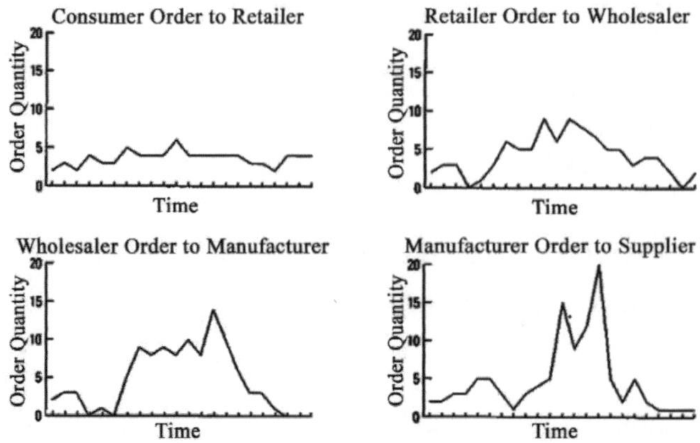

Figure 1. Increasing Orders Variance Amplification from Downstream to Upstream of Supply Chain .

Figure 1 showed that graph is one of the ways to represent bullwhip effect [2]. Bullwhip effect also can be understood and illustrated with physical model, computer program, mathematic equation, and combination of those models. Computer modelling is an effective, efficient, and up-to-date tool for simulation. Bullwhip effect modelling with computer program has been conducted by several previous studies such as Li [3], Lambrecht and Dejonckheere [4], and Boute and Lambrecht [5].

The purpose of the study is analysing impact of forecasting methods and forecasting parameters to the bullwhip effect using Bullwhip Explorer [5]. The remainder of this paper is organized as follows. Section 2 provides theoretical background about bullwhip effect, previous bullwhip effect studies, and Bullwhip Explorer program. Section 3 presents the research method. Section 4 discusses about input data and simulation result. Section 5 concludes the research and presents future research directions.

2. Theoretical Background

Lack of coordination in supply chain occurs either because of different stages of the supply chain have conflicting objectives or because of delayed and distorted information flows between stages [6]. This condition can cause bullwhip effect. Bullwhip effect is a phenomenon where order to the supplier tend to have larger variance than sales to buyer and the distortion propagates upstream of supply chain in amplified form [2]. Chen et al. in [5] stated the bullwhip effect equation as below:

$$\text{Bullwhip effect} = \frac{\text{Variance of Orders}}{\text{Variance of Demand}} \quad (1)$$

There are four causes of the bullwhip effect [2] :

1. Forecasting methods and parameters choice
   If an upstream of supply chain relies only on the order data from its downstream, it will cause lose track of the real demand pattern. In this case there will be double forecasting that can lead to bullwhip effect. Wrong forecasting methods and parameters also can cause bullwhip effect.
2. The rationing and shortage gaming
   Downstream of supply chain sometimes orders to upstream of supply chain more than the actual necessity to defend from lack of goods in the future. When the demand decrease again there will be many cancellations of order and this condition causes bullwhip effect.
3. Order batching
   When inventory of a level of supply chain decreases, that level usually does not order directly because the order is based on batch or accumulated demand. Order batching is done to reduce order cost and transportation cost, but it can result on bullwhip effect because the orders from downstream represent accumulated demand, not the actual demand.
4. Price fluctuation
   One level of supply chain can give promotion in form of discount, bonus, coupon, and many more. This can result on increasing purchase from downstream of supply chain in promotion period

although it has not needed the goods. When the price is returned to normal again, the demand decreases. This condition will cause bullwhip effect and piled-up inventory.

There are several kinds of bullwhip effect research such as empirical study, experiment, analytical study, and mathematical model [7]. Mathematical model can precisely quantify the bullwhip effect and its causes to predict system respond to several kinds of factor. One of the ways to create mathematical model is using simulation. There are some previous bullwhip effect studies using simulation. Li [3] made a controlling mathematical model that show inventory fluctuation resulted from information or connectivity availability rate difference among supply chain parties. Lambrecht and Dejonckhere [4] made a mathematical and graphical model in spreadsheet program for three echelon of supply chain. The input is inventory policy and parameter. The output is demand and order variance of supplier, order variance of wholesaler, total cost, and cycle service level and fill rate.

Boute and Lambrecht [5] made a mathematical and graphical model in a spreadsheet program. Demand parameters that must be defined are mean demand, variance of error, physical lead time, safety factor, unit holding cost per period, unit backlog cost per period, unit switching cost period, and safety stock factor. Forecasting methods that can be used are mean demand, moving average, exponential smoothing, demand signal processing, and minimum expected mean squared error. Forecasting parameters that can be defined are moving average period, smoothing parameter, signalling factor, and safety stock factor. The output is bullwhip effect ratio, net stock amplification, cycle service level, fill rate, average inventory cost per period, average switching cost per period, and bullwhip effect graphs.

3. Research Method

The study used Bullwhip Explorer spreadsheet program to analyse impact of forecasting methods and forecasting parameters to the bullwhip effect. We built two scenarios in this study. Scenario 1 was built to know impact of forecasting methods and their own parameters to bullwhip effect and other output. In this scenario, forecasting method was changed so we would know its impact to bullwhip effect. Some forecasting methods have its own forecasting parameter so the forecasting parameter was also changed to know its impact to bullwhip effect. Scenario 2 was built to know impact of safety stock factor to bullwhip effect and other output. In this scenario, safety stock factor was changed for each forecasting method. There were further explanations about the scenarios in section 4.

Total repetitions for each scenario with each method are thirty-five. Then, there was analysis impact of forecasting methods and parameters to bullwhip effect and other output such as net stock amplification, cycle service level, fill rate, average inventory cost/period, and average switching cost/period. Finally, the conclusion was drawn from result analysis.

4. Simulation and Analysis

4.1 Scenarios

Simulation of some scenarios is needed to understand impact of forecasting methods and forecasting parameters to bullwhip effect and other output. Each scenario and method is simulated thirty-five times. Demand data that are used are hypothetical data which are assumed as cassava chips demand at distributor. For each scenario, there are same input data as shown in Table 1:

Table 1. Same Input Data for Scenario 1 and 2.

Input Type	Input Data
Order policy	Standard Order-Up-To
Demand pattern	Independent and Identically Distributed (IID)
Mean demand	110 kg per day
Variance of error term	1 kg
Physical lead time	3 days
Unit holding cost per period	0.5 (0.5x1,000 = Rp 500)
Unit backlog cost per period	20 (20x1,000 = Rp 20,000)
Unit switching cost per period	2 (2x1,000 = Rp 2,000)

4.1.1. Scenario 1.

The purpose of scenario is to know impact of forecasting methods and their own parameters to bullwhip effect and other output. Safety stock factor that is used in scenario 1 is 1.645. Moving average period is a parameter of moving average. Smoothing parameter is a parameter of exponential smoothing and range from 0 to 1. Signaling factor is a parameter of demand signal processing and range from 0 to 1. Minimum expected mean squared error (MSE) also has autocorrelation coefficient as a forecasting parameter, but the value of that parameter is 0 if the demand pattern is IID. The input data for scenario 1 can be seen in Table 2.

Table 2. Input Data for Scenario 1.

	Method 1	**Method 2**					**Method 3**					**Method 4**					**Method 5**
Forecasting Method	Mean Demand	Moving Average					Exponential Smoothing					Demand Signal Processing					Minimum Expected MSE
Forecasting Parameter	-	Moving Average Period (Tm)					Smoothing Parameter					Signalling Factor					Autocorrelation Coefficient
		1	3	5	10	20	1	0.75	0.5	0.25	0.05	1	0.75	0.5	0.25	0.05	0

4.1.2. Scenario 2.

The purpose of this scenario is to know impact of safety stock factor to bullwhip effect and other output. Five forecasting methods with its own forecasting parameter are also used in this scenario. Each forecasting method is simulated with three different safety stock factor. The input data for scenario 2 can be seen in Table 3.

Table 3. Input Data for Scenario 2.

	Safety Stock Factor 1	**Safety Stock Factor 2**	**Safety Stock Factor 3**
Safety Stock Factor Value	2.326	1.645	0.842
Forecasting Method and Forecasting Parameter	1) Mean Demand 2) Moving Average (moving average period = 5) 3) Exponential Smoothing (smoothing parameter= 1) 4) Demand Signal Processing (signalling factor = 1) 5) Minimum Expected MSE		

4.2 Simulation Result

The average simulation output value of scenario 1 can be seen in Table 4. The average simulation output value of scenario 2 can be seen in Table 5.

Table 4(a). Average Simulation Output Value of Scenario 1.

Output	**Method 1 Mean Demand**	**Method 2 Moving Average Parameter = Moving Average Period (Tm)**					**Method 3 Exponential Smoothing Parameter = Smoothing Parameter (α)**				
		1	**3**	**5**	**10**	**20**	**1**	**0.75**	**0.5**	**0.25**	**0.05**
Bullwhip (analytical)	1.00	41.00	7.22	3.88	2.12	1.48	41.00	21.40	10.33	4.14	1.44
Bullwhip (simulated)	1.00	40.64	7.37	4.00	2.27	1.59	41.51	21.54	10.46	4.29	1.55
NSA (analytical)	4.00	20.00	9.33	7.20	5.60	4.80	20.00	13.60	9.33	6.29	4.41
NSA (simulated)	3.93	19.90	9.42	7.13	5.72	4.82	19.95	13.57	9.61	6.30	4.47
Cycle Service Level (%)	98.60	83.59	90.43	93.81	96.05	97.41	83.40	87.39	90.97	94.90	97.54
Fill Rate (%)	99.98	99.56	99.83	99.90	99.94	99.96	99.56	99.72	99.83	99.92	99.97

Average Inventory Cost/period	2.37	12.01	5.81	4.12	3.18	2.66	12.09	8.36	5.74	3.71	2.59
Average Switching Cost/period	2.24	17.91	6.35	4.67	3.48	2.85	18.32	12.58	8.34	5.12	2.84

Table 4(b). Average Simulation Output Value of Scenario 1.

Output	Method 4 Demand Signal Processing Parameter = Signalling Factor (χ)					Method 5 Minimum Expected MSE Parameter = Autocorrelation Coefficient (ρ)
	1	**0.75**	**0.5**	**0.25**	**0.05**	**ρ=0**
Bullwhip (analytical)	5.00	3.63	2.50	1.63	1.11	1.00
Bullwhip (simulated)	5.00	4.26	2.63	1.42	1.00	1.00
NSA (analytical)	5.00	4.56	4.25	4.06	4.00	4.00
NSA (simulated)	5.01	4.74	4.23	3.99	4.02	3.96
Cycle Service Level (%)	97.26	97.58	97.80	98.35	98.38	95.30
Fill Rate (%)	99.96	99.97	99.97	99.98	99.98	99.93
Average Inventory Cost/period	2.84	2.72	2.55	2.45	2.46	2.87
Average Switching Cost/period	6.20	5.75	4.22	2.71	2.28	2.26

Table 5(a). Average Simulation Output Value of Scenario 2 (Part 1).

Output	Method 1 Mean Demand			Method 2 Moving Average			Method 3 Smoothing		
	2.326	**1.645**	**0.842**	**2.326**	**1.645**	**0.842**	**2.326**	**1.645**	**0.842**
Bullwhip (analytical)	1.00	1.00	1.00	3.88	3.88	3.88	41.00	41.00	41.00
Bullwhip (simulated)	1.00	1.00	1.00	3.98	4.00	4.00	40.97	41.51	41.21
NSA (analytical)	4.00	4.00	4.00	7.20	7.20	7.20	20.00	20.00	20.00
NSA (simulated)	3.98	3.93	3.94	7.31	7.13	7.19	19.86	19.95	19.70
Cycle Service Level (%)	99.61	98.60	89.32	96.46	93.81	78.07	88.39	83.40	70.34
Fill Rate (%)	100	99.98	99.85	99.93	99.90	99.57	99.70	99.56	99.07
Average Inventory Cost/period	2.59	2.37	4.44	3.56	4.12	10.44	9.24	12.09	21.97
Average Switching Cost/period	2.23	2.24	2.25	4.68	4.67	4.63	18.22	18.32	18.39

Table 5(b). Average Simulation Output Value of Scenario 2.

Output	Method 4 Demand Signal Processing			Method 5 Minimum Expected MSE		
	2.326	**1.645**	**0.842**	**2.326**	**1.645**	**0.842**
Bullwhip (analytical)	5.00	5.00	5.00	1.00	1.00	1.00
Bullwhip (simulated)	4.97	5.04	5.02	1.00	1.00	1.00
NSA (analytical)	5.00	5.00	5.00	4.00	4.00	4.00
NSA (simulated)	4.98	4.94	4.98	3.87	3.96	3.84
Cycle Service Level (%)	99.16	97.50	86.05	99.65	95.30	89.38
Fill Rate (%)	99.99	99.97	99.78	100.00	99.93	99.86
Average Inventory Cost/period	2.76	2.80	6.02	2.60	2.87	4.28
Average Switching Cost/period	6.10	6.18	6.25	2.25	2.26	2.31

From scenario 1 result, it can be seen that if the forecasting method is mean demand, bullwhip ratio is one means that there is no bullwhip effect. It is because of the steadiness of the forecasting result, which is average of past demand. The quantity is only added if there is unfulfilled demand. Because of that, variance of order and variance of demand have same value and do not result on bullwhip effect.

Moving average forecasting method has impact to bullwhip effect. Bigger moving average period cause smaller bullwhip effect ratio, smaller net stock amplification, bigger cycle service level, bigger fill rate, smaller average inventory and switching cost per period. It is because forecasting uses more past demand data when the moving average period is bigger so forecasting result will approach mean demand. Exponential smoothing method has impact to bullwhip effect. Smaller smoothing parameter cause smaller bullwhip effect, smaller net stock amplification, bigger cycle service level, bigger fill rate, smaller average inventory and switching cost per period. It is because if the smoothing parameter value near zero, it will give less weight to single past demand.

Demand signal processing method has impact to bullwhip effect. Smaller signalling parameter cause smaller bullwhip effect, smaller net stock amplification, bigger cycle service level, bigger fill rate, smaller average inventory and switching cost per period. It is because if signalling factor value near zero, the order will not be set by the change of demand. Minimum expected mean squared error with zero autocorrelation coefficient does not have impact on bullwhip effect because customer demand variability is reduced, not amplified.

From scenario 2, it can be seen that smaller signalling parameter cause no significant effect to bullwhip effect, no significant effect to net stock amplification, smaller cycle service level, smaller fill rate, bigger average inventory per period and no significant effect to switching cost per period. It is because safety stock factor only affects safety stock quantity and does not affect forecasting result.

5. Conclusion and Future Research

Forecasting methods and forecasting parameters have impact on bullwhip effect. Mean demand forecasting method does not have impact on bullwhip effect because the forecasting result is always steady. Moving average forecasting method has impact on bullwhip effect but can be reduced by using bigger moving average period or more past demand data. Exponential smoothing forecasting method has impact on bullwhip effect but can be reduced by using smaller smoothing parameter because it will give less weight to single past demand. Demand signal processing has impact to bullwhip effect but can be reduced by using smaller signalling factor because the order will not be set by the change of demand. Minimum expected mean squared error with zero autocorrelation coefficient does not have impact on bullwhip effect because customer demand variability is reduced.

Future research about bullwhip effect in more than two echelon using Bullwhip Explorer or other simulation model can be conducted to learn more about the phenomenon. This research only considered one of four bullwhip effect causes which is forecasting methods and parameters choice so in the future, the three other causes should be included. In the future, raw data can also be the input so demand pattern between different situations can be observed and differentiated even though the mean and variance of those situations are the same.

6. References

[1] Makajić-nikolić D, Panić B and Vujošević M 2004 *Bullwhip Effect and Supply Chain Modelling and Analysis Using CPN In: Jensen K.(Ed.) Proc. of the Fifth Workshop and Tutorial on Practical Use of Coloured Petri Nets and the CPN Tools* (Aarhus: Denmark) pp 219–34.

[2] Lee H L, Padmanabhan V and Whang S 1997 *The bullwhip effect in supply chains Sloan Management Review* **38** pp 93–102

[3] Li C 2012 *Controlling the bullwhip effect in a supply chain system with constrained information flows Applied Mathematical Modelling* **37** pp 1897–1909

[4] Lambrecht M R and Dejonckheere J 1999 *A bullwhip effect explorer Research Report 9910* (Leuven, Belgium: Department of Applied Economics, Katholieke Universiteit)

[5] Boute R N and Lambrecht M R 2009 *Exploring the bullwhip effect by means of spreadsheet simulation INFORMS Transaction on Education* **10** pp 1–9

[6] Chopra S and Meindl P 2009 *Supply Chain Management: Strategy, Planning, and Operation* (New Jersey: Pearson Education, Inc.) vol **4** pp 483–5

[7] Wang X and Disney S M 2015 *The bullwhip effect: progress, trends and directions European Journal of Operational Research* **250** pp 691-701

IOP Conf. Series: Materials Science and Engineering **337** (2018) 012027 doi:10.1088/1757-899X/337/1/012027

Integration of three echelon supply chain (supplier-manufacturer-distributor-drop shipper) with permissible delay in payment and penalty contract

M F Ibrahim[1*] and Suparno[1]
[1] Department of Industrial Engineering, Institut Teknologi Sepuluh Nopember, Kampus ITS Sukolilo-Surabaya 60111, Indonesia

*faisalibrahim.ie@gmail.com

Abstract. Supply Chain Management (SCM) has to be considered in the company in order to improve the sustainability and competitiveness. SCM executed to integrating any companies on the supply chain in a way of coordinating the flow of goods, information, and financial. Permissible delay in payment is one of the coordination ways with allowing the costumers delay the payments to a vendor in some certain periods without any interest charges. In the supply chain system, drop-shipping player already familiar in this era. In drop-shipping internet retailing, the supplier will hold supplies and also carry out physical distribution service on behalf of drop-shipper. Drop-shipper will just focus on selling, on the other hand, their supplier will be responsible for the physical process. Generally, drop-shipper have information of the customer demands better than the distributor. But, it is also unrare when the drop-shipper send the estimation of demands which bigger than their own estimation in order to maximize their own interest, so they hope supplies of the distributor will always enough to accommodate their demands. Contributions in this research will be focused on integration of three echelons supply chain, which are the supplier, manufacturer, distributor, and drop-shipper. With considering delay in payment on first and second echelons, and also the contract penalty on third echelon. The problem on this research will be modeled in some kind of cases which can represent the problem of real supply chain system. Sensitivity analysis will be done on certain significant variables toward the changes of total supply chain cost. Coordination with delay in payment success to integrate supply chain. Contract penalty plan success to maintain the profit of distributor and drop-shipper.

1. Introduction

Supply Chain Management (SCM) has to be considered in the company in order to improve the sustainable and competitiveness. According to Seuring [1], cooperation is the only way for a company to increase supply competitiveness. SCM's ran to integrate the companies in a supply chain by meansof coordinating the flow of goods, information, and financial with the aim of fulfilling consumer demands and competitive supply chain [2]. With focusing to the sustainable supply chain management, a company will realize if the inventory in every supply chains will be more efficient if it is maintained with a better cooperation and coordination [3]. It is difficult for share complete information in decentralization supply chain, due to the majority of members are independent company which focus on maximizing their own benefits [4]. Different with centralization which allows every player in supply chain to cooperate together in the decision making. As an example according to Glock [5], coordination decisions in term of filling the supplies in a chain supply system can improve the channel efficiency and the position of the company which involves there. Centralization scenario and joint decision making will run in line to increase the total benefits of supply chain [6].

Delay in payment is one way to coordinate with allowing the costumers to delay their payment to the vendor in some certain period without any interest expenses [7]. Delay in payment will be given by the seller to buyers for anticipating the decreasing amount of inventory in the downstream. With providing them that policy also will attract buyers to buy in a bigger lot from their economic order quantity. With the improvement of buyer order quantity, then the inventory will move to the buyers [8]. For buyers, a benefit exists because buyers get time before they have to pay. According to Goyal **Error! Reference source not found.** when the supplier gives certain period for delaying the payment, indirectly supplier provide their customer loan without any interest expenses until the period ended. In result, generally, costumers will hold that until the end of period.

The online retail sale is estimated will always grow, we know 7% of general retail sales in 2011 will be becoming 9% in 2016 [10]. In drop-shipping internet retailing, the supplier will hold supplies and also carry out physical distribution service on behalf of drop-shipper [11]. Thus, drop-shipper will only focus on sales, and in the other hand physical process will always be handled by their supplier. Drop-shipping channel becomes more efficient rather than traditional channel because generally traditional channel usually lacked coordination [12]. In the previous research Gan et al. [13] stated if generally drop-shipper have the information of customer demands which is better that distributor. It is very common when the drop-shipper send the estimation of demands which bigger than their own estimation in order to maximize their own interest, so they hope supplies of the distributor will always enough to accommodate their demands. For prevent that kind of cases, they have to establish a commitment penalty contract.

In some decades ago, research which did by Goyal **Error! Reference source not found.** success to develop the model of economic order quantity with consideration of delay in payment. The research focused on the decision variables of the optimal quantity. Research on delay in payment developed frequently, the researcher tries to analyze some different situation of supplies and assumption. In some research about delay in payment which focused on the purpose function of deciding optimal lot size already discussed by Chung et al. [14], Nurshanti [15], Glock [5], and Jaber and Osman [7]. Some years ago, Aljazzar et al. [8] tried to develop delay in payment model of one echelon with a long term purpose. Then, Aljazzar et al. [3] tried to develop that delay in payment model into two echelons. Recently, Marchi et al. [6] developed investment sharing model which beneficial to both vendor and buyer. The development of strategic model for determination of price and size order on chain supply with drop-shipping has been done by Chiang and Feng [12]. On the research which did by Gan et al. [13], developed a penalty contract model which has a purpose for the supplier to get demand retailed information that more valid and expected profit supplier can be optimized. In another research, Khouja and Stylianou [16] developed inventory model with drop-shipping option. In the research which did by Edirisinghe and Atkins [17], echelon meant as a relation among two players. For example, supplier and distributor called as one echelon. Different with the research which did by Heydari et al. [18], echelon meant as a number of players which exist inside chain supply system. In this research, the echelon is defined as a relation between two players.

This research based on two different topics which have gap between each other. The first topic is the research on delay in payment and the second topic is the research about drop – shipping. Previous research discussed lots of delay in payment in some cases, explicitly can be seen on Table 1. The concept of one level delay in payment with one-way payment method already developed in some previous research, such as Aljazzar et al. [8]; Chung et al. [14]; Glock [5]; [7]. In the other hand, second concept of delay in payment level with one way of payment also already developed in some previous research, such as Aljazzar et al. [3]; Nurshanti [15]; Moussawi-Haidar et al. [19]. Most of the inventory system which used in previous research is EOQ or EPQ inventory system. Delay in payment model on EOQ inventory system also already developed in some previous research, such as Aljazzar et al. [8]; Aljazzar et al. [3]; Chung et al. [14]; Glock [5]; Jaber and Osman [7]; Jaber et al. [20]; and Nurshanti [15] on EPQ.

Table 1. Input parameters.

No.	Researcher	Delay in Payment		Inventory System		Payment Method
		1 Level	2 Level	EOQ	EPQ	1 Method
1	Aljazzar et al. [8]	√		√		√
2	Chung et al. [14]	√		√		√
3	Glock [5]	√		√		√
4	Jaber and Osman [7]	√		√		√
5	Aljazzar et al. [3]		√	√		√
6	Nurshanti [15]		√		√	√
7	Moussawi-Haidar et al. [19]		√	√		√

This research will refer to research by Aljazzar et al. [3] in model development because having the second system of delay in payment level which are supplier – manufacturer, manufacturer – retailer. That condition can represent chain supply system of first and second echelon on this research. Previous researches on delay in payment like something above generally between traditional channel, for example, the one which given by manufacturer to distributor or distributor to retailer. By looking at the conditions of the channel in the current era which is busy on the entering of new player, drop-shipper, so that things also important to be researched. How the way of coordination on supply chain which also membered by the drop – shipper.

Drops–shipper has some superiority on marketing because they just focus on product disposal without any needed to think about the physical distribution of goods. The supplier that supplies the drop–shipper will become the holder which control the stock. It often happens when the information of demand which given by drop–shipper to supplier create some disadvantages because the information is not valid. In the other hand, it also often when the supplier cannot fulfill the demand of drop – shipper, so it is creating shortage cost and some disadvantages to drop – shipper. From some previous research about drop – shipping, research successfully discussed that topic in some cases, such as:

1. Model concept of stock with drop – shipper as a member Chiang and Feng [12]; Khouja and Stylianou [16].
2. Concept of penalty contract for coordinating supplier and drop–shipper Gan et al. [13].

Some penalty policies Gan et al. [13] can be used to coordinating the third echelon on developing the model in this research. The current research tries to develop penalty contract which can accommodate win–win interactions between supplier and drop – shipper, so two sides can feel benefited. Because drop–shipper does not have any inventory, so the invemtory system of the player one level above drop–shipper will be different compare with any other traditional player. Inventory model Khouja and Stylianou [16] will be used by the player one level above drop–shipper in developing the model in this research.

Focus on some research above as the primary reference, such as Aljazzar et al. [3]; Gan et al. [13]; Jaber et al. [20]; Khouja and Stylianou [16]. The current research will combine some relevant model in that researchers. Thus, will develop the model which can be applied in supply chain system on this research. Focus of this research is on coordination between players in supply chain of three echelons with one of the player is drop–shipper. One first and second echelon will create coordination model with delay in payment as a consideration and on the third echelon will create coordination model with penalty contract as a consideration.

Contribution in this research will be focused on integration of three echelons supply chain, which are the supplier, manufacturer, distributor, and drop-shipper. The consideration which will be used is coordination by delay in payment on first and second echelons, and also coordination by penalty contract on third echelons. The problem on this research will be modeled and resolved by assuming general cycle time. With using simulation of some kind of cases which can represent the problem of supply chain system that already explained previously, will be counted the total cost of supply chain in all cases. Then, the influence of every each independent variables in total cost of supply chain under some scenario cases will be analyzed. At the end, the sensitivity analysis will be applied on certain significant variables toward the changes of total supply chain cost. Thus, the best scenario can identify from each player in supply chain.

2. Notations and assumptions

Based on performance criteria and decision variables which have been formulated, current research define the parameters that will be used to build mathematical models in this research as mentioned below:

i : Supply Chain Player (s: supplier, m: manufacturer, d: distributor, and ds: drop-shipper).
j : Type of supplies level (w: raw material, f: finished goods).
$A_{i,j}$: Setup / order cost which issued by player i toward the item j.
$C_{i,j}$: Production / purchasing cost per item which issued by player i toward the item j.
$h_{i,j}$: Financial holding cost per item which issued by player i toward the item j.
$S_{i,j}$: Physical (storage) holding cost per item which issued by player i toward the item j.
Q : Distributor order quantity.
Q_{ds} : Drop-shipper order quantity.
n_1 : Total of shipments by supplier to manufacture per manufacturing raw material cycle.
n_2 : Total of shipments by manufacture to distributor per distributor cycle.
α : Total of raw materials needed to producing one finished product.
t_i : Period of delay in payments offered by player i.
τ_i : Time of payments which done by player i.
k_i : Return of investment for player i.
P : The annual production rate of manufacturing.
D : The annual distributor demand $D<P$.
T : General cycle length $=\frac{n_2 Q}{D}$.
T_s : Supplier cycle length $=\frac{n_2 Q}{P}$.
T_w : Manufacturing raw materials cycle length $=\frac{n_2 Q}{n_1 P}$.
T_m : Finished product manufacturing cycle length $=\frac{n_2 Q}{D}$.
T_d : Distributor cycle length $=\frac{Q}{D}$.
m : Minimum drop-shipper purchase commitment.
v : Shortage penalty.
w : Less than contract penalty.

In order to simplify the formation of model, several assumptions are defined which will be used for this model development, such as:

- One supplier, one manufacture, one distributor, and one drop-shipper.
- Demands on manufacture and distributor is deterministic and constant over time.
- Supplier production level higher than manufacture demand for raw material, and manufacture production level higher than distributor demand.
- Only consider one product.
- All of the shipment in supply chain have the same size.
- There are two components of holding cost: financial and physical holding costs.

IOP Conf. Series: Materials Science and Engineering **337** (2018) 012027 doi:10.1088/1757-899X/337/1/012027

- The period of delay in payment is offered by supplier to manufacture and manufacture to distributor.
- Period of delay in payment become the decision variables.
- Manufacture and distributor invest what their vendor debt (supplier or manufacture) $C_{i,j}Q$ in free risk investment during the permitted period.
- Distributor and manufacture pay the debt with one single payment τ.
- Manufacture has two warehouse: for raw materials and finished goods.
- Drop-shipper demand level is deterministic.
- Demand on manufacture and distributor is deterministic and constant over time.
- Not consider the product shipment cost.
- Penalty based on time on echelon 1, 2, and based on quantity on echelon 3.

3. Mathematical model

In this part, the mathematical model is formulated which has a purpose to get the optimal decision variables with some certain consideration that has been described in the performance criteria. Other considerations which also be a concern are cost and income component of each player in the supply chain. In this research will be considered the general length of cycle in formulating a mathematical model, where the length of cycle is $T=\frac{n_2Q}{D}$. Manufacture will ship the same amount of interval adjusted for production of items for the distributor through ongoing cycle including during production. Manufacture begins production when the supplies empty. It is also assumed that manufacturer can do the shipment to buyers along with ongoing production processes. The proposed mathematical model will be divided into some cases. Each case is a combination of sub-case of echelon 1 (supplier and manufacturer), echelon 2 (manufacturer and distributor), and echelon 3 (distributor and drop–shipper).

3.1. Scenario 1: $0\leq t_s=\tau_m\leq\frac{n_2Q}{Pn_1}$ and $0\leq t_m=\tau_d\leq\frac{Q}{D}$ and $m=Q>Q_{ds}$

In this case, echelon 1 assumed if supplier provide the manufacture certain period of time t_s for finishing the payment without any interest expenses. Manufacture do the payment τ_m at the last time of the period t_s. On echelon 2, assumed if manufacture provides the distributor certain period t_m to finish the payment without any interest expenses. Distributor do the payment τ_d at the last time of the period t_m. On echelon 3, assumed if in one general cycle $T=\frac{n_2Q}{D}$ distributor demand equal with the commitment quantity which already been agreed by the drop-shipper before. But, the amount of demand from the drop-shipper less than commitment which has been agreed before. The annual total cost of supplier which include setup cost, production, physical holding cost, financial holding cost, and opportunity cost can be formulated as:

$$\psi^1_{s,f}=\frac{A_{s,f}D}{n_2Q}+C_{s,f}\alpha D+\frac{n_1(n_1-1)}{2}\left(h_{s,f}+S_{s,f}\right)\frac{\alpha n_2QD}{Pn_1{}^2}+h_{s,f}\tau_m\alpha D+\left(C_{m,w}-C_{s,f}\right)e^{k_st_s}\alpha D \tag{1}$$

Manufacture invest $C_{m,w}Q$ which haven't paid to supplier in free risk investation. Cost component which related with the annual manufacture raw materials cost include ordering, purchasing, physical holding cost, financial holding cost, and the income of investation interest expenses can be formulated as :

$$\psi^1_{m,w}=\frac{n_1A_{m,w}D}{n_2Q}+\frac{C_{m,w}\alpha n_2QD}{n_2Q}+\frac{h_{m,w}\alpha^2n_2{}^2Q^2D}{2\alpha Pn_1n_2Q}-\frac{h_{m,w}\alpha n_2Qt_sD}{n_2Q}+\frac{h_{m,w}n_1\alpha Pt_s{}^2D}{2n_2Q}+\frac{S_{m,w}\alpha n_2{}^2Q^2D}{2Pn_1n_2Q}-\frac{C_{m,w}\alpha n_2Qe^{k_m\tau_m}D}{n_2Q} \tag{2}$$

Manufacture providing distributor certain period t_m for finishing the payment $C_{d,f}Q$ without any interest expenses. However, it will cause the opportunity cost in manufacture with comparison if distributor total debt invested. Cost component which related with the annual cost finish manufacture product include setup, production, physical holding cost, financial holding cost, and opportunity cost can be formulated as:

$$\psi^1_{m,f}=\frac{A_{m,f}D}{n_2Q}+C_{m,f}D+\left(h_{m,f}+S_{m,f}\right)\left(\frac{Q(2D+(P-D)n_2-P)}{2P}\right)+h_{m,f}\tau_dD+\left(C_{d,f}-C_{m,f}\right)e^{k_mt_m}D \tag{3}$$

Distributor invest $C_{d,f}Q$ which haven't paid to manufacture in free risk investation. Distributor also get compensation from the drop-shipper because of the demands which less than the commitment which has been agreed before. Cost component which relate with the annual cost distributor product include ordering, purchasing, physical holding cost, financial holding cost, investation interest expenses income, and contract penalty compensation income can be formulated as:

$$\psi^1_{d,f}=\frac{A_{d,f}D}{Q}+C_{d,f}D+\frac{h_{d,f}(Q-Dt_m)^2}{2Q}+\frac{S_{d,f}Q}{2}-C_{d,f}D\left(1-e^{k_d t_m}\right)-\frac{w(m-Q_{ds})D}{n_2Q} \quad (4)$$

With the existance of contract penalty which agreed by distributor and drop-shipper, so drop-shipper will charged because in one general cycle $T=\frac{n_2Q}{D}$ demands from the drop-shipper less than commitment which has been agreed before. Drop-shipper doesn't has inventory cost component because the product which sold will directly sent by distributor. Cost component which related with the annual drop-shipper cost product include ordering, purchasing, and penalty charge can be formulated as:

$$\psi^1_{ds,f}=\frac{Q_{ds}A_{ds,f}D}{n_2Q}+\frac{C_{ds,f}Q_{ds}D}{n_2Q}+\frac{w(m-Q_{ds})\,D}{n_2Q} \quad (5)$$

The total annual cost of the system can be searched by summing up the total annual fees of each player:

$$\psi^1_{SC}=\frac{A_{s,f}D}{n_2Q}+C_{s,f}\alpha D+\frac{n_1(n_1-1)}{2}\left(h_{s,f}+S_{s,f}\right)\frac{\alpha n_2QD}{Pn_1^2}+h_{s,f}\tau_m\alpha D+\left(C_{m,w}-C_{s,f}\right)e^{k_s t_s}\alpha D+\frac{n_1A_{m,w}D}{n_2Q}+$$
$$\frac{C_{m,w}\alpha n_2QD}{n_2Q}+\frac{h_{m,w}\alpha^2n_2^2Q^2D}{2\alpha Pn_1n_2Q}-\frac{h_{m,w}\alpha n_2Qt_sD}{n_2Q}+\frac{h_{m,w}n_1\alpha Pt_s^2D}{2n_2Q}+\frac{S_{m,w}\alpha n_2^2Q^2D}{2Pn_1n_2Q}-\frac{C_{m,w}\alpha n_2Qe^{k_m\tau_m}D}{n_2Q}+$$
$$\frac{A_{m,f}D}{n_2Q}+C_{m,f}D+\left(h_{m,f}+S_{m,f}\right)\left(\frac{Q(2D+(P-D)n_2-P)}{2P}\right)+h_{m,f}\tau_dD+\left(C_{d,f}-C_{m,f}\right)e^{k_m t_m}D+\frac{A_{d,f}D}{Q}+C_{d,f}D+$$
$$\frac{h_{d,f}(Q-Dt_m)^2}{2Q}+\frac{S_{d,f}Q}{2}-C_{d,f}D\left(1-e^{k_d t_m}\right)-\frac{w(m-Q_{ds})D}{n_2Q}+\frac{Q_{ds}A_{ds,f}D}{n_2Q}+\frac{C_{ds,f}Q_{ds}D}{n_2Q}+\frac{w(m-Q_{ds})\,D}{n_2Q} \quad (6)$$

3.2. Scenario 2 : $0\le t_s=\tau_m\le\frac{n_2Q}{Pn_1}$ and $0\le t_m<\tau_d\le\frac{Q}{D}$ and $Q<Q_{ds}\le m$

In this case, echelon 1 assumed if supplier provide manufacture relief in certain period of time t_s for finishing the payment which out any interest expenses. Manufacture do the payment at the end of period t_s. On echelon 2, assumed if manufacture provide distributor relief in certain period t_m for finishing the payment without interest expenses. Distributor do the payment τ_d after the end of the period t_m and before receiving the next shipping. On the echelon 3, assumed if in one general cycle $T=\frac{n_2Q}{D}$ distributor demands less than the commitment quantity which has been agreed by drop-shipper before. It turns out that the drop-shipper demand is in line with the commitment and more than the distributor demand. Thus, the total annual cost of the system is :

$$\psi^2_{SC}=\frac{A_{s,f}D}{n_2Q}+C_{s,f}\alpha D+\frac{n_1(n_1-1)}{2}\left(h_{s,f}+S_{s,f}\right)\frac{\alpha n_2QD}{Pn_1^2}+h_{s,f}\tau_m\alpha D+\left(C_{m,w}-C_{s,f}\right)e^{k_s t_s}\alpha D+\frac{n_1A_{m,w}D}{n_2Q}+$$
$$\frac{C_{m,w}\alpha n_2QD}{n_2Q}+\frac{h_{m,w}\alpha^2n_2^2Q^2D}{2\alpha Pn_1n_2Q}-\frac{h_{m,w}\alpha n_2Qt_sD}{n_2Q}+\frac{h_{m,w}n_1\alpha Pt_s^2D}{2n_2Q}+\frac{S_{m,w}\alpha n_2^2Q^2D}{2Pn_1n_2Q}-\frac{C_{m,w}\alpha n_2Qe^{k_m\tau_m}D}{n_2Q}+$$
$$\frac{A_{m,f}D}{n_2Q}+C_{m,f}D+\left(h_{m,f}+S_{m,f}\right)\left(\frac{Q(2D+(P-D)n_2-P)}{2P}\right)+h_{m,f}\tau_dD+\left(C_{d,f}-C_{m,f}\right)e^{k_m t_m}D-C_{d,f}e^{k_m(\tau_d-t_m)}D+$$
$$\frac{A_{d,f}D}{Q}+C_{d,f}D+\frac{h_{d,f}(Q-D_dt_m)^2}{2Q}+\frac{S_{d,f}Q}{2}+C_{d,f}e^{k_m(\tau_d-t_m)}D+\frac{v(Q_{ds}-Qn_2)D}{n_2Q}-C_{d,f}D\left(1-e^{k_d t_m}\right)+\frac{Q_{ds}A_{ds,f}D}{n_2Q}+$$
$$\frac{C_{ds,f}Q_{ds}D}{n_2Q}-\frac{v(Q_{ds}-Qn_2)D}{n_2Q} \quad (7)$$

3.3. Scenario 3: $0 \leq t_s < \tau_m \leq \frac{n_2 Q}{Pn_1}$ and $0 \leq t_m = \tau_d \leq \frac{Q}{D}$ and $m = Q > Q_{ds}$

In this case, echelon 1 assumed if supplier provide manufacture relief in certain period of time t_s for finishing the payment without any interest expenses. Manufacture do the payment τ_m after the end of period t_s and before receiving the next shipping. On the echelon 2, assumed if manufacture provide distributor relief in certain period t_m for finishing the payment without any interest expenses. Distributor do the payment τ_d at the end of period t_m. On echelon 3, assumed if in one general cycle $T = \frac{n_2 Q}{D}$ distributor demands equal with the commitment quantity which has been agreed by drop-shipper before. But demand from the drop-shipper less than the commitment which has been agreed before. Thus, the total annual cost of the system is :

$$\psi_{SC}^3 = \frac{A_{s,f}D}{n_2 Q} + C_{s,f}\alpha D + \frac{n_1(n_1-1)}{2}(h_{s,f}+S_{s,f})\frac{\alpha n_2 QD}{Pn_1{}^2} + h_{s,f}\tau_m \alpha D + (C_{m,w}-C_{s,f})e^{k_s t_s}\alpha D -$$

$$C_{m,w}e^{k_s(\tau_m - t_s)}\alpha D + \frac{n_1 A_{m,w}D}{n_2 Q} + \frac{C_{m,w}\alpha n_2 QD}{n_2 Q} + \frac{h_{m,w}\alpha^2 n_2{}^2 Q^2 D}{2\alpha Pn_1 n_2 Q} - \frac{h_{m,w}\alpha n_2 Q t_s D}{n_2 Q} + \frac{h_{m,w}n_1 \alpha P t_s{}^2 D}{2n_2 Q} +$$

$$\frac{S_{m,w}\alpha n_2{}^2 Q^2 D}{2Pn_1 n_2 Q} + \frac{C_{m,w}\alpha n_2 Q e^{k_s(\tau_m - t_s)}D}{n_2 Q} - \frac{C_{m,w}\alpha n_2 Q e^{k_m \tau_m}D}{n_2 Q} + \frac{A_{m,f}D}{n_2 Q} + C_{m,f}D +$$

$$(h_{m,f}+S_{m,f})\left(\frac{Q(2D+(P-D)n_2-P)}{2P}\right) + h_{m,f}\tau_d D + (C_{d,f}-C_{m,f})e^{k_m t_m}D + \frac{A_{d,f}D}{Q} + C_{d,f}D + \frac{h_{d,f}(Q-Dt_m)^2}{2Q} +$$

$$\frac{S_{d,f}Q}{2} - C_{d,f}D(1-e^{k_d t_m}) - \frac{w(m-Q_{ds})D}{n_2 Q} + \frac{Q_{ds}A_{ds,f}D}{n_2 Q} + \frac{C_{ds,f}Q_{ds}D}{n_2 Q} + \frac{w(m-Q_{ds})D}{n_2 Q} \quad (8)$$

3.4. Scenario 4: $0 \leq t_s < \tau_m \leq \frac{n_2 Q}{Pn_1}$ and $0 \leq t_m < \tau_d \leq \frac{Q}{D}$ and $Q < Q_{ds} \leq m$

In this case, echelon 1 assumed if supplier provide manufacture relief in certain period of time t_s for finishing the payment without any interest expenses. Manufacture do the payment τ_m after the end of period t_s and before receiving the next shipping. On echelon 2, assumed if manufacture provide distributor relief in certain period t_m for finishing the payment without any interest expenses. Distributor do the payment τ_d after the end of period $t_m$and before receiving the next shipping. On echelon 3, assumed if in one general cycle $T = \frac{n_2 Q}{D}$ distributor demands less than the commitment quantity which has been agreed by drop-shipper before. It turns out that drop-shipper demands in line with commitment and more than distributor demands. Thus, the total annual cost of the system is :

$$\psi_{SC}^4 = \frac{A_{s,f}D}{n_2 Q} + C_{s,f}\alpha D + \frac{n_1(n_1-1)}{2}(h_{s,f}+S_{s,f})\frac{\alpha n_2 QD}{Pn_1{}^2} + h_{s,f}\tau_m \alpha D + (C_{m,w}-C_{s,f})e^{k_s t_s}\alpha D -$$

$$C_{m,w}e^{k_s(\tau_m - t_s)}\alpha D + \frac{n_1 A_{m,w}D}{n_2 Q} + \frac{C_{m,w}\alpha n_2 QD}{n_2 Q} + \frac{h_{m,w}\alpha^2 n_2{}^2 Q^2 D}{2\alpha Pn_1 n_2 Q} - \frac{h_{m,w}\alpha n_2 Q t_s D}{n_2 Q} + \frac{h_{m,w}n_1 \alpha P t_s{}^2 D}{2n_2 Q} +$$

$$\frac{S_{m,w}\alpha n_2{}^2 Q^2 D}{2Pn_1 n_2 Q} + \frac{C_{m,w}\alpha n_2 Q e^{k_s(\tau_m - t_s)}D}{n_2 Q} - \frac{C_{m,w}\alpha n_2 Q e^{k_m \tau_m}D}{n_2 Q} + \frac{A_{m,f}D}{n_2 Q} + C_{m,f}D +$$

$$(h_{m,f}+S_{m,f})\left(\frac{Q(2D+(P-D)n_2-P)}{2P}\right) + h_{m,f}\tau_d D + (C_{d,f}-C_{m,f})e^{k_m t_m}D - C_{d,f}e^{k_m(\tau_d - t_m)}D + \frac{A_{d,f}D}{Q} +$$

$$C_{d,f}D + \frac{h_{d,f}(Q-D_d t_m)^2}{2Q} + \frac{S_{d,f}Q}{2} + C_{d,f}e^{k_m(\tau_d - t_m)}D + \frac{v(Q_{ds}-Qn_2)D}{n_2 Q} - C_{d,f}D(1-e^{k_d t_m}) + \frac{Q_{ds}A_{ds,f}D}{n_2 Q} + \frac{C_{ds,f}Q_{ds}D}{n_2 Q} - \frac{v(Q_{ds}-Qn_2)D}{n_2 Q} \quad (9)$$

4. Numerical results and discussion

In this part will be carried out the numerical experiment on model which has been developed previously. The experiment will be made for each case scenario which has been explained previously. From all of

IOP Conf. Series: Materials Science and Engineering **337** (2018) 012027 doi:10.1088/1757-899X/337/1/012027

the case scenarios, will be decided the optimum condition in supply chain system which could produce the most minimum total cost. Before doing the numerical experiment, first, need to set the value of known variables. Variable values which used in this research adopted data from the previous research. Most of the data refer to the research which conducted by Aljazzar et al. [3]. The input parameter which will be used in this research is shown in table 2.

Table 2. Input parameters.

D	P	α	$A_{s,f}$	$A_{m,w}$	$A_{m,f}$	$A_{d,f}$	$A_{ds,f}$	$C_{s,f}$	$C_{m,w}$	$C_{m,f}$	$C_{d,f}$	$C_{ds,f}$
3069	4720	1	441	206	175	384	2.5	20	30	50	70	80
$h_{s,f}$	$h_{m,w}$	$h_{m,f}$	$h_{d,f}$	$S_{s,f}$	$S_{m,w}$	$S_{m,f}$	$S_{d,f}$	n_1	n_2	k_s	k_m	k_d
3	3	12	13,3	3	7,5	9	7,7	1	2	1%	8%	4%
v	e	w										
30	10%	50										

The basic parameter which already explained will be used for the search of the optimal solution from decision variable. In this research, the search of solution did by Solver on Excel software. The use of Solver in Excel chosen because computing with Excel Solver is simpler and produces relatively short computation period. To validate the result of calculations with Excel Solver, then it will also be done manual calculations. From several experiments conducted can show that calculations using Excel Solver or manual have relatively similar results. After doing the counting for all of the cases scenario, a decision is made for each case scenario. The calculation result of all case scenarios can be seen in table 3.

Table 3 shows the result of the numerical experiment showed that supplier identical provide a short period of time of delay the payment. The total of the minimum cost which issued by supplier also occurs when the short delay period is given. Supplier provides time to delay the payment which relatively short because of a small percentage of supplier return on investment (ROI), which is 1%. Minimum total cost of manufacture occurs when an actor who is one level below delay the payment out of time which provided before. And also seen if total cost which issued by manufacturer will become lower when manufacturer delays the payment out of time which provided by supplier than does the payment on time. That kind of thing happened because the percentage of manufacture ROI is very high if it's compared with the percentage of supplier ROI. The total cost of distributor has seen lower when distributor does the payment in accordance with the time which provided by the manufacturer. This kind of thing happened because the percentage of distributor ROI which lower if it's compared with the percentage of manufacture ROI. In the other hand, total cost on drop-shipper become lowest when the distributor can fulfill all of their demands. With the integration of supply chain, best scenario that comes from case scenario which has the least total cost of supply chain system is scenario 1.

Table 3. Result of the numerical example.

Skenario	0	1*	2	3	4
Q	296	326	311	326	311
t_s	0.00	0.01	0.01	0.01	0.01
τ_m	0.00	0.01	0.01	0.02	0.02
t_m	0.00	0.09	0.06	0.09	0.06
τ_d	0.00	0.09	0.07	0.09	0.07
m	592	651	745	651	745
SC	94,355	94,233	94,334	2,276	2,378
MC	221,935	224,662	14,804	316,880	107,021
DC	237,264	233,316	440,702	233,316	440,702
DsC	243,218	243,218	266,082	243,218	266,082
TSC	796,773	795,429	815,922	795,691	816,183

Table 4 shows the test of varying k_d, it can conclude if higher percentage of distributor ROI will also cause higher total benefits. That is happen because the improvement percentage of distributor ROI. Higher percentage of distributor ROI will generate higher Q values. Higher Q values will generate longer t_m value or period of payment delay which provided by manufacture to distributor. When percentage of distributor investment return increase 50% will decreasing the total cost on supply chain as big as 0,13%. The decreasing of total cost in that supply chain always increase in line with the increase of distributor investment return percentage. Order quantity will be increased in line with the improvement of distributor investment return percentage. After further calculation, it can be seen that there is an increase of order quantity by 7% when the percentage of distributor investment is increasing 50%.

Table 4. Sensitivity when varying k_d.

k_d	Q	t_s	τ_m	t_m	τ_d	m	SC	MC	DC	DsC	TSC
0.04	326	0.01	0.01	0.09	0.09	651	94233	224662	233316	243218	795429
0.06	349	0.01	0.01	0.11	0.11	698	94094	225547	231548	243218	794407
0.08	379	0.01	0.01	0.12	0.12	758	93939	226193	229908	243218	793258
0.1	418	0.01	0.01	0.14	0.14	836	93774	227049	227969	243218	792010
0.12	470	0.01	0.01	0.15	0.15	940	93595	228228	225593	243218	790634
0.14	542	0.01	0.01	0.18	0.18	1085	93403	229929	222538	243218	789088
0.16	650	0.01	0.01	0.21	0.21	1300	93196	232519	218369	243218	787302

The increase of return of investment distributor causes an increase in delay in payment period provided by the manufacturer to distributor. That thing can be connected with the increasements on Q quantity of orders, the higher the Q value will result the value or the delay in payment period gived by the manufacturer to the distributor will be longer. If it is seen more detail, when the increasements of order quantity exist as big as 7%, will cause the delay in payment period increased which given by the manufacturer to the distributor by 22%. That increasement will increase in line with order quantity increasements and return of investment distributor percentage.

According to the test of varying, it can be concluded if the higher percentage of distributor ROI will generate lower cost on distributor itself. But, if we see the change of cost issued by the manufacturer, it's actually increased. This kind of thing happened because of the longer period of delay in payment. Thus, that will lead to increase the opportunity cost of manufacture. But, the manufacturer also gets more benefits because of distributor order quantity which increased. Thus, from table 4 we can conclude if the percentage of distributor ROI is one of the parameter which significantly influential in supply chain system which investigated.

From that calculation results can be seen that the total cost of minimum supply chain occurs when the distributor has a greater percentage of return on investment than manufacturer and manufacturer has a greater percentage of return on investment than the supplier. When the percentage of distributor return on investment is between the two other players, which is manufacturer has the highest percentage of return on investment and the lowest is supplier. Then there is an increase in supply chain cost by 0.17%. When the distributors has the lowest percentage of return on investment compared to the other two players, the highest supplier and manufacturing is under the supplier. There is an increasement of total supply chain cost by 0.38% compared to the initial condition when the distributor has the highest percentage of return on investment. Different things happen when the percentage of return on investment distributor is between the two other players, but the highest is supplier and the lowest is manufacturing. There is an increase in total cost of supply chain by 0.26%, lower than the smallest return on investment of distributors. However, in that scenario the increase in total cost of supply chain is higher than the distributor return of investment which is between the other two players, the highest is manufacturer and the lowest is supplier with an increase of 0.17%. When return of investment percentage of that three players is same, also raise the total cost of supply chain by 0.22%. A larger Q value occurs when the return of investment percentage distributor is greater as well. The value of Q is also influenced by the return of investment percentage of all players. Thus, we can conclude if the percentage of return on distributor investment is very important on the improvement of order quantity.

Table 5. Sensitivity when varying (k_d, k_m, k_s).

Scenario					Q	ts	τm	tm	τd	m	SC	MC	DC	DsC	TSC
kd 0.09	> >	km 0.06	> >	ks 0.03	387	0.01	0.01	0.13	0.13	773	93,891	226,660	229,176	243,218	792,945*
km 0.09	> >	kd 0.06	> >	ks 0.03	353	0.01	0.01	0.11	0.11	706	94,059	225,490	231,489	243,218	794,256
ks 0.09	> >	km 0.06	> >	kd 0.03	313	0.01	0.01	0.07	0.07	773	94,259	224,090	234,371	243,218	795,939
ks 0.09	> >	kr 0.06	> >	km 0.03	331	0.01	0.01	0.10	0.10	663	94,140	225,544	232,099	243,218	795,002
ks 0.06	= =	km 0.06	= =	kd 0.06	331	0.01	0.01	0.10	0.10	663	94,161	225,192	232,099	243,218	794,670

The analysis of financial holding cost manufacturer - raw material is also done by increasing the cost of existing as much as 50% stage by stage. The result of variation financial holding cost manufacturer – raw material detail can be seen in table 6. When there is an increase of financial holding costs manufacturer – raw material as big as 50% will cause a decrease in order quantity by 2% from the previous. This decrease of order quantity is due to the cost of the manufacturer which constantly increases. When there is an increase in financial holding cost manufacturer – raw material as big as 50% will result in the increase of manufacturer cost itself by 0.06% from the previous condition. The increased cost to be borne by the manufacturer led to a decline period of delay in payment provided by the manufacturer to the distributor. As can be seen in table 6, manufacturer reduces the period of delay in payment provided to distributors from the previous 0.09 years to 0.08 years. Period of delay in payment was declined when the costs incurred by the manufacturer continue to increase.

Table 6. Sensitivity when varying financial holding cost manufacturer - raw material $(h_{m,w})$.

$h_{m,w}$	*Q*	*ts*	*τm*	*tm*	*τd*	*m*	*SC*	*MC*	*DC*	*DsC*	*TSC*
3	326	0.01	0.01	0.09	0.09	652	94,233	224,662	233,316	243,218	795,429
5	319	0.01	0.01	0.09	0.09	638	94,276	224,804	233,401	243,218	795,699
7	310	0.01	0.01	0.09	0.09	620	94,339	225,012	233,524	243,218	796,093
10	298	0.02	0.02	0.08	0.08	597	94,535	225,199	233,686	243,218	796,638
15	291	0.06	0.06	0.08	0.08	582	94,877	225,174	233,799	243,218	797,068
23	286	0.08	0.08	0.08	0.08	572	95,094	225,162	233,872	243,218	797,347
34	283	0.09	0.09	0.08	0.08	566	95,235	225,157	233,919	243,218	797,529

As can be seen in table 6, the costs were borne by the supplier also increase in proportion to the increase of costs borne by the manufacturer. If the manufacturer decreases the period of delay in payments which given to the distributor when the cost increase. In term of suppliers actually, happen different things, suppliers actually increase the period of delay in payments given to manufacturer. From the results of the experiment can be concluded that suppliers have been trying to stabilize the order quantity by providing a longer delay in the payment period. However, the quantity of order continues to decrease due to the costs that borne by manufacturer continuously increased. The increase of cost causes the shortening of payment delay period given by the manufacturer to the distributor so that the order quantity of distributor decreases. The total cost of supply chain increased in line with the increase of financial holding cost manufacturer–raw material warehouse. Costs issued by suppliers also continue to increase in line with increased financial holding costs manufacturer-raw material warehouse is increasing. The cost which issued by manufacturer initially increased but thereafter declined. That thing happens because the supplier continues to provide additional time limits for payment settlement for the manufacturer, while on the other hand at the same time manufacturer reduces the delay in payment period given to the distributor.

Table 7. Sensitivity when varying drop-shipper demand.

Demand Probabilities	Q	m	Qds	SC	MC	DC	DsC	TSC
0.90	326	651	586	94,233	224,662	233,316	243,218	795,429
0.8	326	651	521	94,233	224,662	248,661	233,244	800,800
0.70	326	651	456	94,233	224,662	264,006	223,270	806,171
0.6	326	651	391	94,233	224,662	279,351	213,296	811,542
0.50	326	651	326	94,233	224,662	294,696	203,321	816,912
0.4	326	651	261	94,233	224,662	310,041	193,347	822,283
0.30	326	651	195	94,233	224,662	325,386	183,373	827,654

Table 7 shows distributor will stand on the contract which already agreed before. Overall total cost of supply chain system will increase. Total cost which issued by drop-shipper seen decreased, this kind of thing caused by the product price which is higher than penalty cost which must be paid by the drop-shipper to the distributor. Even if the distributor doesn't change the minimum quantity in contract penalty and will get the compensation from the drop-shipper but total cost which issued by distributor has seen increased. This is caused by penalty charge which given by drop-shipper lower than purchasing cost of one distributor product. But if we look the overall of total system cost that will be increased because there are some actors who must pay more charge.

5. Conclusions

The model which created on this research can accommodate supply chain system of three integrated echelons. Where on the first and second echelons, coordination between actors did with the consideration of permissible delay in payment. Then, on the third echelon which one of the actors is drop-shipper, coordination between actors did by the consideration of contract penalty. Coordination with consideration of delay payment period success to integrate supply chain. This kind of thing can be seen from overall total cost of supply chain system which lower if we considering the payment delay period than not considering it. With providing the certain period of time for delay the payment by supplier and manufacture, so the actor one level below will have more benefits with investing their income until the payment done. Distributor order quantity influential to order quantity and even production of a player who is in upstream. That optimal quantity of order influenced significantly by the delay in payment period which provided by the seller. The longer period of payment delay which provided by the seller will force the buyer to do ordering above optimal quantity of them. Because buyer will get more benefits with getting the longer payment delay period. Ordering quantity which based on information of drop-shipper demand level becoming a reference for the distributor in order to decide the quantity in contract penalty plan. The distributor will stand on quantity corresponding on contract even if the demand of the drop-shipper decreased. That kind of thing must be done by the distributor to maintain their benefits. Where drop-shipper must guarantee the cost which increases time to time because of excess supplies. Other considerations can be used to develop a model in further research. For example, compensation if the buyer makes a payment faster than the end of the period of time delayed payments provided. Probabilistic demand type and discount when a buyer makes a certain quantity purchase.

6. References

[1] Seuring S 2004 Integrated chain management and supply chain management comparative analysis and illustrative cases *Journal of Cleaner Production* **12** pp 1059-1071

[2] Stadtler H 2005 Supply chain management and advanced planning—basics, overview and challenges *European Journal of Operational Research* **163** pp 575-588

[3] Aljazzar S M, Jaber M Y and Moussawi-Haidar L 2016 Coordination of a three-level supply chain (supplier–manufacturer–retailer) with permissible delay in payments *Applied Mathematical Modelling* **40** pp 9594-9614

[4] Iyer G, Narasimhan C and Niraj R 2007 Information and Inventory in Distribution Channels *Management Science* **53** pp 1551-1561

[5] Glock C H 2012 The joint economic lot size problem: A review *Int. Journal of Production Economics* **135** pp 671-686

[6] Marchi B, Ries J M, Zanoni S and Glock C H 2016 A joint economic lot size model with financial collaboration and uncertain investment opportunity *Int. Journal of Production Economics* **176** pp 170-182

[7] Jaber M Y and Osman I H 2006 Coordinating a two-level supply chain with delay in payments and profit sharing *Computers & Industrial Engineering* **50** pp 385-400

[8] Aljazzar S M, Jaber M Y and Goyal S K 2015 Coordination of a two-level supply chain (manufacturer–retailer) with permissible delay in payments *Int. Journal of Systems Science: Operations & Logistics* **3** pp 176-188

[9] Goyal S K 1985 Economic Order Quantity under Conditions of Permissible Delay in Payments *The Journal of the Operational Research Society* **36** pp 335-338

[10] Yu D Z, Cheong T and Sun D 2017 Impact of supply chain power and drop-shipping on a manufacturer's optimal distribution channel strategy *European Journal of Operational Research* **259** pp 554-563

[11] Rabinovich E, Rungtusanatham M and Laseter T M 2008 Physical distribution service performance and Internet retailer margins: The drop-shipping context *Journal of Operations Management* **26** pp 767-780

[12] Chiang W K and Feng Y 2010 Retailer or e-tailer? Strategic pricing and economic-lot-size decisions in a competitive supply chain with drop-shipping *Journal of the Operational Research Society* **61** pp 1645-1653

[13] Gan X, Sethi S P and Zhou J Commitment-penalty contracts in drop-shipping supply chains with asymmetric demand information *European Journal of Operational Research* **204** pp 449-462

[14] Chung K -J, Goyal S K and Huang Y -F 2005 The optimal inventory policies under permissible delay in payments depending on the ordering quantity *Int. Journal of Production Economics* **95** pp 203-213

[15] Nurshanti I 2011 Kebijakan Persediaan Terintegrasi Pada Sistem Dual Sales Channel Mempertimbangkan Penundaan Pembayaran *Industrial Engineering Master Thesis* (Surabaya: Institut Teknologi Sepuluh Nopember)

[16] Khouja M and Stylianou A C 2009 A inventory model with a drop-shipping option for e-business *Omega* **37** pp 896-908

[17] Edirisinghe C and Atkins D 2017 Lower bounding inventory allocations for risk pooling in two-echelon supply chains *Int. Journal of Production Economics* **187** pp 159-167

[18] Heydari J, Mahmoodi M and Taleizadeh A A 2016 Lead time aggregation: A three-echelon supply chain model *Transportation Research Part E: Logistics and Transportation Review* **89** pp 215-233

[19] Moussawi-Haidar L, Dbouk W, Jaber M Y and Osman I H 2014 Coordinating a three-level supply chain with delay in payments and a discounted interest rate *Computers & Industrial Engineering* **69** pp 29-42
[20] Jaber M Y, Bonney M and Guiffrida A L 2010 Coordinating a three-level supply chain with learning-based continuous improvement *Int. Journal of Production Economics* **127** pp. 27-38

IOP Conf. Series: Materials Science and Engineering **337** (2018) 012028 doi:10.1088/1757-899X/337/1/012028

Modified allocation capacitated planning model in blood supply chain management

A Mansur[1,2*], I Vanany[1], and N I Arvitrida[1]
[1]Department of Industrial Engineering, Institut Teknologi Sepuluh Nopember, Kampus ITS Sukolilo-Surabaya 60111, Indonesia
[2]Industrial Engineering Department, Faculty of Industrial Technology, Universitas Islam Indonesia

*agusmansur@uii.ac.id

Abstract. Blood supply chain management (BSCM) is a complex process management that involves many cooperating stakeholders. BSCM involves four echelon processes, which are blood collection or procurement, production, inventory, and distribution. This research develops an optimization model of blood distribution planning. The efficiency of decentralization and centralization policies in a blood distribution chain are compared, by optimizing the amount of blood delivered from a blood center to a blood bank. This model is developed based on allocation problem of capacitated planning model. At the first stage, the capacity and the cost of transportation are considered to create an initial capacitated planning model. Then, the inventory holding and shortage costs are added to the model. These additional parameters of inventory costs lead the model to be more realistic and accurate.

1. Introduction

Blood management has been considered critical in these past years, resulting in the emergence of blood center as a specialized organization for blood management. The blood center regulates and manages the blood distribution from the donors to hospitals. "One size does not fit all" is an apt description for blood management, since the problems and challenges are uniquely entitled to each country. This indicates that the blood management system in each country requires to be adjusted and customized based on the characteristics of the corresponding country [1][2][3][4][5]. One of the challenges in synchronizing supply and demand of blood management is the difficulties in determining the exact number of demand and accessible supply. This is because the procurement mechanism of blood supply is conducted through volunteering act and unpredictable demand [6]. Meanwhile, blood is a perishable product as the useful lifetime is limited. As a result, when blood passed a certain period of time it becomes unsafe to use and turn to waste [7]. The expired blood will cause a large amount of cost for procurement and processing unit, and also extra expenditure for blood waste disposal [8].

The main risks in blood management are the supply stockout problem and an abundant amount of outdated blood. Blood supply stockout could lead to mortality risk for untreated patients in emergency need of blood. The supply stockout will also lead to the high cost of a sudden blood procurement, since there are costly operational actions to be taken, such as bring in the blood supply from another institution when the shortage occurs. An intensive effort to cover blood stockout will add transportation cost and other extra costs [9]. This is the main reason why blood management is more than just maintaining the availability of blood stock, but also creating efficiency in the blood management operational. Hence, management distribution and other things related to whole cost minimization in blood stock procurement are essential [5].

Blood supply chain (BSC) consists of four stages or echelons. First stage is collection in which the blood is collected from donors. The second phase is production, where blood processed using various process into each product types. The third stage is inventory, which represents special storage in blood centers or hospitals. Then, the distribution echelon, which reflects blood transported to hospitals, so that the transfusion for patients could be conducted [10][7][11], the challenges that are needed to overcome is determining the amount of blood stock to be transported and distribution route from blood processing unit to hospitals. The precision of blood stock, punctuality, and minimum cost become three main consideration for distribution management [12][1][13][14].

This research aims to developed distribution blood model, hence, it can minimize the allocation cost, decrease the stock out and expired date risk. A contribution given from this research is the development of allocation capacitated planning model by consider the variable which affects the blood management.

2. Literature Review

Blood supply chain (BSC) become essentials since blood is a highly valued commodity. It has a strict safety regulation, bear in mind that perishable characteristic of blood could play a big part in patients' health, and its tendency of enlarging the gap between a number of donors with the blood demands [15]. The effective supply chain in blood management is expected to optimize the number of blood stock, as a result, creating balance and match between supply and demand. In order to support this purpose, precise production planning system should match the blood procurement scheduling through blood donor activities [16]. Guaranteeing the balance between demand and supply in blood supply chain is a critical factor to be noticed. It needs good coordination to the supply source (vendor) in order to lower blood supply chain risk. Since the early step of the supply chain, coordination should have been planned by reinforcing the importance of it to all stakeholder, including the vendor, so that the scheduling could have done right [5].

The distribution activities can happen between blood centers, or from blood center to hospital blood bank. Good distribution management will ensure short lead time and minimum cost. In the strategic level, distribution policies include vehicle type selection, capacities and staff number involved in the activities. While at operational level, the decision is focused on transported blood stock, picking point, and time constraint.

Hosseinifard & Abbasi **Error! Reference source not found.** have attempted to compare decentralization system to centralization system effectivity in distribution chain from blood centers to hospitals. The scenario, in this case, is that the hospitals which were located in close proximity were asked to conduct collaborative planning and ordering (collaboration and sharing information), by assuming that all orders could be fulfilled by blood center. As a result, centralized model is deemed to fit this scenario more, by implementing the system that the hospital with the most strategical position is appointed to be the blood bank for other hospital coalition members. Yet, this research only focused on blood inventory number and have not considered the efficiency nor distribution cost.

Allocation capacitated planning is planning model that will allocate a number of the source with certain capacities to a number of the consumer with certain demand level, where its purpose is minimizing production distribution cost from the source to destination [17]. Allocation capacitated planning is modelled by Chopra & Meindl [18] as follow:

$$Min \sum_{i=1}^{n} \left[f_i Y_i + \sum_{j=1}^{m} c_{ij} X_{ij} \right] \quad (1)$$

Subject to:

$$\sum_{i=1}^{n} X_{ij= D_j} \quad \forall j \quad (2)$$

$$\sum_{j=1}^{m} X_{ij} \leq K_i y_i \ \forall i \tag{3}$$

$$yi \in \{0,1\} \tag{4}$$

The above model has the objective to minimize the costs of allocation of a number of products from multiple sources to multiple destinations with which based on fixed costs for production and shipping costs, the assumptions used in inventory carrying costs are ignored and all requests can be met. Where in case of blood management, the model can be explained as follows.

Xijl = number of blood bag l to be transported from PMI i to Hospital Blood Bank (BDRS) j

Yi = the value will be equal to 1 if PMI i is chosen and 0 if it is not chosen.

The parameters could be defined as follows.

i = index for PMI

j = index for Hospital Blood Bank (Bank Darah Rumah Sakit/BDRS)

l = index for blood bag

Dj = demand per period from BDRS j.

Ki = blood capacity that is available to be supplied by PMI i.

Fi = fixed cost for PMI i operational.

Cijl = production and distribution cost for 1 blood bag l in PMI i to BDRS j.

3. Proposed Model

This research is the development of Chopra & Meindl [18] model, where there is a consideration that the initial model has not been able to accommodate the decreasing a value of perishable products and the penalty when the products are out of stock. Whereas both matters are important to notice in blood management.

Centralization and decentralization model were studied in this research because both model already implemented in many countries. Differences of both models are related to the planning system, production and distribution. However, the focus of this research is in distribution management.

3.1. Decentralization Model

Blood management in Indonesia officially conducted under authorization of Palang Merah Indonesia (PMI). This model giving an opportunity to each independent PMI to distribute blood bags for several hospitals. The function of the modified model could be formulated as decentralization model, as follows:

$$Min \sum_{i=1}^{n}\left[\sum_{l}^{4}\sum_{t}^{5} f_{ilt}Y_{ilt} + \sum_{j=1}^{m}\sum_{l}^{4}\sum_{t}^{5} c_{ijlt}X_{ijlt} + \sum_{l}^{4}\left(\sum_{t}^{5} U_{ilt}H_{ilt} + Z_{ilt}S_{ilt}\right)\right] \tag{5}$$

Subject to:

$$\sum_{i=1}^{n} X_{ijlt} = D_{jlt} \qquad \forall\, j,l,t \tag{6}$$

$$\sum_{j=1}^{4} X_{ijlt} \leq K_{ilt}Y_{ilt} \ \forall\, i,l,t \tag{7}$$

$$S_{ilt} = \begin{cases} 0, if\ X_{ijlt} > D_{jlt} \\ X_{ijlt} - D_{jlt},\ if\ X_{ijlt} \leq D_{jlt} \end{cases},\quad \forall\, i \tag{8}$$

$$H_{ijlt} = X_{ijlt} - D_{jlt},\quad \forall\, i \tag{9}$$

$$Y_{ilt} \in \{0,1\} \tag{10}$$

$$X_{ijlt}\ integer \tag{11}$$

This model represents the decentralized system. This is because of each PMI has impartial decentralization which expects more opportunity to provide supply the same to the hospital, so that a solution will be chosen based on the minimal cost of each hospital.

3.2 Centralized Model

The next model is developed with a centralized model, where certain PMI will serve as a central storage of blood from another PMI and distribution will be done centralized of PMI designated as the center. In the centralized model, there are two suggested scenarios, which are centralization in blood center and centralization in BDRS. This paper focused in blood center centralization, the proposed model is formulated below:

$$Min \sum_{i=1}^{n} \left[f_i Y_i + \sum_{j=1}^{m} c_{ijlt} X_{ijlt} + \sum_{l}^{4} \left(\sum_{t}^{5} U_{ilt} H_{ilt} + Z_{ilt} S_{ilt} \right) \right] + C_T \tag{12}$$

Subject to:

$$\sum_{t=1}^{5} X_{jlt} = D_{jt} \quad \forall\, j, l, t \tag{13}$$

$$\textstyle\sum_{j=1}^{m} \sum_{t=0}^{5} X_{jt} \leq K'_t\ \text{, Where } K'_t = \sum_i K_{it} \tag{14}$$

$$S_{lt} = \begin{cases} 0, if\ X_{ijlt} > D_{ijlt} \\ X_{jlt} - D_{jlt},\ if\ X_{jl} \leq D_{jlt} \end{cases},\quad \forall\, j, l \qquad H_{jlt} = X_{jlt} - D_{jlt},\quad \forall\, j, l \tag{15}$$

$$\text{Xi, Di, } yi \in \{0,1\} \tag{16}$$

Where:

Z_{ilt} = stock out cost for blood inventory l in PMI I at period t.
S_{ilt} = number of a shortage of blood inventory l in PMI I at period t.
U_{ilt} = holding cost for blood inventory l in PMI i on t period (U_(il,t-1)≪ U_(il,t))
H_{ilt} = number of blood inventory l in PMI i on t period
l = blood type (1 = A, 2 = B, 3 = AB, 4 = O)
t = blood life span (platelet) (integer: 0,1,2,3,4,5)
C_T = Transportation Cost, from PMI i to central PMI

4. Result and Discussion

Research about optimization cost for product allocation often only based on transportation cost. While model that developed in this research offers related variables to be included in the distribution, such as inventory cost, stock out and expired cost. Terminology that was developed are decentralization and centralization model. It is based on consideration that in Indonesia the blood distribution is using decentralization concept, where each PMI has the same opportunity to gather, produce and distribute the blood to several hospitals. Whilst, the model system will be compared with the centralization concept that implemented in several countries. In centralization planning model, production and distribution conducted by one blood banking, hence this definitely giving different impact for the cost. Disadvantage from this research are, has not provided an in-depth study of the probability of shortages or expired date in any deliveries made and has not yet developed an alternative supply between PMI and hospitals.

Illustration of centralization model can be used in the following scenario: in which one PMI is selected as a distribution center. The arrival of blood to PMI occurs in every period 1 and will be sent to BDRS for 5 periods. The holding cost should be calculated when blood inventory in PMI surpasses 1 or more period. In this research, the time frame used is 5 period (t) with the assumption that expiration date for blood is equal or more than 5 (t). The value of inventory cost is different in each period, the longer blood stored, the more expensive the inventory cost is, at t = 1 inventory cost is 0 and will be increasing every period with formula t+n. This caused by the degenerating blood quality over time, so the declining value of bloodstock charged toward inventory cost. The stockout cost happened when PMI could not be able to fulfill the blood demand. This scenario pictured and simulated one blood center to serve two BDRS.

Table 1. Scenario of blood allocation problem at centralization model.

	Type		**Deliv. + hold cost to BDRS1**				**Deliv. + Hold cost to BDRS2**				
Stock		**Period (t)**	**A**	**B**	**AB**	**O**	**A**	**B**	**AB**	**O**	**Demand/ period**
10000	A	1	3	-	-	-	3	-	-	-	5000
		2	4	-	-	-	4	-	-	-	3000
		3	5	-	-	-	5	-	-	-	1000
		4	6	-	-	-	6	-	-	-	2000
		5	7	-	-	-	7	-	-	-	2000
8000	B	1	-	3	-	-	-	3	-	-	3000
		2	-	4	-	-	-	4	-	-	2000
		3	-	5	-	-	-	5	-	-	2000
		4	-	6	-	-	-	6	-	-	3000
		5	-	7		-	-	7	-	-	2000
9000	AB	1	-	-	3	-	-	-	3	-	3000
		2	-	-	4	-	-	-	4	-	1000
		3	-	-	5	-	-	-	5	-	3000
		4	-	-	6	-	-	-	6	-	2000
		5	--	-	7	-	-	-	7	-	1000
10000	0	1	-	-	-	3	-	-		3	3000
		2	-	-	-	4	-	-		4	4000
		3	-	-	-	5	-	-		5	3000
		4	-	-	-	6	-	-		6	2000
		5	-	-	-	7	-	-		7	3000
		Demand	10000	8000	3000	5000	3000	4000	7000	10000	

From Table 1, stock of blood in PMI is delivered to BDRS that have a demand at period (t), inability to fulfill the demand will be given a penalty stock out cost and every blood bag that has not been distribute will give a holding cost.

5. Conclusion

The conclusions for this research is that it has succeeded in developing blood distribution model with a minimum cost under consideration of delivery cost, inventory cost and the possibility of incurred stockout cost. This result is different with the previous allocation capacity model. The previous model did not facilitate cost adjustment on the inventory and stock out. Moreover, for the future research suggested could be approached by incorporating blood span or lifetime as consideration in advancing the blood distribution model.

References

[1] Blake J T and Hardy M 2014 A generic modelling framework to evaluate network blood management policies: The Canadian Blood Services experience *Operations Research for Health Care* **3** pp 116–128 https://doiorg/101016/jorhc201405002

[2] Lowalekar H and Ravichandran N 2013 Blood bank inventory management in India *Opsearch* **51** pp 376–399 https://doiorg/101007/s12597-013-0148-z

[3] Zahraee S M, Rohani J M, Firouzi A and Shahpanah A 2015 Efficiency Improvement of Blood Supply Chain System Using Taguchi Method and Dynamic Simulation *Procedia Manufacturing* **2** pp 1–5 https://doiorg/101016/jpromfg201507001

[4] Bahrami H 1997 *How Can Canadian Blood Services Product And Hospital Services In Bc And Yukon Region Implement A More Effective Blood Supply Chain Model To Achieve Its Vision ?* Microbiology UBC 1994 RT Medical Laboratory Technology BCIT

[5] Stanger S H W 2013 *Vendor managed inventory in the blood supply chain in Germany Evidence from multiple case studies* https://doiorg/101108/17538291311316054

[6] Beliën J and Forcé H 2012 Supply chain management of blood products: A literature review *European Journal of Operational Research* **217** pp 1–16 https://doiorg/101016/jejor201105026

[7] Dillon M, Oliveira F and Abbasi B 2017 A two-stage stochastic programming model for inventory management in the blood supply chain A Two-Stage Stochastic Programming Model for Inventory Management in the Blood Supply Chain *Journal of Production Economics* **187** pp 27–41 https://doiorg/101016/jijpe201702006

[8] Nahmias S 1982 Perishable inventory theory: a review *Operations Research* **304** pp 680–708 https://doiorg/101287/opre304680

[9] Lowalekar H and Ravichandran N 2013 Blood bank inventory management in India *Opsearch* **513** pp 376–399 https://doiorg/101007/s12597-013-0148-z

[10] Osorio A F, Brailsford S C and Smith H K 2015 A structured review of quantitative models in the blood supply chain: a taxonomic framework for decision-making *International Journal of Production Research* **11** pp 1–22 https://doiorg/101080/00207543201510057 66

[11] Hosseinifard Z and Abbasi B 2016 The inventory centralization impacts on sustainability of the blood supply chain *Computers and Operations Research* pp 1–7 Elsevier https://doiorg/101016/jcor201608014

[12] Coleman A and Akinsola O S 2012 Effective blood distribution in rural hospitals through ICT service oriented architecture SOA framework: A case study in rural hospitals in South Africa *Studies on Ethno-Medicine* **6** pp 141–147 https://doiorg/101007/s10287-011-0133-z

[13] Stanger S H W, Wilding R ,Hartmann E, Yates N and Cotton S 2013 Lateral transshipments: an institutional theory perspective *International Journal of Physical Distribution and Logistics Management* **439** pp 747–767 https://doiorg/101108/IJPDLM-09-2011-0155

[14] Sarkar B 2013 A production-inventory model with probabilistic deterioration in two-echelon supply chain management *Applied Mathematical Modelling* **37** pp 3138–3151 Elsevier Inc https://doiorg/101016/japm201207026
[15] Reynolds E, Wickenden C and Oliver A 2001 The impact of improved safety on maintaining a sufficient blood supply *Transfusion Clinique et Biologique* **83** pp 235–239 https://doiorg/101016/S1246-78200100139-2
[16] Ghandforoush P and Sen T K 2010 A DSS to manage platelet production supply chain for regional blood centers *Decision Support Systems* **50** pp 32–42 Elsevier BV https://doiorg/101016/jdss201006005
[17] Pujawan N and Mahendrawati 2010 *Supply Chain Management* Guna Widya
[18] Chopra S and Meindl P 2013 *Supply Chain Management: Strategy, Planning, and Operations* (New Jersey:Prentice Hall)

Acknowledgements:
Foremost, I would like to express my very profound gratitude to the Ministry of Research, Technology and Higher Education of the Republic of Indonesia and the Ministry of Finance, Republic of Indonesia, for generous support and I am honored to be one of the recipients of BUDI-DN scholarship.

The relationship among the resiliency practices in supply chain, financial performance, and competitive advantage in manufacturing firms in Indonesia and Sierra Leone

I Musa[1,*] and I Nyoman Pujawan[1]
[1]Department of Industrial Engineering, Institut Teknologi Sepuluh Nopember, Kampus ITS Sukolilo-Surabaya 60111, Indonesia

*ibmusa58@gmail.com

Abstract. Current supply chain management (SCM) has become a potentially treasured way of safeguarding competitive advantage and improving organizational performance since competition is no longer between organizations, but among supply chains. This research conceptualizes and develops four resiliency practices (Flexibility, Redundancy, Collaboration and Agility) and tests the relationships between organizations' financial performance and competitive advantage in manufacturing firms. The study involves manufacturing firms in Indonesia and Sierra Leone. The study used stratified random sampling to pick a sample size of 95 manufacturing firms, which represented different industrial sectors. The respondents were mainly managers of different manufacturing companies. The relationships proposed in the conceptual framework were tested using correlation analysis. The results indicate that higher levels of resilience practices in manufacturing firms can lead to enhanced competitive advantage and improved financial performance.

1. Introduction

In today's competitive business setting, outsourcing and globalisation have made stretched and more difficult supply chains, which exposed business to disruptions in a stormy environment. The amount of supply chain vulnerability is largely determined by the extent of complexity of the supply chain. Over the past years, many supply chains have encounter a wide range of disruptive situations such as transportation delays, port strikes, terrorist attacks, cyber-attacks, economic crises, accidents, natural disasters, poor communication, shortages of parts, quality operational and quality problems . The cost of these supply chain disruptions to a company can be of significance. Disruptions of any nature in the supply chain have the possibility to cause the entire supply chain to be unsuccessful **Error! Reference source not found.**. As noted by Sheffi and Rice **Error! Reference source not found.**, disruptions can result to loss in supply chain organisations and can cause a range of problems such as long lead-times, stock-outs, inability to meet customer demand and increases in costs. When major disruptions occur, many supply chains tend to stop and take a reasonable time to recover. Evidence that demonstrate the importance of even minor disruptions is the example of the automotive manufacturing supply chain in Japan. On July 16, 2007, a magnitude 6.8 earthquake in central Japan severely damaged the facilities of Riken Corp. a supplier of automobile components that is specialized in piston rings. Riken located all its plants in a single area of Japan to raise efficiency, making the whole production capacity vulnerable to a catastrophic incident. The earthquake damage to Riken facilities and its utilities completely shut down production for one week and required another week of repairs to return to full output. Because of carrying limited inventories, Toyota, one of Riken's many customers, was highly exposed to production and transportation disruptions. Because Toyota sourcing strategy stressed on close relationships with a limited number of suppliers, but in this case, Toyota has no substitute but to shut down all 12 of its domestic assembly plants, delaying production of approximately 55,000 vehicles [2]. We intend to derive specific supply chain resiliency practices that either contribute to or conversely impact a firm's disruption mitigation capabilities. Specifically, our objectives are to:

1. Identify resilience factors that affect a firm's supply chain performance.
2. Propose an empirically tested model that represents the relationships between the supply chain resilience factors, firm's financial performance and competitive advantage
3. Offer guidelines for managers to enhance disruption mitigation capabilities through understanding the role of resilience factors in a supply chain.

2. Supply Chain Risk, Disruptions, and Resilience

Mensah et al. [3] Argue that Supply chain risk can be conceptualized as an event that adversely affects supply chain operations and hence its desired performance measures, such as chain-wide service levels and responsiveness, as well as costs. There are wide ranges of events that can affect supply chain operations, from environmental, and intra-organisational to inter-organisational. Their impact is divided from short-term to long-term, as highlighted in various literatures by practical examples. From the logistics point of view, supply chain member's interaction become very complex, especially in the context of growing uncertainty, where the main drivers of uncertainty are new business models that will applied in order to escalate both the logistics efficiency and competitiveness. Hence, as the main sources of risk in supply chain could be identified in both the relations between members in supply chain and its environment. Risks to supply chain range from short delays to catastrophic disasters and one of the key stages in proactively managing them is to visualize and understand these various types of risks [4].

Supply disruptions stem from a broad range of risk sources and can emerge from within the supply chain or from external events [5]. In recent years, supply chains have become extended and more complex, while this severity and frequency of supply chain disruptions seems to be increasing **Error! Reference source not found.**[6]. According to the World Economic Forum report launched in close cooperation with supply chain researcher, it indicates that significant supply chain disruptions reduce the share price of affected companies by as much as seven percent on average. Natural disasters and extreme weather conditions are not the only threats to supply chains. There are also Systemic vulnerabilities, such as oil dependence and information fragmentation, that pose serious risks, as do political unrest, cybercrime and the rising cost of insurance and trade finance [6].

From an organisational perspective, resiliency has been termed as a dynamic capacity of adaptability, which grows and develops over time. It reflects an organisations capacity to adjust and maintain required functions under challenging situations. Thus, resiliency studies within an organisation identify both the ability to absorb shocks in the form of extreme events and an adaptive capability to adjust to new conditions. Organisations with better resiliency practices are more likely to deal with everyday problems, as well as those coming from a crisis [7]. Therefore, resiliency is a source of competitive advantage [8].

3. Development of Theoretical Framework

Arani et al. [9] has developed and advance Strategic Choice Theory (SCT). According to this theory the goal of the organizations is to achieve high performance standards and increase the efficiency to the limits of economic constraints. Anew priority has emerged for business planning that is a higher degree of resilience [10]. However, no existing study provides a complete framework that incorporates the scale of issues both internal and external to the supply chain. This research will propose a model in the form of hypothesis to illustrate the relationship between resilience factors specifically flexibility, redundancy, collaboration and agility to that of a firm's financial performance and competitive advantage in manufacturing firms as shown in' figure 1' below [11].

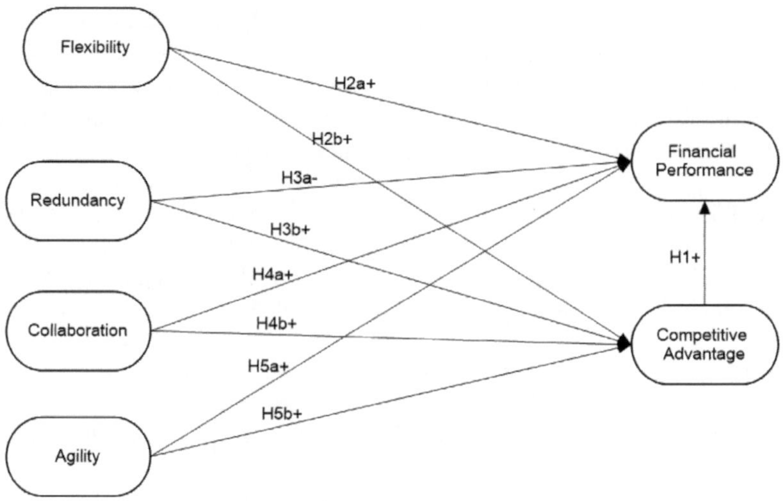

Figure 1. Proposed model.

Figure 1 above represents a framework developed to show the relationship between resilience factors and a firm's financial performance and competitive advantage in a manufacturing supply chain. The framework proposes that the four resilience factors flexibility, redundancy, collaboration and agility have a positive impact on manufacturing firms supply chain financial performance and competitive advantage.

4. Methodology

The questions in the questionnaire were designed based on a review of the literature. For constructs, which had not been well documented and tested in the literature, we developed new items based on our understanding of the constructs. The initial survey instrument was revised based on feedback by and experts in supply chain management for content, clarity and understanding. Based on their feedbacks, we modified, added and deleted some questions, making them more understandable and relevant to resiliency practices in Indonesia and Sierra Leone. All the scales used in the study employed a five-point Likert scale. The ends points were labelled 'Strongly disagree' (1) to 'Strongly agree' (5). The mid-point (3) was labelled 'Neither agree nor disagree'.

Data was randomly selected from manufacturing companies in Indonesia and Sierra Leone. Questionnaires were distributed electronically via mail, WhatsApp and other social media applications. We sent the questionnaire to key informants, who were CEOs, managers in charge of operations and supply chain, along with a cover letter highlighting the study's objectives. Follow-up emails or calls were made to improve the response rate. After the follow-up calls or emails, 95 usable questionnaires were received for analysis.

5. Data Analysis

The first common step to consider when examining reliability is factor analysis in order to assess dimensionality [12]. Three measures of reliability or internal consistency were examined in this research: composite reliability, average variance and Cronbach's alpha. A lower bound of .7 is recommended for composite reliability. It is recommended that the average variance extracted (AVE) be at least .50, as this indicates that the indicators of the latent variable [13] explain 50% or more of the variance. For Cronbach's alpha, a minimum value of .70 is considered acceptable [12]. 'Table 1' provides a summary of Convergent validity for all the constructs. The minimum acceptable values for each construct on each of the measures are met. Therefore, the reliability of the measurement model is adequate. Convergent validity, the ability of items in a scale to converge or load together as a single construct, is measured by examining individual loadings for each indicator. The items with the lowest loadings are generally candidates for removal from the scale. Most of the loadings loads at least 0.60 and above indicating that each measure is accounting for 50% or more of the variance of the underlying latent variable [12]. Table 1 provides a list of standardised loadings for each construct, and it is shown that they are all above the acceptable minimum value [13].

Table 1. Shows convergent validity of the various construct and their loadings

Construct	Indicator	Mean	SD	Loading
Flexibility	FLex2	3.58947	1.18945	0.81691
	Flex3	3.63158	1.14903	0.76481
	Flex5	3.43158	1.04833	0.82247
Redundancy	Redu3	3.47368	0.92071	0.85071
	Redu4	3.47368	0.76967	0.85636
	Redu5	3.52632	0.96582	0.80562
Collaboration	Colla2	3.54737	1.16481	0.8796
	Colla3	3.34737	1.11821	0.84095
	Colla4	3.46316	0.98726	0.63382
	Colla5	3.38947	0.98169	0.62346
Agility	Agil4	1.09442	1.07834	0.85033
	Agil3	1.07834	1.09442	0.91572
	Agil5	1.03824	1.03824	0.87442
Financial Performance	Finper2	0.9936	0.9936	0.89835
	Finper4	3.56842	0.84618	0.88287
	Finper5	3.36842	0.88789	0.7522
Comppetitive Advantage	ComAd2	3.96842	0.99416	0.84844
	ComAd3	3.70526	1.09032	0.79093
	ComAd4	3.82105	0.93375	0.90507
	ComAd5	3.57895	1.05769	0.79165

6. Hypothesis Testing

As shown in ‘Table 2’ below competitive advantage has a significantly positive correlation on a firm's financial performance (0.01211; $p< 0.05$). Thus, H1 is supported. The correlation coefficient between flexibility and financial performance is not statistically significant (0.1480; $p> 0.05$) thus, the null hypothesis H2a is rejected. However, the correlation coefficient from flexibility to competitive advantage is significant, in this case the null hypothesis is accepted (0.0119 $p<0.05$). Thus, supply chains with high level of flexibility exhibits a lower level of financial performance and a higher level of competitive advantage.

Likewise, for H3a, the correlation coefficient from redundancy to a company financial performance is significant thus, the null hypothesis is accepted (0.0339; $p <0.05$). However, for H3b the correlation coefficient from redundancy to competitive advantage is not significant, thus the null hypothesis is rejected (0.1167; $p> 0.05$). Supply chain collaboration has significantly positive correlation to both financial performance (0.3431; $p<0.05$) and competitive advantage (0.0007; $p<0.05$) of a firm. Thus, supply chains with high collaboration exhibits high level of financial performance and competitive advantage. Hence, both H4a and H4b are supported. Finally, the correlation between supply chain agility to that of financial performance (H5a) is not significant (0.0910; p >0.05), and the correlation coefficient from supply chain agility to competitive advantage is found to be significant H5b (0.0282; p<0.05). Thus, supply chains with high level of agility found to exhibit a low level of financial performance and high level of competitive advantage. Hence, H5a is rejected and H5b is supported.

Table 2. Correlation matrix of latent variables.

Pearson Correlation Coefficients, N = 95
Prob > |r| under H0: Rho=0

	Flexibility	Redundancy	Collaboration	Agility	Finper	Compadv
Flexibility	1	0.41204	0.35813	0.02179	0.14955	0.25718
Flexibility		<.0001	0.0004	0.834	0.148	0.0119
Redundancy	0.41204	1	0.17984	0.24205	0.21793	0.16202
Redundancy	<.0001		0.0812	0.0181	0.0339	0.1167
Collaboration	0.35813	0.17984	1	0.35284	0.34319	0.2702
Collaboration	0.0004	0.0812		0.0005	0.0007	0.0081
Agility	0.02179	0.24205	0.35284	1	0.1744	0.2252
Agility	0.834	0.0181	0.0005		0.091	0.0282
Finper	0.14955	0.21793	0.34319	0.1744	1	0.25645
Finper	0.148	0.0339	0.0007	0.091		0.0121
Compadv	0.25718	0.16202	0.2702	0.2252	0.25645	1
Compadv	0.0119	0.1167	0.0081	0.0282	0.0121	

7. Discussion and Implications

Supply chain with low capabilities corresponds to moderate or high vulnerabilities that can dramatically degrade the supply chain's resilience. For example, a supply chain with a high vulnerability to connectivity can face disastrous consequences if it has poor visibility and collaboration. In addition, the framework provides managerial guidance for setting priorities to create a strategy for improving supply chain resilience. The strategy is based on assessment of the firm's pattern of vulnerabilities and its competitive advantages, weighed against the potential returns on investment. In doing so, corporate strategy will focus on resource investments to fill the gaps which enhances a company's disruption recovery capability. This study represents an important contribution for managers. It gives insights on the kind of lean, resilient practices with significant impact on supply chain sustainability. In the presence of this information, it is easier for companies to choose the set of practices that should be employed to improve social, economic and environmental sustainability. However it is also important for companies to formulate appropriate strategies for improving resilience. Understanding the source and types of uncertainty and disruptions would be critical for this purpose. As suggested Angkiriwang et al. [2], approaches could be reactive or proactive in nature. Likewise, Pujawan et al. [14] also suggest two different broad strategies in dealing with uncertainty. Obviously, being able to handle uncertainty is also an important characteristic of resilient companies.

8. Conclusion

From the study findings, it could be concluded that resiliency factors had a positive significant influence on manufacturing firm's financial performance and competitive advantage. The study showed that there was a strong relationship between various resiliency factors and firm's financial performance and competitive advantage. Hence, it could also be concluded that if management of manufacturing firms embraces resilience practices, it could increase supply chain resilience in manufacturing firms and improve competitive advantage. In addition, based on the results of this study, it could be concluded that manufacturing firms collaborate frequently with their key supply chain partners through the various platforms like sharing of information, synchronizing of decisions, aligning of incentives in a form of co-developing systems, sharing costs, risks and benefits, sharing of resources and constant communication.

Because the concept of supply chain resiliency is a fast emerging topic it is very beneficial to expand the scope of the research. Further research should be undertaken to investigate the concept of resiliency practices in other fields of studies. For example petroleum industry, mining industry, building and construction industry etc. In future investigations, it might be possible to use different resiliency factors such as robustness, leanness etc.

References

[1] Sheffi Y and Rice Jr J B 2005 A supply chain view of the resilient enterprise *MIT Sloan management review*, **47** p 41
[2] Angkiriwang R, Pujawan IN and Santosa B 2014 Managing uncertainty through supply chain flexibility: reactive vs. proactive approaches *Prod Manuf Res* **2** pp 50–70. doi:10.1080/21693277.2014.882804.
[3] Mensah P, Merkuryev Y, and Manak S. 2015 Developing a Resilient Supply Chain Strategy by Exploiting ICT *Procedia Comput Sci* **77** pp 65–71 doi:10.1016/j.procs.2015.12.360.
[4] Habermann M, Blackhurst J, and Metcalf AY 2015 Keep Your Friends Close? Supply Chain Design and Disruption Risk. *Decis Sci* doi:10.1111/deci.12138.
[5] Jüttner U and Maklan S 2011 Supply chain resilience in the global financial crisis: an empirical study. *Supply Chain Manag An Int J* 2011**16** pp 246–59 doi:10.1108/13598541111139062.
[6] Shao X -F 2013 Supply chain characteristics and disruption mitigation capability: an empirical investigation in China *Int J Logist Res Appl* **6** pp 277–95 doi:10.1080/13675567.2013.815695.
[7] Gualandris J and Kalchschmidt M 2015 Supply risk management and competitive advantage: a misfit model. *Int J Logist Manag* **26** pp 459–78 doi:10.1108/IJLM-05-2013-0062.
[8] Purvis L, Spall S, Naim M and Spiegler V 2016 Developing a resilient supply chain strategy during "boom" and "bust." *Prod Plan Control* **7287** pp 1–12 doi:10.1080/09537287.2016.1165306.
[9] Arani W, Mukulu E, Waiganjo E, and Wambua J Waiganjo E and Wambua J 2016 Strategic Sourcing an Antecedent of Supply Chain Resilience in Manufacturing Firms in Kenya *International Journal of Academic Research in Business and Social Science* **6** pp 1–18. doi:10.6007/IJARBSS/v6-i10/2317.
[10] Christopher M and Peck H 2004 Building the Resilient Supply Chain *Int J Logist Manag* **15** pp 1–14. doi:10.1108/09574090410700275.
[11] Erol O, Sauser B J and Mansouri M A 2010 A Framework for investigation into extended enterprise resilience. *Enterp Inf Syst* **4** pp 111–36 doi:10.1080/17517570903474304.
[12] Clark L A and Watson D 1995 Constructing validity: Basic issues in objective scale development. *Psychol Assess* **7** pp 309–19 doi:10.1037/1040-3590.7.3.309.
[13] Samuels P 2016 *Advice on Exploratory Factor Analysis* doi:10.13140/RG.2.1.5013.9766.
[14] Pujawan I N, Er M, Somboonwiwat T, Kritchanchai D 2014 Uncertainty and schedule instability in supply chain: insights from case studies *Int J Serv Oper Manag* **19** pp 468 doi:10.1504/IJSOM.2014.065670

Acknowledgements

We would like to thank the sponsors of the research especially my supervisor and the Department of Industrial Engineering ITS Surabaya. In addition, we would like to thank the referees for their valuable suggestions and comments.

Managing risks in the fisheries supply chain using House of Risk Framework (HOR) and Interpretive Structural Modeling (ISM)

T L T Nguyen[1*], T T Tran[1], T P Huynh[1], T K D Ho[1], A T Le[1] and T K H Do[1]
Department of Industrial Management, Can Tho University, Can Tho, Viet Nam

*ntlethuy@ctu.edu.vn

Abstract. One of the sectors which contributes importantly to the development of Vietnam economy is fishery industry. However, during recent year, it has been witnessed many difficulties on managing the performance of the fishery supply chain operations as a whole. In this paper, a framework for supply chain risk management (SCRM) is proposed. Initially, all the activities are mapped by using Supply Chain Operations Reference (SCOR) model. Next, the risk ranking is analyzed in House of Risk. Furthermore, interpretive structural modeling (ISM) is used to identify inter-relationships among supply chain risks and to visualize the risks according to their levels. For illustration, the model has been tested in several case studies with fishery companies in Can Tho, Mekong Delta. This study identifies 22 risk events and 20 risk agents through the supply chain. Also, the risk priority could be used for further House of Risk with proactive actions in future studies.

1. Introduction

The Mekong Delta is one of the great regions contributing to Vietnam's economy. According to Can Tho University' estimation, this region takes account for 70 % of nation's aquaculture areas and 60 % of nation's fish. Nevertheless, the growth rate is low, variable and unsteady. In fact, the common factors affecting the fishery industry are climate change, temperature, uncontrollable weather, flood-tide, disaster and disease. Also, the majority of fishery households in Mekong Delta have small extent of culturing lands, do not aware of cooperating with others as well as how to accommodate with the changing of climate. Moreover, the concerns of protecting environment restricts the using of modern technologies. Besides, some other factors such as quality requirements, production process, transportation, etc., can affect the supply chain as well.

Supply chain risks management is the implementation of strategies to manage both daily and extra-ordinary among supply chain based on continuous risk assessment. Each partner of supply chain has its risks linking from backward or forward one in supply chain adversely affecting the effectiveness of a whole chain. With the objective of reducing vulnerability and ensuring continuity, SCRM is collaboratively with partners in a supply chain or on your own as well as applies risk management tools to deal with risks and uncertainties caused by logistics related activities and resources in supply chain.

In this approach, we will analyze the activities of partners in supply chain as well as identify problems and essential risks. Some of risks can be solved, diminished, transferred whereas others are unavoidable. First, SCOR model (Supply Chain Operation Reference) is applied for the purpose of analyzing the activities according to five main stages including plan, resource, make, deliver, and return, of all partners among the fisheries supply chain. Second, using HOR (House of Risk) to assess risks and their roots as well as analyzing the relationship between risks and root causes because they can be caused by different sources and otherwise. Next, the inter-relationships among risks are defined and mapped by ISM (Interpretive Structural Model) to obtain the visual relationship of the problem. To sum up, the purpose of this research is to identify the risk and the source of agent of any risks that k affect the process of fisheries in Can Tho, Mekong Delta so that a sustainable fisheries supply chain can be achieved.

2. Literature review

2.1 The supply chain operations reference model (SCOR model)

The supply chain operations reference model (SCOR model) was developed in 1996 by the management consulting firm PRTM, now part of PricewaterhouseCoopers LLP (PwC) and AMR Research, and endorsed by the Supply-Chain Council (SCC). SCOR is a process reference model describes the business activities associated with satisfying a customer's demand, which include plan, source, make, deliver, and return [1][2]. Use of the model includes analyzing the current state of a company's processes and goals, quantifying operational performance, and comparing company performance to benchmark data. This reference model enables users to address, improve, and communicate supply chain management practices within and between all interested parties in the extended enterprise.

2.2 House of risk

House of risk is developed upon foundation of well-known model House of Quality (HOQ) but in sense of determining which risk actions to be tackled first and selecting a set of proactive actions deemed cost-effective to be prioritized. It is divided into two phases, House of risk 1 (HOR1) is used to determine which risk agents are to be given priority for preventive actions [3][4][5] whereas House of risk 2 (HOR-2) is to give priority to those actions considered effective but with reasonable money and resource commitments.

HOR first stage as the stage for the data input work has 8 steps as follows:

- Step 1 Identify the activities in the supply chain based on the SCOR model, with a view to facilitate the detection process in which the risk of potentially emerge (where are the risk).
- Step 2 Identify the entire incident risks that may appear on any activity in the supply chain.
- Step 3 Identify Severity Level or degree of impact of each risk event using a scale of 1-10.
- Step 4 Identification result (potential causes) an occurrence of the activity of the supply chain process, as a result will help to describe what disorders arising from any risk.
- Step 5 Identify the agency risk (risk agent), which detects any factors which may cause the occurrence risks identified in step.
- Step 6 Identification of correlation between an event to trigger agent risk. If an agency risk of causing a risk, it can be said there is a correlation. If a strong correlation is weighted 9; correlations are given weights 3 and a weight of 1 to the value of the correlation is weak.
- Step 7 Identify opportunities emergence (occurance) of each agent risks, to determine the risk of chance occurrence of an agent using a scale of 1-10.
- Step 8 Determination of the risk priority index value, priority will be used benchmark index for recommendation selecting agent which risks need to design a risk mitigation strategy.

2.3 Interpretive structural modeling (ISM)

Interpretive structural modeling (ISM) is a well-established methodology for identifying relationships among specific items, which define a problem or an issue [6][7][8]. This approach has been increasingly used by various researchers to represent the interrelationships among various elements related to the issue. ISM has several steps as follow [9][10][11][12]:

(1) Identification of variables: The key variables of the system are identified using literature study and brain storming sessions with the industry experts and academicians.
(2) Contextual relationship: A contextual relationship is identified among each variable (identified in step 1) with respect to which the pairs of variables would be examined. The contextual relationship is in the form of a matrix called the structural self-interaction matrix (SSIM).
Notations used to develop the SSIM:
V: Risk variable i leads to variable j A: Risk variable j leads to variable i
X: Risk variable i leads to variable j and vice versa O: No relationship between the variables
(3) Initial Reachability Matrix (IRM): The SSIM is then converted into a binary matrix, called initial reachability matrix by substituting V, A, X and O by 1 and 0 as per the following rules:
Rule 1: If the (i, j) entry in the SSIM is V, then the (i, j) entry in the reachability matrix becomes 1 and the (j, i) entry is 0.
Rule 2: If the (i, j) entry in the SSIM is A, then the (i, j) entry in the reachability matrix is 0 and the (j, i) entry becomes 1.
Rule 3: If the (i, j) entry in the SSIM is X, then the (i, j) entry in the reachability matrix becomes 1 and the (j, i) entry also becomes 1.
Rule 4: If the (i, j) entry in the SSIM is O, then the (i, j) entry in the reachability matrix becomes 0 and the (j, i) entry also becomes 0.
(4) Transitivity check: The reachability matrix is developed from the SSIM and the matrix is checked for transitivity. The transitivity of the contextual relation is a basic assumption made in ISM. It states that if variable A is related to B and B is related to C, then A is necessarily related to C.
(5) Levels: The transitivity matrix obtained in step (4) is converted into the canonical matrix format by arranging the elements according to their levels.
(6) Building the ISM model: Variables in each level are then connected based on their relationships as defined in the structural self-interaction matrix.

3. Methods

The data was collected from the companies through questionnaires formed based on the result of SCOR Model and House of risk framework. The sample size consist of 8 small and medium fishery companies located in Can Tho, Mekong Delta from the total of 50 companies (16%). The risk analysis was constructed by risk mapping, risk classifying based on SCOR and HOR. Also, ISM is proposed to identify the relationship among any events of risk for ranking the level of risks. Which risk directly affects to partner of fisheries supply chain will be positioned at level one while the lowest level is the risk that has less affection to partner. It helps us to select which risk should be tackled first in the constraint of budget and resource.

4. Results

4.1. The SCOR model and HOR framework for fisheries companies

Risks and causes found from activities in the planning, source, making, delivery and return of the fisheries supply chain are listed in the Table 1. According to survey, we obtain the aggregated of severity of risks, the probabilities of causes and the correlations among them are shown in Table 2.

Table 1. SCOR model with risks and causes of risks at five stages through fisheries supply chain

SCOR model	Activities in supply chain	Risks in supply chain (E_i)	Causes of risks (A_j)
Planing	– Planning product development – Planning for seafood procession and product quality – Understanding the contract or requirement of customers	– Cancel the contract (E1) – Depend on a supplier (E2)	– Do not have long- term plan (A1) – Manage the finance ineffectively (A2) – Weakness in suppliers selection (A3)

IOP Conf. Series: Materials Science and Engineering **337** (2018) 012030 doi:10.1088/1757-899X/337/1/012030

Table 1. SCOR model with risks and causes of risks at five stages through fisheries supply chain

SCOR model	Activities in supply chain	Risks in supply chain (E_i)	Causes of risks (A_j)
Source	– Build a supplier selection strategy – Forecast the seafood market in the coming time – Predict the cycle of entering new material sources – The process of sharing information on fishery quality requirements and production lead time for suppliers – Consolidate invoices, pay for the cost of orders – Maintain the relationship with suppliers and sub-contract company	– The price of input materials fluctuate high (E3) – The quality of input materials does not match standardization (E4) – Lack of high skilled workers (E5) – Lack of capital (E6)	– Natural disaster (drought, flood,...) (A4) – Environmental pollution (A5) – Economic crisis (A6) – Quantity limit (materials, products) from suppliers or sub-contract company (A7) – Difficult to compare the suppliers (A8) – Suppliers or sub-contractors went bankrupt (A9) – Production techniques are limited (A10)
Making	– Prepare for production schedule – Planning the amount of workers – Maintaining human resources attached to the company – Assign human resources among parts of company – Check the quality of products and production process – Controlling unexpected interruption in production system (devices, human...)	– Production process is delayed (E7) – Devices are out of order in production process (E8) – Regularly increasing production time (overtime) (E9) – Strike (E10) – Lack of materials (E11) – Errors in marking components which are used in production process (E12) – Products are contaminated, exceeded of proportion of heavy metal (E13) – Closed the company (E14)	– Changing production plan (A11) – Weakness in controlling system (quality of material, product, check hygienic of workers before production...) (A12) – Strict requirements for product (A13) – Low workers salary (A14)
Delivery	– Managing orders (amount, delivery date, received date) – Build shipping and distribution schedules – Managing the co-operative relationship with distributors	– Means of transportation are out of order regularly (E15) – Error in delivery (date, amount, type of product, address) (E16) – Delivery time of suppliers change many times (E17) – Reserved products/ materials are spoiled, increasing inventory cost (E18) – The risks of trade or negotiation with international ports (E19) – Exchange rate risks (E20)	– Less maintenance of machinery (A15) – Regularly late delivery (A16) – Long-term shortage of products in stock (A17) – Lack of collaboration with outside organizations (A18)
Return	– Adapting to the changes of customers' needs, time, age... – Managing the co-operative relationship with customers	– Do not meet the expectation of customers (E21) – Products are refunded (E22)	– Do not note the orders in detail (wrong date, amount, type of product) (A19) – Quality of products does not match requirements (A20)

For examples, we compute ARP_1 as follow:

$$ARP_1 = O_1 * (R_{11} * S_1 + R_{51} * S_5 + R_{61} * S_6 + R_{14;1} * S_{14} + R_{19;1} * S_{19})$$
$$= 5*(3*8 + 9*7 + 3*5 + 1*1 + 1*7) = 550$$

Based on this calculation, the aggregate risk potentials (ARPj) are obtained in Table 2.

Table 2. HOR model

Risks	Causes of risk																				S_i
	A1	A2	A3	A4	A5	A6	A7	A8	A9	A10	A11	A12	A13	A14	A15	A16	A17	A18	A19	A20	
E1	3			9			3						3			9					6
E2			3					9													4
E3						3															6
E4			9		1							9									5
E5	9	3																			7
E6	3	9																			5
E7		1		9	3						3	3									3
E8															9						5
E9							3				3										4
E10														9							1
E11			9				3		3												6
E12										9		3									8
E13										9		9									9
E14	1	9																			1
E15												1			9						6
E16												3							9		4
E17			1																		5
E18												3					9				4
E19	1																	3			7
E20				3		9															3
E21													9								4
E22																3				9	5
O_j	5	2	1	1	2	2	3	1	3	2	4	2	4	4	1	2	3	2	1	1	
ARP_j	550	160	116	90	40	90	162	36	54	306	108	378	144	36	99	30	108	42	36	45	
P_j	1	5	7	11	16	12	4	17	13	3	8	2	6	17	10	20	8	15	17	14	

4.2. Building the ISM model

At the beginning, based on personal insights and experts' opinions, we have developed the structural self-interaction matrix (SSIM). Then, we convert the SSIM to the IRM matrix based on the principles developed in step 3 and step 4.

Table 3. Transitional IRM matrix.

	E22	E21	E20	E19	E18	E17	E16	E15	E14	E13	E12	E11	E10	E9	E8	E7	E6	E5	E4	E3	E2	E1
E1	1	0	0	0	0	0	0	0	1*	0	0	1	0	0	0	1*	0	0	0	0	0	1
E2	1*	1*	0	1*	1*	1	1	0	1*	1	0	1	0	0	0	1*	0	0	1	1	1	1
E3	1*	0	0	0	0	0	0	0	1*	0	0	1	0	0	0	1*	0	0	0	1	0	1
E4	1	1*	0	1*	1*	0	0	0	1*	1	0	1	0	0	0	1*	0	0	1	0	0	1
E5	0	0	0	0	0	0	0	0	0	0	0	0	1*	1	0	1	0	1	0	0	0	1*
E6	0	0	0	0	0	0	0	0	1	0	0	1	1	1*	0	1	1	0	0	0	0	1*
E7	1*	0	0	0	0	0	0	0	0	0	0	1*	1*	1	0	1	0	0	0	0	0	1
E8	0	0	0	0	0	0	0	0	0	0	0	0	1*	1	1	1	0	0	0	0	0	1*
E9	0	0	0	0	0	0	0	0	1*	0	0	0	1	1	0	0	0	0	0	0	0	0
E10	0	0	0	0	0	0	0	0	1	0	0	0	1	0	0	0	0	0	0	0	0	1*
E11	1*	0	0	0	0	0	0	0	1	0	0	1	0	1*	0	1	0	0	0	0	0	1
E12	1*	1	1*	1	1	0	0	0	0	1	1	1*	0	0	0	1*	0	0	0	0	0	1*
E13	1	1	1*	1	1	0	0	0	0	1	0	1*	0	0	0	1*	0	0	0	0	0	1
E14	1*	0	0	0	0	0	0	0	1	0	0	1*	0	0	0	0	0	0	0	0	0	1
E15	1*	1*	0	0	0	1	1	1	0	0	0	1*	0	1*	0	1	0	0	0	0	0	1*
E16	1	1	0	0	0	1	1	0	0	0	0	1*	0	1*	0	1	0	0	0	0	0	1

	E22	E21	E20	E19	E18	E17	E16	E15	E14	E13	E12	E11	E10	E9	E8	E7	E6	E5	E4	E3	E2	E1
E17	1*	1	0	0	0	1	0	0	1*	0	0	1	0	1*	0	1	0	0	0	0	0	1
E18	0	0	0	0	1	0	0	0	1*	0	0	1	0	1*	0	1	0	0	0	0	0	1*
E19	1	0	1	1	0	0	0	0	0	0	0	1*	0	0	0	0	0	0	0	0	0	1
E20	1*	0	1	1	0	0	0	0	0	0	0	0	0	0	0	0	0	0	0	0	0	1*
E21	1	1	0	0	0	0	0	0	0	0	0	1*	0	0	0	0	0	0	0	0	0	1
E22	1	0	0	0	0	0	0	0	0	0	0	1*	0	0	0	0	0	0	0	0	0	1

From transitional IRM matrix in Table 3, we determine the level of each risk based on the principles developed in step 5. By dropping the row of risks which have been chosen and these risks in others Reachability set, Antecedent set and intersections simultaneously. Next, keep doing step by step until the risks are filled up. The following table shows the degree of risks.

Table 4. Result of level partitioning of reachability matrix.

Risk	Reachability set	Antecedent set	Intersection set	Level
E1	1,7,11,14,22	1,2,3,4,5,6,7,8,10,11,12,13,14,15,16,17,18,19,20,21,22	1,7,11,14,22	I
E22	1,11,22	1,2,3,4,7,11,12,13,14,15,16,17,19,20,21,22	1,11,22	I
E14	1,11,14,22	1,2,3,4,6,9,10,11,14,17,18	1,11,14	II
E20	1,19,20,22	12,13,19,20	19,20	II
E10	1,10,14	5,6,7,8,9,10	10	III
E9	9,10,14	5,6,7,8,9,11,15,16,17,18	9	IV
E7	1,7,9,10,11,22	1,2,3,4,5,6,7,8,11,12,13,15,16,17,18	1,7,11	V
E11	1,7,9,11,14,22	1,2,3,4,6,7,11,12,13,14,15,16,18,19,21,22	1,7,11,14,22	V
E3	1,3,7,11,14,22	2,3	3	VI
E5	1,5,7,9,10	5	5	VI
E6	1,6,7,9,10,11,14	6	6	VI
E8	1,7,8,9,10	8	8	VI
E18	1,7,9,11,14,18	2,4,12,13,18	18	VI
E19	1,11,19,20,22	2,4,12,13,19,20	19,20	VI
E21	1,11,21,22	2,4,12,13,15,16,17,21	21	VI
E13	1,7,11,12,13,18,19,20,21,22	2,4,12,13	12,13	VII
E17	1,7,9,11,14,17,21,22	2,15,16,17	17	VII
E4	1,4,7,11,13,14,18,19,21,22	2,4	4	VIII
E12	1,7,11,12,13,18,19,20,21,22	12,17	12	VIII
E16	1,7,9,11,16,17,21,22	2,15,16	16	VIII
E2	1,2,3,4,7,11,13,14,16,17,18,19,21,22	2	2	IX
E15	1,7,9,11,15,16,17,21,22	15	15	IX

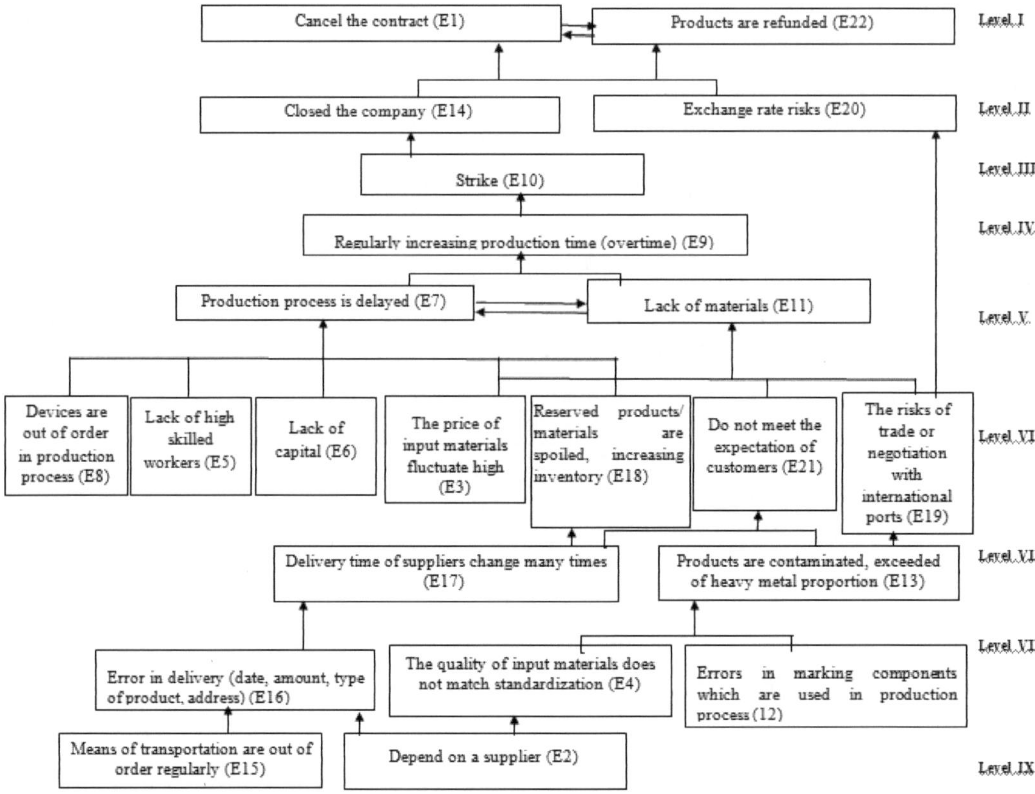

Figure 1. Interpretive structural model (ISM) for risks relationship in fisheries companies

Based on the results of level partitioning, we developed the interpretive structural model presented in Figure 1. In this diagram, the risks "Depend on a supplier" (E2) and "Means of transportation are out of order regularly" (E15) are classified in the lowest level (level 9) of the system, with the risks E1 "Cancel the contract" and E22 "Products are refunded" having the highest influence on the others.

5. Discussion and conclusions

There are many types of risks that affect the operation of aquaculture supply chain, but we just focus on the most common types of risks that regularly occur. Results showed that 22 risk events and 20 risk agents are identified, and the two most important risks are A1 "Do not have long term plan" and A13 "Strict product requirement". This risk priority information will be used effectively for further House of Risk with proactive actions. Moreover, ISM model is used to identify the relationship among risks and draw a relationship diagram so that companies may have a general knowledge of which risks will directly or indirectly influence on their supply chains.

In conclusion, the paper proposes a model for the Risk mapping and Risk priority calculation using ISM and SCOR-HOR for the applications in fisheries supply chain. The findings would help managers to analyse and to take actions for managing the risk factors to improve the performance of their organizations effectively. However, the sample size of experts is small and to remove the biasness of opinion, the model can be further validated using Structural Equation Modelling (SEM) in the future.

In addition, the dependencies between risk events or relative severity of risk events can be further analysed by combining between Analytical Network Process (ANP) and Interpretive Structural Modelling (ISM).

References

[1] Hessam Z H and Ava S 2011 *Int.J. Eco. Management Sci.* **3** pp 60 – 72

[2] Ntabe E N, Lebel L, Munson A D and Santa-Eulalia L A 2015 A systematic literature review of the supply chain operations reference (SCOR) model application with special attention to environmental issues *Int. J. of Product. Eco.* **169** pp 310–332

[3] Pujawan I N and Laudine H G 2009 House of Risk: a model for proactive supply chain risk management. *Business Process Management Journal.* **15** (6) pp 953 – 976

[4] Titik K, Adi D G and Heru C R Mapping of Supply Chain Risk in Industrial Furniture Base on House of Risk Framework *Euro. J. of Business and Management* **7** (34) 104 – 115

[5] Dewanti A, Putu D K and Martian S 2015 Managing quality risk in a frozen shrimp supply chain: a case study *Procedia Manuf.* **4** pp 252 – 260

[6] Arvind J and Mohd A 2014 Analysis of Barriers to Implement Green Supply Chain Management (GSCM) Practices: An Interpretive Structural Modeling (ISM) Approach *Procedia Engineering* **97** pp 2157-66

[7] Venkatesh V G, Snehal R and Sriyans P 2015 Analysis on supply chain risks in Indian apparel retail chains and proposal of risk prioritization model using Interpretive structural modelling *J. of Retail. and Consu. Serv* **26** 153–167

[8] Devendra K Y and Akhilesh B 2015 Analysis of critical success factors of humanitarian supply chain: An application of Interpretive Structural Modeling *Int. J. Disaster Risk Reduction* **12** pp 213 – 225

[9] Sage A P 1977 *Interpretive Structural Modeling: Methodology for Large-scale Systems* (New York: McGraw-Hill)

[10] Faisal M N, Banwet D K and Shankar R 2006 Supply chain risk mitigation: modelling the enablers *J. of Bus. Process Manag.* **12** (4) pp 535–552

[11] Sushil 2012 Interpreting the Interpretive Structural Model *Global J. of Flexib. Sys. Manage.* **13** (2) pp 87–106

[12] Jharkharia S and Shankar R 2004 IT enablement of supply chains: modeling the enablers *Int. J. Prod. Perform. Manag.* **53** (8) pp 700–712

Identifying risk event in Indonesian fresh meat supply chain

H.C. Wahyuni[1,2*], I.Vanany[1], U. Ciptomulyono[1]
[1]Department of Industrial Engineering, Institut Teknologi Sepuluh Nopember, Kampus ITS Sukolilo-Surabaya 60111, Indonesia
[2]Industrial Engineering Department, Universitas Muhammadiyah Sidoarjo

*hanacatur@umsida.ac.id

Abstract. The aim of this paper is to identify risk issues in Indonesian fresh meat supply chain from the farm until to the "plate". The critical points for food safety in physical fresh meat product flow are also identified. The paper employed one case study in the Indonesian fresh meat company by conducting observations and in-depth three stages of interviews. At the first interview, the players, process, and activities in the fresh meat industry were identified. In the second interview, critical points for food safety were recognized. The risk events in each player and process were identified in the last interview. The research will be conducted in three stages, but this article focuses on risk identification process (first stage) only. The second stage is measuring risk and the third stage focuses on determining the value of risk priority. The results showed that there were four players in the fresh meat supply chain: livestock (source), slaughter (make), distributor and retail (deliver). Each player has different activities and identified 16 risk events in the fresh meat supply chain. Some of the strategies that can be used to reduce the occurrence of such risks include improving the ability of laborers on food safety systems, improving cutting equipment and distribution processes

1. Introduction

The food supply chain is one of supply chain domain which suffers a lot of uncertainty in its functioning [1]. Structure of the supply chain relatively difference between one country to others countries [2], such as in the United Kingdom [3], Japan [4], and Australia [5]. Besides that, types of food products also have uniqueness based on types of food products such as agrifood [6] and beef products [7]. In the flow of food products in the supply chain, there is a risk that food contaminated with hazardous materials for human health, and food cannot be consumed. A risk is defined as consumer perceptions of uncertainties and events to be received as a result of purchasing a good or service [8]. Basically, there are two main components of risk, namely: (1) risk category of product, which is the perception of a person who must be accepted because of buying a certain product and (2) risk to product specification (product specific risk), which is the perception of the association to the risk that will be accepted on a product [9].

In the context of the supply chain of food products, the possible risks are the contamination of the food from the chemical, biological and physical aspects of the food security indicator or the risk that occurs due to the behavior that causes the food not to be consumed.This condition is driven by several factors that influence, among others, lack of institutional support (government policy) in terms of law enforcement on food safety, lack of knowledge and awareness of farmers, as a source in the supply chain process of microbiology as a whole [10] and less optimal implementation of food safety in the process of making food [11] resulted in growth rate, improvement of quality and increase of product competitiveness of food industry product experiencing obstacles [12] .The objective of risk management in the supply chain is to increase the capability so as to reduce the probability of occurrence of risk [13]. Quantitatively, the risk can be measured so that it can be estimated its severity [1].

The contamination of hazardous substances in the supply chain is also affected by the negative manipulations perpetrated by the actors in the supply chain resulting in anxiety and inconvenience to consumers, requiring information that ensures that the products consumed to meet the quality standards and halal requirements [14]. Therefore, in the supply chain, companies should be able to identify the activities that are the source of risk so that food is unsafe to eat. Some sources of the risks if not managed properly will cause risks that have a negative impact on food products, such as contaminated Salmonella ssp, S.aureus, E Coli, Shigella spp, Fecal coliform [15].

A fresh meat is one of the food sources of livestock are indispensable as a source of functional protein and growth, especially in children of an early age where at that age the rate of growth and development of brain cells is very high [16]. However, meat is also food that is at high risk of being contaminated if not managed properly. In Indonesia, several incidents due to meat poisoning occurred in Purbalingga resulting in 93 people in hospital [17], and 70 people poisoned in Nunukan District [18]. Therefore, it is necessary to identify activities at risk in the supply chain of meat. This condition indicates that fresh meat needs to be managed in order to avoid the risk of harm to health. Several preliminary studies on fresh meat supply chains are associated with improved meat quality with traceability [19] .With the knowledge of various kinds of the risk event, it is expected to be a good meat management so that the negative impact of the risk can be avoided.

2. Research Method

SCOR (Supply Chain Operations References) model are utilized to develop the research framework which consists of the source, process, and delivery. Figure 1 shows the proposed research framework based on the SCOR business process model. Each stage of the business process is carried out by each actor/player with different activities and different risks. Any activity that does carry risks that can be used as a basis for improvement activities

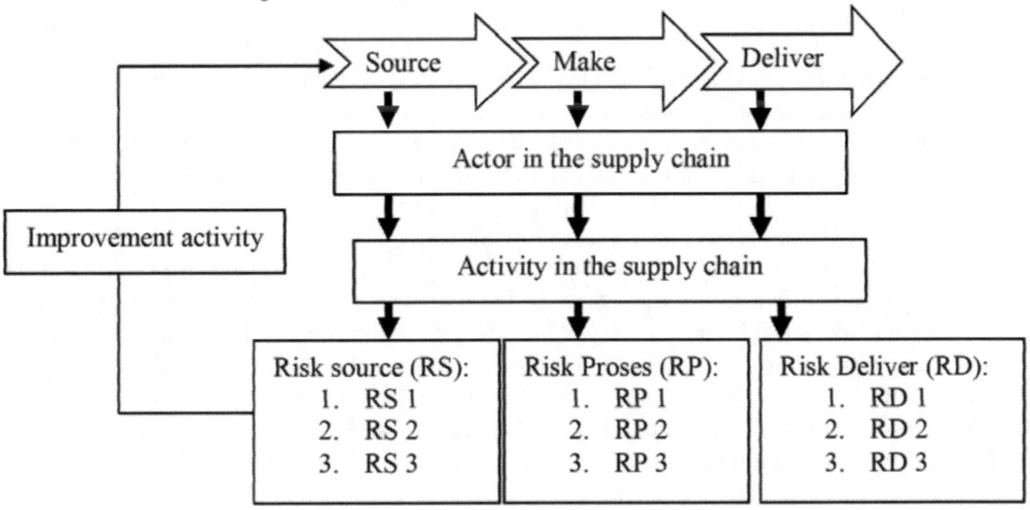

Figure 1. Research framework.

The research stages were conducted in three (3) stages, namely (1) risk event identification, (2) risk measurement, (3) determination of Risk Priority Value (RPV). The detail of stages is described in below.

Stage 1. The identification of risk events

This phase aims to: (1) identify the scope of food safety and halal integration in the supply chain, (2) identification of risk factor types at each stage of the supply chain so that information about risk activities for each Player (risk event), (3) the impact of each risk event on the risk agent (see Table 1)

Stage 2. Risk measurement

This phase aims to measure the various types of risks identified in phase 1. The risk assessment conducted in this phase will generate risk value for each risk event and serve to prepare risk priority and risk classification for each player in the supply chain. The measurement of risk value for each risk event is based on the Likelihood (L), Impact (I) and Detection (D) values. The technique of data retrieval is done by using questionnaire with scale 1 s / d 10 for S, I and D.

Stage 3. Determination of RPV

This phase aims to establish an integrated risk priority in the food supply chain. The preparation of the RPV is based on the value gained during the risk assessment. Data processing is done by ranking risk value for each risk event from the biggest value to the smallest value. The largest risk value indicates the risk level of the activity is large, meaning that the implementation of the activity is at risk of contamination of harmful substances.

Table 1. The identification of risk events template.

Process business	**Player**	**Activity**	**Risk Event**	**Likelihood (L)**	**Impact (I)**	**Risk Score (LxI)**	**Detection (D)**	**RPV (LxIxD)**
Source	PS	PS1	SR1	....	...	....	...	...
		PS2	SR2	...	....	...	...	...
		PS3	SR3	....	...	...	....	....
Make	PM	PM1	MR1	...	...	...	...	...
		PM2	MR2	...	...	...	...	...
		PM3	MR3	...	...	...	..	...
Deliver	PD	PD1	PD1	...	...	...	...	...
		PD2	PD2	...	...	...	...	...
		PD3	PD3	...	...	...	...	...

Indonesian food industries has experienced several accidents of risk such as SARS, bird flu, and other food contaminations [20][21]. Indonesian manager believed that the implementation of supply chain risk is important to be conducted in order to reduce cost, loss of profit and others [22]**Error! Reference source not found.** and the relationships among risks should be identified to manage the risk [23].

The case study in this study is a fresh beef located in Sidoarjo, East Java, Indonesia. The data collection method was carried out by using survey and interview in February 2017. The survey aims to determine the flow of processes and information flow of fresh meat supply chain management. The interviews were conducted by multiple various important information related to the management of fresh meat supply chain with the manager who understands the process and quality of the product. Three stages were implemented from this instrument: Firstly, the players, process, and activities in the fresh meat industry were identified. In the second interview, the critical points for food safety were also identified. In the third interview, the risk events of each player and processes were identified.

3. Results and Discussions

In general, meat is food that is easily obtained in the market. Meat supply chain shows the process of meat procurement from the beginning until the meat is received by consumers. The length of the supply chain is strongly influenced by spatial and vertical marketing reach, where cattle sold out to the island have long supply chains than those sold in local markets [24]. The results showed that the model of fresh meat supply chain in Sidoarjo can be seen in Figure 2.

Figure 2 shows 4 actors in the fresh meat supply chain. Firstly, farmers are the beginning of the supply chain that serves as a provider of raw materials, namely cattle. Secondly, distributor of fresh meat, a part that will distribute fresh meat to retailers or small businesses such as food vendors meatballs, sausage industry, and others. Thirdly, fresh meat retailers, a part that will distribute fresh meat to the final consumer. Finally, retailers are the last part of the fresh meat supply chain process.Each stage of the supply chain poses a risk that has a negative impact on the safety of fresh meat food.The results of identification of risks in fresh meat can be seen in Table 2:

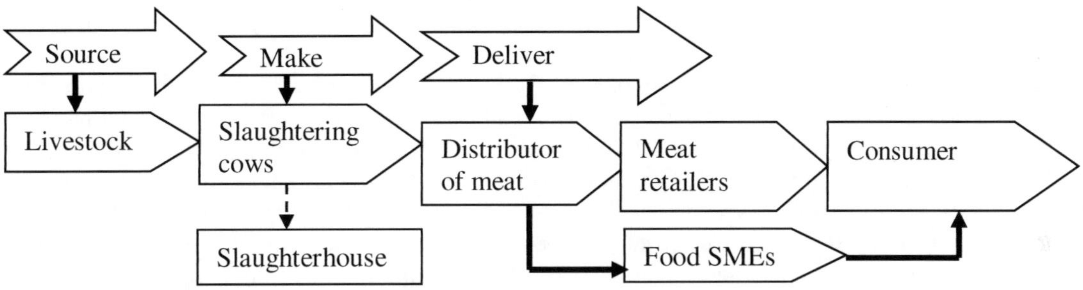

Figure 2. Fresh meat supply chain.

Table 2. Identification of risks in fresh meat supply chains.

Player of supply chain	Activity	Risk Event
Livestock	Providing cattle as the raw materials of fresh meat.	Animals are not legal (L1)
	Delivering cattle to slaughterhouses.	Depressed animals (L2)
Slaughter	Animal antemortem examination	Animals in sickness (S1) Found infectious diseases (S2)
	Slaughter	Animals in a state of stress when cut (S3) Knife is not sharp (S4) Officer is not disciplined (S5)
	Postmortem examination	Found infectious diseases in animals (S6)
	Maintenance of sanitation	The drains are not well maintaned (S7).
Distributor of fresh meat	Delivery of fresh meat to distributors	Non standard transport system (D1)
	Meat sales to retailers	Non standard storage system (D2) Meat marketed under open conditions (D3)
Fresh meat retailers	Sales of meat to end consumers	Meat marketed under open conditions (Fr1) Meat is often held by consumers (Fr2) Not using plastic packaging type PE (polyethylene) and PP (polypropylene) (Fr3) Stored in high air temperature (Fr4)

The table 2 shows the risks occurring at each stage of the supply chain. Totaly, the total of the risks events is 16 risk events in the fresh meat supply chain, consisting of 2 risk events on livestock, 7 risk events on slaughter, 3 risk events on distribution and 4 risk events in retailers. This condition indicates that every player in the supply chain has different characteristics. In order to improve the effectiveness and efficiency of risk prevention on fresh meat, it is necessary to conduct various food safety related measures. Some strategies that can be used to mitigate risk events including improving workers' understanding of food security systems, complying with government requirements on slaughtering animals, slaughter equipment, physical environment and equipment used for cutting and distribution processes. Therefore, to prevent risks that have a negative impact on human health, it is necessary to implement the application of food safety system and the improvement of integrated infrastructure for each actor in the supply chain. A large number of actors in the fresh meat supply chain encourages inter-agency coordination. This coordination must ensure sustainability in the risk prevention process in the fresh meat supply chain.

4. Conclusion

Fresh meat is a food source of the protein important for human health. Fresh meat supply chain is the process of providing raw materials to consumers. The results showed that there were found 4 actors in the fresh meat supply chain, namely: breeders, slaughterers in the slaughterhouse, distributors, and retailers. Any activity perpetrated by the offender poses a risk that has a negative impact on human health. In overall, found 16 risk events that will be a source of risk in the supply chain of fresh meat. To reduce the impact of risks to human health, it is necessary to implement an integrated food security system among actors in the fresh meat supply chain. For example, to maintain the quality of fresh meat in delivery, it is necessary to use the type of vehicle in accordance with food safety standards, in terms of temperature and packaging.

The study was limited to the risk identification process in fresh meat. This study shows the type of risk activities performed on the process of fresh meat supply chain. However the magnitude of the risks posed by these activities and other aspects that affect food security in the supply chain has not yet included. For future research, it is necessary to consider the aspects of food safety, and halal in the supply chain. The relationships among risks will be investigated in the future.

Reference

[1] Diabat A, Govidan K and Panicker V V 2012 Supply chain risk management and its mitigation in a food industry *Int. Journal of Production Research* **50** (11) pp 3039-50

[2] Georgise F B, Thoben K D and Seiferi M 2014 Identifying the characteristics of the supply chain processes in developing country: A manufacturing industry perspective *Wseas Transaction on Business and Economics* **11**.

[3] Bourlakis M and Weightman P W H 2004 *Food Supply Chain Management* Blackwell Publishing Ltd.

[4] Saidon I M, Radzi R M and Ghani N A 2015 Food Supply Chain Integration: Learning From The Supply Chain Superpower, *Int. Journal of Managing Value and Supply Chains (IJMVSC)* **6** (4)

[5] Australian Goverment 2012 *Resilience in the Australian food supply chain* Departement of Agriculture, Fisheries, and Forestry

[6] Akhtar P, Tse Y K, Khan Z and Rao-Nicholson R 2016 Data-driven and adaptive leadership contributing to sustainability: Global agri-food supply chains connected with emerging markets. *Int. Journal of Production Economics* **181** pp 392-401

[7] King R P, Digiacomo G and Ortmann G F 2014 Beef Case Studies in the Minneapolis– St. Paul– Bloomington MSA *Growing Local, Case Studi On Local Food Supply Chain*, University Of Nebraska Press, Lincoln And London.

[8] Dwoling G and Staelin R 1994 A Model Of Perceived Risk And Intended Risk Handling Activity *Journal of Cunsumer Research* **21** pp 119-134.

[9] Fearne A, Hornibrook S and Dedman S 2001 The Management of Perceived Risk In The Food Supply Chain: A Comparative Study of Retailer Led Beef Quality Assurance Schemes In Germany And Italy *Int. Food And Agribusiness Management Review* **4** pp 19-36.

[10] Kirezieva K, Luning P A, Jacxens L, Allende A, Johanessen G S, Tondo E C, Rajkovic A, Uyttendaele M, and Boekel M A J S 2015 Factors affecting the status of food safety management systems in the global fresh produce chain *Food Control* **52** pp 85-97

[11] Herzman J and Barrash D 2007 An assessment of food safety knowledge and practices of catering employees *British Food Journal* **109** (7) pp 562-576

[12] Raharja S, Munarso S, and Puspitasari S 2012 Perbaikan dan Evaluasi Penerapan Sistem Manajemen Mutu Pada Industri Pengolahan Tahu (Studi Kasus di UD Cinta Sari, DIY) *Manajemen IKM* **7** (1) pp 28- 36.

[13] Pujawan N and Geraldin L H 2009 House of risk: a model for proactive supply chain risk management, *Business Process Management Journal* **15** (60 pp 953-967

[14] Purnomo D, Pujianto T, and Maulana R 2015 Information System Development on Halal Supply Chain Traceability (Case Study: Supply Chain of Beef Meat Balls) *Journal Of Halal Research* **1** (1) pp 19-24

[15] Shabbithi A, Reddi L S G D N, Kumr R N 2017 Identifying Critical Risk Practices Among Street Food Handlers *British Food Journal* **119** (2) pp 390-400

[16] Bahri S, Sani Y, and Indraningsih 2006 Beberapa Faktor Yang Mempengaruhi Kemanan Pangan Asal Ternak Di Indonesia *Wartazoa* **16** (1)

[17] Republika co.id *Korban Keracunan Daging Ayam Di Purbalingga Bertambah Jadi 93 Orang* accessed on May 26, 2017

[18] Tribun Kaltim *70 orang yang kerancunan diduga karena santap rendang daging asal Malaysia* accessed on May 26, 2017

[19] Pizzuti T, Mirabelli G, Grasco G, and Paldino G 2017 MESCO (MEat Supply Chain Ontology): An ontology for supporting traceability in the meat supply chain *Food Control,* Vol **72**, February pp 123-133.

[20] Vanany I and Zailani S 2010 Urgency In managing the risk in supply chain amongst Indonesian manufacturing companies *Int. Business Management* **4** (4) pp 199-208.

[21] Vanany I, Zailani S, and Pujawan N 2009 Supply chain risk management: Literature review and future research *Int. Journal of Information System and Supply Chain Management (IJISSCM)* **2** (1) pp 16-22.

[22] Vanany I, Zailani S and Rusdiansyah A 2007 Supply Chain Risk Management (SCRM) in the Indonesian Manufacturing Companies: Survey from Manager's Perspectives. *Paper presented at the The 2nd Int. Conference Operations and Supply Chain Management* (OSCM)

[23] Indrawati C D, Karningsih P D, and Vanany I 2015 Development of supply chain risks interrelationships model using interpretive structural modeling and analytical network process, *Journal of Proc. Series* (1). IPTEK

[24] Kruse H 2015 Food safety in an international perspective *Journal of Consumer Protection and Food Safety*, **10**, pp 105-107.

IOP Conf. Series: Materials Science and Engineering **337** (2018) 012032 doi:10.1088/1757-899X/337/1/012032

Shifting orders among suppliers considering risk, price and transportation cost

C Revitasari[1*] and I N Pujawan[1]
[1]Department of Industrial Engineering, Institut Teknologi Sepuluh Nopember, Kampus ITS Sukolilo-Surabaya 60111, Indonesia

*revitasaricindy@gmail.com

Abstract. Supplier order allocation is an important supply chain decision for an enterprise. It is related to the supplier's function as a raw material provider and other supporting materials that will be used in production process. Most of works on order allocation has been based on costs and other supply chain performance, but very limited of them taking risks into consideration. In this paper we address the problem of order allocation of a single commodity sourced from multiple suppliers considering supply risks in addition to the attempt of minimizing transportation costs. The supply chain risk was investigated and a procedure was proposed in the risk mitigation phase as a form of risk profile. The objective including risk profile in order allocation is to maximize the product flow from a risky supplier to a relatively less risky supplier. The proposed procedure is applied to a sugar company. The result suggests that order allocations should be maximized to suppliers that have a relatively low risk and minimized to suppliers that have a relatively larger risks.

1. Introduction

In today's competitive environment, companies have to optimize their business processes and improve overall performance of their supply chain. Every company is attempting to meet demand, improving the quality and reducing costs to optimize the business processes [1]. In most industries, cost of raw materials and components formed as a major part of production cost until up to 70% [2]. An efficient and effective supply chain depends on how good the company in selecting the suppliers that will provide materials. The right decisions would result in supplier being able to supply quality materials at the right time, and thus would be able to reduce not only purchasing cost but also enhance the competitiveness of the company [3]. This is related to the function of the supplier itself as a provider of raw materials and supporting materials that will be used in the production process. In procurement, companies have to pay attention on various costs including the procurement cost as well as transportation cost [4]. Transportation costs should be considered in determining the amount of the order and also the removal of order to improve the efficiency of the overall supply chain [5].

Some published works only focused on creating an optimal order allocation for each supplier to fulfill demands of each manufacturer [6][3][4][7][2]. Most of these works however only focus on minimizing overall supply chain cost, but neglect the importance of reducing risks. Among the few authors, Oguzhan and Erol [8] proposed a model that takes risk into account when allocating orders to suppliers. Risk factors are used to determine an optimal order allocation that can maximize orders to suppliers that have a relatively low risks. Each supplier has different risks, therefore it is necessary to analyze risk for each supplier, then optimize the order allocation by entering the risk factor in the model [9].

In this paper we present a model that take into account supply chain risks when making order allocation to suppliers. A two-stage model is applied. First, the optimization model is developed for order allocation without considering risk. The model is simply to optimize order allocation with an objective function of cost minimization. Second, a risk profile for each supplier is assessed. These risk profile values will then become the basis for shifting order allocation, mainly from higher risk suppliers to lower risk suppliers as long as the capacity is still available. To do risk assessment, a simple risk assessment tool is developed. This includes such risk as delivery delay, quality problems, etc. There are various risks that could happen in the supply chain. For an example of risk associated with supply chain, see for example Pujawan and Geraldin [10] who list various supply chain risks in a fertilizer company and Parenreng et al. [11] that present risks of tuna supply chain. We apply the procedure to a sugar company.

2. Proposed Procedure of Procurement Plan and Shifting Orders Among Suppliers

In this research, a case study has been conducted in a company producing sugar. One of the materials needed is soda caustic which is used in this paper as a case example. Tables 1 – 4 present some basic data that we will need as input parameters for the model. Table 1 presents capacity and unit price. Table 2 is demand data which represents three manufacturers. Table 3 presents' transportation costs and table 4 is the results of risk assessments. The higher the risk profile values, the higher is the risk associated with the corresponding supplier.

Table 1. Capacity and unit purchasing price of each supplier

Suppliers	**1**	**2**	**3**	**4**	**Total**
Capacity (C_i) per kg	31800	30200	29450	28800	120250
Unit Price (P_i) per kg	8300	8100	8400	8650	33450

Table 2. Demand of each manufacturer/ assembler

Manufacturers/assemblers	**1**	**2**	**3**	**Total**
Demand (D_i) –kg	38400	34600	29650	102650

Table 3. Unit Transportation Costs from suppliers to manufacturers/assemblers (T_{ij})

		Manufacturers/assemblers (j)		
		1	**2**	**3**
Suppliers (i)	**1**	63.64	55.45	70.45
	2	97.73	102.27	95.45
	3	65.91	56.82	72.73
	4	59.09	56.82	68.18

Table 4. Risk Profile of each supplier.

	Supplier 1	Supplier 2	Supplier 3	Supplier 4
Risk profiles(R_{Ti})	58	24	44	32

2.1 Initial Procurement Plan

The first stage is optimizing the procurement plan without consideration of risk profile. The procurement quantity of each supplier is obtained by the cost criterion.

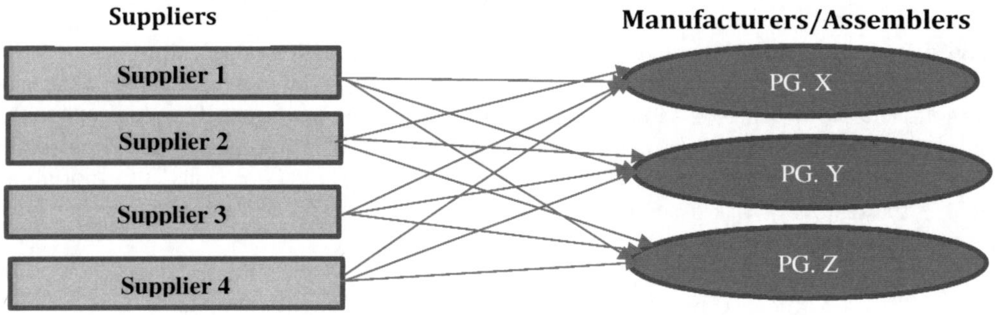

Figure 1. Supply Chain Network

Figure 1 shows the network where suppliers can supply any plant and each plant may allocate orders to the available suppliers. The optimization model below is used to find the optimum allocation decision. The objective function is to minimize the total cost which consists of purchasing costs and transportation costs. The constraints include consideration of capacity and demand. Supply from a supplier to any plant is limited by the capacity (2) and supply will not exceed the demand.

$$Mincost = \sum_{ieV_1} P_i \sum_{jeV_2} y_{ij} + \sum_{ieV_i} \sum_{jeV_2} T_{ij} * y_{ij} \quad (1)$$
$$\sum_{jeV_2} y_{ij} \leq C_i i \in V_1 \quad (2)$$
$$\sum_{ieV_1} y_{ij} \geq D_j i \in V_2 \quad (3)$$
$$y_{ij} \geq 0 \quad (4)$$

Where :

i	: suppliers	P_i	: unit purchasing price of supplier i
T_{ij}	: unit transportation cost from supplier i to manufacturer j	y_{ij}	: quantity to be transported from supplier i to manufacturer j
P_i	: unit purchasing price of supplier i	C_i	: capacity of supplier i
J	: manufacturers/assemblers	Dj	: demand of manufacturer j

After imposing the input parameters for both objective function and constraints, the model will have the following form.

Mincost = $8364y_{11}+ 8355y_{12}+ 8370y_{13}+ 8198y_{21}+ 8202y_{22}+ 8195y_{23}+ 8466y_{31}+ 8457y_{32}+ 8473y_{33} + 8709y_{41}+ 8707y_{42} + 8718y_{43}$

Subject To

- $y_{11} + y_{12} + y_{13} \leq 31800$
- $y_{21} + y_{22} + y_{23} \leq 30200$
- $y_{31} + y_{32} + y_{33} \leq 29450$
- $y_{41} + y_{42} + y_{43} \leq 28800$
- $y_{11} + y_{21} + y_{31} + y_{41} \leq 38400$
- $y_{12} + y_{22} + y_{32} + y_{42} \leq 34600$
- $y_{13} + y_{23} + y_{33} + y_{43} \leq 29650$
- $y_{ij} \geq 0$

Optimal Solution of processing the formulation using LINGO software, as follows.

Table 5. Optimal Solution that minimizes total costs

	Manufacturers/Assemblers (j)			
Suppliers (i)	**1**	**2**	**3**	**Total (kg)**
1	0	31800	0	31800
2	550	0	29650	30200
3	26650	2800	0	29450
4	11200	0	0	11200
Total (kg)	38400	34600	29650	102650

Table 5 shows the optimal order allocation for each supplier to each manufacturer in accordance to the objective function, constrains and data of the case company. However, the order allocation in Table 4 has not considered the risk factor. In this study, risk factor is included in the procurement plan to reduce risk that will happen in the order allocation decision between supplier and manufacturer. Shifting of orders from higher risk suppliers to lower risk suppliers will be described in the following section.

2.2 Shifting Orders Based on Risk Profile Value

According to Table 4, Supplier 2 is most reliable and Supplier 1 is the most risky supplier. Product Transfer will be performed from a risky supplier to a relatively less risky supplier by using value in Table 4. As shown in Table 4, supplier 2 has the risk profile value of 24, which is the lowest. This value is then used to subtract risk profile values of other suppliers. The remaining risk profile values are then normalized as shown in Table 6. Since there will be no product transfer from supplier 1 to others, zero is the base value and the differences between the risk profiles of suppliers remain the same.

The product transfer network based on risk profiles of suppliers is presented in Figure 2 and parameters used in the model are presented in Table 6 and Table 7.

Table 6. Normalized Risk Value

Suppliers	Total Risk Value	Relative total risk values	Normalized values
Suppliers - 1	58	58-24=34	R_{N1} = (34-0)/63=0.54
Suppliers - 2	24	0	R_{N2} = 0
Suppliers - 3	44	44-24=20	R_{N3} = (20-0)/63=0.32
Suppliers - 4	32	32-24 = 9	R_{N4} = (9-0)//63= 0.14
	Total	63	1

Table 7. Parameters used in the model

Suppliers	Number of Product procured according to min cost	Normalized risk Values	Number of Product to be transferred	Product to be kept in the supplier	Remaining capacity of the supplier
Suppliers -1	31800	0.54	17172	14628	0
Suppliers -2	30200	0	0	30200	0
Suppliers -3	29450	0.32	9424	20026	0
Suppliers -4	11200	0.14	1568	9632	17600

Illustration in Figure 2 below was made based on value and calculation in Table 6 which is the existing parameters on the model that including the risk. Shifting order illustration began from maximizing order for supplier that has the lowest risk profile value (Supplier 2). The same steps repeated for reallocating the excess supply order from Supplier 2 to the next supplier that has larger value of risk profile than Supplier 2 (in this case supplier 4).

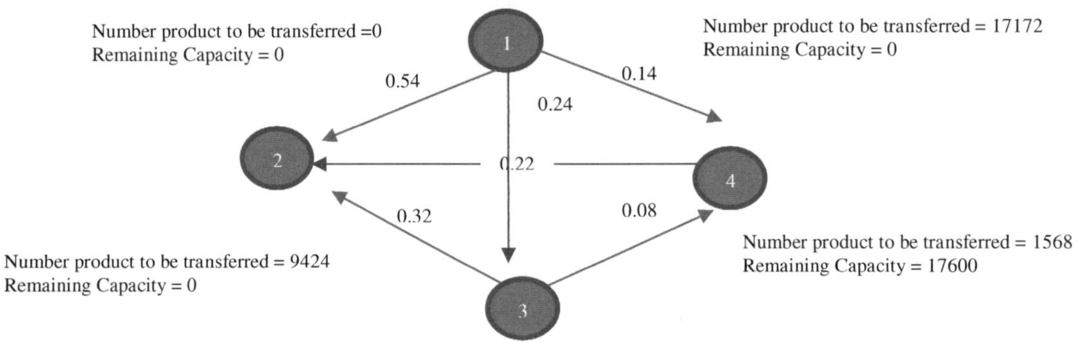

Figure 2. Illustration of shifting order between suppliers

Table 8. Recapitulation The Difference between the normalized risk values of suppliers (R_{ij})

R_{12}	R_{32}	R_{42}	R_{14}	R_{34}	R_{13}
0.54	0.32	0.24	0.14	0.08	0.22

The objective function of formulation (5) does not represent any quantity but since the objective function is maximization, it determines condition of transfer from a risky supplier to a less risky supplier. It is basically to maximize the transfer quantity based on the normalized risk value. The model to transfer the product from a risky supplier to a relatively less risky supplier is given by the model below. Constraint (6) is to determine the lowest and highest risky nodes for each supplier and (7) is condition that the difference between the quantity entering and leaving the node cannot be greater than the remained capacity of that supplier.

$$Maxz = \sum_{ij} N_{ij} * Y_{ij} \quad (5)$$

$$\sum_j^J X_{ij} \leq Q_{Ti} \forall i \neq j \quad (6)$$

$$\sum_k^K X_{ki} - \sum_j^J X_{ij} \leq C_{Ri} \quad (7)$$

Where :

IOP Conf. Series: Materials Science and Engineering **337** (2018) 012032 doi:10.1088/1757-899X/337/1/012032

N_{ij}	: Positive difference between the normalized risk	C_{Ri}	: remained capacity of supplier i
J	: indicates all suppliers less risky than supplier i	X_{ij}	: number of products to be transferred from supplier i to supplier j
Q_n	: quantity to be transferred less risky than supplier i		
K	: all suppliers more risky than supplier i		

Below is the transfer model incorporating the parameter values from the case company.

Max Z = $0.54*X_{12} + 0.32X_{32} + 0.24*X_{42} + 0.14*X_{14} + 0.08*X_{34} + 0.22*X_{13}$

Subject To

- $X_{12} + X_{13} + X_{14} \leq 17172$
- $X_{12} + X_{32} + X_{42} \leq 0$
- $X_{13} - X_{32} - X_{34} \leq 0$
- $X_{32} + X_{34} \leq 9424$
- $X_{14} + X_{34} - X_{42} \leq 17600$
- $X_{42} \leq 1568$
- $X_{ij} \geq 0$

The formulation is solved via LINGO and the optimal solution is presented in Table 9 below.

Table 9. Optimal Solution

R_{12}	R_{32}	R_{42}	R_{14}	R_{34}	R_{13}
0	0	0	17172	9424	0

Modified procurement plan in Table 10 was a comparison between optimal solution of data processing that not considering risk and optimal solution of data processing that including risk. Table 10 also shown order allocation changing in supplier 1, 3 and 4. Supplier 1 should decrease 54% of total supply order, supplier 3 should decrease 32% of their supply order; while supplier 4 should increase 42% of supply order. This condition occurred because supplier 1 and supplier 3 has relatively higher risk profile value so both of supplier 1 and 3 have to decrease their order allocation and reallocate their order to the supplier that has relatively less risk profile value (supplier 4), therefore supplier 4 should increase their supply order. There was no changing order allocation for suppliers 2 because the number of supply is already reach the maximum capacity.

Table 10. Modified Procurement Plan

Current Procurement Plan		Modified Procurement Plan		Decrease	Increase	Percentage (%)
Supplier 1	31800 kg	**Supplier 1**	14628 kg	17172 kg	0	(–) 54%
Supplier 2	30200 kg	**Supplier 2**	30200 kg	0	0	0%
Supplier 3	29450 kg	**Supplier 3**	20026 kg	9424 kg	0	(–) 32%
Supplier 4	11200 kg	**Supplier 4**	37796 kg	0	26596 kg	(+) 42%
Total	102650 kg	**Total**	102650 kg			

3. Model Verification

In this research, verification process was done by analyzing the output of the lingo. Referring to formula (1) until (7), the validity was obtained if the output of the model meets two criteria below:

1. The number of orders to be delivered by each supplier does not exceed the capacity of suppliers.
2. The number of orders for each supplier have to fulfill the demand of each manufacturer.

4. Conclusions

In this paper we present a model that can be used to allocate order to suppliers by considering both total costs and the risk. The first step is to allocate order to minimize cost. This initial solution is then used as a basis for transferring orders to lower risk suppliers by looking at both capacity constraints and risk profile of each supplier. The model was applied to a case study. The results suggest that the model has been able to include both objectives in making allocation decisions.

This procedure can be extended to multi-period, multi commodity and multi-echelon Supply Chain in further research. Risk analysis as parameter in determining order allocation for suppliers can use more parameters as supporting data to determine risk profile value. Processing data and analysis of order allocation can be further developed by using the integration of fuzzy method, AHP, revised analytic hierarchy process or TOPSIS fuzzy.

References

[1] Chen P and Wu M 2013 A modified failure mode and effects analysis method for supplier selection problems in the supply chain risk environment : A case study *Comput. Ind. Eng.* **66** pp 634–642
[2] Jadidi O, Zolfaghari S and Cavalieri S 2014 A new normalized goal programming model for multi-objective problems : A case of supplier selection and order allocation *Intern. J. Prod. Econ.* **148** pp158–165.
[3] Nazari-shirkouhi S, Shakouri H, Javadi B and Keramati A 2013 Supplier selection and order allocation problem using a two-phase fuzzy multi-objective linear programming *Journal of Purchasing & Supply Management* pp 849–860
[4] Pazhani S, Ventura J A and Mendoza A 2016 A serial inventory system with supplier selection and order quantity allocation considering transportation costs **40** pp 612–634
[5] Yu F, Xue L, Sun C and Zhang C 2016 Product transportation distance based supplier selection in sustainable supply chain network *J. Clean. Prod.* **137** pp 29–39
[6] Sodenkamp M A, Tavana M and Di D 2016 Modeling synergies in multi-criteria supplier selection and order allocation : An application to commodity trading *European Journal of Operational Research* **254** pp 859–874
[7] Guo C and Li X 2014 A multi-echelon inventory system with supplier selection and order allocation under stochastic demand *Intern. J. Prod. Econ.* **151** pp 37–47
[8] Oguzhan K and Erol S 2016 A proactive approach to supply chain risk management : Shifting orders among suppliers to mitigate the supply side risks *Journal of Purchasing & Supply Management* pp 842–843
[9] Prasannavenkatesan S and Goh M 2016 Multi-objective supplier selection and order allocation under disruption risk *Transp. Res. Part E* **95** pp124–142
[10] Pujawan I N and Geraldin L H 2009 House of risk: a model for proactive supply chain risk management *Business Process Management Journal* **15** pp 953-967
[11] Parenreng S M, Pujawan I N, Karningsih P D and Engelseth P 2016 Mitigating Risk in the Tuna Supply through Traceability System Development *International Food and Agribusiness Management Review* **19** pp 59 - 82

Failure mode effect analysis and fault tree analysis as a combined methodology in risk management

N A Wessiani[1], F Yoshio[2]
[1]Department of Industrial Engineering, Institut Teknologi Sepuluh Nopember, Kampus ITS Sukolilo-Surabaya 60111, Indonesia

*wessiani@gmail.com

Abstract. There have been many studies reported the implementation of Failure Mode Effect Analysis (FMEA) and Fault Tree Analysis (FTA) as a method in risk management. However, most of the studies usually only choose one of these two methods in their risk management methodology. On the other side, combining these two methods will reduce the drawbacks of each methods when implemented separately. This paper aims to combine the methodology of FMEA and FTA in assessing risk. A case study in the metal company will illustrate how this methodology can be implemented. In the case study, this combined methodology will assess the internal risks that occur in the production process. Further, those internal risks should be mitigated based on their level of risks.

1. Introduction

Failure Mode Effect and Analysis (FMEA) were formally introduced in the mid-1960. It initially focused on safety issues in the aerospace industry [1]. Further development showed that FMEA are widely used across the research under the topic of risk management. It indicates that this methodology robust enough to be applied in managing risk. Moreover, FMEA methodology is also one of the risk analysis techniques recommended by international standards [2].

Many research that used FMEA as the risk analysis technique has demonstrate the successful implementation. Kurt and Ozilgen could found the major risks that occurred in the dairy product manufacturing by utilizing FMEA methodology. Their finding might help large number of dairy product manufacturers in producing safe products since almost all dairy products share common manufacturing stages [3]. Kumru and Kumru also reported a corrective actions recommendation for improving purchasing process in the public hospital after adopting fuzzy FMEA [4]. The corrective actions recommendation was the result of failures identification in the purchasing process using fuzzy FMEA methodology. Kumru and Kumru's research was the first authentic example of how fuzzy FMEA approach might be helpful to the management processes [4]. There are still many other researches that also adopt the FMEA methodology as the risk analysis technique [5][6][7].

On the other side, Fault Tree Analysis (FTA) has also been reported as an extensively used technique in risk analysis [8]. This technique was formerly developed by Watson in 1962 at Bell Telephone Laboratories in USA for safety evaluation of the Minuteman launch control system. Then, followed by Haasl et. al. that introduced the basic rules for the construction of the fault trees in 1965 [9]. This technique was then widely used in various field of study. Hyun et. al. implemented this technique combined with Analytical Hierarchy Process (AHP) for identifying potential risk of undesirable events occurring during tunneling with application of a shield tunnel boring machine (TBM) method. The result showed that the proposed method reliable and indicates its applicability to risk management for shield TBM tunneling [10]. Park and Lee published the application of FTA in the field of hygiene management. They applied the FTA method to investigate the causes for faults in handwashing process. They concluded that FTA, with qualitative and quantitative analyses, was a good alternative approach to hazard analysis in HACCP system implementation [11]. Liu et al employed FTA method combined with quantitative analysis to investigate high-speed railway accidents. They reported that the proposed method was effective to be applied since they can give conclusions and suggestions to decrease the occurrence possibilities of similar accidents [12].

The discussions above are shown how both techniques, FMEA and FTA, reliable to be applied for risk management. Each technique has their own advantages. Whiteley et. al. mentioned the advantage of FMEA is a comprehensive way to analyses all of the potential component failure modes in a system [13]. A system, design, process, or service may usually have multiple failure modes or causes and effects. In this situation, FMEA methodology enable the analyst to assess each failure mode or cause and prioritized in terms of their risks so that high risky (or most dangerous) failure modes can be corrected with top priority [14]. Moreover, conducting FMEA technique properly and appropriately will provide the practitioner with useful information that can reduce the risk (work) load in the system, design, process, and service. FMEA is one of the most important early preventive actions in system, design, process, or service which will prevent failures and errors from occurring and reaching the customer [15]. On the other hand, FTA also offers its advantages. FTA is a deductive technique that can be used to classify the instrumental relationships leading to a specific failure mode [13]. FTA is a systematic method for analyzing the risks that are sourced from various combinations of failure events [8][10]. This technique also enables to estimate the probability of an accident from both qualitative and quantitative perspectives. Qualitative analysis of a fault tree helps to find out its minimal-cut sets while its quantitative analysis computes the failure probability of top event using its basic events' exact failure probabilities [8].

Besides their advantages, both techniques actually also own their disadvantages. Some literatures mention that using FMEA may be difficult and tedious for the case of complex systems that have multiple functions consisting of a number of components so that compound failure effects cannot be analyzed [2][15]. Whitley et. al. also states that the main drawback when using only FMEA for reliability analysis, is that the technique is geared towards analyzing individual component failure mode occurrences. Because these failure modes are considered one by one, the interaction of multiple failure mode occurrences are often not listed using this method. Additionally, this type of reliability process is not fully quantitative. Severity and risk rankings can be made, however overall reliability levels cannot be deduced using FMEA [13]. On the other side, the main drawbacks of using FTA is that it cannot take into account dependencies between failure modes. They also don't tend to consider the cause of failure modes, rather just tackling the knock-on effect of the failure's occurrence. Also, a FT must be undertaken individually for each failure mode of interest. The entire range of operating conditions that the system can operate in are not considered in one tree. A tree must be made for each operating condition to be fully accurate [13].

This paper aims to combine FMEA and FTA methodology for managing risks in order to reduce the shortcomings of both methodologies. Risk to be identified are internal risks that occurred in production process. This study uses a production process on a metal company as the case study to implement the combined method. It will choose one product of the metal company to show how this combine method will work to identify the internal risk.

2. Literature Review

2.1. Failure Mode and Effect Analysis (FMEA)

As stated earlier, initially FMEA focuses on safety issues, but further development shows that FMEA becomes one of many other tools in risk management. Applying FMEA as a tool for risk management involving ten steps to be conducted as shown in Table 1 below [1].

Table 1. Ten Steps for Implementing FMEA in Risk Management

Steps	Description
Step 1	Review the process or product.
Step 2	Brainstorm potential failure modes.
Step 3	List potential effects of each failure mode.
Step 4	Assign a severity ranking for each effect.
Step 5	Assign an occurrence ranking for each failure mode.
Step 6	Assign a detection ranking for each failure mode and/or effect.
Step 7	Calculate the risk priority number for each effect.
Step 8	Prioritize the failure modes for action.
Step 9	Take action to eliminate or reduce the high-risk failure modes.
Step 10	Calculate the resulting RPN as the failure modes are reduced

Source: [1]

FMEA method defines the level of risk based on the value of risk priority number (RPN) from each risk. The higher the value of RPN will be the higher the risk. The formula for obtaining the RPN can be written as below.

$$RPN = S \times O \times D \quad (1)$$

Where RPN is risk priority number, S is severity ranking for each effect, O is occurrence ranking for each failure mode, and D is detection ranking for each failure mode and/or effect. Each of these three rankings is based on a 10-point scale, with 1 being the lowest ranking and 10 the highest. Based on the RPN, each failure mode could be ranked in order from the highest RPN to the lowest. This prioritization will show which failure mode should be prioritize first to mitigate

2.2. Fault Tree Analysis (FTA)

Fault tree analysis (FTA) is an analytical technique that performed deductively from the occurrence of undesired event to the identification of root cause. The fault tree itself is a graphic model of the various parallel and sequential combinations of faults that will result in the occurrence of the predefined undesired event [9]. The faults can be events that are associated with component hardware failures, human errors, or any other pertinent events which can lead to the undesired event. A fault tree thus depicts the logical interrelationships of basic events that lead to the undesired event-which is the top event of the fault tree.

Developing fault tree will involve six steps as following [16]:

1. Define undesired event
2. Acquire understanding of system
3. Construct fault tree
4. Collect quantitative data
5. Evaluate fault tree probability
6. Analyze computer results

IOP Conf. Series: Materials Science and Engineering **337** (2018) 012033 doi:10.1088/1757-899X/337/1/012033

As fault tree is a graphical representation of the relationship between certain specific events and the ultimate undesired event, constructing the fault tree will need several common symbols as depicted in picture 1 below.

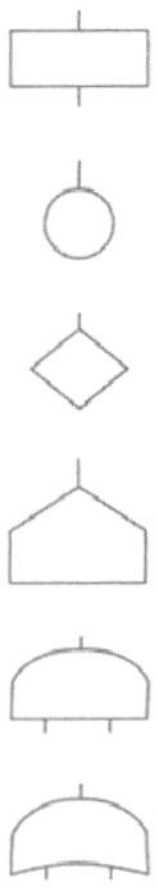

Intermediate Events	An event that happens between two other events
Basic Events	Failure
Undeveloped Events	Event with not enough information
External Events	Event that expected to happen
AND gate	Output happens if both of the branch happens
OR gate	Output happens if one of the branch happens

Figure 1. Several Common Symbol of FTA [16]

3. Combining the FMEA and FTA Methodology for Risk Assessment

FTA methodology will be used to identify the potential failure mode as stated in the FMEA methodology. Rather than using brainstorming itself, using the FTA methodology for generating the failure mode is expected to have better results. If FMEA is an inductive approach for finding the fault, the FTA itself is a deductive approach. Using FTA for generating the fault, it means attempt to find out what modes of system/component behavior contribute to the failure. Having this deductive approach for identifying the fault (risk), it can be known the root cause of those risks. Mitigation effort that being developed will base on the root cause of the risks hence it would have better results. Recalling the table 1 and modified the steps with the FTA steps, we will have the steps of combining the FMEA and FTA methodology for assessing the risk. The modified methodology can be seen in Table 2 below.

Table 2. Steps for Implementing FTA and FMEA in Risk Management

Steps	Description
Step 1	Review the process or product.
Step 2	Define undesired events
Step 3	Construct the fault tree
Step 4	Collect quantitative data
Step 5	Assign a severity ranking for each fault/failure.
Step 6	Assign an occurrence ranking for each fault/failure.
Step 7	Assign a detection ranking for each fault/failure.
Step 8	Calculate the risk priority number for each fault/failure.
Step 9	Prioritize the fault/failure for action.
Step 10	Take action to eliminate or reduce the high-risk fault/failure.
Step 11	Calculate the resulting RPN as the fault/failure are reduced

4. Case Study

Metal Company is one of metal work and engineering company in Indonesia. This metal company has 3 main business lines. There are, engineering procurement and construction (EPC), industrial tools manufacturing, and foundry. Some products of this metal company are bogie, mill roll, pipeline, road rollers, storage tank, and also ball tank. PT X is a leading bogie supplier in Indonesia with 100% local demand fulfillment. Moreover, this metal company has been penetrated United States market with 1,500 bogies delivered since 2012. Bogie's supply both local and worldwide shows a positive trend. This condition forced this metal company to maintain the commitments to give a competitive service in terms of quality, price, and delivery [17].

In this research bogie production process will be the case study for implementing the combining method of FTA and FMEA in assessing the risk.

Step 1. Review the process or product

Reviewing the process or product is conducting by making the operation process chart (OPC) of bogie production process. The picture below shows the OPC of bogie production process.

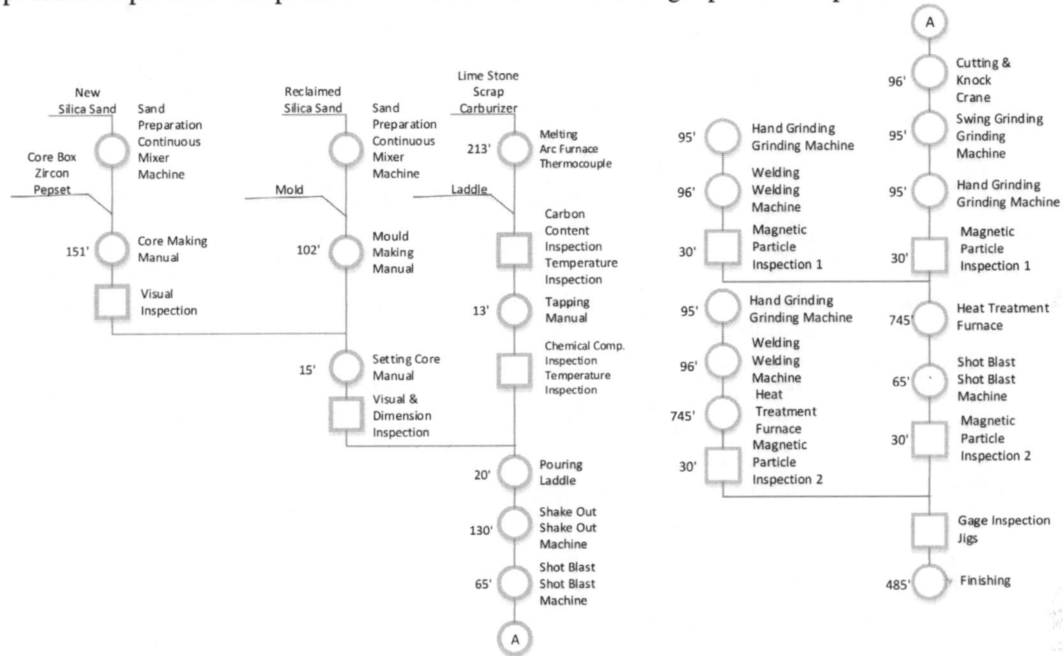

Figure 2. Operation Process Chart of Bogie Production Process

Further the activities involved in each production process are identified to acquire detailed understanding of the system. The example of detailed activities in moulding process can be shown in picture 3 below.

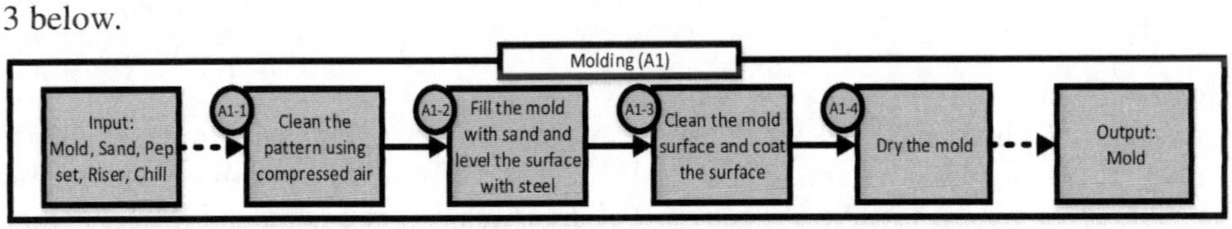

Figure 3. The Detailed Activities of Molding Process

Step 2. Define undesired events

Undesired event is identified through the activities that have been identified in the step 1. Brainstorming is conducted to define undesired events that probably happen through the production process. Those undesired events further will be analyzed by the FTA to find the root cause.

Step 3. Construct the fault tree

Fault tree analysis is conducted in order to find the root cause of each undesired events. Those root causes that have identified will be categorize as the internal risks that could occurred in the production process. Picture 4 shows the example of fault tree in the molding process.

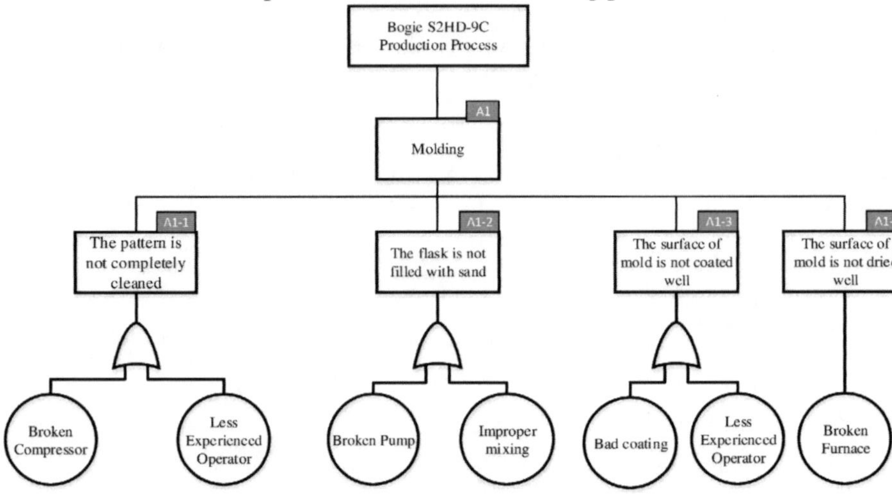

Figure 4. Fault Tree of Molding Process

Step 4. Collect quantitative data
Quantitative data of each internal risk should be collected as much as possible. The frequency of occurrence, the impact, the effort to handle the risk is kinds of data that should be collected. This data is important for the next step that will apply FMEA methodology for defining the severity, occurrence, and detection level of each internal risk.

Step 5. Assign a severity ranking for each fault/failure
Severity ranking will show the level of impact that would be happened if the internal risks occurred. The result of severity ranking for each internal risk is shown in Appendix 1.

Step 6. Assign an occurrence ranking for each fault/failure
Occurrence ranking reflects how often each fault could occurred in the production process in certain period. Quantitative data that are collected in the previous step could help to determine in which ranking each fault would be fallen. The result of occurrence ranking for each internal risk is shown in Appendix 1.

Step 7. Assign a detection ranking for each fault/failure
Detection ranking will show the existence of current control inside the company to detect the failure or effect of failure. If company has no current control, the likelihood of detection will be low and the failure will have high ranking for the detection. The result of detection ranking for each internal risk is shown in Appendix 1.

Step 8. Calculate the risk priority number for each fault/failure
The calculation of risk priority number for each internal risk will follow equation number 1. The higher the RPN, the higher risk the failure. The failure mode can be ranked based on the RPN further the failure mode that has higher RPN should be prioritized for the mitigation effort. Appendix 1 shows the severity, occurrence, detection and RPN for each internal risk.

Step 9. Prioritize the fault/failure for action.

Pareto diagram is used to prioritize which failures should be mitigated first. Pareto diagram shows 80 percent of the total RPN for the FMEA comes from just 20 percent of the potential failures and effects. Picture 5 shows the pareto diagram of internal risks that occurred in the production process.

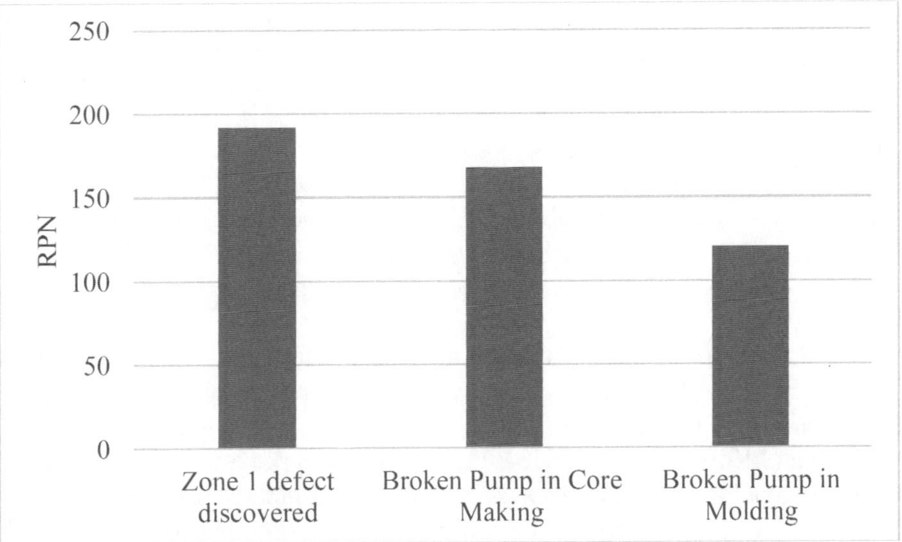

Figure 5. Pareto Diagram of Risk Ranking

Step 10. Take action to eliminate or reduce the high-risk fault/failure
Mitigation effort could be the alternative of eliminate the risk, transfer the risk, reduce the severity, reduce the occurrence, and/or accept the risk. Recommendation of mitigation effort for the prioritize risk could be seen in table 3 below.

Table 3. Recommendation of Mitigation Effort

Risk	Risk Treatment				
	Eliminate	**Transfer**	**Reduce the Occurrence**	**Reduce the Impact**	**Accept**
Zone 1 defect discovered	Repair	-	-	-	-
Broken Pump in Core Making	Scheduled and preventive maintenance	-	Allocate idle time for machine	Provide spare part	-
Broken Pump in Molding	Scheduled and preventive maintenance	-	Allocate idle time for machine	Provide spare part	-

Step 11. Calculate the resulting RPN as the fault/failure are reduced
When the recommendation efforts are implemented by the company, further the resulting RPN is calculated. A mitigation effort is effective when it could reduce significant RPN. The resulting RPN could be compared with the initial RPN to analyze how significant the mitigation effort that already done by the company. This research could not provide the resulting RPN since the company does not implement yet the recommendation of mitigation efforts.

5. Conclusion

This research tries to combine fault tree analysis (FTA) and failure mode and effect analysis (FMEA) for managing risk that could be occurred inside the company. The result is expected better since by having FTA the company could define the root cause of failure (risk) and further the risk could be prioritized by mapping the risk priority number (RPN) from the FMEA methodology. This prioritization is important since certain costs will exist to take the mitigation efforts. Case study is taken in this research for showing how the combined methodology could be implemented.

6. Reference

[1] McDermott R E, Mikulak R J and Beauregard M R (2009) *The Basic of FMEA* (2nd ed) New York: Productivity Press Taylor and Francis Group

[2] Dyadem Press (2003) *Guidelines for Failure Mode and Effects Analysis for Automotive Aerospace and General Manufacturing Industries* Ontario Canada: CRC Press

[3] Kurt L and Ozilgen S (2013) Failure mode and effect analysis for dairy product manufacturing: Practical safety improvement action plan with cases from Turkey *Safety Science*(55) 195-206

[4] Kumru M and Kumru P Y (2013) Fuzzy FMEA application to improve purchasing process in a public hospital *Applied Soft Computing*(13) 721-733

[5] Feili H R, Akar N Lotfizadeh H Bairampour M and Nasiri S (2013) Risk analysis of geothermal power plants using Failure Modes and Effects Analysis (FMEA) technique *Energy Conversion and Management*(72) 69-76

[6] Wessiani N A and Sarwoko S O (2015) Risk analysis of poultry feed production using fuzzy FMEA *Procedia Manufacturing*(4) 270-281

[7] Yang Z and Wang J (2015) Use of fuzzy risk assessment in FMEA of offshore engineering systems *Ocean Engineering*(95) 195-204

[8] Komal (2015) Fuzzy fault tree analysis for patient safety risk modeling in healthcare under uncertainty *Applied Soft Computing*(37) 942-951

[9] Vesely W E, Goldberg F F Roberts N H and Haasl D F (1981) *Fault Tree Handbook* Washington D C: U S Government Printing Office

[10] Hyun K -C, Min S Choi H Park J and Lee I -M (2015) Risk analysis using fault-tree analysis (FTA) and analytic hierarchy process (AHP) applicable to shield TBM tunnels *Tunnelling and Underground Space Technology*(49) 121-129

[11] Park A and Lee S J (2009) Fault tree analysis on handwashing for hygiene management *Food Control*(20) 223-229

[12] Liu P, Yang L Gao Z Li S and Gao Y (2015) Fault tree analysis combined with quantitative analysis for high-speed railway accidents *Safety Science*(79) 344-357

[13] Whiteley M, Dunnett S and Jackson L (2016) Failure Mode and Effect Analysis and Fault Tree Analysis of Polymer Electrolyte Membrane Fuel Cells *International Journal of Hydrogen Energy*(41) 1187-1202

[14] Wang Y -M, Chin K -S Poon G K and Yang J -B (2009) Risk evaluation in failure mode and effects analysis using fuzzy weighted geometric mean *Expert Systems with Applications*(36) 1195-1207

[15] Stamatis D H (2003) *Failure mode and effect analysis : FMEA from theory to execution* (2nd ed) Milwaukee: Quality Press

[16] Hixenbaugh A F (1968) *Fault Tree for Safety* Seatle Washington: The Boeing Company
[17] Metal Company in Indonesia (2013) *Metal Company* Retrieved from Metal Company

7. Appendix

Appendix 1 Severity, Occurrence, Detection and RPN for Each Internal Risk

Risk Code	Risk	Severity	Occurrence	Detection	RPN
	MOLDING				
R1	Broken Compressor	2	2	2	8
R2	The pattern is not completely cleaned	4	2	2	16
R3	Broken Pump	8	5	3	120
R4	Improper mixing	3	2	1	6
R5	Bad coating	8	2	2	32
R6	Broken Furnace	3	2	4	24
	CORE MAKING				
R7	Broken Compressor	2	2	2	8
R8	The pattern is not completely cleaned	3	2	2	12
R9	Broken Pump	8	7	3	168
R10	Improper mixing	3	2	1	6
R11	Broken core box	2	6	1	12
R12	Broken core	2	5	1	10
R13	Bad coating	6	4	2	48
	SETTING CORE				
R14	Incomplete core	2	3	1	6
R15	Broken core	2	5	1	10
R16	The core and parting lines is not completely cleaned	4	2	1	8
R17	Loose material left behind	6	5	3	90
R18	Broken core	2	2	1	4
	MELTING				
R19	Raw material doesn't meet standard	2	4	1	8
R20	Broken thermocouple	2	2	1	4
R21	Broken Furnace	8	5	2	80
R22	Improper chemical composition	4	2	1	8
R23	Improper chemical composition	4	3	2	24
R24	Improper ladle setting	5	2	2	20
R25	Temperature higher than standard	3	2	1	6
	POURING				
R26	Temperature higher than standard	8	2	1	16
R27	Pouring time too long	3	2	1	6
	SHAKE OUT				
R28	Broken Belt Conveyor	8	7	2	112
R29	The casting was shake when the temperature is high	6	2	1	12

Risk Code	Risk	Severity	Occurrence	Detection	RPN
R30	Malfunctioned crane	8	6	2	96
R31	Casting cannot be pulled out	3	5	2	30
R32	Broken Belt Conveyor	8	7	2	112
R33	The casting was shake when the temperature is high	6	2	1	12
R34	Malfunctioned crane	8	6	2	96
R35	Casting cannot be pulled out	3	2	2	12
R36	Malfunctioned crane	3	5	2	30
	SHOT BLAST				
R37	Setting need a longer time to finish	4	2	2	16
R38	Malfunctioned impeller	8	7	2	112
R39	The casting still has a rough surface	3	4	3	36
	CUTTING				
R40	Setting need a longer time to finish	2	2	1	4
R41	Cutting result doesn't meet standard	3	4	1	12
	GRINDING				
R42	Setting need a longer time to finish	2	2	1	4
R43	Line full	5	5	2	50
R44	The casting need a lot of grinding	4	6	2	48
R45	The casting still need to be processed	4	5	2	40
	MP 1				
R46	The fluid is not applied well	3	4	2	24
R47	Improper inspection	6	4	2	48
R48	Broken MPI tools	6	4	2	48
R49	Zone 1 defect discovered	8	8	3	192
	WELDING				
R50	Setting need a longer time to finish	2	2	2	8
R51	Line full	4	5	2	40
R52	Casting need a lot of welding	4	5	2	40
R53	The casting still need to be processed	4	2	1	8
	HEAT TREATMENT				
R54	Broken furnace	2	4	2	16
R55	Undisciplined method	4	3	1	12
R56	Thermocouple broken	2	4	2	16
R57	The casting cannot be pulled out from the machine	2	3	2	12
R58	The casting need a longer time to cool down	2	2	1	4
	MP 2				
R59	Casting need heat treatment process	8	3	2	48
	MACHINING				
R60	The clamp cannot hold the work piece	3	2	1	6
R61	Line full	6	5	2	60

Risk Code	Risk	Severity	Occurrence	Detection	RPN
R62	Broken CNC machine	6	4	2	48
R63	work piece cannot meet the dimensional standard	4	3	1	12
	GAGE INSPECTION				
R64	Setting need a longer time to finish	4	2	1	8
R65	The casting dimension doesn't meet requirement	4	2	1	8
	FINISHING				
R66	Nuts and bolts cannot be set	4	2	1	8
R67	Wrong arrangement of product	6	2	1	12

IOP Conf. Series: Materials Science and Engineering **337** (2018) 012034 doi:10.1088/1757-899X/337/1/012034

University building safety index measurement using risk and implementation matrix

A Rahman[1*], F Arumsari[1] and A Maryani[1]
[1]Department of Industrial Engineering, Institut Teknologi Sepuluh Nopember, Kampus ITS Sukolilo-Surabaya 60111, Indonesia

*arief@its.ac.id

Abstract. Many high rise building constructed in several universities in Indonesia. The high-rise building management must provide the safety planning and proper safety equipment in each part of the building. Unfortunately, most of the university in Indonesia have not been applying safety policy yet and less awareness on treating safety facilities. Several fire accidents in university showed that some significant risk should be managed by the building management. This research developed a framework for measuring the high rise building safety index in university The framework is not only assessed the risk magnitude but also designed modular building safety checklist for measuring the safety implementation level. The safety checklist has been developed for 8 types of the university rooms, i.e.: office, classroom, 4 type of laboratories, canteen, and library. University building safety index determined using risk-implementation matrix by measuring the risk magnitude and assessing the safety implementation level. Building Safety Index measurement has been applied in 4 high rise buildings in ITS Campus. The building assessment showed that the rectorate building in secure condition and chemical department building in beware condition. While the library and administration center building was in less secure condition.

1. Introduction

High-rise building management has a responsibility to identify and evaluate any risk in order to avoid injury and damage or losses of building facilities [1]. Safety inspection as regular program must be conducted as part of building management responsibilities [2]. The number of occupant who live or work and number of floors of the high-rise building is the complicated factors to be managed during evacuation process and may lead to the big number of death or losses [3].

In last decades, many high-rise building constructed in several University in Indonesia. Land utility, energy efficiency, and green environment mission become most of the consideration to build the high-rise building in University. However, safety management is not implemented and supporting facilities for the emergency condition also not installed follow the minimum safety standard for the high-rise building. Some fire accident was made big losses in the university such as agriculture technology building in Bogor Agricultural Institute (2017), the chemical laboratory in University of Indonesia (2015), and industrial engineering building in Bandung Institute of Technology (2013). These facts showed that the potential hazards inside the high rise building must be controlled. Some security issues in university high rise building in Indonesia such as student murderer and robbery should warn the management that safety regulation must be implemented in order to reduce some possible security threat inside the building.

Safety assessment or audit tools have been published by many different techniques and most of them have similar phases i.e.: hazard identification, risk assessment, and risk control [1]. Risk assessment should be able to provide the future projection of safety status or condition [2]. Considering the historical safety data does not enough to analyze the threat of the hazard. The present condition of safety implementation must be audited and should be considered as the important aspect in evaluating the building safety condition [3].

Safety checklist techniques may become the most practical tools apply by safety auditor to meet the ease to use and generalization [4]. Unfortunately, specific safety checklist for University high-rise building only present partially, still not completely covered the part of university rooms such as chemical laboratory, mechanical laboratory, library, etc. This paper introduces the university high-rise building safety index measurement tools by considering not only the risk mapping but also the safety implementation level. A modular checklist based on the specific room in University has been developed where the entire building safety index can be measured aggregately.

2. Methodology

Risk assessment result and evaluation of safety implementation in observed high-rises building are considered to determine the building safety index. The building safety index measure based on the aggregate index of each room in the observed building. Building Safety Index (BSI) = Average (Safety index room (1 to i)) if the observed building has i number of rooms. Figure 1 presents the general phases of high-rise building safety index measurement. Matrix and checklist models are applied in this safety audit framework in order to simplify the measurement.

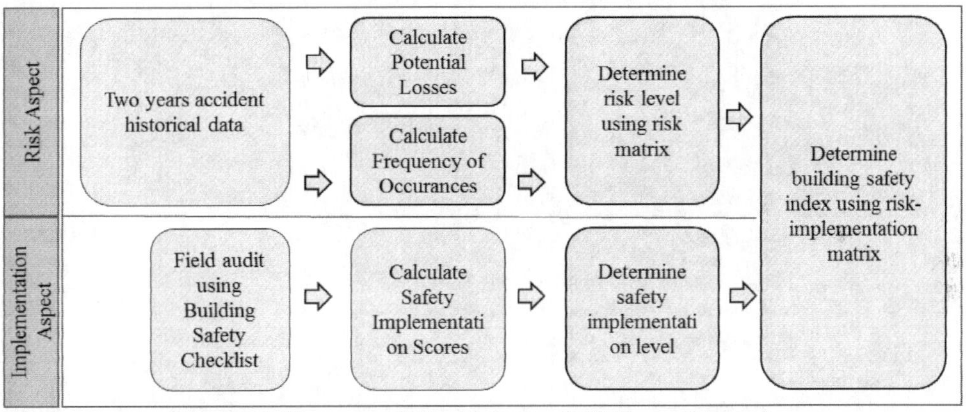

Figure 1. Framework for high-rise building safety index

2.1. Risk Aspect

Risk aspect measured by identifying the previous actual accidents occurred in the building. Two years historical data are required to determine the risk magnitude. The fundamental formula of risk magnitude in any risk assessment exercise is the multiplication of the frequency (how often) and the consequences (how big the impact) of each risk [1]. The accident occurrence rate table is defined to determine the frequency of occurrence as presented in table 1.

Table 1. Accident occurrence rate classification

Extremely Frequent	Frequent	Occasional	Rare	Very Rare
Number of accident more 6 times in a year	Number of accidents 2-6 times in a year	Number of accident only 1 in 1 year	No accident in 1 year	No accident in 2 years

The potential losses if the accident occurred may be measured by identifying the severity received by the victims in the accident. Besides, the losses also may be measured by measuring the duration losses or production/services opportunity losses as the effect of an accident. Five grades of potential losses are presented in accident losses classification as provided in Table 2.

By determining the potential losses grades and the accident occurrence rate, the risk level of each room can be defined using risk matrix. There are five level of risk based on below risk matrix i.e.: very low (5), low (4), moderate (3), high (4), and very high (5). The risk matrix is depicted in table 3.

Table 2. Accident losses classification

	Catastrophic	**Major**	**Moderate**	**Minor**	**Insignificant**
Human	1 or more fatality or death accident	1 or more accident made loss part of body	1 or more accident made loss 10 of day of works	1 or more accident made loss 1-3 of day of works	1 or more incident and without any working hour interruption
Production/ Service Opportunity	All line production or services must be stop in several days	All Production or services must be stop about 1 day production or 1 shift of work	One line of production or service must be stopped	One machine or facility stop and the others run normally	One machine or facility has been stopped for a short of time

Potential Losses	Frequency of occurrence: Extremely Frequently	Frequently	Occasionally	Rarely	Very Rarely
Catastrophic	1 Extreme Risk	1 Extreme Risk	1 Extreme Risk	2 High Risk	3 Moderate Risk
Major	1 Extreme Risk	1 Extreme Risk	2 High Risk	2 High Risk	3 Moderate Risk
Moderate	1 Extreme Risk	2 High Risk	2 High Risk	3 Moderate Risk	4 Low Risk
Minor	2 High Risk	2 High Risk	3 Moderate Risk	3 Moderate Risk	4 Low Risk
Insignificant	3 Moderate Risk	3 Moderate Risk	4 Low Risk	4 Low Risk	5 Very Low Risk

Figure 2. Risk matrix (potential losses against frequency of accident)

2.2. Implementation Aspect

The implementation of safety management in university high-rise building must be audited in order to evaluate the safety regulation and standard fulfillment. The proposed building safety checklist has been reconstructed from some regulation and standard including Indonesia National Safety and Health law and regulation [5][6][7][8]. Table 4 presents the detailed university building safety checklist.

IOP Conf. Series: Materials Science and Engineering **337** (2018) 012034 doi:10.1088/1757-899X/337/1/012034

Table 3. General university high-rise building safety checklist

	SAFETY ASPECTS	Office	Classroom	Library	Canteen/Cafe	Laboratory			
						Chemical	Mechanical	Electrical	Computer
Organization and Administration for Safety									
A1	Official safety team	☐	☐	☐	☐	☐	☐	☐	☐
A2	Yearly budget for safety programs	☐	☐	☐	☐	☐	☐	☐	☐
A3	Accident and incident reporting and recording	☐	☐	☐	☐	☐	☐	☐	☐
A4	Safety training and socialization programs	☐	☐	☐	☐	☐	☐	☐	☐
A5	Emergency preparedness and mitigation plan	☐	☐	☐	☐	☐	☐	☐	☐
A6	Safety facilities inspector	☐	☐	☐	☐	☐	☐	☐	☐
Emergency Equipment and Hazard Detectors									
B1	Fire alarm and sensors detectors	☐	☐	☐	☐	☐	☐	☐	☐
B2	Automatic sprinkler system	☐	☐	☐	☐	☐	☐	☐	☐
B3	Fire extinguisher	☐	☐	☐	☐	☐	☐	☐	☐
B4	First aid kits	☐	☐	☐	☐	☐	☐	☐	☐
B5	Water hydrant installation	☐	☐	☐	☐	☐	☐	☐	☐
Emergency Planning									
C1	Evacuation routes	☐	☐	☐	☐	☐	☐	☐	☐
C2	Assembly points and signs	☐	☐	☐	☐	☐	☐	☐	☐
C3	Emergency call center and evacuation procedure	☐	☐	☐	☐	☐	☐	☐	☐
C4	Electrical power for emergency condition	☐	☐	☐	☐	☐	☐	☐	☐
Building Safety Facilities									
D1	Smoking room	☐	☐	☐	☐	☐	☐	☐	☐
D2	Stairs	☐	☐	☐	☐	☐	☐	☐	☐
D3	Corridors	☐	☐	☐	☐	☐	☐	☐	☐
Exit and Communication									
E1	Exit doors	☐	☐	☐	☐	☐	☐	☐	☐
E2	Evacuation map and room location	☐	☐	☐	☐	☐	☐	☐	☐
E3	Communication system and facilities inside the building	☐	☐	☐	☐	☐	☐	☐	☐
E4	Road access for fire truck to reach the building	☐	☐	☐	☐	☐	☐	☐	☐
Occupant knowledge about emergency preparedness									
F1	Knowledge about operating the fire extinguisher and hydrant	☐	☐	☐	☐	☐	☐	☐	☐
F2	Knowledge about performing the first aid	☐	☐	☐	☐	☐	☐	☐	☐
F3	Knowledge about emergency response plan	☐	☐	☐	☐	☐	☐	☐	☐
Procedure and Equipment for personal protection									
G1	Operating procedure to use the equipment and materials					☐	☐	☐	☐
G2	Personal Protective Equipment					☐	☐	☐	☐
G3	Hazard warning signs	☐	☐	☐	☐	☐	☐	☐	☐
G4	Access control and limitation for each rooms	☐	☐	☐	☐	☐	☐	☐	☐
G5	Electrical socket, cable, and switch	☐	☐	☐	☐	☐	☐	☐	☐
G6	Electrical and gas stove equipment				☐	☐	☐	☐	
G7	Explosive and flammable gas or materials management					☐	☐		
G8	Toxic materials management					☐	☐		
G9	Chemical Inventory management system					☐	☐		
G10	Mechanical equipment in Laboratory					☐	☐		
G11	Noise and vibration					☐	☐		
G12	Air conditioning at server room							☐	☐
G13	Material handling devices for transportation of good/materials	☐	☐	☐	☐	☐	☐	☐	☐
G14	Staircase for working at height	☐	☐	☐	☐	☐	☐	☐	☐

The safety implementation checklist divided into 8 modules which represented the common university room. Those checklist modules are office [7], classroom, library, canteen/cafe, chemical laboratory, mechanical, laboratory, electrical laboratory, and computer laboratory. Each modules considered some safety aspects on each checklist i.e.: Organization and Administration for Safety [3] [5][7][9][10], Emergency Equipment and Hazard Detectors [5][7][9][10][11], Emergency Planning [4][5][7][9][10][11], Building Safety Facilities [5][7][9][10][11], Exit and Communication [5][7][9][10][11], Occupant knowledge about emergency preparedness [12][13], and Procedure and Equipment for personal protection [4] [7][9][10][11] [13].

Building safety scorecard has been developed as detailed guidelines for auditor during building assessment. Each safety aspects scaled with 5 confirmatory question in building safety scorecard. There were 2 assessed conditions for each safety aspect which presented the requirement for each aspect to be fulfilled. The most complete or ideal condition would be marked as 5 points and the incomplete or worse condition would be marked as 1 points. Table 5 shows the 5 grades design for each aspect on the scorecard. An example of the question in building safety scorecard is shown in table 6.

Table 4. Grades of each safety aspect on building safety scorecard

1	2	3	4	5
Both condition has not been achieved	One Condition has been achieved partially, other condition has not been achieved.	Both condition has been achieved partially	One Condition has been achieved perfectly, other condition has been achieved partially	Both condition has been achieved perfectly

Table 5. Example of building safety scorecard

No	Aspect	1	2	3	4	5
E.1	Exit Doors	All the width of exit door < 70 cm. All exit door open against the direction of escape corridor	Not all the width of exit door ≥ 70 cm. All exit door open against the direction of escape corridor	Not all the width of exit door ≥ 70 cm. Not all exit door open in the direction of escape corridor	All the width of exit door ≥ 70 cm. Not all exit door open in the direction of escape corridor	All the width of exit door ≥ 70 cm. All exit door open in the direction of escape corridor

The level of building safety implementation can be determined by calculating the average scores of building safety assessment result. There are also 5 level of safety implementations i.e.: Neglect (scores: 0.00 – 1.00), Poor (1.01-2.00), Growth (2.01-3.00), Advance (3.01-4.00), and Sustain (4.01-5.00).

2.3. Building safety Index (BSI)

Safety status of each observed room in the building whether in secure or dangerous condition can be determined using risk-implementation matrix. Table 123 is depicted the safety status by plotting the previous risk level and the safety implementation level results. There are 5 statuses of safety i.e.: very dangerous, dangerous, beware, less secure, secure, relief.

3. Building Assessment Results

Building safety measurement has been applied in 4 high-rise building in ITS Campus. The building safety assessment conducts by 2 observers and conducted the audit for 2 days. Th result of building safety exercises depicted in table 7. The assessment result showed that the Rectorate building under secure condition. However, Chemical department building must improve their building safety management. The rectorate building as the head quarter of university provides with proper safety facilities, get higher attention from the building safety manager and even have no high-risk type of room. Chemical department manages several chemical laboratories where the potential hazards have a significant contribution to building safety awareness. These assessment results have shown that the proposed building safety index delivered appropriate assessment result.

Table 6. Four buildings safety assessment in ITS Campus, Surabaya

Building	No of floors	Number and type of rooms	Implementation		Risk		BSI	
			score	level	score	level	score	Status
Rectorate	4	16 offices, 4 classes, 3 pantries,	3,2	Advance	5	Very low	4.1	Secure
Library	6	5 offices, 7 classes, 2 canteens, 5 library, 3 computer lab.	2.9	Growth	5	Very low	3.9	Less Secure
Administration Centre	2	16 offices, 5 classes	2,2	Growth	5	Very low	3.6	Less Secure
Chemical Department	4	1 office, 3 classes, 1 pantry, 1 library, 7 chemical lab	2,7	Growth	3	Moderate	2.8	Beware

The auditors in this assessment able to finish their observation less than 2 days for visiting 2 big building in the university. Based on the auditor's experiences, the building safety scorecards is easy to apply and very fast to get the final result by following the BSI matrix.

	IMPLEMENTATION LEVEL				
RISK LEVEL	1 Neglect	2 Poor	3 Growth	4 Advance	5 Sustain
1 Extreme Risk	1 Very Dangerous	1 Very Dangerous	2 Dangerous	2 Dangerous	2 Dangerous
2 High Risk	1 Very Dangerous	2 Dangerous	2 Dangerous	3 Beware	b Beware
3 Moderate Risk	2 Dangerous	2 Dangerous	3 Beware	4 Less Secure	4 Less Secure
4 Low Risk	3 Beware	3 Beware	4 Less Secure	4 Less Secure	5 Secure
5 Very Low Risk	4 Less Secure	4 Less Secure	4 Less Secure	5 Secure	5 Relief

Figure 3. Risk-Implementation Matrix

4. Conclusions

High rise building safety index in a university can be measured modularly by assessing each typical room with specific safety scorecards. Implementation aspect on high rise building safety assessment consists of the safety organization, safety facilities, occupant's knowledge, and personal protection from any hazard. Risk-Implementation matrix can be applied to determine the building safety index. The safety checklist module may be extended to other laboratories types such medical laboratory, biological laboratory, etc.

5. References

[1] Al-Homoud M S and Khan M M 2014 Assessing Safety Measures in Residential Buildings in Saudi Arabia Building Research and Information pp 32-4 300-305

[2] Bamber L 2003 Risk Management: Techniques and Practices [book auth] J Ridley and J Channing Safety at Work (Burlington: Elsevier Ltd pp 227 Chapter 2 4)

[3] Peraturan Menteri Kesehatan RI No 48 2016 Standar Keselamatan dan Kesehatan Kerja Perkantoran (Jakarta: Kementerian Kesehatan RI)

[4] Peraturan Menteri Tenaga Kerja RI No 5 1996 Sistem Manajemen Keselamatan dan Kesehatan Kerja (Jakarta: Kementerian Tenaga Kerja RI)

[5] Department of Environmental Health & Safety Pasific Lutheran University Building Safety Inspection Checklist [Online] [Cited: 06 05 2017] https://www plu edu/

[6] Texas Department of State Health Services Building Safety and Environment Health Checklist [Online] [Cited: 06 04 2017] http://www dshs texas gov/

[7] Office of Research Office National Institutes of Health Office Safety Checklist [Online] [Cited: 06 05 2017] https://www ors od nih gov/Pages/home aspx

[8] National Research Council Prudent Practices in the Laboratory Washignton D C: The National Research Council Press, 2011 978-0-309-21158-1
[9] Morales R and Paula S 2011 Biosafety Guide Cambridge Campus: Harvard Environmental Health and Safety Biosafety Office
[10] Keputusan Menteri Tenaga Kerja RI No 51 1999 Nilai Ambang Batas Faktor Fisika di Tempat Kerja (Jakarta: Kementerian Tenaga Kerja RI)
[11] Rahman A 2004 Measurement of Safety and Health Implementation Level National Seminar on Applied Ergonomics in Industry (Yogyakarta: Forum Komunikasi Teknik Industri p 647)
[12] Craighead G 2009 High-Rise Security and Fire Life Safety Elsevier Inc
[13] Peraturan Menteri Tenaga Kerja dan Transmigrasi RI No 1 2017 Pedoman Pemberian Penghargaan Keselamatan dan Kesehatan Kerja (Jakarta: Kementerian Tenaga Kerja dan Transmigrasi RI)

IOP Conf. Series: Materials Science and Engineering **337** (2018) 012035 doi:10.1088/1757-899X/337/1/012035

Mapping Risks of Indonesian Tuna Supply Chain

P D Karningsih[1,3*], D Anggrahini[1,3], N Kurniati[1,3], M Suef[1,3], A R Fachrur[1,3], and N Syahroni[2,3]
[1]Manufacturing System Laboratory, Industrial Engineering Department, Institut Teknologi Sepuluh Nopember, Kampus ITS Sukolilo-Surabaya 60111, Indonesia
[2]Ocean Engineering Department, Institut Teknologi Sepuluh Nopember, Kampus ITS Sukolilo-Surabaya, 60111, Indonesia
[3]Marine Research Centre, Institut Teknologi Sepuluh Nopember, Kampus ITS Sukolilo-Surabaya 60111, Indonesia

*dana@ie.its.ac.id

Abstract. Due to its high economic value and is produced by many countries, Tuna is considered as one of the world's popular fish. Demand for Tuna species are very high and it usually sells in three form: fresh, frozen or canned. Competition in Tuna trading is challengin with the potential risk of price and supply fluctuations. With recent focus of Indonesia government that see the future of Indonesia civilization depend on the oceans and as the three biggest Tuna producing country, Ministry of Marine Affairs and Fisheries should ensure sustainability and competitiveness of Indonesian tuna. Therefore, there is a great need to develop a proper and effective strategy to manage potential risks in Indonesian Tuna supply chain. This paper is aimed at identifying and mapping potential Tuna supply chain risks and its interrelationships that would assist government in determining proper strategies to manage Indonesian Tuna. A framework for identifying Tuna supply chain risks is proposed. Generic risk structure of Supply Chain Risk Identification System is adopted and modified to match with particular object, which is Indonesian Tuna. The proposed model consists of hierarchical and causal structure that encompass potential risks of Tuna supply chain operations from fishing, trading, processing and distribution. The causal structure consist of risk events and its risk agents which is the cause of risk events. To ensure the root cause of risk events are identified properly, five why's analysis is utilized to obtain risk agents. This proposed model also captures risk interrelationship between internal and external environment of Tuna supply chain. Preliminary result of this study identifies 15 risk events and 13 risk factors on fishing and trading operations and maps their interrelationships.

1. Introduction

Indonesian Government has a vision to set the oceans, the seas, the straits and the bays as the future of Indonesia civilization. In the next five years, Ministry of Marine Affairs and Fisheries (MMAF) of Indonesia plan "to achieve sovereign, independent and sustainable management of marine and fisheries resources for the people's prosperity". Strategic plan of MMAF is based on three pillars they are: sovereignty, sustainability and prosperity. One of MMAF target is to increase Gross Domestic Product for fisheries sector to 12% on 2019. However, there are many challenges for MMAF to achieve its target. Some of the problems are as follows: illegal, unreported and unregulated fishing, unresolved maritime territorial, lack of clear zoning regulations, low competitiveness and quality of fisheries products.

Therefore, MMAF is aiming to optimize marine resources management and improve fisheries competiveness and sustainability [1] [2] [3] [4].

Fish is not only important sources of food, many people also depend heavily on fisheries as source of their income. There are many stakeholders involved in Indonesian fisheries, from fishermen, traders, processing industries, distributors, retailers, to end of customers. To secure its competitiveness, Indonesian fisheries stakeholders (supply chain players) should be managed effectively. Furthermore, any undesire events which could make negative impact to competitiveness of Indonesian fisheries supply chain should be identified and managed properly. Therefore, supply chain risk management (SCRM) approach is required to support MMAF in accomplishing their goal. In general, SCRM consists of four steps, namely: risk identification, risk analysis, risk evaluation and risk treatment.

Tuna, shrimp and crab are top three Indonesia's capture fisheries products [1]. Tuna is considered as one of the world's popular fish which has a high economic value [5]. Majority of tuna species in international market are Skipjack tuna (Katsuwonus pelamis), Bigeye tuna (Thunnus obesus), Yellowfin tuna (Thunnus albacares), Albacore tuna (Thunnus alalunga), North Atlantic bluefin tuna (Thunnus thynnus), Pacific bluefin tuna (Thunnus thynnus orientalis) and Southern bluefin tuna (Thunnus maccoyii). Tuna is produced by approximately 70 countries. Positions of Indonesia, which is surrounded by Indian and Pasific Ocean, provides proper environment for tuna stock. Thus, it makes Indonesia as the largest tuna producing nations [6]. Most of Indonesia tuna catch is from Western Central Pacific Ocean and the rest is from Eastern Indian Ocean. In 2014, Indonesia has the biggest tuna landing for more than 600.000 metric tonnes (MT), then follows by Japan and Taiwan [7]. Skipjack and Yellow fin are two major species of tuna that mainly caught in Indonesia. Most of Tuna catch in Indonesia goes to further process domestically such as canneries or cured methods such as smoked, dried and salted. Small proportion of Tuna catch goes to domestic fresh markets and the smallest portion are exported to overseas canneries industries as loins [7]. As the largest tuna landing nation and the importance of Tuna in Indonesian fisheries, it is critical for MMAF to have a effective strategy to ensure sustainability and competitiveness of Indonesian tuna. For that reason, this research is aiming at developing strategy to manage potential risks in Indonesian Tuna supply chain. While, the focus of this paper is the first step of SCRM, that is to identify and provide a list of potential SC risks so it would guide MMAF in determining proper risk treatment strategies.

Study about Indonesian Tuna supply chain is not new but only few available, as can be seen in [8], [9], [10], and [11]. However, there is fewer number of research on managing risk of Indonesian Tuna supply chain [12] develops a model for managing risk in Tuna supply chain and use Bitung (Western Central Pacific Ocean) Indonesia as the case study. The model is developed using House of Risk (HOR) which is proposed by [13]. Since there are different interests from different supply chain players in Tuna SC, HOR is then modified to accomodate this issue. This paper uses case study of Tuna SC located in Eastern Indian Ocean, in order to give MMAF a more comprehensive view about Indonesian Tuna SC risks. Moreover, the structure of supply chain risk identification developed by [14] and combined with five why's analysis is adopted on this paper to identify Tuna SC risks and to map their interrelationships. To generate SC risks strategies, SC risk that has been identified is then analyzed futher by using HOR for multistakeholders proposed by [15].

The rest of this paper is divided into four subchapters. The second subchapter is literature review of supply chain risk management and its related tools. Then, it is followed by proposed framework for identifying SC risks of Tuna. Next, preliminary case study findings is presented. Finally, conclusions and future works is elaborated in the last subchapter.

2. Literature Review

Supply Chain (SC) can be defined as the network of entities, which consist of suppliers, carriers, manufacturing sites, distribution centres, retailers, and customers, through which material flows [16]. While [17] propose that SC also not only comprise product and material flow but also the flow of services, finances, and information. Therefore, SC includes financial providers, market research firms, logistics providers, carriers and transport firms as part of SC partners besides suppliers, manufacturer, retailers, and customers.

Supply Chain Risk Management (SCRM) is "an inter-organizational collaborative endeavor utilizing quantitative and qualitative risk management methodologies to identify, evaluate, mitigate and monitor unexpected macro and micro level events or conditions, which might adversely impact any part of a supply chain" [18]. Managing supply chain (SC) risks are very important to be able to formulate suitable strategy to minimized or avoid these risks. If SC risks is not managed properly and promptly, it may leads to failures to the whole SC and usually involving financial losses and/or business discontinuity. On the other hand, when SC risks is managed properly, it would lead to a higher competitiveness of the SC. Risks in supply chain may arise from both internal and external SC environments. Internal SC is represented by SC network that consist of several players/actors that are: focal organization and its suppliers and customers including logistic providers. External SC is represented by external factors linked with SC network such as man made and natural disaster, economic and political stability, etc [19].

SCRM consists of four steps [20] [21] [22], they are: (1) risk identification: to determine potential SC risks and their sources, (2) risk assessment: to measure the magnitude of SC risks, (3) risk evaluation: to prioritize SC risks according to certain criteria and magnitude, and (4) risk mitigation: to formulate suitable action to deal with SC risks. Risk identification should provides most (if not all) of possible risks on the SC in order to enable correct mitigation strategy formulation [23] [24] According to [25], risk has several attributes namely: (1) source of risks, (2) consequences of risks, (3) time, (4) location, and (5) person/factor/activity that may involved. Furthermore, it is also crucial to understand interrelationships between risks to recognize their impact on the entire SC network.

[14] propose utilization of Knowledge Based System (KBS) to support SC risk identification which is called Supply Chain Risk Identification System (SCRIS). For identifying SC risks as well as their interrelationships, combination of hierarchical and causal structure is used. First, hierarchical structure represents risk attribute "time" (level 0), "location" (level 1) and "risk owner" (level 2). Risk attribute "time" divided according to five phases of SC operations (i.e. Plan, Source, Make, Deliver, and Return) [26]. Risk attribute "location" consist of manufacturing (focal) organization, its SC partners and SC external environment which is adopted from [19]. Every risk "time" stages at level 0 are connected to each risk "locations" at level 1. Risk "owner" (level 2) is located under each risk "location" (level 1) which represents two risk attributes (i.e. persons or factors) and consists of different members. Under risk "location" manufacturing organization, there are six generic functional areas/divisions namely: Operational, Technical, Organisational, Communication, Resource and Financial . While, there are three players (i.e. Suppliers, Customers, and Transportation carriers / logistics providers) under risk "location" SC partners, there are nine factors under risk "location" SC external environment, they are: Political-governmental, Macroeconomic, Socio-cultural, Competitor, Complementary product organisation, Other related organisations, Technology development, Nature and Accidents. Second, Causal structure represents risk attributes "risk sources" which are represented by risk factors and risk sub factors and "risk consequences" which are represented by risk events. This structure is located under and linked to level 2 of hierarchical structure. With this causal structure, interrelations between and within risk events, risk factors and risk sub factors can be depicted.

House of Risk (HOR) is an approach to manage SC risks by identifying risk event and its relationship with source of risk event (risk agents). By adopting Failure Mode and Effect Analysis (FMEA) and House of Quality (HOQ) framework, HOR prioritizes the most influential risk agents and the best preventive actions for those risk agents accordingly [13]. Tuna SC consists of multiple actors, namely: traditional and industrial fishermen, traders, traditional and modern processors, domestic and overseas distributors. Naturally, different actors have different interests and perspectives about risks that could happened in their supply chain. Thus, [12] suggest HOR for multistakeholders to accomodate this issue.

To ensure effectiveness of preventive action in reducing or eliminate risk events, root cause of risk events should be identified properly. One technique that have been utilized largely as root cause analysis tool is Five Why's analysis. This technique aims to discover and differentiate between the real root cause with the symptom by asking "why?" five times. Then, after the answer (risk agent) are obtained, a corrective solution are proposed accordingly [27].

3. Proposed Framework for Tuna Supply Chain Risk Identification

To develop framework for managing risk in Tuna supply chain, the potential SC risks are identified and mapped using the proposed framework which is elaborated in this section.

Firstly, supply chain of Tuna is defined. According to [15], Tuna supply chain operations consist of fishing, trading, processing and distribution. [8] presents similar supply network and suggests that fishermen, traders (middlemen), processors, distributors, retailers, end customers including transport firms are the main players/stakeholders of SC Tuna. Whereas, in this research, supporting players, such as government institution (MMAF), and local or international non government institutions are also taken into account.

Secondly, Tuna SC risks event are identified through literature review and interview with SC players including experts from government institutions. These risks are then organized using modified risk structure of [14]. Thus, this research also applies combination of hierarchical and causal risk structure but with some modification on risk attributes. On hierarchical structure, risk attribute "time" (level 0) is divided according to four operations of Tuna SC that are aligned with SCOR but do not include "plan" and "return", they are: (1) fishing and trading are considered as "source" operations, (2) processing is a "make" operation, and (3) distribution is a "deliver" operation. Risk attribute "location" (level 1) divided into internal and external environment of Tuna SC. In this research, internal Tuna SC is not divided into focal organization and its SC network partners, but it comprises all players ofTuna SC. Thus, risk "owner" (level 2) that located under internal Tuna SC consists of all Tuna SC players from fishermen to end customers including transportation firms. SC player at fishing, trading and processing operations are only one player each (e.q. fishing operations is conducted by fishermen). While, for deliver operations consists of four players, they are: distributors, transport firms (third party logistics), retailers, and end customers. External environment consist of similar factors as is in [14] with two modifications: (a) Addition of "Tuna International Trade Market" and "Government/MMAF", (b) "Complementary product organisation" is omitted. Next, for causal structure also use similar arrangement but only take into account risk events and risk agents/factors (risk sub factors are omitted) which is located under each SC players. To determine risk agents (factors) which is the origin source of risk events, the five why's analysis is adopted. By using this structure any risk agent/risk event that is occured at "Source" or Fishing which then iniates risk agent/risk event at later stage (e.q. at "Make" or Processing stage) can be mapped. Thus, interrelationships between risk events and risk agents within and between operations stages can also be depicted in this structure. As a result, the proposed Tuna Supply Chain Risk Identification (SCRI) structure can be seen in figure 1. After the risk events and risk agents are obtained and depicted in Tuna-SCRI, these risks are incorporated into modified House of Risk that take into account multistakeholder.

4. Preliminary Case Study Findings

4.1. Eastern Indian Ocean Tuna Supply Chain

Tuna supply chain and its potential risks for Eastern Indian Ocean is developed and identified based on literatures and interview with respondents who are SC players of Eastern Indian Ocean Tuna. They are several fishermen and middlemen at Sendang Biru Port - East Java, top management of a Tuna processor company at Pasuruan – East Java, a small size Tuna processor industry, experts from MMAF of East Java province and Trenggalek region, and Institute For Mariculture Reseach and Development Gondol – Bali. Figure 2 shows Tuna supply chain for Eastern Indian Ocean.

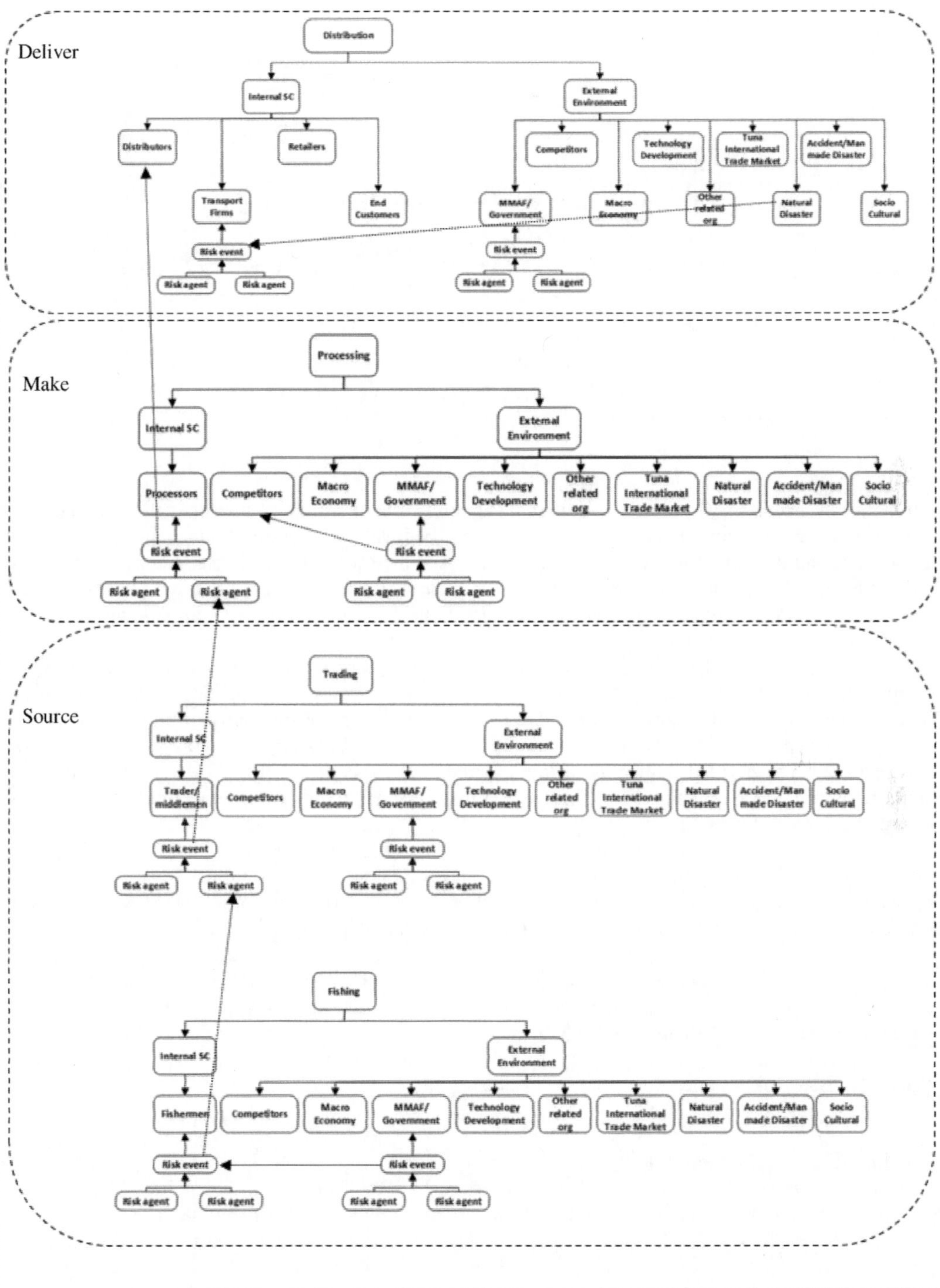

Figure 1. Tuna Supply Chain Risk Identification Framework

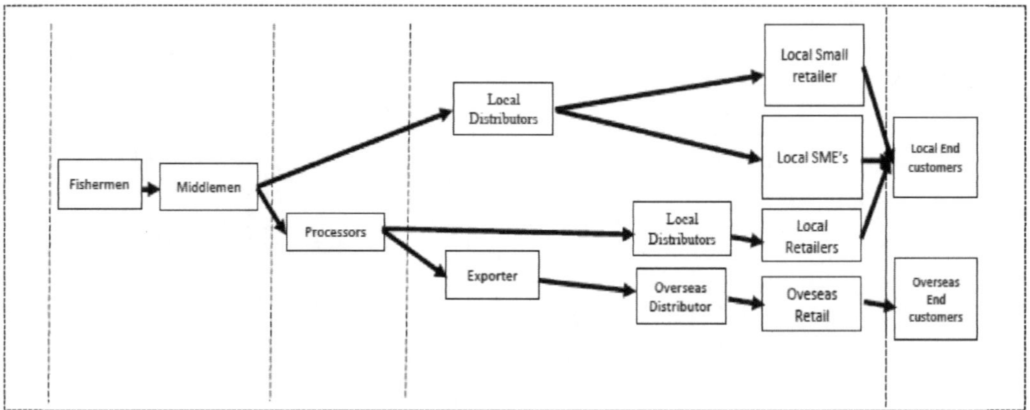

Figure 2. Eastern Indian Ocean Tuna SC

4.2. Supply Chain Risk Mapping of Eastern Indian Ocean Tuna

Potential risk events of SC Tuna are initially gathered from literature review and then validated by interviewing respondents. After risk events are identified, risk agents are obtained using five why's analysis. Then finally, risk events and risk agents are validated with the same respondents. All potential risk is then mapped on proposed framework of Tuna Supply Chain Risk Identification (Tuna-SCRI). A partial part of identified Tuna SC risks at Eastern Indian Ocean at Fishing and Trading operations (Source stage) can be seen on figure 3.

In general, most of SC risks that identified in figure 3 also is also occured on Tuna at Western Central Pacific Ocean. For example, improper fish quality due to lack of cold storage facilities during fishing. The same risk also applies on trading. In this case, due to unavailability of proper cold storage on the fishermen ships, they rely on ice block/pack vendor. Similarly, unpredictable weather in the last two years (2015 to 2016) also affect to lack of captured fish. Risk of unfair fish price is also occured. It could happened due to unregulated trading system in ports, lack of law enforcement of bidding scheme or declining price of world tuna.

Since 2015, Indonesian government/MMAF applies new regulations including war on illegal and unreported (IUU) fishing by foreign vessels in Indonesian waters and illegal transshipments (i.e. offloading of catch to foreign reefer vessels and processors at sea). Therefore, few SC risks listed on [12] are no longer applicable in this research (e.q. illegal transshipments)

One of interesting finding of this paper is when the price of tuna is too low (at trading operations), it could affects back to fishermen unwillingness to go fishing (at fishing operations). The reason for this, when the price of fish too low, fishermen could not get any revenue or even suffer a loss since they should provides funding around US$ 1500 each time they go fishing while usually spend 2 weeks in the offshore. This study also acknowledges the instution that responsible for regulating Tuna world price (located in Bangkok) as one of external environment.

5. Conclusion and future works

This paper shows development and application of framework for Tuna Supply Chain Risk Identification, which is then called as Tuna-SCRI, to identify and map supply chain risks of tuna. This framework is adopted from SCRIS by [14]. Some changing in the framework is conducted to consider particular characteristics of Tuna SC operations. Tuna-SCRI consists of three stages (i.e. Source, Make and Deliver), while on Source stages there are two operations that are: fishing and trading.

IOP Conf. Series: Materials Science and Engineering **337** (2018) 012035 doi:10.1088/1757-899X/337/1/012035

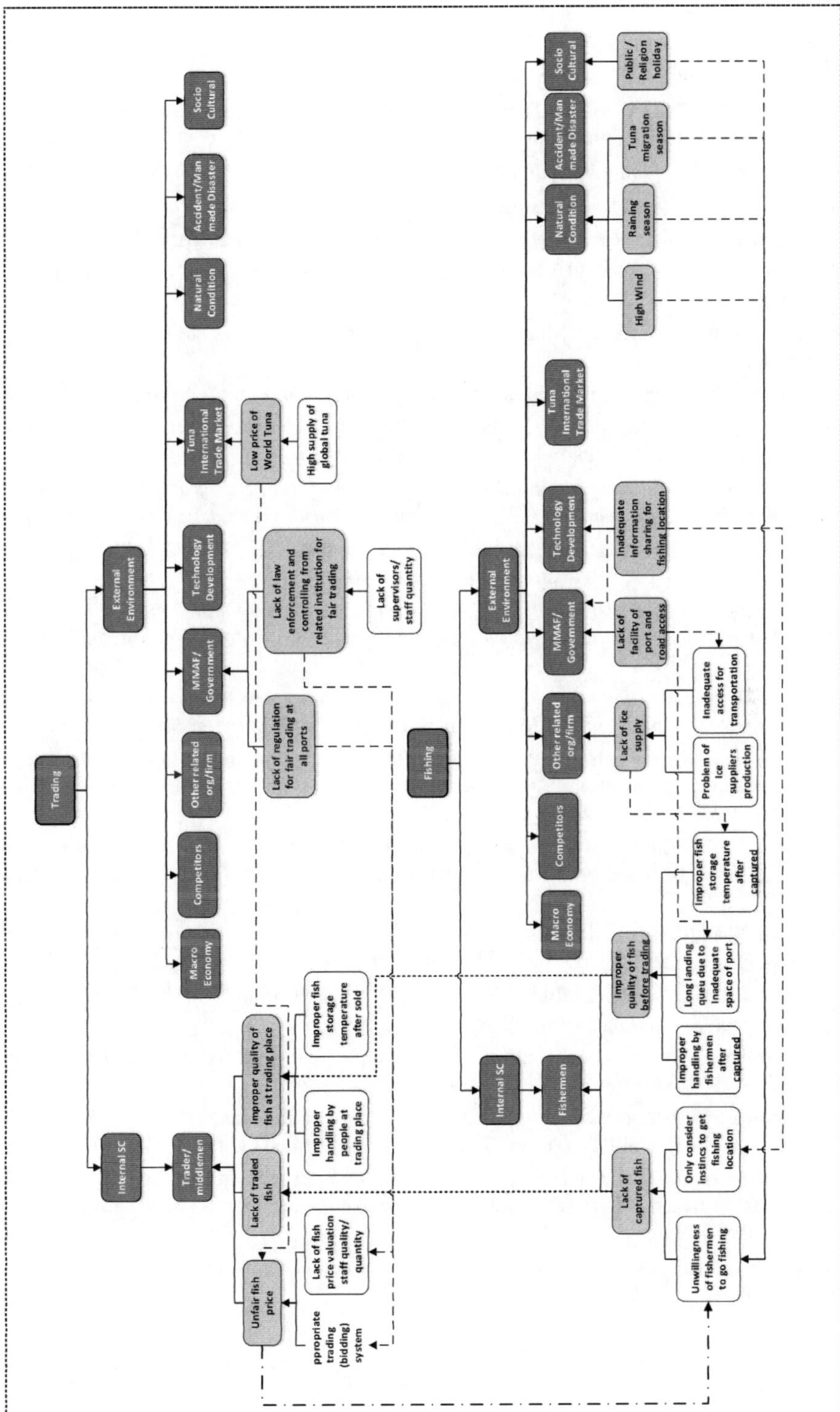

Figure 3. Eastern Indian Ocean Tuna SC

The next steps of this study are to identify and map SC risks events on Make and Deliver stages and then determine the risk agents. Modified House of Risk will be utilized to prioritize risk agents and determine the proper preventive actions by considering more than one SC stakeholders.

6. References

[1] CEA 2016 *Indonesia Fisheries: 2015 review*

[2] Noegroho A 2016 *Improving Traceability and Chain of Custody to Meet Market and Export Requirement* 2nd Bali Tuna Conference

[3] Pudjiastuti S 2016 *Manajemen dan Konservasi Sumber Daya Kelautan* (Surabaya: ITS Surabaya);

[4] IFAD 2016 *Republic of Indonesia Country strategic opportunities programme*

[5] FAO 2016 *The State of World Fisheries and Aquaculture* (Rome: United Nations)

[6] Sunoko R and Huang H W 2014 Indonesia tuna fisheries development and future strategy *Marine Policy* 43 p 174-83

[7] Pseudon Aquatic Resource Management Ltd 2016 *Estimates of Global Sales Tuna Fisheries* (Hampshire)

[8] Sustainable Fisheries Partnership 2010 *Indonesian Tuna Supply Chain Analysis*

[9] Ministry of Foreign Affair of the Netherlands 2012 *The Indonesian Seafood Sector: A Value Chain Analysis* (Ministry of Foreign Affair of the Netherlands)

[10] Maulida J F 2014 *Deskripsi Rantai Pasok Ikan Tuna (Studi Kasus PT Awindo International di Pelabuhan Perikanan Samudera Nizam Zachman Jakarta* (Bogor: Institut Pertanian Bogor)

[11] Soeratno D and Jan A H 2016 Analisis Model Supply Chain Ikan Cakalang Di Kota Manado (Studi Kasus Pada TPI PPP Tumumpa) *Jurnal EMBA* **4** (2) p 602-12

[12] Parenreng S M 2016 *Pengelolaan Risiko Rantai Pasok Komoditas Tuna Untuk Menciptakan Keberlanjutan Bisnis Dan Ketahanan Pangan* (Surabaya: Institut Teknologi Sepuluh Nopember, Teknik Industri)

[13] Pujawan I N and Geraldin LH. House of risk: a model for proactive supply chain risk management. Business Process Management Journal. 2009; 15(6): p. 953-967.

[14] Kayis B and Karningsih P D 2012 SCRIS: A knowledge-based system tool for assisting manufacturing organizations in identifying supply chain risks J*ournal of Manufacturing Technology Management* **23** (7) p 834–52.

[15] Parenreng S M, Engelseth P, Pujawan N and Karningsih P D 2016 Mitigating risk in the tuna supply through traceability system development *International Food Agribusiness Management Rev* **19** (1) p 59–82

[16] Lummus R and Vokurka R 1999 Defining Supply Chain Management: A Historical Perspective and Practical Guidelines *Industrial Management & Data Systems* **99** p 11-17

[17] Mentzer J T, Dewitt W, Keebler J S, Min S, Nix N and Smith C 2001 Defining Supply Chain Management *Journal of Business Logistics* **22** (1) p 1-24

[18] Ho W, Zheng T, Yildiz H and Talluri S 2015 Supply chain risk management: a literature review. I*nt. J. Prod. Res* **53** (16) p 5031–69

[19] Chapman P, Christopher M, Jüttner U, Peck H and Wilding R 2002 Identifying and Managing Supply Chain Vulnerability *Logistic Transportation Focus* **4** (4) p 59–70

[20] Li J and Hong S 2007 Towards a New Model of Supply Chain risk Management: The Cross-Functional Process mapping Approach *International Journal of Electronic Customer Relationship Management* **1** p 91-107

[21] Tuncel G and Alpan G 2010 Risk assessment and management for supply chain networks: A case study *Computers in Industry* **61** (3) p 250-59

[22] Trkman P and McCormack K 2009 Supply chain risk in turbulent environments - A conceptual model *Int. J. Production Economics* **119** p 247-58

[23] Waters C D J 2007 *Supply Chain Risk Management: Vulnerability and Resilience in Logistics* (London and Philadelpia: Kogan)

[24] Hallikas J, Karvina I, Pulkkinen U, Virolainen V M and Tuominen M 2004 Risk management process in supplier networks. International *Journal of Production Economic* **90** (1) p 47-58

[25] Australian/New Zealand Standar 2004 *Risk management* 3rd ed

[26] SCOR 2012 *Supply Chain Operations* (Supply Chain Council, Inc)

[27] Nicholas J M 2011 *Lean Production for Competitive Advantage: A comprehensive guide to lean methodologies and management practices* (New York: Productivity Press Taylor & Francis Group)

Acknowledgments
This study is funded by Lembaga Penelitian dan Pengabdian kepada Masyarakat (LPPM) and Pusat Studi Kelautan - Institut Teknologi Sepuluh Nopember (ITS) Surabaya Indonesia.

IOP Conf. Series: Materials Science and Engineering **337** (2018) 012036 doi:10.1088/1757-899X/337/1/012036

The effect of occupational health and safety, work environment and discipline on employee performance in a consumer goods company

D O Putri[1*], B Triatmanto[2], and S Setiyadi[2]
[1]Department of Industrial Engineering, University of Merdeka, Malang, Indonesia
[2]Department of Economy and Business, University of Merdeka, Malang, Indonesia

*digitha.oktaviani@unmer.ac.id

Abstract. Employee performance can be the supporting factor of company performance. However, employee performance can be affected by several factors. Employees can have optimal performance if they feel safe, have good working environment and have discipline. The purposes of this research are to analyze the effect of occupational health and safety, work environment and discipline on the employee performance in PPIC Thermo section in a consumer goods company and to find the dominant variable which primarily affects employee performance. This research was conducted by taking data from 47 respondents. The data were collected using questionnaire. The techniques in data analysis is multiple linear regression with SPSS software. The result shows that occupational health and safety, work environment and discipline are simultaneously significant to the employee performance. Discipline holds the dominant factor which affects employee performance.

1. Introduction

Human resources are important part in an organization. Company as an organization can increase their organizational performance if there is a good relation between company and their employee. Employees are important resource in a company which contribute to every company's activity. If there is an obstacle when they are working, it can affect their performance. Job performance is a work achieved by a person in carrying out the tasks assigned to him based on skills, experience, and sincerity and time [1].

Job performance in PPIC Thermo section in this company can be known based on the accuracy, amount of inventory availability that is based on stock opname. The amount of stock opname can indicate the similarity between the last stock and stock opname. Based on data stock opname, the differences between last stock and stock opname can be depicted and it indicates whether the employee performance is in good condition or not.

Job performance can be affected by several factors. While working, it is important for the employee to feel safe and comfortable with their working environment. The feeling of safety and comfort in a work place can be affected if there are health and safety insurance. Besides health and safety, there are other factors which can affect job performance. Author of [2] said that job performance can be known from several factors; one of them is discipline. Employees' discipline level can be indicated from their attendance data.

This study aims to analyze the effect of occupational health and safety, work environment and discipline to the employee performance and to find the dominant variable which primarily affects employee performance.

2. Research Framework

2.1 Job Performance
Work performance is a work achieved by a person in carrying out the tasks assigned to him based on his skills, experience, and sincerity and time [1]. In this study, the researcher used four indicators in the variable of work performance, i.e. job quality, job quantity, work attitude and work interest.

2.2 Occupational Health and Safety
Under occupational safety legislation no.1 of 1970 Article 2, occupational safety protection covers all aspects of hazardous work of any workplace, whether on land, in the soil, on the surface of water, in water or in the air within the jurisdiction of the Republic of Indonesia. By this regulation, then every worker in various fields of work is expected to be given guaranteed protection for safety in carrying out his work. There are five indicators in occupational safety and health variables, which are knowledge of occupational safety and health, health conditions, occupational health and safety, completeness of work equipment and the availability of protective equipment.

2.3 Work Environment
Work environment is the physical environment in which employees' works affect their performance, safety, and quality of work life. Illumination, noise and vibration, temperature, humidity, and air quality are work-environment factors under the control of the organization and the operations manager. The manager must approach them as controllable [3]. The indicators in this variable are workplace conditions, job security, peer relation, and relationship between superiors and subordinates.

2.4 Discipline
Discipline is the awareness and willingness of a person to comply with all corporate rules and prevailing social norms. Consciousness is the attitude of a person who voluntarily obeys all rules and is aware of his duties and responsibilities. Therefore, the employee will obey or do all the work well, not under coercion. Willingness is an attitude, behavior and action of a person in accordance with company regulations, whether written or not [1]. In this study, the researcher used indicators of compliance with regulations, adherence to working hours, compliance in the use of office facilities and infrastructure and work according to procedures.

2.5 *Research Hypothesis*

Author of [4] examined the effect of work environment and work discipline on the work performance. The result of the research is that work environment and work discipline, respectively, have significant effects on work performance; and of these two variables, work discipline is the most dominant influence. Furthermore, author of [5] found that occupational health and safety have a significant effect to the job performance, while author of [6] found that occupational health and safety have a significant effect, directly and indirectly, through motivation variable to employee performance. Based on these arguments, the hypotheses that can be prepared are as follows:

H_1 : Occupational health and safety, work environment and discipline has simultaneous partial effect on the employee peformance.

H_2 : Occupational health and safety have dominant effect on employee performance.

The theoretical framework of this research is given in Figure 1.

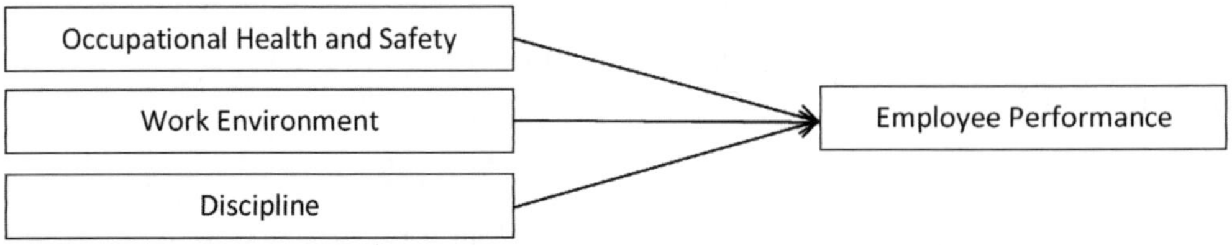

Figure 1. The Theoretical Framework.

3. Research Methodology

This study used survey method and surveyed all employees in PPIC Thermo section as respondents. Forty-seven employees took part as respondents. All respondents answered all thirty-five questions. There are 11 questions for occupational health and safety variable, 8 questions for work environment variable, 8 questions for discipline variable and 8 questions for employee performance. Each question describes the indicators that exist on each variable. Every question provides several choices that respondents have to response. Five-points Likert was used for this questionnaire.

Variables used in this study are independent variable and dependent variable. Dependent variable is employee performance, while the independent variable consists of occupational health and safety, work environment and discipline. Questionnaire that had been collected then processed with SPSS software to analyze the instrument test that consists of validity and reliability analysis. After that, multiple linear regressions were used to analyze data.

4. Results

4.1 Validity test

Validity test was conducted by using Pearson product moment correlation. From thirty-five questions, one question was not valid. This question was deleted and not included in the next analysis.

4.2 Reliability test

Reliability test was conducted to measure whether an instrument is reliable enough to be used as data collection tool by testing the score between items using Cronbach's alpha. After analyzed, all variables were declared reliable because all Cronbach's alpha were greater than 0.6.

4.3 Multiple Linear Regressions

After the data were analyzed with SPSS software, the results can be seen in the table below:

Table 1. The Resume of Results Multiple Linear Analysis.

Variable	Unstandardized Coefficients B	t	Sig	Note
Occupational Health and Safety	0.231	2.274	0.028	**Significant**
Work Environment	0.241	2.302	0.026	**Significant**
Discipline	0.285	2.311	0.026	**Significant**
Coefficient Determination (R^2)	= 0.734			
F	= 39.494			
F Sig	= 0.000			

From the table above, it can be described that:

- Simultaneous hypothesis analysis (F test)
  This analysis is to describe if there is an effect of all independent variables on the dependent variable. Based on the results of regression model analysis, the value of F is 39.494. This value is greater than the value of F table (2.821). Therefore, each independent variable is simultaneously significant at $p<0.05$.
- Partial hypothesis analysis (t test)
  The analysis of partial regression model is used to identify if there is an effect of every independent variable on the dependent variable. The results are presented in this table below:

Table 2. The Results of Partial hypothesis analysis (t test).

Variable	t	Sig
Occupational Health and Safety	2.274	0.028
Work Environment	2.302	0.026
Discipline	2.311	0.026

  With t table 2.012 , t value of every variable is greater than the value of t table. Thus, every independent variable significantly affects employee performance at $p <0,05$.
- Dominant test
  To determine one independent variable that has the dominant effect on the dependent variable, comparison of standardized coefficients among all independent variables was carried out. Independent variable is said to have dominant effect on the dependent variable if it has the greatest standardized coefficient. The comparison of all standardized coefficients of all independent variables is presented in the table below:

Table 3. The Resume of Standardized Coefficients of all Independent Variables

Variable	**Standardized Coefficients**
Discipline	0.330
Occupational Health and Safety	0.308
Work Environment	0.296

  From this table, discipline is the independent variable which has the greatest standardized coefficient among the other independent variables. This dominant test did not match the second hypothesis. It shows that employee performance is more affected by discipline factor than the other independent variables in this study. Standardized coefficients have positive value; it means that the higher the level of discipline, the higher the performance of employee.

5. Conclusion

After analysis with multiple linear regressions, occupational health and safety, work environment and discipline have simultaneous partial effect on the employee performance. From the dominant test, it can be seen that discipline has a dominant effect on employee performance. Upgrading employee performance can be done by improving their level of discipline in addition to making sure that they get health and safety insurance in their work place and have a good working environment (including their relationship with their peer, superior and subordinates). Good employee performance can better support company performance.

6. References

[1] Hasibuan M 2011 *Human Resources Management* (Jakarta-Bumi Aksara)

[2] Simamora H 2004 *Human Resources Management* (Yogyakarta-Penerbit Andi)

[3] Heizer J and Render B 2012 *Operations Management book1*, 9th editions

[4] Nurhaida 2010 *The Effect of Work Environment and Discipline to Employee Performance Kopertis I Nanggroe Aceh Darussalam North Sumatera* (Graduated Thesis of Graduated School of University of North Sumatera Medan)

[5] Husni 2011 *The Analyze of Occupational Helath and Safety and Work Environment to Employee Performance of PT Indonesia Asahan Aluminium (Inalum) Kuala Tanjung* (Graduated Thesis of University of North Sumatera Medan)

[6] Paramita C and Wijayanto A 2012 The Effect of Occupational Health and Safety to Employee Performance of PT. PLN (Persero) APJ Semarang *Journal of Business Administration* **1** (1)

SME Worker Affective (SWA) index based on environmental ergonomics

M Ushada[1*], G Kusuma Aji[2], T Okayama[3], M Khidir[4]
[1]Department of Agro-industrial Technology, Universitas Gadjah Mada, D. I. Yogyakarta, Indonesia
[2]Department of Bioresource Technology and Veterinary, Vocational College, Universitas Gadjah Mada, D. I. Yogyakarta, Indonesia
[3]Department of Regional and Environmental Science, Ibaraki University, Ibaraki, Japan
[4]PT. Stechoq Robotika Indonesia, D. I. Yogyakarta, Indonesia

*mirwan_ushada@ugm.ac.id

Abstract. Small-Medium sized (SME) is a focal type of Indonesian industry which contributes to national emerging economies. Indonesian goverment has developed employee social security system (*BPJS Ketenagakerjaan*) to support worker quality of life. However, there were limited research which could assist BPJS Ketenagakerjaan in evaluating worker quality of life. Worker quality of life could be categorized as the highest worker needs or affective states. SME Worker Affective (SWA) index is being concerned as a basic tool to make balance between worker performance and quality of life in workstation of SMEs. The research objectives are: 1) To optimize the environmental ergonomics in SMEs; 2) To quantify SME Worker Affective (SWA) index based on optimized environmental ergonomics. The research advantage is to support Indonesian goverment in monitoring SMEs good practices to its worker quality of life. Simulated annealing optimized the heart rate and environmental ergonomics parameters. SWA index was determined based on comparison between optimized heart rate and environmental ergonomics parameters. SWA index were quantified for 380 data of worker. The evaluation indicated 51.3% worker in affective and 48.7% in non-affective condition. Research results indicated that stakeholders of SMEs should put more attention on environmental ergonomics and worker affective.

1. Introduction

Small-Medium sized (SME) is a focal type of Indonesian industry which contributes to national emerging economies. The characteristics of Indonesian SMEs were [1]: 1) Application of appropriate technology; 2) Labour intensive-based production; 3) Significant economic contribution; 4) Not yet total ergonomics-oriented. Indonesian government has developed employee social security system (*BPJS Ketenagakerjaan*) to support worker quality of life [2]. However, there were limited research which could assist *BPJS Ketenagakerjaan* in evaluating worker quality of life. Worker quality of life could be categorized as the highest worker needs, or known as affective states. Affective or implementation of Kansei has relationship with ergonomic as psychological and emotional response from worker satisfaction [3]. Heart rate was proven as reliable non-verbal parameter to represent affective states [4]. Environmental ergonomics influenced significantly worker affective states [5][6]. Environmental ergonomics is the effects of indoor environment on the health, convinience and productivity of human work [7]. SME Worker Affective (SWA) index is being concerned as a basic tool to assess quality of life in workstation of SMEs.

There were some research which developed ergonomic asessment based on environmental ergonomics. Gläser et al. [8] have developed the ergonomic assesment based on simulation of digital human model and data sensor. Fallahi et al. [9] have developed ergonomic assesment for tractor based on algometer. Ahmadi et al.[10] have developed indicator of ergonomic assessment based on standard of international labour organization and international ergonomic association for the application in assembly-packaging industry. However, none of them is applicable for assessment of SWA index.

The research objectives are: 1) To optimize the environmental ergonomics in SMEs; 2) To quantify SME Worker Affective (SWA) index based on optimized environmental ergonomics. The research advantage is to support Indonesian government in monitoring SMEs good practices to its worker quality of life. In industrial practice, SWA index could be used as an indicator to assess the ergonomic feasibility of a production system.

2. Materials and Method

2.1. Database

The Six (6) SMEs of different food products were used for the database, i.e., crackers, nuggets, fish chips, *bakpia* (a traditional pastry), tempeh and herbal instant beverages. Three hundred and eighty (380) data sets were analysed. The environmental ergonomics data consisted of heart rate after working, temperature before and after working, Relative Humidity (RH) before and after working, light intensity after working. The work measurement before and after working was suggested by Ushada et al. [5].

2.2. Research Methodology

Research methodology was defined in Fig. 1. Environmental ergonomics data consisted of affective states (Heart rate after working) and workplace environment (Workplace temperature and relative humidity before and after working and light intensity after working). The environmental ergonomics were modeled using simulated annealing algorithm. The temperature before working was set as target for the optimization. The temperature was measured before working since it was the closest one to the convinient temperature in the morning time. The temperature ranged from 24.8 to 33.1^{0}C. The other parameters were optimized based on target of temperature before working. The index of each parameters were determined based on the comparison between measured and optimized values. SWA index was determined based on standardization of each parameter index.

2.3. Simulated Annealing

The algorithm for simulated annealing were developed using MATLAB R2015a (8.5.0.19763) and consists of following steps:

1. Set the cooling scheme (Value of annealing parameter):
   a. Initial temperature (T_0) and final temperature (T_1)
   b. Factor of decreasing temperature (F)
   c. Amount of iteration in each temperature value (N)
2. Select the initial solution X by random or using specific heuristic method.
3. Determine the value of objective function: $E = f(X)$.
4. Set the value of temperature parameter: $T = T_0$.
5. Generating the new solution X_{new}
6. Determine the value of objective function: $E_{new} = f(X_{new})$.
7. Determine $X = X_{new}$ and $E = E_{new}$ based on probability (Metropolis criterion) in equation 1:

$$\min\left(1, \exp\left\{\frac{(E-E_{new})}{T}\right\}\right) \quad (1)$$

IOP Conf. Series: Materials Science and Engineering **337** (2018) 012037 doi:10.1088/1757-899X/337/1/012037

8. Repeat the step 5 to 7 for the *N*
9. Reduce the temperature parameter by setting: $T = T \times F$
10. Stop the algorithm if termination criterion is satisfied. If it is not, back to step 5.

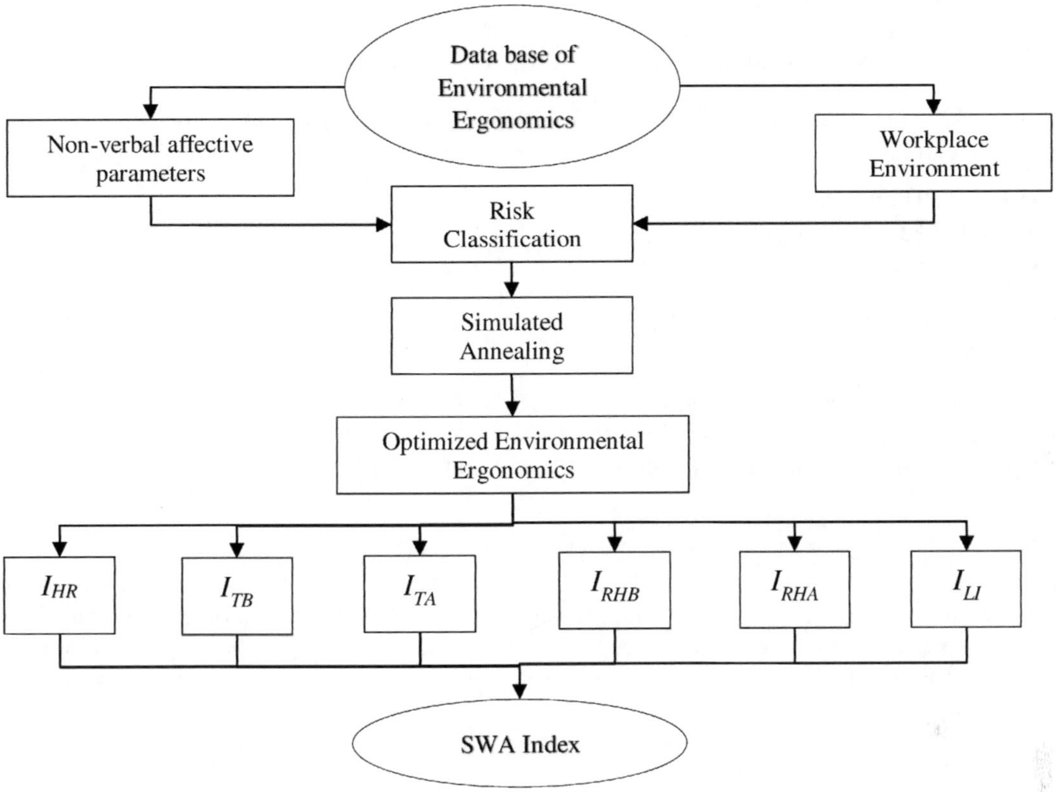

Figure 1. Research methodology.

2.4. SME Worker Affective

Index of heart rate is determined based on equation 2:

$$I_{HR} = \frac{HR_0}{HR_1} \times 100 \quad (2)$$

I_{HR} = Index of heart rate
HR_0 = Measured heart rate (Pulse/minutes)
HR_1 = Optimized heart rate based on simulated annealing (Pulse/minutes)

Index of workplace temperature before working is determined based on equation 3:

$$I_{TB} = \frac{TB_0}{TB_1} \times 100 \quad (3)$$

I_{TB} = Index of temperature before working
TB_0 = Measured temperature before working (^{0}C)
TB_1 = Target of temperature before working (^{0}C)

Index of workplace temperature after working is determined based on equation 4:

$$I_{TA} = \frac{TA_0}{TA_1} \times 100 \quad (4)$$

I_{TA} = Index of temperature after working
TA_0 = Measured temperature after working (^{0}C)

TA_1 = Optimized temperature after working based on simulated annealing (^{0}C)

Index of workplace RH before working is determined based on equation 5:

$$I_{RHB} = \frac{RHB_0}{RHB_1} \times 100 \quad (5)$$

I_{RHB} = Index of RH before working
RHB_0 = Measured RH before working (%)
RHB_1 = Optimized RH before working based on simulated annealing (%)

Index of workplace RH after working working is determined based on equation 6:

$$I_{RHA} = \frac{RHA_0}{RHA_1} \times 100 \quad (6)$$

I_{RHA} = Index of RH after working
RHA_0 = Measured RH after working (%)
RHA_1 = Optimized RH after working based on simulated annealing (%)

Index of light intensity after working is determined based on equation 7:

$$I_{LI} = \frac{LI_0}{LI_1} \times 100 \quad (7)$$

I_{LI} = Index of light intensity after working
LI_0 = Measured light intensity after working (Lux)
LI_1 = Optimized light intensity after working based on simulated annealing (Lux)

SWA index are determined based on equation 8:

$$SWA = ((I_{HR} \times 0.4) + (I_{TB} \times 0.4) + (I_{TA} \times 0.05) + (I_{RHB} \times 0.05) + (I_{RHA} \times 0.05) + (I_{LI} \times 0.05)) \quad (8)$$

Workers were regarded as affective if SWA Index $\leq$ 112. The value of 112 was determined using the average value from population of 380 data. The weights of SWA index was determined based on the importance of each index. The highest weight was determined on I_{HR} due to representative of non-verbal affective parameters. The highest weight was determined on I_{TB} due to target value of simulated annealing optimization.

3. Results and Discussion

3.1. Risk Classification

The risk score was determined based on classification of affective and environmental parameter. The heart rate was classified by AIHA in Kolus et al. [11] (Table 1). This kind of classification was reviewed by Kolus et al. [12]. The temperature was classified based on Regulation of Ministry of Energy and Mineral Resources (RMEMR) No.13 Year of 2012 [13]. RH range was classifed based on Regulation of Ministry of Health No.1077 Year of 2011 [14]. Light intensity was classified based Regulation of Ministry of Health No.1405 Year of 2002 [15].

Table 1. Risk classification

No	Parameters	Classification	Range	Score
1.	Heart rate (Pulse per minute) [11][12]	Sitting	60-75	0
		Very light	65-75	0
		Light	75-100	1
		Moderate	100-125	2
		Heavy	125-150	3
		Very heavy	150-175	4
		Extremely heavy	>175	5
2.	Workplace Temperature (^{0}C) [13]	Recommendable	$\leq$ 30	0
		Not recommendable	>30	1
3	Workplace RH (%) [14]	Recommendable	40 - 60	0
		Not recommendable	>60	1

4.	Light intensity (Lux) [15]	Recommendable Not recommendable	<3000 >3000	0 1

3.2. Simulated Annealing
Simulated annealing was conducted on 17 trials and various targets of before working temperatures as shown in Table 2.

3.3. Optimized Environmental Ergonomics
Optimized parameters were generated using simulated annealing and risk classification as following (Table 3):

Table 2. Target of before working temperatures based on simulated annealing

Trials	Simulated Annealing (^{0}C)	Measured (^{0}C)	Absolute Error	Risk Score
1.	26,94	27,50	0,56	5
2.	29,27	29,30	0,03	4
3.	30,12	31,10	0,98	5
4.	26,74	27,50	0,76	4
5.	29,02	29,30	0,28	5
6.	30,19	31,10	0,91	4
7.	29,06	29,10	0,04	5
8.	28,67	30,20	1,53	5
9.	28,71	29,00	0,29	5
10.	28,59	29,00	0,41	4
11.	27,64	28,20	0,56	5
12.	28,63	28,80	0,17	1
13.	28,52	29,60	1,08	5
14.	28,80	29,10	0,30	1
15.	29,51	30,20	0,69	6
16.	29,36	29,40	0,04	2
17.	28,27	29,00	0,73	1

Table 3. Optimized environmental ergonomics

No	Parameter	Metric	Values
1.	Heart rate	Pulse/minutes	69,58
2.	Workplace temperature after working	^{0}C	28,46
3.	Workplace RH before working	%	59,80
4.	Workplace RH after working	%	40,64
5.	Light intensity after working	Lux	64357,93
6.	Target of temperature before working	^{0}C	28,80
7.	Simulated temperature before working	^{0}C	28,63

3.4. Quantification SWA Index
Based on the optimized parameters in Table 3, the index of environmental ergonomics were determined using equations (2), (3), (4), (5), (6) and (7). Subsequently, SWA index was quantified for 380 worker data using equation (8). The evaluation indicated 51.3% workers were in affective condition while 48.7% others were in non-affective condition. The research results indicated that stakeholders of SMEs should put more attention on environmental ergonomics and worker affective. SWA index could be used to assess feasibility of workplace environment to generate worker quality of

life. Further research is suggested for the development of national affective assessment system based on integration of SWA index and employees social security system.

4. Conclusion

Simulated annealing optimized the heart rate and environmental ergonomics parameters based on temperature before working. SWA index was determined based on comparison between optimized heart rate and environmental ergonomics parameters. SWA index were quantified for 380 workers. The evaluation indicated 51.3% workers were in affective state while the rest of 48.7% workers were in non-affective condition. The research results indicated that stakeholders of SMEs should put more attention on environmental ergonomics and worker affective. SWA index could be used to assess feasibility of workplace environment to generate worker quality of life.

5. References

[1] Ushada M, Suyantohadi A, Khuriyati N and Okayama 2017 Identification of environmental ergonomics control system for Indonesian SMEs *Proc. of IEEE-ICCAR 2017* Nagoya Japan.

[2] Anonym 2017 *Badan Penyelenggara Jaminan Sosial Ketenagakerjaan/BPJS Ketenagakerjaan (National employees social security system)* http://www.bpjsketenagakerjaan.go.id/, accessed on June 7, 2017

[3] Kwong C K, Huimin J and Luo X G 2016 AI-based methodology of integrating affective design, engineering, and marketing for defining design specifications of new products. *Eng. Appl. Artificial Intelligence* **47** pp 49-60

[4] Ushada M, Okayama T, Khuriyati N and Suyantohadi A 2015 *Kansei's physiological measurement in small-nedium sized enterprises using profile of mood states and heart rate. Springer Lecture Notes in Electrical Engineering series (LNEE).*

[5] Ushada M, Okayama T and Murase H 2015 Development of kansei engineering-based watchdog model to assess worker capacity in Indonesian small-medium food industry *Eng. in Agric., Environ. Food* **8** pp 241-250

[6] Ushada M, Okayama T, Suyantohadi A, Khuriyati N and Murase H 2017b Kansei Engineering-based artificial neural network model to evaluate worker performance in small-medium scale food production system. *Int. J. Indust. Sys. Eng.*, In Press.

[7] Parsons K C 2000 Environmental ergonomics: a review of principles, methods and models. *App. Ergon.* **31** pp 581-594

[8] Gläser D, Fritzsche L, Bauer S and Sylaja V J 2016 *Ergonomic assessment for DHM simulations facilitated by sensor data. Procedia CIRP* **41** pp 702-705

[9] Fallahi H,Abbaspour-Fard M H, Azhari A, Khojastehpour M and Nikkhah A 2016 Ergonomic assessment of drivers in MF285 and MF399 tractors during clutching using algometer. *Inform. Process. in Agric.* **3** pp 54-60

[10] Ahmadi M, Zakerian S A and Salmanzadeh H 2017 Prioritizing the ILO/IEA ergonomic checkpoints' measures; a study in an assembly and packaging industry. *Int. J. Indust. Ergon.* **59** pp 54-63

[11] American Industrial Hygiene Association (AIHA) Ergonomic guide to assessment of metabolic and cardiac costs of physical work (1971) in Kolus A, Imbeau, D, Dubé, V, Dubeau, D. Classifying work rate from heart rate measurements using an adaptive neuro-fuzzy inference system *App. Ergon.* **54** pp.158-168

[12] Kolus A, Imbeau D, Dubé P-A and Dubeau D 2016 Classifying work rate from heart rate measurements using an adaptive neuro-fuzzy inference system *App. Ergon.*, vol. **54** (2016), pp.158-168

[13] Anonym 2012 *Regulation of Ministry of Energy and Mineral Resources of Republic Indonesia No.13: Electricity Saving (In Bahasa Indonesia: Peraturan Menteri Energi dan Sumber*

Daya Mineral Nomor 13 tentang Penghematan Pemakaian Listrik). Ministry of Energy and Mineral Resources of Republic Indonesia

[14] Anonym 2011 *Regulation of Ministry of Health Republic Indonesia No. 1077: Healty Indoor Air at Home (In Bahasa Indonesia: Pedoman Penyehatan Udara dalam Rumah).* Ministry of Health Republic Indonesia.

[15] Anonym 2002 *Regulation of Ministry of Health Republic Indonesia No. 1405: Requirement of occupational health for office and industry (In Bahasa Indonesia: Persyaratan kesehatan lingkungan kerja perkantoran dan industri).* Ministry of Health Republic Indonesia.

Acknowledgments

Authors wishing to acknowledge financial support from Ministry of Research, Technology and Higher Education of the Republic of Indonesia by 2017 Research Grants of International Collaboration Competitive Research Grant for International Publication-Universitas Gadjah Mada 'Penelitian Kerjasama Luar Negeri dan Publikasi Internasional Tahun 2017" No: SP DIPA-042.06.1.401516/2017, December 7, 2016 and No: 2273/UN1.P.III/DIT-LIT/LT/2017, 19 April 2017

Manual handling methods evaluation based on oxygen consumption

E Nurmianto[1*], U Ciptomulyono[1], Suparno[1], S Kromodihardjo[2], H Setijono [3], and N A Arief [3]
[1] Department of Industrial Engineering, Institut Teknologi Sepuluh Nopember, Kampus ITS Sukolilo-Surabaya 60111, Indonesia
[2] Department of Mechanical Engineering, Institut Teknologi Sepuluh Nopember, Kampus ITS Sukolilo-Surabaya 60111, Indonesia
[3] Achilles Laboratory of Sport Sciences – Faculty of Sport Science, Universitas Negeri Surabaya, Surabaya, Indonesia 60213

*nurmieko@gmail.com

Abstract. Mining industry has become one of the largest industries in Indonesia, now competing in billions dollar market, with numbers people employed. Deliveries of a Return Rolls (RR) involve the use of a hand truck and, in many cases, a shoulder/elbow-mode of carriage. Workers usually prefer to the *Gendong* (carrying on the small of the back or the hip, supported by the waist and arm) mode or *Manggul* (carrying on some stuff shoulder) mode, because they feel safer by carrying RR on the shoulder/elbow. In this study, the physiological workload involved in shoulder/elbow-mode carrying was investigated, especially focusing on the effects of load weight and inclination. To measure heart rate and oxygen uptake while carrying on the shoulder/elbow, a laboratory experiment was conducted and safety guidelines for such tasks were proposed, based on the experimental results. Four healthy male subjects performed shoulder/elbow-mode carrying, weight between 20 and 24 kg: (1) on inclination of 10°, (2) 20° and (3) 30°. The results showed that inclination involved an increased physiological burden, and that a load of 24 kg entailed a significantly higher physiological cost than carrying a load of 20 kg. Although shoulder/elbow-mode carrying has some advantages, the worker should be advised to carry a load of less than 20 kg, to avoid a high physiological load. During shoulder/elbow-mode carrying, it is also recommended that a person prepare more training in order to have muscular strength.

1. Introduction

Coal mining continues to be an important source of musculoskeletal stress and injuries. The primary aim of this paper described here was to reduce injury risks associated with manual tasks performed by miners [1]. Coal business process is started from overburden stripping, coal mining, coal loading coal from the stockpile to hauler, coal hauling, crushing and sizing, barge loading, and ended up in coal barging. In coal mining industry, discrete-event simulation has been widely used to support decisions in material handling system (MHS) [2] while Bow-tie analysis combines aspects of fault-tree analysis and event-tree analysis to identify an initiating event [3].

The mining industry in this study is important to investigate because of its relatively on high incidence and severe muscle injury [4][5][6][7]. Besides working in various minerals industries (such as coal mining) is rarely considered as safe and comfortable as working in the office [3][8]. This will cause Musculoskeletal Disorders (MSDs) or muscle bone injuries. It is this injury that will increase the dangers and diseases caused by work. More importantly, the mining industry is an industry with many potential hazards that can cause injury and death [9]. Accordingly, some of the major causes of hazards in the mining industry are dust, vibration, noise, work shift and other manual tasks [8].

Several studies have been conducted related to the mining industry [10][11][12], and research on the ergonomic work environment has been done [11]. Activities in the mining industry are full of risks for both underground and over ground mines. These activities lead to high risk of occupational accidents due to lifting especially on the back of the rear. Such work accidents are more common in the mining industry than in other industries.

The formulation of the problem or research question (RQ) of this research is how the RR load and inclination conveyor affects the VO_2 value when lifting RR based on VO_2. The purpose of this research is to determine the effect of conveyor load and inclination on VO_2 value when lifting RR based on VO_2. This research recommend to the company on how the mechanics should carry the RR (Return Rolls). Theoretical implications obtaining the significance value of each factor's influence on VO_2 obtaining a lift method that impacts yields lower VO_2. The proposed managerial implications are as follows.

- Providing recommendations to companies regarding appropriate lifting methods and priority factors that are important and should be considered to minimize the possibility of injury to workers.
- Providing recommendations to companies regarding the value of the burden and inclination, appropriate to minimize the possibility of injury to workers.
- Considering lifting methods as one of the important safety factors in MMH design work as it has an effect on the physiological aspects of the respondents.

Contribution of this study contribute to the different RR lift styles and mechanical capabilities shown through VO_2 as well as the maximum time and distance traveled. The different lift styles give consideration to how RR should be raised in the mine area. In addition to the consideration, this lift style difference also makes it easier in planning the necessary actions in case of errors in lifting.

2. Definition

Manual handling of objects in an industrial setting has been a significant concern to occupational health professionals who attempt to prevent injury. Tasks that demand frequent and heavy lifting are associated with an increased risk of low back pain [4].

Manual Material Handling (MMH) defined by several researchers [16][17][18] that Manual Material Handling and movement is a part of many activities, usually it includes involves lifting, lowering, pushing, pulling, gripping, pinching, carrying and holding objects by hand. For instance of how manual material handling and movement diffuses all facet of our work life in health care involves loading and unloading trucks, carts, boxes or wooden container; lifting bags, instrument trays; moving parts, food or assemblies from one place to another; loading paper in the copier or picking binders from an overhead shelf; and handling a mop, polisher or vacuum. Equal with technology available today, manual material handling and movement will always be with us. The one thing all these tasks have in common is the potential to result in some contrary effect such as personal hurt or property loss.

3. Methodology

The current research methodology generally consists of three stages: the preparatory phase of the research, the MMH experimental stage, and the analysis and discussion stage of the MMH experiments using statistical analysis. In the preparation phase of the study discussed about the background of the research. Having obtained the background of the next stage of research is to formulate the problems to be researched followed by preparing research objectives that answer the problems that have been formulated. Also drawn up the benefits and contributions of the current research. The next step is to study the theories related to MMH from various literatures so that from the theoretical studies that have been done, obtained the responses to be measured that is VO2 and HR, the factors that are considered to give influence to the response that is load, inclination and how to lift, As well as methods to be performed for MMH experiments and analysis of MMH experimental data i.e. MANOVA (Multivariate Analysis of Variance) if correlation between high value response variable and ANOVA if otherwise. Preliminary theoretical and experimental studies are also conducted to get the right level on each factor.

After getting the response variable, the factors as well as the level, as well as the experimental method and MMH analysis, the next step is to start doing MMH experiment. Before the experiment begins, the thing that needs to be prepared is the sampling of the operator in the experiment. After that the MMH experiment was done in accordance with the general factorial design that has been made. During the experiment the VO_2 and HR values were measured with the Ergometer FitmatePro Cosmed tool and the results were recapitulated.

4. Results

4.1. The average difference of Oxygen Consumption for each lift position

Based on Figure 1 it can be seen that for the lifting method with the holding position (G) is better used when the weight conditions are 20 kg inclination 20, the weight of 20kg inclination 30°, the weight of 24kg inclination 20° and the weight of 24kg inclination 30°. As for the lifting method with the *Manggul* position (M) is more suitable when used in heavy conditions 20 inclinations 10, weight 24kg inclination 10° this is because it has a smaller value VO_2.

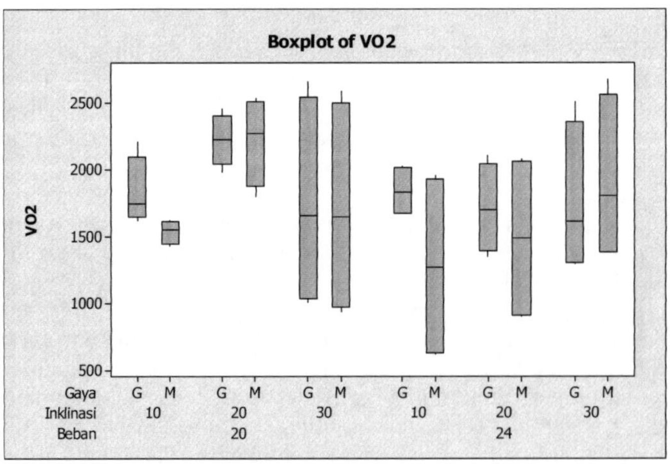

Figure 1 Boxplot for Consumption VO_2.

4.2. Distribution of plot data from oxygen consumption for each individual

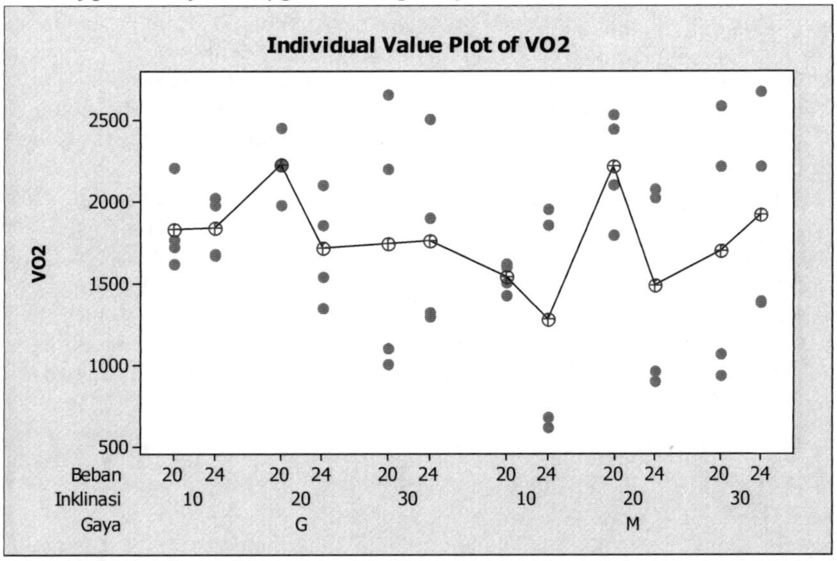

Figure 2 Differences VO_2 for each respondent and different load conditions.

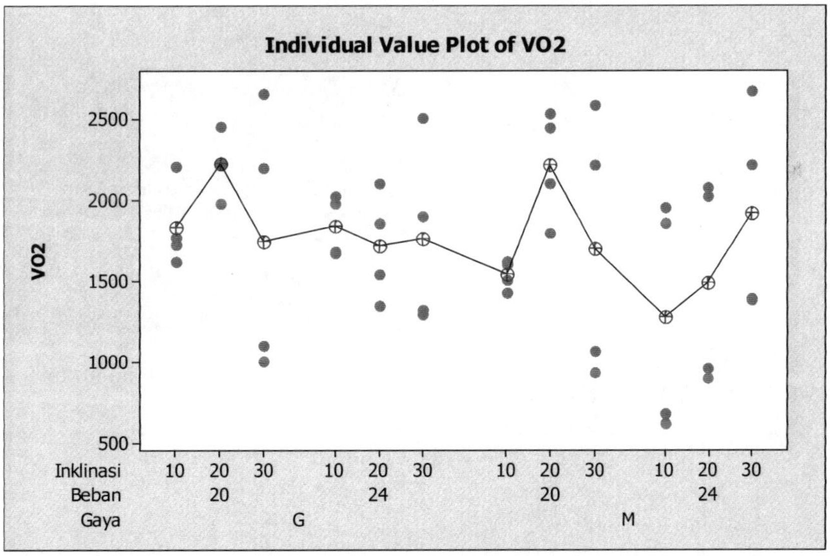

Figure 3 Difference VO_2 for each respondent and different inclination conditions.

When the VO_2 ratio is compared when the load is different, for 24 kg loads it has a higher VO_2 when compared to a 20 kg load, but when the activity with the 20 kg incubation position of the *Manggul* for 20kg and the inclination of 20° *Manggul* positions for load 20kg gives a higher VO_2 value From the 24kg load. Meanwhile, when viewed from inclination, for inclination 10° always gives higher VO_2 value, the next highest is inclination 30°. So it can be concluded that the higher the inclination the VO_2 value will be higher also. Furthermore it will also be known for which factors influence VO_2.

5. Discussion

Figure 4.a and Figure 4.b shows the output of Ergometer FitmatePro. At the output it appears 2 graphs are the red color HR and the blue color VO_2. One's lifting condition is indicated by the increase in HR and VO2 which ends with the discharge of the AT line. AT is the limit of Anaerobic Threshold where its oxygen reserves are getting thinner or almost exhausted. At that time oxygen is taken from the muscle so it will feel tired in anaerobic conditions. At that time someone doing the activity should rest to get O_2 back followed by a decline in HR.

Based on Figure 5, it can be explained that when using α by 10% then the factors affecting Heart Rate when appointment of return roll are load factor, lift / position, and inclination. In the calculation of Anova also obtained R-sq value of 39.95% which means very low so it can be concluded that for the Heart Rate is influenced by the interaction between load, lift / position of 39.95% the rest entered in other factors. Based on Figure 4 it can be explained that when using α by 10% then the factor that influences VO_2 when the appointment of return roll is inclination. In the calculation of Anova also obtained R-sq value of 69.45% which means that factors used in this study can only explain / affect VO2 of 69.45%.

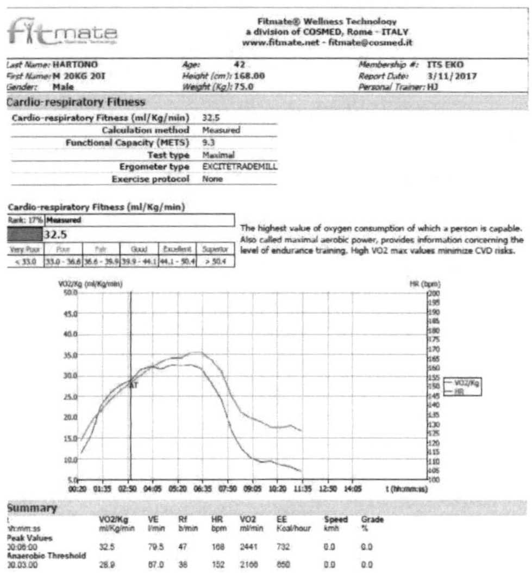

Figure 4.a. Example output FitmatePro (Hartono-*Manggul*-20kg-20°).

Figure 4.b. Manual handling in stock pile area.

Analysis of Variance for VO2, using Adjusted SS for Tests

Source	DF	Seq SS	Adj SS	Adj MS	F	P	
Load	1	0.213	0.213	0.213	0.02	0.880	NS
Inclination	2	681.626	681.626	340.813	37.07	0.000	S
Position	1	0.301	0.301	0.301	0.03	0.857	NS
Load*Inclination	2	47.675	47.675	23.838	2.59	0.089	NS
Load*Position	1	2.341	2.341	2.341	0.25	0.617	NS
Inclination*Position	2	18.725	18.725	9.363	1.02	0.371	NS
Load*Inclination*Position	2	1.638	1.638	0.819	0.09	0.915	NS
Error	36	331.020	331.020	9.195			
Total	47	1083.540					

S = 3,03233 R-Sq = 69,45% R-Sq(adj) = 60,12%

Figure 5. Anova for VO_2 with 4 respondents.

6. Conclusion

Anova analysis result by using Minitab showed that factors influencing Heart Rate when appointment of return roll are load factor, lift / position, and inclination. For all activities with the *manggul* position have a lower HR when compared with the position of carrying. This can be used as a suggestion for the company that raising the *Manggul* way better than the way *Gendong*. Whereas when the VO_2 ratio is compared when the load is different, for 24kg loads it has a higher VO_2 when compared to a 20kg load, but when the activity with the 20kg incoordination position of the *Manggul* for 20kg load and the inclination of 20° *Manggul* positions for load 20kg gives a higher VO_2 value High from 24kg load. Whereas when viewed from inclination, for inclination 10° always gives higher VO_2 value, the next is highest when inclination 30° *Gendong*. The results of this study contribute to the different RR lift styles and mechanical capabilities shown through VO_2 as well as the maximum time and distance traveled. The different lift styles give consideration to how RR should be raised in the mine area. In addition to the consideration, this lift style difference also makes it easier in planning the necessary actions in case of errors in lifting

7. References

[1] Burgess-limerick R, Straker L, Pollock C, Dennis G, Leveritt S, and Johnson S 2007 Implementation of the Participative Ergonomics for Manual tasks (PErforM) programme at four Australian underground coal mines *Int. J. Ind. Ergon.* vol **37** pp 145–155

[2] Meng C, Srinivas S, Maghsoudi A, Son Y, and Dessureault S 2013 Computers & Industrial Engineering Data-driven modeling and simulation framework for material handling systems in coal mines *Comput. Ind. Eng.* vol **64**(3) pp 766–779

[3] Burgess-limerick R, Horberry T, and Steiner L 2014 Bow-tie analysis of a fatal underground coal mine collision *Ergon. Aust.* vol **10**(2) pp 1–5

[4] Plamondon A, Delisle A, Trimble K, Desjardins P, and Rickwood T 2006 Manual materials handling in mining: the effect of rod heights and foot positions when lifting 'in-the-hole' drill rods. *Appl. Ergon.*, vol **37**(6) pp 709–18

[5] Nurmianto E, Ciptomulyono U, Suparno, and Kromodiharjo S 2015 Manual handling problem identification in mining industry: an Ergonomic perspective *Procedia Manuf.* vol **4** pp 89–97

[6] Widanarko B, Legg S, Devereux J, and Stevenson M 2015 Interaction between physical and psychosocial work risk factors for low back symptoms and its consequences amongst Indonesian coal mining workers *Appl. Ergon.* vol **46** pp 158–167

[7] Widanarko B, Legg S, Devereux J, and Stevenson M 2015 Interaction between physical and psychosocial risk factors on the presence of neck / shoulder symptoms and its consequences *Work* vol **139**

[8] Horberry T, Burgess-limerick R, and Fuller R 2013 The contributions of human factors and ergonomics to a sustainable minerals industry *Ergonomics* vol **56** pp 37–41

[9] Delisle A, Trimble K, Desjardins P, and Rickwood T 2006 Manual materials handling in mining : The effect of rod heights and foot positions when lifting "in-the-hole" drill rods *Applied Ergonomics* vol **37** pp 709–718

[10] Abrahamsson L 2000 Production economics analysis of investment initiated to improve working environment *Appl. Ergon.* vol **31** pp 1–7

[11] Parsons K C 2000 Environmental ergonomics: a review of principles, methods and models *Appl. Ergon.* vol **31**(6) pp 581–94

[12] Torp S and Moen B E 2006 The effects of occupational health and safety management on work

environment and health: A prospective study *Applied Ergonomics* vol **37** pp 775–783

[13] Gallagher S and Hamrick C A 1992 Acceptable workloads for three common mining materials *J. Ergon.* vol **35** pp 1013–1031

[14] Widanarko B, Legg S, Stevenson M, Devereux J, Eng A, Cheng S, Douwes J, Ellison-loschmann L, Mclean D, and Pearce N 2011 Prevalence of musculoskeletal symptoms in relation to gender, age, and occupational / industrial group *Int. J. Ind. Ergon.* vol **41**(5) pp 561–572

[15] Widanarko B, Legg S, Stevenson M, Devereux J, and Jones G 2013 Prevalence of Low Back Symptoms and Its Consequences in Relation to Occupational Group *J. Ind. Med.* vol **589** pp 576–589

[16] Bridger R S 2003 *Introduction to Ergonomics* Taylor & Francis Group

[17] Lehto M R and Buck J R 2005 *Introduction to Human Factors and Ergonomics for Engineers* Taylor & Francis Group

[18] Ferguson S A, Marras W S, and Burr D 2005 Workplace design guidelines for asymptomatic vs low-back-injured workers *Applied Ergonomics* vol **36** pp 85–95

Acknowledgments
Authors wishing to acknowledge that this research was financially supported from Directorate General of Higher Education, Ministry of Technology Research and Higher Education, Republic of Indonesia

IOP Conf. Series: Materials Science and Engineering **337** (2018) 012039 doi:10.1088/1757-899X/337/1/012039

Musculoskeletal disorder survey for pond workers

A Maryani[1*], S G Partiwi[1] and H N F Dewi[1]
[1]Department of Industrial Engineering, Institut Teknologi Sepuluh Nopember, Kampus ITS Sukolilo-Surabaya 60111, Indonesia

*annymaryani@ie.its.ac.id

Abstract. Mucsuloskeletal disorder will affect worker performance and become serious injury when ignored, so that workers cannot work normally. Therefore, an effective strategy plan is needed to reduce the risk of musculoskeletal disorder. A pond worker is profession with high risk of physical complain. Four main activities are ponds preparation, seed distribution, pond maintenance, and harvesting. The methods employed in this current musculoskeletal disorder survey are questionnaire and interview. The result from 73 questionnaires shown that most of pond workers were working for 7 days a week. Prevalence physical complain are on neck, shoulders, upper back, lower back, and knees. The level of perceived complaint is moderate pain. However, most of them do not contact therapists or physicians. Therefore it is necessary to improve the working methods to be able to reduce physical complains due to musculoskeletal disorder.

1. Introduction

Aquaculture can be defined as a human intervention to increase aquatic productivity through cultivation activities. What is meant by cultivation activities is the maintenance activities to reproduce, growth, and improve the quality of aquatic biota so that it can be profitable [1]. Specifically, the aims of aquaculture are for (1) food production, (2) natural stock improvement, (3) fish production for recreation, (4) fish production, (5) ornamental fish production, (6) organic materials recycling and (7) production of industrial materials.

The fishery businessmen who are usually fishermen, pond workers, fishery processors and product marketers are expected to increase productivity, added value, and competitiveness, as well as build a modern and integrated production system from upstream to downstream. Like most of pond workers in Indonesia, the work process is conducted manually. A variety of manual activities carried out are lifting/lowering, pushing/pulling, twisting the body, carrying objects and holding/restraining objects. Work manuals that rely on musculoskeletal are very susceptible to health problems. Repetitive work and work position errors lead to the accumulation of pain that will ultimately have a bad effect for workers.

Based on the information from [2], the musculoskeletal disorders affect workers in various industries. Musculoskeletal disorders may affect a worker's performance, or even if they are not seriously aware, it can result in a musculoskeletal injury so that workers cannot work normally. Therefore, it is necessary to have an effective strategy plan that can reduce the risk of musculoskeletal injury. However, until now there has been no integrated data of musculoskeletal disorders in Indonesia, including on pond workers. The problem formulation that is solved is how the condition of the fish farmer's complaints against the disorders (pain, agony, and discomfort) in the part of the body caused by the manual work that they do.

The various positions that are observed are lifting/lowering, pushing/pulling, twisting the body, carrying objects and holding/restraining objects. The results of the survey that will be conducted produce the prevalence data of musculoskeletal disorders on pond workers in Surabaya and its surrounding areas. These data can be used as a basis for consideration in designing an effective ergonomic intervention to reduce the risk of musculoskeletal injury.

2. Literature Review

Currently, aquaculture is responsible for the livelihood and nutrition of millions of people [3]. What is meant by cultivation activities is the maintenance activities to reproduce, growth, and improve the quality of aquatic biota so that it can be profitable [1]. Aquaculture is divided into three parts, namely freshwater aquaculture, brackish water aquaculture, and sea water (mariculture). [4] explained that aquaculture development is important because aquaculture provides a high quality of animal protein and other essential nutrients at an affordable price. Aquaculture is also important because it provides job opportunities, cash income and is a valuable export commodity. [5] reinforced that aquaculture in rural areas contribute to poverty alleviation directly through small scale fisheries cultivation for consumption and income. Their indirect contributions are through the provision of employment for the poor in commercial cultivation. Shrimp is one of the commodities in aquaculture. The Vannamei shrimp have been domesticated for several generations, and have become the most important shrimp species in the world [6].

Physical work - often referred to as a heavy work, manual labor, muscle work or manual work - can be defined as (1) activities that require a strong physical (muscle) effort during the work process, and (2) work that requires energy consumption - Kcal/min (measured by heart rate or O2 consumption); which is then used as a measure to determine the weight of the work activities. Mechanization and automation will be considered to be able to accelerate the work process, increase productivity and more importantly reduce the physical workload that must be done by humans. Manual work activities that are mostly done by many pond workers are lifting/lowering, pushing/pulling, twisting the body, carrying objects and holding/restraining objects.

Musculoskeletal disorders are also called as MSDs. [7] suggested that musculoskeletal disorders (MSDs) include various inflammatory and degenerative conditions affecting the muscles, tendons, ligaments, joints, peripheral nerves, and supporting blood vessels. These conditions include clinical syndromes such as tendon inflammation and its related conditions (tenosynovitis, epicondylitis, bursitis), nerve compression disorders (carpal tunnel syndrome, sciatica), and osteoarthritis, and also less standard conditions such as in myalgia, lower back pain and the other regional areas with the condition of the pain syndrome caused by a known pathology.

The most commonly affected areas of the body are the back, neck, shoulders, arms, and hands. Although recently the lower extremities have received more attention, the accurate data on the incidence and prevalence of musculoskeletal disorders are difficult to compare between countries.

However, MSDs are the largest category of occupational illness, representing one-third or more of all occupational diseases listed in the US, Nordic countries, and Japan [8]. Standardization is required in the analysis and recording of musculoskeletal disorder symptoms [9]. Assessment of pain is needed to confirm the level of pain felt by the respondents. This is done to see the severity as a result of working in a non-ergonomic position. Assessment of pain is based on [10] using 0-10 numeric pain rating scale.

3. Research Methodology

3.1. Identification of Observed Objects

The reason of selecting pond workers as an observation object is because they are currently do their job manually, consisting of the activity of lifting/lowering, pushing/pulling, twisting the body, carrying objects and holding/restraining objects. The entire works rely on musculoskeletal strength in its implementation. Mismatching of movements, repetitive and lengthy movements, and performing a job that exceed human capabilities will potentially lead to musculoskeletal disorders that cause pain and decrease productivity. At this first stage of data collection, Tlocor Village is the observed object that is selected. It is located in Jabon District, Sidoarjo, East Java. The location of the selected pond is Vannamei Shrimp Pond. The pond is one of the incentive ponds that has quite sophisticated equipment compared to the ponds in general.

3.2. Preparation of Survey Instrument
The next step is the preparation of a musculoskeletal disorder survey instrument. The Nordic Body Map questionnaire has been commonly used in the identification of pain in body parts, particularly in the musculoskeletal. However, for its implementation in Indonesia, it is still necessary to make adjustments. PEI (Association of Indonesian Ergonomics) has conducted an initial preparation of a musculoskeletal disorder survey instrument that can be used by researchers to obtain the data of pain complaints in various industries.

The survey instrument consists of two parts, namely:

Part A is demography, containing:

- The respondents' self-data including gender, place/date of birth, height and weight, latest education and ethnicity.
- Company data consisting of the industry type, number of employee (workers) and company location.
- The data of work in the form of specialization/division of work, duration of work, average working hour per day, tendency of using left hand and working shift.

Part B is the form consisting of:

- Perceived complaints (such as pain, agony, or discomfort) in nine parts of the body. Those nine parts of the body are (1) neck, (2) shoulders, (3) upper back, (4) elbow, (5) lower back, (6) hand wrist, (7) buttocks/thighs, (8) knee, and (9) ankle.
- Pain assessment using a scale of 0 to 10. Where the value 0 indicates no pain at all and 10 indicates the highest degree of pain that is felt.

3.3. Selection of Respondents
Respondents who became the object in the survey questionnaires and interviews were workers in intensive ponds in Sidoarjo district. In this study, the observed activities were pond preparation, seed distribution/seedling, cultivation maintenance, and harvesting. For those respondents, we explained about the purpose of the survey, the questions in the questionnaire and the interviews about the activities that they do. After determining the respondents, the data were collected.

3.4. Data Collection
The data were collected on Saturday, June 17, 2017 at 08:30 – 15:00 WIB at Vannamei Shrimp Pond and Semi Incentives Pond in Tlocor Village, Jabon Subdistrict, Sidoarjo District, East Java. The data collection was carried out through direct interviews to the determined respondents. Respondents were not allowed to complete the questionnaire independently. This was to ensure that the choice of answers given by respondents were in accordance with the existing conditions in the questionnaire. Each surveyor asked questions based on the questionnaire. Then each respondent gave the answer. Furthermore, if the respondent had difficulty in giving their answers, then the surveyor would provide more detailed explanation dealing with the questions and answer choices in the questionnaire. Figure 1 shows the respondent's working position and data collection process.

Figure 1. The respondent activities that were observed

3.5. Data Processing and Analysis
Data processing was conducted after collecting the entire data. Data processing was done in two types of questions that exist in the questionnaire, namely biodata and content identification of pain in the body. Statistical methods were used to process the data and obtain the most vulnerable body parts to musculoskeletal. Then, the results of data processing were analysed. The analysis was addressed to the part of the body that predominantly got the high scores for pain complaints.

4. Result and Discussion
4.1. Respondents Characteristics
At the end of the data collection, 73 questionnaires filled by respondents were obtained. The data processing of respondents based on their age, working period, weight, height, number of working days per week, and working hours per day were conducted by category in [11].

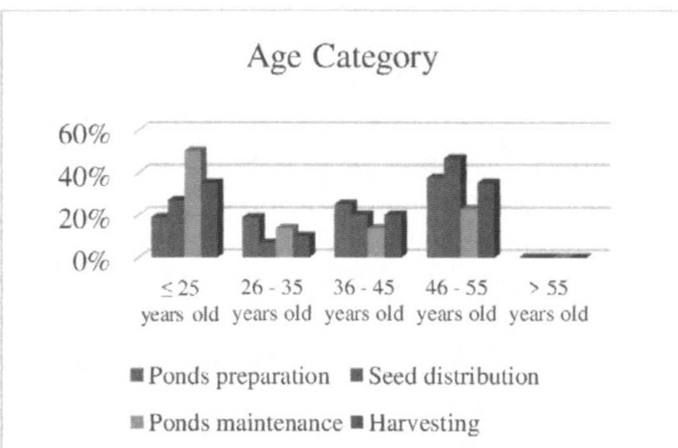

Figure 2. Respondents based on age

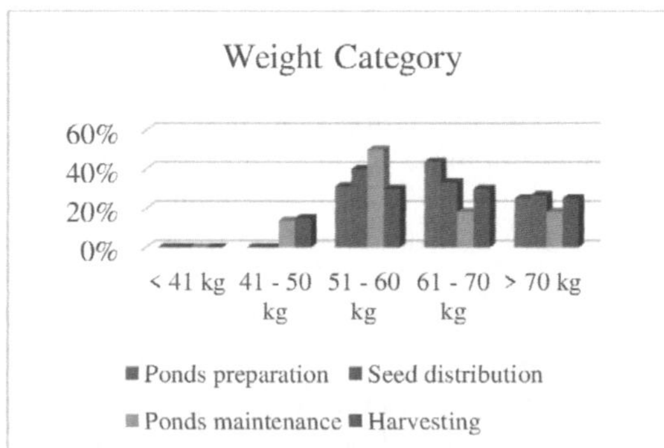

Figure 3. Respondents based on weight

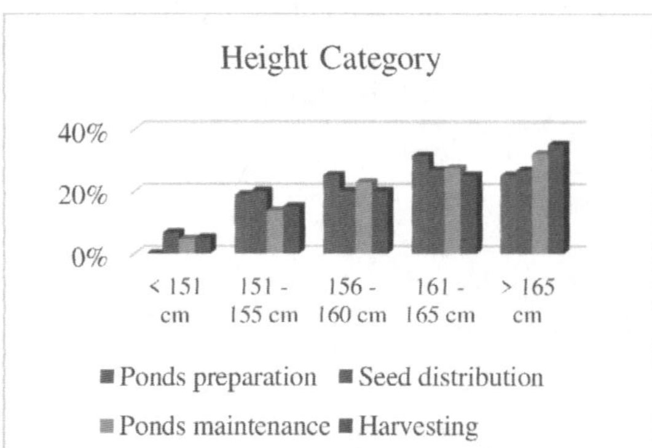

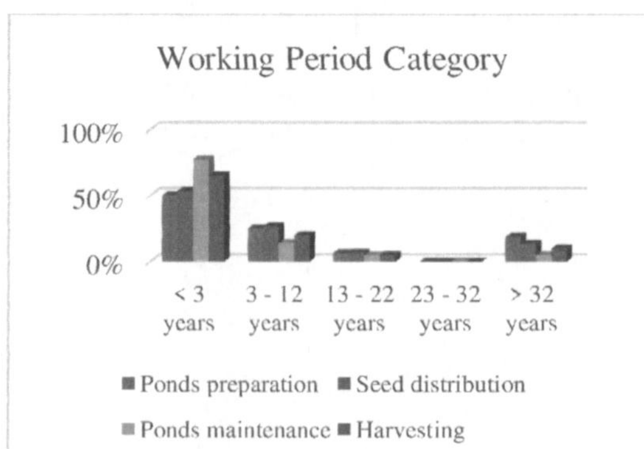

Figure 4. Respondents based on height

Figure 5. Respondents based on working periods

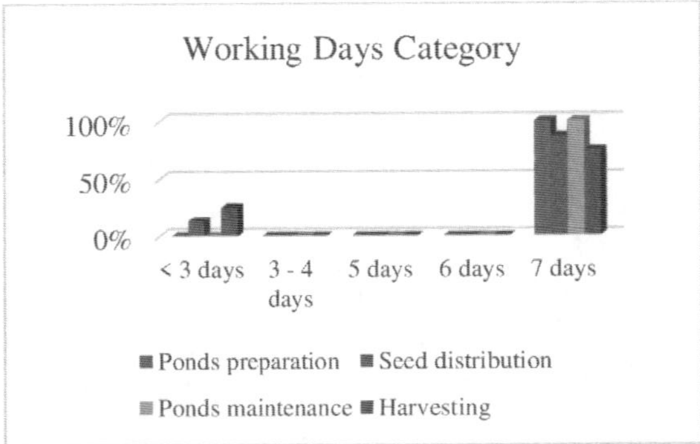

Figure 6. Respondents based on working days

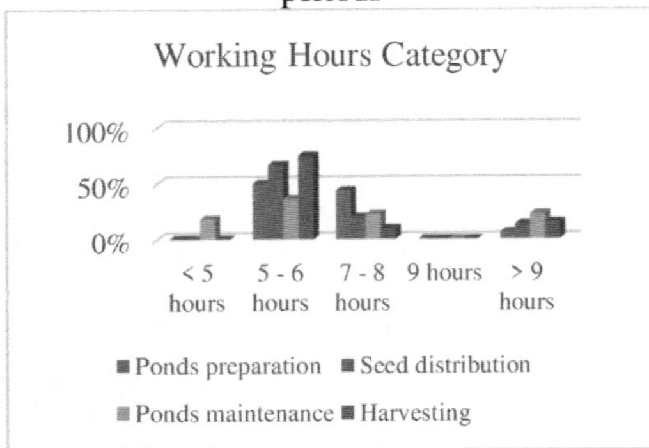

Figure 7. Respondents based on working hours

Figure 2 shows the respondents who are dominated by age less than 25 years. Figure 3 shows that the highest percentage of weight is at body weight 41-60 kg. Figure 4 shows the highest percentage of height more than 165 cm. Figure 5 shows the percentage of working duration dominated less than 3 years. Figure 6 shows that the working days is 7 days for almost all responden. Figure 7 shows the working hours for 5-6 hours per day.

4.2. The Result of Questionnaire
Further, there was an identification of the body parts that had complaints. The data were obtained from the second part of the questionnaire. Figure 8 shows the body parts that experience pain with a time category for last 12 month and last 7 days.

Those body parts was the neck, shoulders, back, waist (lower back), and knee. The highest percentage of pain/complaints on the neck was at the preparation stage. The highest percentage of pain/complaints on the shoulders was on both shoulders at the preparation stage. The highest percentage of pain/complaints on the back was at the harvesting stage. The highest percentage of pain /complaints at the waist (lower back) was at the preparation stage, while the highest percentage of pain/complaints on the knee was on both knees in the preparation stage.

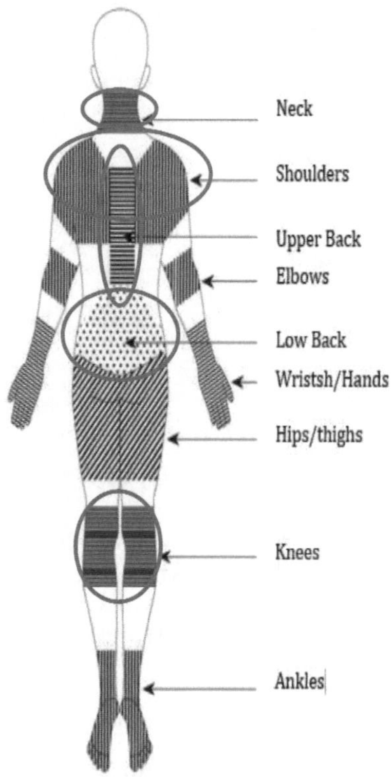

Figure 8. Parts of body that have pain complaints

Based on the questionnaire about the pain experience, it did not interfere the workers in doing their work activities. Furthermore, the results of identification of the level of pain felt by the respondents were conducted based on the given answers. The pain was divided into 10 categories. Most respondents assessed the pain that they felt in "moderate pain" degree, shown in figure 9.

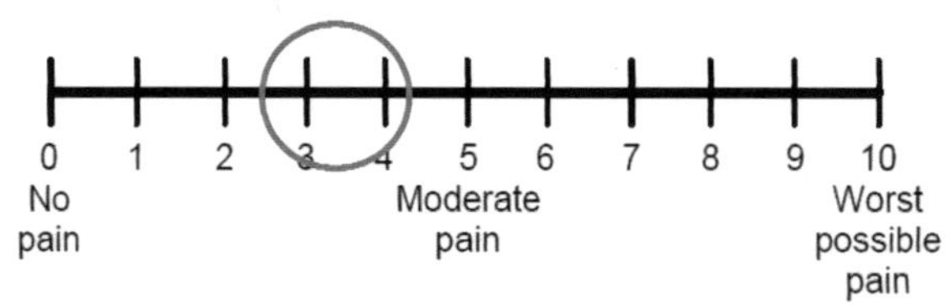

Figure 9. The result of pain assessment

Most respondents did not follow up the pain complaints that they experienced. They did not contact therapists or physicians (doctors, medics, etc.) to solve the problem.

5. Conclusion

There has been a survey and we obtained 73 questionnaires filled by respondents of pond workers in the village of Tlocor, District Jabon, Sidoarjo, East Java. The results of a musculoskeletal disorder survey in the pond workers revealed five body parts with the main complaints; they were, the neck, shoulders, back, waist and ankle. The level of pain was a "moderate pain" with no follow-up to therapists or physicians (doctors, nurses, etc.).

Next, it is necessary to observe in detail the working position that is not ergonomic. This is aimed at obtaining the correlation between the pain and work position. Further, the advice and recommendations will be based on the aspects of ergonomics and design of work systems in the implementation of work by humans. The improvements of working methods are expected to create a working condition that is called as ENASE (effective, convenient, safe, healthy and efficient) and reduce the pain.

6. Acknowledgments

Thanks to the pond workers who were willing to be a respondent in the distribution of musculoskeletal disorder questionnaires.

7. References

[1] Effendi M I 2004 *Pengantar Akuakultur* (Jakarta : Penebar Swadaya)

[2] Perhimpunan Ergonomi Indonesia 2016 Instrumen Survei Gangguan Otot-Rangka

[3] Rivera A, Unibazo J, León P, Vásquez-Lavín F, Ponce R, Mansur L and Gelcich S 2017 *Stakeholder Perceptions Of Enhancement Opportunities In The Chilean Small And Medium Scale Mussel Aquaculture Industry* (Elsevier – Aquaculture 479 (2017)) p 423-431

[4] Tacon A G J 2001 *Increasing the Contribution of Aquaculture for Food Security and Poverty Alleviation* (FAO Corporate Document Repository)

[5] Edwards P 1999 *Towards Increased Impact of Rural Aquaculture* (A Discussion Paper prepared for the First Meeting of the APFIC Ad Hoc Working Group of Experts on Rural Aquaculture FAO Regional Office for Asia and the Pacific (RAP) Bangkok Thailand 20-22 October 1999)

[6] Nolasco-Alzaga H R, Perez-Enriquez R, Enez F, Bestin A, Palacios-Mechetnov and Haffray P 2017 *Quantitative Genetic Parameters Of Growth And Fatty Acid Content In The Hemolymph Of The Whiteleg Shrimp Litopenaeus Vannamei* (Elsevier – Aquacuture 482 (2018)) p 17-23

[7] Punnet L and Wegman D H 2004 *Work-related musculoskeletas disorders: the epidemiologic evidence and the debate* (Journal of Electromyography and Kinesioloy 14) p 13-23

[8] Sjøgaard G, Sejersted O M, Winkel J, Smolander J, Westgaard K and Westgaard R H 1993 *Exposure assessment and mechanisms of pathogenesis in work-related musculoskeletal disorders: significant aspects in the documentation of risk factors* (Johanesen S Johanses (Eds) Work and health: Scientific basis of progress in the working environment) p 22-25

[9] Kuorinka I, Jonsson B, Kilbom A, Vinterberg H, Biering-Sørensen F, Andersson G and Jørgensen K 1987*Standardised Nordic Questionnaires for The Analysis of Musculokeletal Symptoms* (Applied Ergonomics 18 3) p 233-237

[10] McCaffery M and Pasero C 1999 *Pain: Clinical Manual – 0 –10 Numeric Pain Rating Scale* (St Louis Copyrighted by Mosby Inc) p 16

[11] Maryani A, Partiwi S G and Ratnasanti D A 2016 *Improving Cashew Peelers Workstation to Reduce Physical Complaints* (4th SEANES International Conference on Human Factors and Ergonomics in South-East Asia)

Bio-mechanical assessment toward throwing and lifting process of i-LOCA (Innovative Lobster Catcher)

A Sudiarno[1], D S Dewi[1], and M A Putri[1]

[1]Department of Industrial Engineering, Institut Teknologi Sepuluh Nopember, Kampus ITS Sukolilo-Surabaya 60111, Indonesia

*adithya.sudiarno@gmail.com

Abstract. Indonesia is the country rich in marine resource, one of which is lobster. East java, one of Indonesian province, especially in Region of Gresik and Lamogan, has very huge potential of lobster. Current condition shown that lobster catch by the fisherman mostly depend on lucky factor, which the lobster unintentionally trapped in fisherman's fish net. By using this mechanism, the number of lobster catch cannot be optimum. Previous researches have produced two versions of i-LOCA, Innovative Lobster Catcher, a special tool for catching the lobster. Although produce more lobster catch, second version of i-LOCA still needs to be scrutinized, one of that is bio-mechanical assessment. The second version of i-LOCA still has no tool to ease throwing and lifting it into the sea. This condition cause Musculoskeletal Disorder (MSD) toward the fisherman. This research perform bio-mechanical assessment toward throwing and lifting process in order to suggest improvement for i-LOCA as the third version. Based on body moment calculation, we found that throwing and lifting process of third version of i-LOCA, each was 3 times and 2 times better than second version of i-LOCA. Meanwhile, Rapid Entire Body Assessment (REBA) score of throwing and lifting process for third version of i-LOCA can be reduced by 5 points compared to second version of i-LOCA.

1. Introduction

One of the Indonesia fishing industry is lobster. Based on the Indonesia statistic fishing data in 2012, it shown that lobster exist in fourth as the highest export commodity from Crustacea family [1]. Besides, based on Balitbang KKP in 2013 report, it shown that increasing of lobster catching result from 2005 to 2012 is 19,23 % or 0,16 million tons per year from total of catching result in Indonesia territorial. And the increasing of lobster utilization for the north of Java Island, especially East Java is 27,4% or 0,98 million tons per year.

But, the lobster product of fishermen is mostly depend on lucky factor or because lobster has been trapped in the catcher tool. It is because the main goal of fishermen is to catch the fish and the tool which is used to catch the lobster is *bubu* (crab's catcher tool) and net so the lobster catching result is not maximal [2]. From this fact, the research on lobster catcher tool i-LOCA was conducted by Felayati [2] and Nafiah[3].

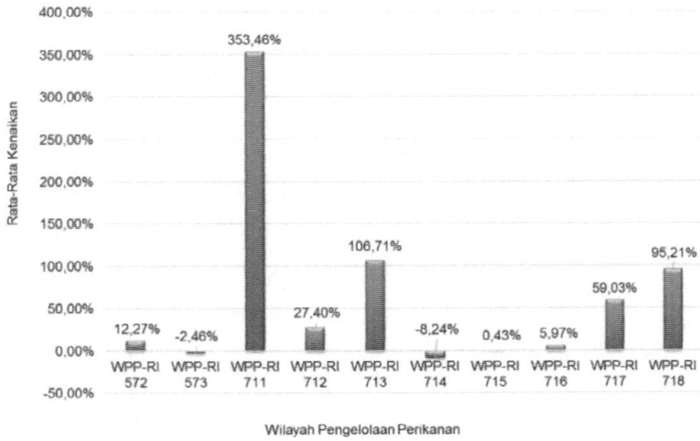

Figure 1. Increasing of lobster utilization in 2005-2012

The lack of the research which conducted by Nafiah that produce the second version of i-LOCA is the absence of bio-mechanical assessment when the fishermen throwing and lifting i-LOCA [3]. The throwing and lifting process has hazard potential like low back pain, so the maximal weight and quantity of lobster has to predicted to avoid *musculosceletal disorder* (MSD) which can influence the performance of fishermen. From this fact, the researcher conducted bio-mechanical assessment toward throwing and lifting process of i-LOCA for the third version of i-LOCA.

Figure 2. Throwing position using the second version of i-LOCA

Figure 3. Lifting position using the second version of i-LOCA

2. Research Methodology

2.1. Data Collection and Data Processing

Data collection is the collection of data that related to this research. Data in this research consist of primary data and secondary data. Primary data in this research is the fishermen body measurement result and the position of throwing and lifting process from the fishermen in Gresik and Lamongan. And the secondary data is the percentage of each body segment.

2.2. Data Analysis and Data Interpretation Phase

After data is processed, the next step is analyze and interpret the result of data processing. The first is biomechanic calculation analysis of throwing and lifting process for existing (the second version of i-LOCA) and improved condition (the third version of i-LOCA). And the second is REBA analysis of throwing and lifting process for each version of i-LOCA.

3. Result and Discussion

3.1. Biomechanics Calculation

To calculate biomechanics, there are some of the influence factors such as mass point of each link segment, gravity, angle, mass of the body, and mass of each body segment. The percentage of each body segment and mass center point is developed by Brown at the Laboratory Department of Kinesiology at Michigan State University [4]:

Table 1. The mass center point and the percentage of body segment

Segment	Mass center position from proximal (%)	Mass center position from distal (%)	Body segment proportion (%)
Trunk	0.562	0.438	0.486
Head and Neck	0.567	0.433	0.079
Thigh	0.433	0.567	0.097
Shank	0.433	0.567	0.045
Foot	0.500	0.500	0.014
Arm	0.436	0.564	0.027
Forearm	0.430	0.570	0.014
Hand	0.506	0.494	0.006

The data in this calculation obtained from the experiment of the second version of i-LOCA which conducted by Mr. Sama'un as the fishermen sample whose weight is 53 kg and height is 165 cm.

Table 2. Measurement point for the second and third version of i-LOCA

Segment	Length (m)	Second Version of i-LOCA				Third Version of i-LOCA			
		Throwing		Lifiting		Throwing		Lifiting	
		Weight (kg)	Angle (0^0)	Weight (kg)	Angle (0^0)	Weight (kg)	Angle (0^0)	Weight (kg)	Angle (0^0)
Fishermen weight	-	53	-	53	-	53	-	53	-
Object mass	-	12 (without lobster)	-	16 (with lobster)	-	3.5 (without lobster)	-	7.5 (with lobster)	-
Hand	0.18	0.318	15^0	0.318	12^0	0.318	20^0	0.318	22^0
Forearm	0.24	0.742	45^0	0.742	25^0	0.742	50^0	0.742	49^0
Arm	0.29	1.431	45^0	1.431	32^0	1.431	38^0	1.431	38^0
Trunk	0.48	25.758	50^0	25.758	40^0	25.758	81^0	25.758	79^0

Development of biomechanic model in both two or three dimencion and static or dinamic is shown in the following equation [5]. Equation (1) is to calculate the static moment in the sequent of joints

$$M_{joint} = M_{joint-1} + \left(L_{link}\ x\ F_{joint-1}\right) + (CM_{link}\ x\ m_{link}\ G) \quad (1)$$

and equation (2) is to calculate the dinamic moment

$$M_{joint} = M_{joint-1} + \left(L_{link}\ x\ F_{joint-1}\right) + (CM_{link}\ x\ m_{link}\ G) \\ (CM_{link}\ x\ m_{link} A_{link} + (\theta_{joint}\ x\ I_{link}) \quad (2)$$

Following is the example of the biomechanics calculation of the i-LOCA second version in throwing process for hand segment

$$\sum F_x = F_w = 0$$

$$\sum F_y = F_{yw} - W_o - W_h = 0$$

$$F_{yw} = W_o + W_h$$

$$117\ 6$$

$$W_o = m \cdot g$$
$$= 12\ kg \cdot 9{,}8\ m/s^2$$
$$= 117{,}6\ N$$
$$W_h = berat\ segmen\ tangan \cdot g$$
$$= 0{,}318\ kg \cdot 9{,}8\ m/s^2$$
$$= 3{,}1164\ \text{N}$$

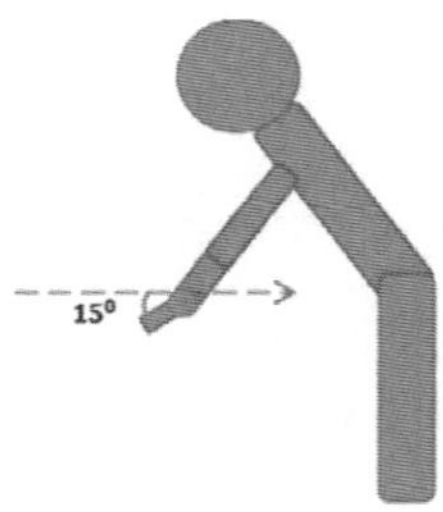

Figure 4. Free body diagram of throwing process for hand segment

$$\sum M_w = M_w - (W_o + W_h)\ x\ SL_1\ x\ \cos\theta_1 = 0$$
$$M_w = (F_{yw})\ x\ SL_1\ x\ \cos\theta_1$$
$$= (61{,}9164\ N)\ x\ 0{,}18\ m\ x\ \cos 15^0$$
$$= 61{,}9164\ N\ x\ 0{,}18\ x\ 0{,}9659$$
$$= 10{,}7649\ Nm$$

Based on the calculation, following is the moment value of using the second and third version of i-LOCA.

Table 3. The moment value of using the second and third version of i-LOCA

Segment	The second version of i-LOCA		The third version of i-LOCA	
	Throwing	Lifting	Throwing	Lifting
Hand	10.76 Nm	14.35 Nm	3.43 Nm	6.65 Nm
Forearm	21.80 Nm	32.76 Nm	7.04 Nm	13.42 Nm
Arm	37.24 Nm	56.10 Nm	14.73 Nm	25.59 Nm
Trunk	169.63 Nm	239.96 Nm	46.34 Nm	75.39 Nm

Based on biomechanics calculation, the moment value for throwing and lifting process of third version of i-LOCA, each was 3 times and 2 times better than second version of i-LOCA. So that the hazard level or musculoskeletal disorder of using the third version of i-LOCA can be reduce until 3 times and 2 times more safety than the second version of i-LOCA

3.2. REBA Analysis

After conducted biomechanics calculation, the next step is running software REBA. This step is used to know position error and risk level when the fishermen is throwing and lifting i-LOCA

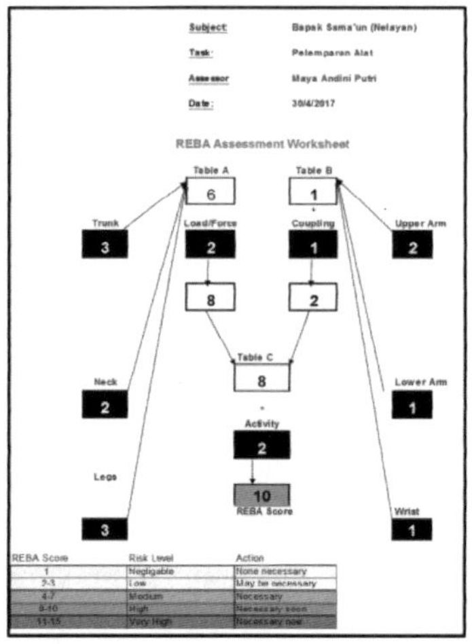

Figure 5. The result from running software REBA when the fishermen is throwing the second version of i-LOCA.

Based on the running result, the following is the REBA score for the both of i-LOCA.

Table 4. REBA score for the second and third version of i-LOCA

Condition	REBA Score	
	Throwing	Lifting
The second version of i-LOCA	10	12
The third version of i-LOCA	5	7

The result from REBA is equivalence with the result of biomechanics calculation which shown the moment value. Based on the running for the second version, throwing and lifting process still need improved. While for the third version, it shown that risk level for each process is in the medium level. It means, it can be reduce until 5 points, but the action level still in necessary level.

After conducted analysis and interview with the fishermen, the first improvement is adding the foothold in the pulley pole to hold the pillar form an angle 60^0. It is the best angle for both process based on the product trial that conducted by fishermen. In addition, the pole is made of stainless steel, so the weight of the pole is not too heavy but the pole is too strong enough. The second is change all of the material with stainless steel, so the weight of i-LOCA is not too heavy. Based on the product trial by fishermen, they said that the weight of the third version of i-LOCA is not too heavy because it made from stainless steel, so they don't need a pulley to throw or lift i-LOCA.

4. References

[1] WWF 2015 Indonesia *Better Management Practice - Perikanan Lobster Laut - Panduan Penangkapan dan Penanganan* (Jakarta Indonesia: WWF Indonesia)

[2] Felayati M A A 2011 *Perancangan Alat Tangkap Lobster dengan Pendekatan Quality Function Deployment (QFD) dan Function Analysis System Technique (FAST) serta Kontribusinya terhadap Kinerja Klaster Industri Perikanan (Studi Kasus : Komunitas Nelayan Paciran)* (Surabaya Indonesia: Institut Teknologi Sepuluh Nopember)

[3] Nafiah R 2016 *Perancangan Alat Tangkap Lobster i-LOCA dengan Pendekatan Teoriya Resheniya Izobretatelskikh Zadatch (TRIZ) dan Business Model Canvas (BMC) (Studi Kasus*

: Komunitas Nelayan Gresik dan Lamongan) (Surabaya Indonesia : Institut Teknologi Sepuluh Nopember)
[4] Miswari N 2016 *Perancangan Peralatan Manual Material Handling Pemuatan Kelapa Sawit untuk Mengurangi Keluhan Muskuloskeletal dan Tingkat Kelelahan (*Surabaya Indonesia: Institut Teknologi Sepuluh Nopember)
[5] Ayoub M M and Woldstad J 1999 *Models in Manual Material Handling (*University of Nebraska-Lincoln: Industrial and Management System Engineering Faculty Publication)
.

5. Acknowledgments

The authors would like to thank the fishermen in Gresik and Lamongan for allowing the research to be conducted. And the special thanks to Kementerian Riset Teknologi dan Pendidikan Tinggi Republik Indonesia for the financial support to conduct this research.

IOP Conf. Series: Materials Science and Engineering **337** (2018) 012041 doi:10.1088/1757-899X/337/1/012041

Targeting overall equipment efficiency for small medium enterprises with irregular production system

Y Prasetyawan[1*], M Suef[1], L Claudia[1] and F D Handayani[1]
[1]Department of Industrial Engineering, Institut Teknologi Sepuluh Nopember, Kampus ITS Sukolilo-Surabaya 60111, Indonesia

*yudhaprase@gmail.com

Abstract. Overall Equipment Effectiveness (OEE) is widely used to measure the maturity of a production system. The company will be considered as World Class Manufacturing if it reaches more than 85% value, with near perfect value for availability, performance and quality factor. This assessment is usually taken on industries with regular production times named shift system. A typical 8 hours shift system is used in OEE measurement and performance monitoring. There are few Small to Medium Enterprise (SME) perform regular production times with shift systems, others using irregular production systems. The irregular production time in the SME production system is used because of demand fluctuations. This paper shows a quantitative analysis as a part of manufacturing system design to achieve a specific value of OEE for SME with irregular production systems, for individual businesses as well as collective business systems (some companies use the same production facilities for several processes). The results of experiments on several companies are presented, as a basis for determining the technical strategy of achieving OEE values.

1. Introduction

In today's current competition between manufacturing companies, world class manufacturing status is no longer a guarantee to become a winner. Some companies that already have that status has been forced to close their production facilities because they think their competitors are small companies. Innovation is becoming the key to winning the competition. In fact, efficiency is also a major factor run the business for a sustainable development path[1][2]. Thus, innovation is used as a breakthrough competition, while efficiency (such as lean manufacturing implementation) is a fuel for businesses to achieve the true benefits of product development in numbers and variants [3]-[5]. SMEs in current tight business competition contribute to economic development through an innovation [7]-[10].

OEE has been widely used by various manufacturing industries. Some companies use OEE as the main indicator of production system performance. Availability is used to ensure that the various equipment is ready to be used when needed or ready to be used at any time. Performance is used to ensure that the production system is running at the expected level or even faster speed to undertake other problems that can decrease availability. Quality is used to ensure that all raw materials are in good condition and can be converted into finished goods. These three factors are not easily achievable indicators but need to be attempted in a systematic and integrative way [11]. Otherwise, a decrease in one value may have a direct or indirect impact on other indicators.

The calculation of indicators in OEE is commonly used in regular shift-based production systems. SMEs generally have an irregular production system, with various conditions. The objective of measurement is the same as OEE in a regular production system. However, some modifications need to be made so that the measurement can provide a genuine picture of the availability, performance, and quality. SMEs do not use all the available time (for example 8 hours of work) to conduct production activities and not all the weekdays are used. In addition, SME performs production under varying conditions between a process and the subsequent process [12]. Thus, the unequal processing time becomes a problem. Moreover, the availability of raw materials based on intermittent order (not continuous order) produce another serious problem [13]-[14].

It is important to have an OEE target values in SME with the irregular production system. Simple technical strategy by placing maintenance activities in time lags between productions is also a good way to increase availability [15]-[17]. Other factors such as precise estimation of demand, good production scheduling and the availability of raw materials, supporting materials, reliable operators are the success of achieving OEE target [18]-[22]. The identification of other factors in SME is also important to reduce unproductive factor [23]-[25]. The pyramid of business orientation needs to be designed with the target value of OEE as its peak and man-money-material-machine-method element as the basis [26]. Followed by the middle part - a technical strategy that needs to be created in managing the basic elements [27]-[28].

This research presents the results of a technical strategy designed based on OEE targets. Technical strategies are made for SMEs that have an irregular production system. The approach of irregular production time (non-full-time) is done to measure availability. The time as described in Standard Operating Procedure along with various modifications is measured to acquire performance. While the quality measurement is approached with the ability maximizing the conversion of raw materials into main and additional products. Such efforts to achieve the highest level of availability, performance, and quality is done by finding the best combination of various factors that have been commonly done in a manufacturing system. Thus, the technical strategy can be done quickly without major manufacturing configuration changes.

2. Methods

First, the OEE formula is acquired in order to have the main expected parameter when performing the manufacturing system design. The value of OEE consist of availability (A), performance (P) and quality (Q) element. The variable of Running Time (RT), Planned Production Time (PPT), Theoretical Cycle Time (TCT), the quantity of Finished Goods (FG) and Good Product (GP) were involved to seek the value of each element. Thus, the typical OEE formula will be as follows.

$$OEE = A\,.\,P.\,Q \quad (1)$$

$$A = RT/PPT \quad (2)$$

$$P = (TCT.\,FP)/RT \quad (3)$$

$$Q = GP/FP \quad (4)$$

It can be seen that A, P, and Q are connected. It is difficult if we are only focus on increasing RT to have greater A as the consequences of getting lower P. Having greater number of finished product might increase P but reduce Q. For SME, bigger PPT means having more demand to be finished and gain more profit. However, RT will depend on the availability of all resources – man as the problem of operator skill, machine as the problem of facilities, material as the problem of accessibility, method as the problem of technology capability and money as the problem of investment. Further, producing zero defect product might the biggest challenge of all as the low confidence of doing the better production system is still a problem in SME. Then, irregularity can be described as the proportion of doing manage process (MP) as the strategic business activities, doing core process (CP) to fulfil the demand (main activity to build OEE) and support process (SP) to enhance the overall capability.

Second, the OEE element should be further explored towards the production consideration within SME. It is proposed that SME should do the following simple steps

- PPT, it depends on how the SME can get the order from the customer. The number of product variant will also contributes to complexity of calculating overall production time. Moreover, creating an accurate Standard Operating Procedure (SOP) contains production activities and estimated cycle time is mandatory.
    - PPT1 – capability of get order
    - PPT2 – capability of innovation and diversification (product variant)
    - PPT3 – capability of creating accurate SOP to estimate cycle time
- RT, the availability of man, machine, material, method and money is important. Such availability of man can be derived from manpower working hours, availability of machine can be derived from time between failure and time to repair within reliability aspect, availability of material can be derived from service level of inventory and supplier connection, availability of method can be derived from degree of adaptation and innovation, and availability of money can be derived from revenue generation or capital inquiries
    - RT1 – increase the availability of resources
    - RT2 – decrease the unavailability of resources
    - RT3 – increase operation continuity
- FP, reaching the production target should be done within the PPT. However, create product faster will be beneficiary in order to get higher performance.
    - FP1 – follow the SOP instruction to reach TCT
    - FP2 – seek for efficiency to reach faster cycle time
    - FP3 – possibility of involving external resources
- GP, it involves engineering drawing that contains product specification. Further, implementation of good quality control (including quality assurance and quality management) is an obligation to reduce defect.
    - GP1 – clear engineering drawing equipped with product specification
    - GP2 – good quality control procedures (can be advanced with quality assurance and quality management)
    - GP3 – reliability of material and manufacturing process

Considering those supporting factors, a weighted proportion (range 0 – 1) is created to have clearer description. SME then can use the simple equation to analyse OEE target.

$$W_{PPT} = (W_{PPT1}+W_{PPT2}+W_{PPT3})/3 \quad (5)$$
$$W_{RT} = (W_{RT1}+W_{RT2}+W_{RT3})/3 \quad (6)$$
$$W_{FP} = (W_{FP1}+W_{FP2}+W_{FP3})/3 \quad (7)$$
$$W_{GP} = (W_{GP1}+W_{GP2}+W_{GP3})/3 \quad (8)$$

Where,
W_{PPT} is weighted factor for PPT, so does W_{PPT1}, W_{PPT2}, W_{PPT3} is related steps as described
W_{RT} is weighted factor for RT, so does W_{RT1}, W_{RT2}, W_{RT3} is related steps as described
W_{FP} is weighted factor for FP, so does W_{FP1}, W_{FP2}, W_{FP3} is related steps as described
W_{GP} is weighted factor for GP, so does W_{RT1}, W_{RT2}, W_{RT3} is related steps as described.

Third, a technical strategy should be built based on all the supporting factors of OEE element. The strategy should be just determining the target of an individual factor and describing how to achieve. Moreover, the scheme of a complete business process will also be included (27). These are manage process (MP), core process (CP), and support process (SP). The proportion of doing each process within a day is defined as an irregularity. The proportion of CP is considered as production time. Thus, SME is not always doing 8 hours production process like a regular manufacturer. However, a weekly basis will be used to simplify the term. A typical path of technical strategy can be seen in Figure 1.

$$WH_{MP} = \sum_{i=1}^{n} WH_{MP_i} . W \quad (9)$$
$$WH_{CP} = \sum_{i=1}^{n} WH_{CP_i} . W \quad (10)$$
$$WH_{SP} = \sum_{i=1}^{n} WH_{SP_i} . W \quad (11)$$

Where,

IOP Conf. Series: Materials Science and Engineering **337** (2018) 012041 doi:10.1088/1757-899X/337/1/012041

WH_{MPi} : Working hours of MP in day i
WH_{CPi} : Working hours of CP in day i
WH_{SPi} : Working hours of SP in day i
W : number of week within a year
n : number of operation day within a week

Finally, the proposed method is also considering a collective production system. In fact, several SME can share a particular production process in order to have better capabilities and flexibility. In collaboration, the predictive result can be determined by a pessimistic value such as several SME who shares production facilities will depend on the slowest production time within members. Thus, for all the elements the weighted factor (range 0 – 1) for collective operation will be the minimum value of all SME member.

$$W_{RTall} = \text{Min}\,(W_{RTSME1}, W_{RTSME2}, \ldots, W_{RTSMEn}) \tag{12}$$
$$W_{PPTall} = \text{Min}\,(W_{PPTSME1}, W_{PPTSME2}, \ldots, W_{PPTSMEn}) \tag{13}$$
$$W_{FPall} = \text{Min}\,(W_{FPSME1}, W_{FPME2}, \ldots, W_{FPSMEn}) \tag{14}$$
$$W_{GPall} = \text{Min}\,(W_{GPSME1}, W_{GPSME2}, \ldots, W_{GPSMEn}) \tag{15}$$

Where,
W_{RTall} is collective weighted factor for RT
W_{PPTall} is collective weighted factor for PPT
W_{FPall} is collective weighted factor for FP
W_{GPall} is collective weighted factor for GP
n is number of SME within the shared production facilities.

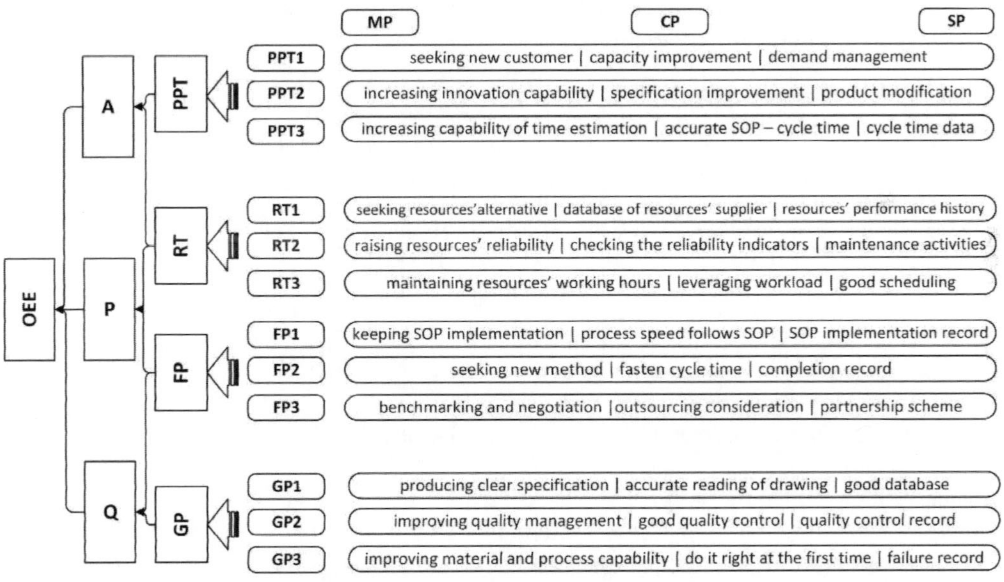

Figure 1. Technical strategy path.

3. Results and Discussion

A survey has been done in several SME, gathering data for testing the proposed method. A various result of OEE has been found in different SME (Figure 2). Testing the improvement scenario trough technical strategy shows significant improvement. Shared production facilities will be beneficiary mostly for SME with lower capability. Thus, the cooperation scenario should be then discussed and improved in order to have more mutual benefit. However, testing the different scenario of MP, CP and SP proportion showing a relatively unpredictable result. This might can be a future reference of improvement.

Considering all the involved factors and capturing the observation within several SME, some consideration of targeting OEE value can be described as follows.

- To increase availability, it is necessary to have a focus on increasing the availability of man, machine, material, and money in order to have more RT. Avoiding unavailability is difficult because maintenance and repair activity requires a high cost. SME should focus on ensuring the sufficient operational funds to guarantee the accessibility of required resources. Thus, providing the effort to reach maximum irregularity is crucial.
- To increase performance, it is better to have a focus of reducing the processing time. Running the production process in its fastest way should be carefully done to avoid a lower quality rate. SME should focus to find alternatives of a better operator, better machine and better method in order to have shorter processing time. Thus, seeking the most flexible option to face irregularity is necessary.
- To increase quality, it is important to have a focus on increasing the proportion of good product rather than increasing the number of the finished product. Expecting high demand must be followed by a near defect free production system for the whole process. SME should focus on producing the number of product based on the normal operation time. Thus, leveraging irregularity of operation time is important.

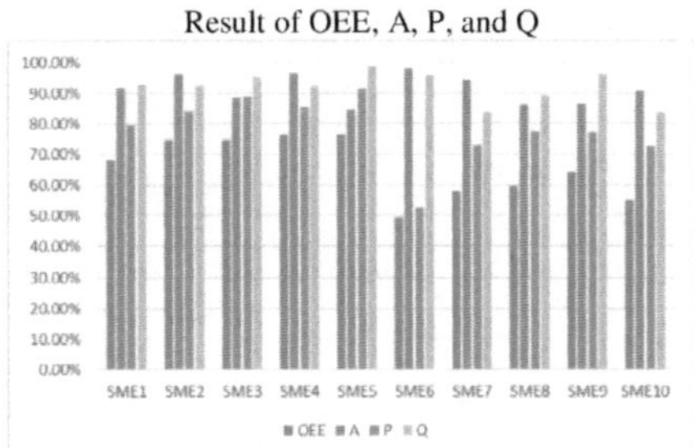

Figure 2a. OEE and its element.

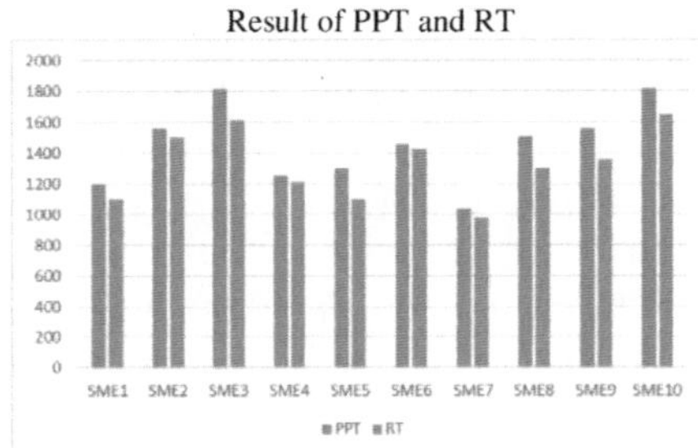

Figure 2b. Value of PPT and RT.

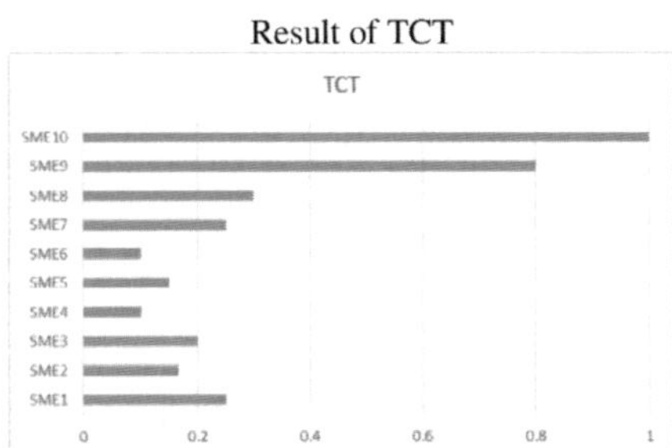

Figure 2c. Value of TCT.

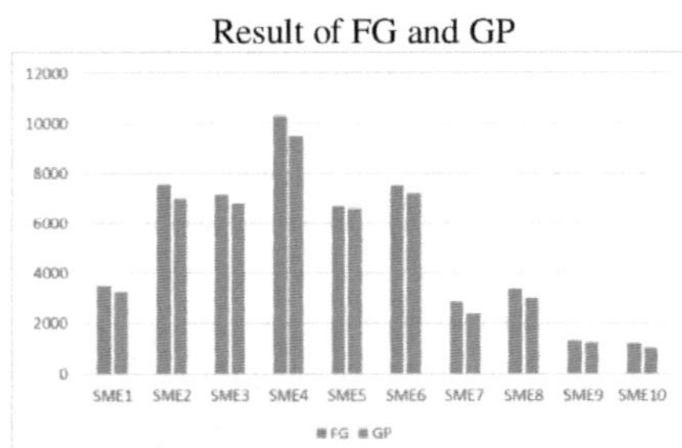

Figure 2d. Value of FG and GP.

4. Conclusion

The advantage of an irregular production system in SME is about to face the problem of limited resources. The proportion of production time (core process) can be somehow altered, in order to have sufficient time for doing manage process and support process. The targeting process is easily done by controlling the number of expected finished good, ideal cycle time, planned production time and running time as the main variables of OEE. The approach of collective production process should be done very carefully as it can decrease the individual OEE of SME. The more combination of both individual and collective production system can be handled by the proposed method. Therefore, seeking the best value of OEE by optimizing the involved variable can be used to determine the required technical strategy.

5. References

[1] Verbano C and Venturini K 2013 Managing Risks in SMEs: A Literature Review and Research Agenda *Journal of Technology Management and Innovation* 8(3) pp. 186-197

[2] Radam A, Abu M L and Abdullah A M 2008 Technical Efficiency of Small and Medium Enterprise in Malaysia: A Stochastic Frontier Production Model *Int Journal of Economics and Management* 2(2) pp 395-408
[3] Castro F P and Araujo F O d 2012 *P*roposal for OEE (Overall Equipment Effectiveness) Indicator Deployment in a Beverage Plant *Brazilian Journal of Operations and Production Managemen*t 9(1) pp 71-84
[4] Pingyu Y and Yu Y 2010 The Barriers to SME's Implementation of Lean Production and Countermeasures-- Based on SME's In Wenzhou *Int Journal of Innovation Management and Technology* 1(2) pp 220-225
[5] Belhadi A, Touriki F E and Fezazi S E 2016 A Framework for Effective Implementation of Lean Production in Small and Medium-Sized Enterprises *Journal of Industrial Engineering* and Management` 9(3) pp 786-810
[6] Prakash N and Prasad C 2014 Lean Practices in Small and Medium Manufacturing Enterprises- A Structured Literature Review *Journal of Basic and Applied Engineering Research* 1(4) pp 78-84
[7] Palanimally Y R 2016 The Growth of Small and Medium Enterprises in Malaysia: A Study on Private Limited Companies in Perak Malaysia *IOSR Journal of Economics and Finance* 7(3) pp 55-60
[8] Martin M S and Namusonge M J 2014 Influence of Innovation on Small and Medium Enterprise (SME) Growth- A Case of Garment Manufacturing Industries in Nakuru County *Int Journal for Innovation Education and Research* 2(6) pp 102-112
[9] Bigliardi B, Colacino P and Dormio A I 2011 Innovative Characteristics of Small and Medium Enterprises *Journal of Technology Management and Innovation* 6(2) pp 83-93
[10] Vasilenko L and Arbaciauskas V 2012 Obstacles and Drivers for Sustainable Innovation Development and Implementation in Small and Medium Sized Enterprises *Environmental Research Engineering and Management* 2(60) pp 58-66
[11] Garza-Reyes J A, Eldridge S, Barber K D and Soriano-Meier H 2010 Overall Equipment Effectiveness (OEE) and Process Capability (PC) Measure: A Relationship Ana*l*ysis *Int Journal of Quality and Reliability Management* 27(1) pp 48-62
[12] Irianto D 2016 Collaborative Manufacturing for Small-Medium Enterprises *IOP Conf Series* 114 pp 1-6
[13] Baluch N, Abdulla C S and Mohtar S 2012 Measuring OEE in Malaysian Palm Oil Mills *Interdisciplinary Journal of Contemporary Research In Business* 4(2) pp 733-743
[14] Carlo F D, Arleo M A and Tucci M 2014 OEE Evaluation of a Paced Assembly Line through Different Calculation and Simulation Methods: A Case Study in the Pharmaceutical Environment *Int Journal of Engineering Business Management* 6(27) pp 1-11
[15] Prasetyawan Y and Sentosa A H 2017 Manufacturing Process Design for Multi Commodities in Agriculture *American Institute of Physics Conf Proceedings* 1855 pp 1-7
[16] Samat H A, Kamaruddin S and Azid I A 2012 Integration of Overall Equipment Effectiveness (OEE) and Reliability Method for Measuring Machine Effectiveness *South African Journal of Industrial Engineering* 23(1) pp 92-113
[17] Ron A D and Rooda J 2006 OEE and Equipment Effectiveness: An Evaluation *Int Journal of Production Research* 44(23) pp 4987-5003
[18] Ngubane N et.al 2015 Inventory Management System used by Manufacturing Small Medium and Micro Enterprises in Cape Town *Mediterranean Journal of Social Science* 6(1) pp 382-390
[19] Paik S -K 2011 *Supply Management in Small and Medium-Sized Enterprises: Role of SME Size* Supply Chain Forum Vol **12** pp 10-21
[20] Puvanasvaran P, Teoh Y and Tay C 2013 Consideration of Demand Rate in Overall Equipment Effectivences (OEE) on Equipment with Constant Process Time *Journal of Industrial Engineering and Management* 6(2) pp 507-524

[21] Vijayakumar S and Gajendran S 2014 Improvement of Overall Equipment Effectiveness (OEE) in Injection Moulding Process Industry *Journal of Mechanical and Civil Engineering* pp 47-60
[22] Low S -N et.al 2014 Measurement of Overall Performance Effectiveness in Setup Improvement *Journal of Industrial Engineering* pp 1-7
[23] Musa M A et.al 2015 Improvement of Overall Equipment Effectivenee (OEE) through Implementation of Autonomous Maintenance in Crankcase Line *Applied Mechanics and Materials* Vol **761** pp 165-169
[24] Kasim N I et.al 2015 Improvement of Overall Equipment Effectiveness (OEE) through Implementation of Total Productive Maintenance (TPM) in Manufacturing Industries *Applied Mechanics and Materials* Vol **761** pp 180-185
[25] Nayak D M N V K M , Naidu G and Shaker V 2013 Evaluation of OEE in a Continuous Process Industry on an Insulation Line in a Cable Manufacturing Unit *Int Journal of Innovation Research in Science* 2(5) pp 1629-1634
[26] Nagdiya M N N P B and Sengar N 2017 Analysis and Implementation of Chute System to Obtain World Class OEE *Int Journal for Science Research and Development* 5(2) pp 1608-1613
[27] Prasetyawan Y, Agustiani E and Jumayla S 2017 *Aligning Business Strategy of Incubator Center and Tenants* American Institute of Physics Conference Proceeding 1855 pp 1-8
[28] Dutta S and Dutta A K 2017 *The Improvement of Overall Equipment of Machines Using TPM Study in a Small Scale Industry* Int Journal of Mechanical Engineering Special Issue pp 1-12

Cost-effective strategy to mitigate transportation disruptions in supply chain

G Albertzeth[1]* and I N Pujawan[1]
[1]Department of Industrial Engineering, Institut Teknologi Sepuluh Nopember, Kampus ITS Sukolilo-Surabaya 60111, Indonesia

*gustav.albertzeth@gmail.com

Abstract. Supply chain disruptions have gained significant attention by scholars. But, even though transportation plays a central role in supply chain, only few studies address transportation disruptions. This research demonstrates a real case of an order delivery process from a focal company (FC) to a single distributor, where transportation disruptions stochastically occurs. Considering the possibility of sales loss during the disruption duration, we proposed a redundant stock, flexible route, and combined flexibility-redundancy (ReFlex) as mitigation strategies and a base case as a risk acceptance strategy. The objective is to find out the best strategy that promotes cost-effectiveness against transportation disruptions. To fulfill this objective, we use simulation modeling and cost-effectiveness analysis (CEA) as our research method. We simulate the delivery process of 5 brands using each strategy to produce two different responses: loss of sales percentage and the incurred costs. Next, using these responses, we evaluate and compare the cost-effectiveness ratio of each strategy using CEA. We found that redundant stock gave the best effectiveness on all brands, ReFlex as the second best, while flexible route gave the least effectiveness. Finally, we recommend which strategy should be applied based on the decision maker willingness to pay.

1. Introduction

Indonesia that geographically prone to natural disasters is facing the risk of having its supply chain crippled. Furthermore, the tardiness of transportation infrastructure development shall lead to the vulnerability against transportation disruptions. In addition, the negligence of most companies to invest in risk management program will consequently worsen the negative impact of transportation disruptions. One example of transportation disruptions that crippling a supply chain was the catastrophic event of Iceland's Eyjafjallajokull volcano eruption in 2010 that disrupting the air transportation from and going to Europe, it had forced Nissan and BMW to stop their production. Wilson [1], Yang & Wu[2] and Figliozzi & Zhang [3] studied the impact of transportation disruptions in supply chain and found that it may lead to drop in supply chain performance. Transportation disruptions as one aspect of the risk management of supply chain disruptions [4][5] has received less attention [1][6][7]. Although several papers have been published in respect of transportation disruptions, we found out that none of them was trying to apply strategies (in a real situation) to protect the supply chain from the harmful effect of transportation disruptions. Talluri et al. [8] mentioned that simply proposing mitigation strategies is not adequate, hence they urged the effectiveness of a strategy must be judged with respect to its cost and non-cost factors. This research was trying to fill the gap of research in transportation disruptions by evaluating cost-effectiveness of the proposed strategies against transportation disruptions. We chose simulation modeling as a tool to measure the cost and effectiveness of each strategy against transportation disruptions because it has the capability to analyze complex model when the analytical techniques are difficult to implement [9], especially if the system incorporates stochastic variables [10]. Finally, a cost-effectiveness analysis (CEA) was chosen as a multi-criteria decision making tool for its proven capability in assessing and comparing the cost-effectiveness of alternate strategies, especially when the effectiveness measure is difficult to monetized [11]. The rest of this paper is structured as follows: Section 2 presents the literature review; Section 3 describes the brief process of building the simulation model; Section 4 shows and discusses the findings; Section 5 presents our conclusion and the potential further studies.

2. Literature review

2.1. Transportation disruptions

Transportation disruptions defined by Sheffi et al. [5]as delay or unavailability of the transportation infrastructure, leading to the impossibility to move goods, either inbound and outbound. The uniqueness of transportation disruptions that differentiate it from supply disruptions lays in the availability of the both parties, consigner and consignee. To the best of our knowledge, published research that studying phenomena of transportation disruptions in supply chain are as follows: Wilson [1], Yang & Wu [2], Bai & Wang [12], and Houshyar et al. [13]found that transportation disruptions may lead to drop in supply chain performance. Chen & Zhang [14] proposed a dispatching vehicle policy that optimize vehicle capacity and dispatching time along a route should a transportation disruptions occurred. Figliozzi & Zhang [3] found that disruptions costs include lost sales, expediting costs, intangibles such as loss of reputation, and financial impacts on companies' cash flows are very difficult to quantify even with full access to a company's proprietary operational, financial, and sales data. Husdal & Brathen [15] found that industry and business are the immediate sector that experience the severe impact of transportation disruptions since the first day after disruptions occurred. Ishfaq [16] found that flexibility in transportation could provide opportunity to build resilience concept into the supply chain. Hishamuddin et al. [6] found that optimal recovery schedule is highly dependent on the relationship between the backorder cost and the lost sales cost parameters. Cui et al. [17] found that their proposed model could yield a supply chain system design that minimizes the impacts from probabilistic disruptions and leverages expedited shipments and inventory management to balance tradeoffs between transportation and inventory costs. Zhen et al. [18] found that the backup transportation is very efficient to reduce the profit loss, causing a less insurance coverage to be purchased in advance of disruptions. Liu et al. [4] found that the improved model of grey neural networks is a feasible prediction method for transportation disruptions.

2.2. Proposed strategies

The risk acceptance strategy roots from the fact that most managers know the inherently risky supply chain is, but chose to do nothing either to avoid the incurred extra costs and/or do not know what to proactively or reactively deal with the risk. This option is commonly adopted by the manager (base case), but Chopra & Sodhi [19] mentioned that doing nothing will likely cost more. The mitigation strategies are aimed to build a resilient supply chain that able to maintain its desired performance level even tough disruptions occurred. As mentioned by Sheffi et al. [5] , creating a resilient supply chain commonly through two approaches: building flexibility and redundancy. Furthermore, they stated that redundancy is easy to build and less expensive in short term, while building flexibility is difficult and costly but appears to be more cost-effective compared to building redundancy in long term. In addition, according to Pujawan [20], one should not pursuing high degree of flexibility unless the market indicates the need for it. Furthermore, Rice & Caniato [21] suggested that ultimately it is possible for the company to adopt a combination of flexibility and redundancy.

2.3. Simulation in risk management

Simulation has been widely used as a tool in risk management, particularly with supply chain disruptions. Schmitt & Singh [22] used Monte Carlo simulation to assess the vulnerability of supply chain against disruption risk and quantity the impact on customer service. Son & Orchard [23] measured the effectiveness of two mitigation strategy: reserved stock (R-policy) and larger order quantity (Q-policy) using simulation. They found that R-policy had better performance that Q-policy. In transportation disruptions, simulation also had been proven as a valuable tool. Wilson [1] used simulation to measure the impact of transportation disruptions in supply chain performance. Yang & Wu [2] also used simulation to investigate the impact of transportation disruptions on performance of e-collaboration supply chain. Bai & Wang [12] Applied discrete-continuous combined simulation to simulate supply chain under transportation disruptions using VMI and traditional structure.

2.4. Cost-effectiveness analysis

Cost-effectiveness analysis (CEA) is preferred over cost benefit analysis (CBA) when monetizing all the benefit of the proposed alternative of strategies is difficult [11]. CEA is used commonly in medicine research area, particularly measuring the effectiveness of the prescribed drugs. In risk management, CEA was used by Schmit & Roth [24] to evaluate the effectiveness of various risk management practices. Sherwin et al. [25] also used CEA to evaluate cost-effective mitigation strategy with a low volume high value supply chain against delay risk. Incremental cost-effectiveness ratio is a method to help measuring cost-effectiveness when more than one strategy need to be evaluated.

3. Problem statement and solution methods

We consider a single supplier – single distributor relationship in a supply chain which doing business in wheat flour commodity. The supplier, which is the focal company, is a wheat flour manufacturer, while the distributor buys products from the focal company. Distributor act as a reseller who fills a known end customer demand from its inventory, which in turn is replenished by the supplier. We assume that natural-catastrophic events may occur at any point in time causing unavailability of road infrastructure (main route) that connecting supplier site with distributor side according to an empirical probability distribution function with a frequency per year. Once transportation disruptions occur, the ability of supplier to replenish distributor's inventory stops temporarily (for the duration of the disruption); however, end customers' demand can continue to be met from inventory until it is depleted. Depending on the duration of the disruptions and on-hand inventory at the time of the beginning of a disruption, distributor could very likely experience stockouts. Here we assume that any unmet demand considered as loss of sales (no backorder). The recovery process (fixing the road) will be executed as soon as a disruption begins. Hence, the disruptions duration is equal to the recovery process which expressed as a lognormal distribution. Replenishment (we call it delivery now on) process from the supplier resumes as soon as recovery process is finished.

We build the simulation model based on the delivery activity of year 2016, hence our model was a terminating simulation with one-year period. We focus onto 80% of total delivery volume which represented in 5 brands (named A, B, C, D, E). Each brand has its own delivery date, volume, and unit price. We deploy 4 strategies: (1) base case/do nothing, (2) redundant stock, (3) flexible route, and (4) redundant-flexibility (ReFlex) into each brand and run the simulation by 4000 replications to get the proper accuracy confidence. Each strategy consists of 3 simulation logic: delivery logic, demand logic, and disruption logic. Input for delivery logic and demand logic are using actual delivery date specific for each brand and read directly. While for disruption logic the frequency and duration of disruptions using the input as explained early in this section. Demand and disruption logic construction are the same for each strategy, except for delivery logic that differ for each strategy. For strategy 1, when disruption occurred, the delivery will do nothing and wait until the recovery finished. Strategy 2 similar with strategy 1, except for a certain amount of redundant stock placed initially at distributor warehouse. For strategy 3, instead of waiting until disruption over, we use alternative route which take 2 days longer than using the regular route. For strategy 4, we combine strategy 2 and 3, that using alternative route plus protection from redundant stock. At the end of simulation run, we record 2 responses: (1) the average percentage of loss of sales (acts as our effectiveness measure) and (2) the average impact cost of executing certain strategy (acts as our cost measure).

We conduct verification and validation (V&V) test and found that the developed simulation model is worked as intended and represents the system under study. After the V&V test have passed, we used these responses as input in CEA to compare and evaluate cost-effectiveness ratio of each strategy. A sensitivity analysis (SA) is conducted to investigate the sensitivity of the factors on responses. Finally, we provide recommendation of the best strategy to be adopted by manager in order to cost-effectively mitigate transportation disruptions based on willingness to pay (WTP).

4. Findings and discussion

4.1. Initial findings

Before conducting sensitivity analysis, some initial findings for all brands were given. These findings are summarized in Table 1 below.

Table 1. Cost and effectiveness before sensitivity analysis.

Cost		Brand				
		A	B	C	D	E
Strategy	Base case	#1	#1	#1	#1	#1
	Flexible route	#2	#2	#2	#2	#2
	Redundant stock	#4	#4	#4	#4	#4
	ReFlex	#3	#3	#3	#3	#3
Effectiveness		Brand				
		A	B	C	D	E
Strategy	Base case	#4	#4	#4	#4	#4
	Flexible route	#3	#3	#3	#3	#3
	Redundant stock	#1	#1	#1	#1	#1
	ReFlex	#2	#2	#2	#2	#2

Table 1 above shows 2 findings: (1) cost and effectiveness oppose each other. For example, redundant stock gives the highest effectiveness (rank 1) of reducing the percentage of loss of sales but gives the highest impact cost (rank 4); (2) Focusing on mitigations strategies, redundant stock gives the best effectiveness on all brands, ReFlex as the second best, and flexible route as the least.

4.2. Sensitivity analysis

Initially, we have 4 strategies on 5 brands that simulated in 4000 replications. When conducting the sensitivity analysis, we produced 5 additional strategies as the result of varying factors of the initial strategies, depicted in Table 2 below.

Table 2. Varying factors on the initial strategies.

Initial Strategy	After varying factors	Modified value: Alternative route travel time	Modified value: Redundant stock
Base Case	BC	n/a by default	n/a by default
Flexible Route	FR-0	4 days (initial value)	n/a by default
	FR-1	3 days	n/a by default
Redundant Stock	RS-0	n/a by default	initial value
	RS-1	n/a by default	down by 50%
ReFlex	RF-0	4 days (initial value)	initial value
	RF-1	3 days	initial value
	RF-2	3 days	up by 400%
	RF-3	4 days (initial value)	up by 400%

From conducting sensitivity analysis in Table 1, we found that: (1) speeding up travel time of alternative route gives no significant changes on decreasing the percentage of loss of sales and not increasing the impact cost; and (2) adding up quantity of redundant stock gives significant changes on decreasing the percentage of loss of sales and significantly increasing the impact cost.

4.3. Final findings and recommendation
We simulate these additional strategies to measure the effectiveness and cost of these strategy and use these measurements to evaluate and compare the cost-effectiveness ratio for all brands. We summarized the findings in Table 3 below.

Table 3. Cost and effectiveness after sensitivity analysis.

Cost		Brand				
		A	B	C	D	E
Strategy	BC	#1	#1	#1	#1	#1
	FR-0	#2	#2	#2	#2	#2
	FR-1	#3	#3	#3	#3	#3
	RS-0	#7	#9	#9	#9	#9
	RS-1	#6	#6	#6	#6	#6
	RF-0	#4	#4	#4	#4	#4
	RF-1	#5	#5	#5	#5	#5
	RF-2	#9	#8	#8	#8	#8
	RF-3	#8	#7	#7	#7	#7
Effectiveness		Brand				
		A	B	C	D	E
Strategy	BC	#9	#9	#9	#9	#9
	FR-0	#8	#8	#8	#8	#8
	FR-1	#7	#7	#7	#7	#7
	RS-0	#3	#1	#1	#1	#1
	RS-1	#4	#4	#4	#4	#4
	RF-0	#6	#6	#6	#6	#6
	RF-1	#5	#5	#5	#5	#5
	RF-2	#1	#2	#2	#2	#2
	RF-3	#2	#3	#3	#3	#3

By comparing Table 3 above with Table 1, it shows that the effectiveness of redundant stock strategy does not change for brand B, C, D, and E, either before or after SA. An exception applied for brand A where the redundant stock strategy no longer becomes the general strategy to mitigate transportation disruptions. On brand A, upgrade its rank by 1 ReFlex strategy (RF-2) from #2 to #1 and RF-3 from #3 to #2. The behavior peculiarity of brand A derived by its nature of having the demand rate higher that the other four brands. Hence, the final findings from Table 3 is that a single strategy is no longer adequate to effectively mitigate transportation disruptions on product with higher demand rate. Table 4 below shows the recommended strategy that cost-effectively mitigate transportation disruptions in supply chain, based on the manager's WTP.

Table 4. Final recommendation for managers.

BRAND	Willingness to Pay (WTP)			Recommended Strategy
A	WTP	<	Rp2,868,009.48	BC
	Rp2,868,009.48	to	Rp3,325,808.03	FR-1
	Rp3,325,809.03	to	Rp5,609,661.03	RF-1
	Rp5,609,662.03	≥	WTP	RF-2
B	WTP	<	Rp3,138,618.04	BC
	Rp3,138,618.04	to	Rp3,314,988.62	RS-1
	Rp3,314,989.62	≥	WTP	RS-0
C	WTP	<	Rp2,035,290.82	BC
	Rp2,035,290.82	to	Rp2,653,694.49	RS-1
	Rp2,653,695.49	≥	WTP	RS-0
D	WTP	<	Rp2,098,035.95	BC
	Rp2,098,035.95	to	Rp2,927,781.42	RS-1
	Rp2,927,782.42	to	Rp7,761,557.26	RF-2
	Rp7,761,558.26	≥	WTP	RS-0
E	WTP	<	Rp1,666,380.06	BC
	Rp1,666,380.06	to	Rp2,013,297.23	RS-1
	Rp2,013,298.23	≥	WTP	RS-0

For example: on brand E, if manager is only willing to pay 2 million for each 1 % of effectiveness, then the best strategy that cost-effectively mitigate transportation disruptions is RS-1 (redundant stock variant 1).

5. Conclusion and further study

This study presents a simulation model to evaluate different strategies in dealing with transportation disruptions. We model a case of a company producing and distribution flour products in Indonesia. The four strategies, namely the base case which is the current situation, the redundant stock, the flexible route, and the combination between flexible route and redundant stock, were simulated and the impacts were compared in terms of service level and total costs. We demonstrated that all alternative strategies are able to negate the impacts of transportation disruptions with different levels of effectiveness. Furthermore, redundant stock strategy was the most expensive alternative but gave the best effectiveness while flexible route strategy was the least expensive but worst in terms of effectiveness. The combined strategy is in the middle between the two proposed strategies in terms of both effectiveness and costs. Overall, after the cost effectiveness analysis, the combined strategy was considered the best.

The issue of transportation disruptions is an interesting research topic which should receive more attention in the field of supply chain management. Various strategies maybe proposed and evaluated, but we believe that the best strategy in one situation may not work well in other situation. An interesting topic should to identify under which condition a certain strategy is preferable. A broad classification of strategies such as reactive vs proactive [26] may be useful as a general framework. Another interesting issue is how the risks maybe collaboratively mitigated by more than one parties in a supply chain.

References

[1] Wilson M C 2007 The impact of transportation disruptions on supply chain performance. *Transp Res Part E Logist Transp Rev* **43** pp 295–320 doi: 10.1016/j.tre.2005.09.008

[2] Yang T, and Wu J 2007 The Impact of Transportation Disruptions on Performance of E-Collaboration Supply Chain *Res. Pract. Issues Enterp. Inf. Syst. II* **254**, Boston, MA: Springer US; 2007, pp 663–7. doi:10.1007/978-0-387-75902-9_74

[3] Figliozzi M A and Zhang Z A 2010 Study of Transportation Disruption Causes and Costs in containerized maritime transportation *Transp Res Board* pp 1–17

[4] Liu C, Shu T, Chen S, Wang S, Lai K K and Gan L 2016 An improved grey neural network model for predicting transportation disruptions *Expert Syst Appl* 2016 **45** pp 331–40 doi: 10.1016/j.eswa.2015.09.052

[5] Sheffi Y, Rice J B, Fleck J M, and Caniato F 2003 Supply Chain Response to Global Terrorism: A Situation Scan *EurOMA POMS Jt Int Conf* pp 1–6

[6] Hishamuddin H, Sarker R A, and Essam D 2013 A recovery model for a two-echelon serial supply chain with consideration of transportation disruption *Comput Ind Eng* **64** pp 552–61 doi: 10.1016/j.cie.2012.11.012

[7] Ho W, Zheng T, Yildiz H, and Talluri S 2015 Supply chain risk management: a literature review. *Int J Prod Res* **7543** pp 1–39 doi:10.1080/00207543.2015.1030467

[8] Talluri S S, Kull T J, Yildiz H and Yoon J 2013 Assessing the Efficiency of Risk Mitigation Strategies in Supply Chains *J Bus Logist* **34** pp 253–69 doi:10.1111/jbl.12025

[9] Law A M 2007 *Simulation Modeling & Analysis* 4th ed McGraw-Hill

[10] Pujawan N, Arief M M, Tjahjono B, and Kritchanchai D 2015 An integrated shipment planning and storage capacity decision under uncertainty *Int J Phys Distrib Logist Manag.***45** doi:10.1108/IJPDLM-08-2014-0198

[11] Boardman E A, Greenberg D, Vining A, and Weimer D 2010 *Cost-Benefit Analysis. Concepts and Practice* 4th ed. (New Jersey: *Pearson Education* Inc.)

[12] Bai S Z and Wang W L 2008 The impact of transportation disruption on adaptive supply chain: A hybrid simulation study. *Proc 2007 Int Conf Manag Sci Eng* ICMSE'07 pp 844–9 doi:10.1109/ICMSE.2007.4421951

[13] Houshyar A N, Baghdadabad M R T, Hoshyar A N and Sulaiman R B 2013 Simulating Effects of Transportation Disruption on Supply Chain Based on Vendor Managed Inventory Approach. *Int J Model Optim* **3** pp 116–9 doi:10.7763/IJMO. 2013.V3.248

[14] Chen X and Zhang J 2009 Optimal dispatching policy under transportation disruption. *50th Annu Transp Res Forum 2009* **2** pp 634–43

[15] Husdal J and Brathen S 2010 Bad Locations, Bad Logistics? How Norwegian Freight Carriers Handle Transportation *World Conf Transp Res 2010* pp 1–17

[16] Ishfaq R 2012 Resilience through flexibility in transportation operations *Int J Logist Res Appl* **15** pp 215–29 doi:10.1080/13675567.2012.709835

[17] Cui J, Zhao M, Li X, Parsafard M and An S 2016 Reliable design of an integrated supply chain with expedited shipments under disruption risks *Transp Res Part E Logist Transp Rev* **95** pp 143–63 doi: 10.1016/j.tre.2016.09.009

[18] Zhen X, Li Y, Cai G (George) and Shi D 2016 Transportation disruption risk management: business interruption insurance and backup transportation. *Transp Res Part E Logist Transp Rev* doi: 10.1016/j.tre.2016.01.005

[19] Chopra S and Sodhi M S 2014 Reducing the Risk of Supply Chain Disruptions *MIT Sloan Manag Rev* **55** pp 73–80

[20] Pujawan I N 2004 Assessing supply chain flexibility: a conceptual framework and case study *Int J Integr Supply Manag* **1** pp 79–97 doi:10.1504/IJISM.2004.004599

[21] Rice J B and Caniato F 2003 Building a secure and resilient supply network *Supply Chain Manag Rev* **7** pp 22–30 doi:10.1080/13675560600717763

[22] Schmitt A J and Singh M 2009 Quantifying supply chain disruption risk using Monte Carlo and discrete-event simulation *Proc. 2009 Winter Simul. Conf.* pp 1237–48 IEEE doi:10.1109/WSC.2009.5429561

[23] Son J Y and Orchard R K 2013 Effectiveness of policies for mitigating supply disruptions *Int J Phys Distrib Logist Manag* **43** pp 684–706 doi:10.1108/IJPDLM-04-2012-0109

[24] Schmit J T and Roth K 1990 Cost effectiveness of risk management practices *J Risk Insur* **57** pp 455–70

[25] Sherwin M D, Medal H and Lapp S A 2016 Proactive cost-effective identification and mitigation of supply delay risks in a low volume high value supply chain using fault-tree analysis *Int J Prod Econ* **175** pp 153–63 doi: 10.1016/j.ijpe.2016.02.001
[26] Angkiriwang R, Pujawan I N and Santosa B 2014 Managing Uncertainty through Supply Chain Flexibility: Reactive vs. Proactive Approaches *Prod Manuf Res* **2** pp 50–70 doi:10.1080/21693277.2014.882804

Acknowledgments
The authors would like to thank the reviewers for the constructive feedback to improve and strengthen the contributions of the paper. The authors are also giving their gratitude to the officers of the focal company for their full support which ensuring this research continuity.

A Literature Review On Multimodal Freight Transportation Planning Under Disruptions

E E Rosyida[1,2*], B Santosa[1] and I N Pujawan[1***]**
[1] Department of Industrial Engineering, Institut Teknologi Sepuluh Nopember Kampus ITS Sukolilo-Surabaya 60111, Indonesia
[2] Department of Industrial Engineering, Universitas Islam Majapahit Mojokerto, Jawa Timur, Indonesia

*erly_ekayanti@yahoo.co.id, **budi_s@ie.its.ac.id, ***pujawan@ie.its.ac.id

Abstract. This paper reviews publication that focuses on multimodal freight transportation planning under disruptions. In this paper, disruptions are specified by the level of the disruptions occurs and the scope of its effect. This becomes an important distinction since the cause and effect that may occur at different levels. The failure to make this distinction has implications for how we understand and manage. The reviewed papers include those that develop framework, model, and technical procedure for freight transportation. Finally, we provide an outlook of future research directions on the domain of transportation planning.

1. Introduction

Transportation development is directed towards the integration of effective and efficient service and intermodal transportation facilities networks. This integration is in form of interconnection on transportation node that functioned as a meeting point of mode interchange, which is named as intermodal terminal that can provides additional values. The service network is created through the inter-route integration of road, train, sea, air, river and lake transportations regarding the specialty of respective transportation mode. This can be based on the concept of combination of main, feeder, and advanced modes. This intermodal integration provides special challenge to logistic service providers, especially in facing the uncertainty factor that always causes losses.

Uncertainty in supply chain is one of the factors that influence freight transportation movement. Each company has different level of uncertainty [1] [2]. This level of uncertainty depends on the driver of the uncertainty itself, including: delivery lead time uncertainty, price uncertainty, quality uncertainty, and uncertainty in availability. Angkiriwang et al.[3] explained that supply uncertainty is one type of uncertainty in the supply chain context and its can disrupt the supply network.

Disruption is one form of uncertainty. Disruptions in terms of supply network may come from internal and external sources. Internal disruptions may be caused by engine damage, disturbance in import or export, transportation failure, disruption on shipping chain, change on demand, technological innovation, change on material price, and many more. External disruption includes supplier failure, supplier quality problems, oil crisis, accident, and natural disaster [4]. There are many researchers that studied disruptions in the context of supply network. Author [5] reviewed disruption on multimodal transportation network with regard on external factor as optimum point of reference, which was the cost minimization. Authors [6] [7] [8] also reviewed papers on disruption on supply network. Author [6] criticized literature on multimodal transportation network based on transportation planning level (strategic, tactical, and operational levels). Author [7] investigated literature on disruption and resilience on supply network. They used graphic theory reference to differentiate the type of disruption by dividing disruptions into three levels: line, node, and network. While author [8] analyzed literature on disruption with regard to recovery. The topic of disruptions on supply network is highly interesting. This is evident from the high number of research regarding this, as well as there are still many research area in this field that need to be explored further [5] [6] [7] [8].

This paper aims to classify papers on disruptions on multimodal transportation network, specifically for freight transportation by classifying the level and behaviour of disruptions, models solution method, as well as objectives of the model. The final objective is to identify research gap that can be used as the reference for future research.

The rest of this article is organized as follows. Section 2 reviews the literature on multimodal freight transportation under disruption that is broken down into description of the conceptual, models and the solution methods. Section 3, discusses the opportunities for future research based on the identified gaps.

2. Literature On Multimodal Freight Transportation Network Under Disruptions

According to [7], disruptions on supply chain network can be described based on the perspective of network graph theory. They illustrated graph as collection of nodes connected by a link (figure 1). Disruption was analyzed based on 3 levels: node, link, and network, featured also on Figure 1. On multimodal transportation network, node is a transportation facility, such as: intermodal terminal, factory, warehouse, depot, or retailer outlet. Link is the line used to transport goods from one node to another by using trucks, trains, ships, airplanes. Figure 1 shows the location/level of disruptions and its effect on the network. Figure 1b explains disruptions occurs on line level (depicted on line a14), and this disruption hampers the line a14 only. Node disruptions level is explained by Figure 1c, occurs on node n7 and damages node 7 itself, as well as line a7 and a13. Meanwhile the network level disruption is different with the former two that still have the alternative line for freight shipping. Disruptions on network level may not provide alternative line for freight shipping. In Figure 1d, disruptions occur on node 3 and line a2. These disruptions trigger, or hamper disruptions in network level.

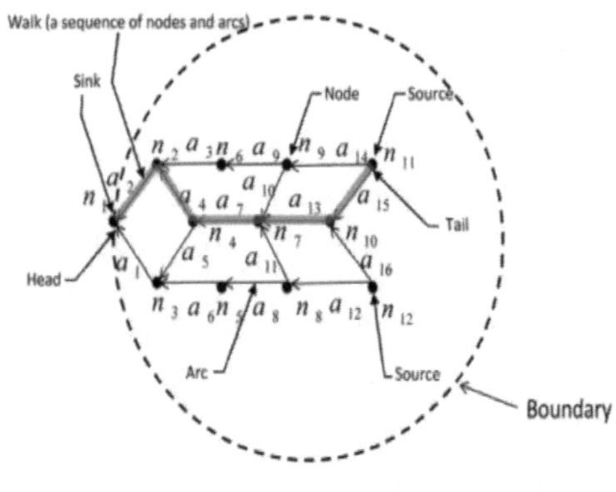

Figure 1. Supply Network[7].

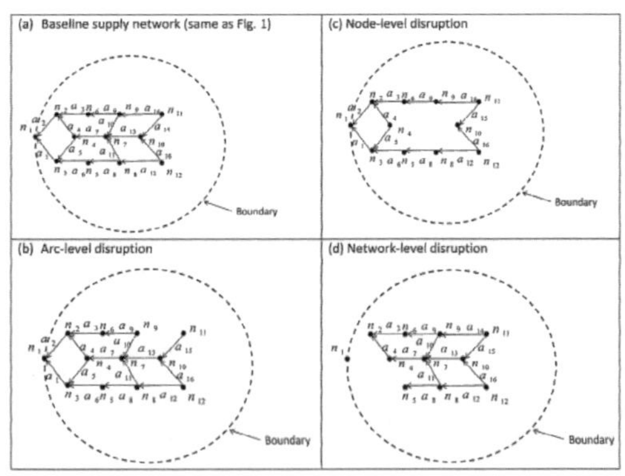

Figure 2. Examples of supply disruptions at links-, nodes-, and network-level [7]

Table 1 explains research progress based on the scope of disruption effect on the network (line, node, or network). Most of research studied disruption on line and node levels. There were only a small number of researches that studied disruption on network level. In addition, some researches were not explained in detail regarding the scope disruption being studied.

Author [7] stated that a clear border must be created between disruptions occurs on line, node, and network levels. This is related to the strategy that must be prepared to face and/or to deal with the disruptions. Some journals clearly stated the scope of the disruption. Nevertheless, there were also some journals did not. From the perspective of graphic theory, if disruptions occurs on line or node level, those line or node only need to remove line and node from the network supply graphic. This is since the disrupted line or node may no longer in function, so that materials cannot pass through those areas, and other alternative routes may probably be found. On the other side, if disruptions occurs on network level, alternative routes may probably unable to be made, since the core line or node cannot be used. This probably means that network level disruption occurs when there are more than one type of disruption take place on particular node or line that may causes disruptions on network level.

Table 1. Matrix Supply Network Disruptions.

Disruptions Level	References
Link	Morlok&Chang[9], Gedik *et al.*,[10] , Narayanaswami and Rangaraj[11], Zilko, Kurowicka and Goverde[12], Huang, Hu and Zhang[13], Burgholzer *et al.*, [14], Ishfaq[15], Gedik *et al.[10]*, Azad, Hassini & Verma[16], Uddin and Huynh[17], Miller-hooks, Zhang and Faturechi[18], Pant, Barker and Landers [19]
Node	Di, Lai and Zuddas[20], Liu, Zheng and Zhang[21], Fialkof et al.[22], Uddin and Huynh[17], Miller-hooks, Zhang and Faturechi[18], Pant, Barker and Landers [19]
Network	Daskin & Snyder [23], Starita et al. [24]

Scope of disruption occurs is also depends on the type of disruption itself. Some journals described the definition of disruption clearly, as well as its size. Nevertheless, some journals did not. Those journals only explained that the occurring disruptions intrude the node or line. Table 2 explains the definition and size of disruption. Each disruption has different effect on network. An operational measure of disruptions is used to find out the effect. Clear explanation on how to measure the disruptions will give the good guidance for the paper. Some paper used an operational measure of disruptions clearly, for example delay as a variable used to show the disruptions performance. In addition, Table 2 also explains network performance measure. In the literature, network performance measured through economic goals that defined as a multimodal network cost. This cost usually consists of transfer cost, transportation cost, delay cost, penalty cost, recovery cost etc. On the other hand, some literatures considered other performance measures in this topic area; time (variation of delay, length of disruptions).

Strategies used in facing disruptions are aimed to make the supply network more effective, efficient, resilient, flexible, and adaptive. Planning on facing disruptions divided into three levels: strategic, tactic, and operational planning level [6]. Strategic level is the stage of risk mitigation; tactical level is a preparation stage; and operational level is the stage of stability and recovery [8]. Many paper use flexibility strategy that used in their transportation planning models. This strategy to maintain service level and responsiveness in facing disruptions [3]. Challenge on implementing flexibility is the increasing cost, even though, on the other side, service level increases. Some studies implemented flexibility strategy on transportation route and mode to face disruption. This is concluded in Table 3.

Table 2. Existing Research on Multimodal Transportation Network.

No.	References	Definition of Disruption		Performance Measures
		Conceptual Definition	Operational Measures	
1	Morlok&Chang [9]	Disruption not formally defined.	Level of delay	Total cost
2	Gedik et al[10]	Disruption not formally defined.	Level of delay	Total transportation & delay cost
3	Narayasnawami &Rangaraj[11]	Disruption as an incident involving track unavailability between a pair of a stations in the operational domain and that can lead to violations of some of the critical and/or preferred constraints of the system.	A small time window of delay (delay of service) and disturbance locations.	Delay
4	Zilko, Kurowicka&Goverde [12]	Disruption defined as an unexpected accident that disrupt the railway timetable (case study in track circuit failure in Netherland).	Length of disruption measured from latency time and repair time. Latency time is the length of time the mechanics need to get disrupted site and the repair time is the length of time they need to repair the problem.	Mean of prediction length of disruption.

No.	References	Definition of Disruption		Performance Measures
		Conceptual Definition	Operational Measures	
5	Huang, Hu&Zhang [13]	Disruption defined as an unexpected events (hurricane, the snow disaster, traffic accident etc) happening in one link of the multimodal chain could result in the disturbance of pre-decided transportation activities.	The duration of a disruption event (delay of current transport activity). The duration of disruption events measured by collect historic statistic data of typical disruption events in the specific transport mode.	disturbance in system level
6	Burgholzer et al [14]	Disruption defined as a disturbance in individual sections (links) that can decrease the performance level of a network.	Disruption as an amount of performance level of a network drops when a link disrupted. The model use network performance indicator to measure it.	Transport time
7	Ishfaq [15]	Disruption defined as a incident that caused by external source in supply chain	Disruption measured by delay time.	Transportation cost.
8	Azad et al. [16]	Disruption not formally defined.	No operational measure for disruption	Transportation cost.
9	Uddin&Hyunh [17]	Disruptions defined as incident that disrupt node, link or terminal	Disruptions measured based on link capacity in node, link and terminal. Capacity will be reducing 50% in link and 80% in node and terminal 50% when this point was disrupt. It make the travel time increase.	Transportation cost (transportation cost, transfer cost, penalty cost)
10	Miller-Hooks, Zhang&Faturechi [18]	Disruption defined as a disaster scenario.	Five disruption scenario that define their level of disrupt (bombing, terrorist attack, flood, earthquake, and intermodal terminal attack).	Resilience level
11	Pant, Barker & Landers [19]	Disruption defined as a incident in inland waterway that caused economic losses (supply and demand shortage) for the industries using facilities along this waterways.	Disruption metric used to measure the disruption effect.	Transportation costs
12	Di, Lai&Zuddas [20]	Disruption not formally defined.	No operational measure for disruption	Transportation Cost
13	Liu, Zheng & Zhang [21]	Disruption defined as a machinery breakdown of QCs in work. QCs are the major tools to unload and load containers from the vessels.	Disruption period is multiples time of the unit time. The disruption duration not be calculated accurately but repair workers can give an approximate recovery time according to their experience.	Negative deviation from the originally planned schedule
14	Fialkof et al.[22]	Natural Disaster	Level of delay	Transportation cost.
15	Daskin&Snyder [23]	Disruption defined as a incident that disrupt the system (natural disaster, manmade, etc)	No operational measure for disruption	Transportation cost.
16	Starita et al.[24]	Natural disaster	Banjir	Transportation cost.
17	Chen&Miller-Hooks[25]	natural or human caused disaster	No operational measure for disruption	Transportation cost.
18	Udenta et.al [26]	Disaster and external effects caused by vehicles entering the system	Beta variable to describe whether disruption occurs or not. Beta is a distribution function. If normal, its value is close to zero. While if disruption occurs, its value will be close to 1.	Transportation time and vehicle flow on line
19	Sun&Sconfeld[27]	Disruption on vehicle movement on intermodal terminal	No operational measure for disruption	Holding time
20	Jiang, Wang & Ding [28]	Disruption not formally defined	No operational measure for disruption	Deviation between planning and real condition

No.	References	Definition of Disruption		Performance Measures
		Conceptual Definition	Operational Measures	
21	Ahmadi-Javid&Seddighi[29]	Disruption occurs on production process/capacity disruption (influences location allocation) as well as on delivery process (influences routing)	No operational measure for disruption	Transportation cost.
22	Cui et al.[30]	Disruption not formally defined	No operational measure for disruption	Transportation cost.
23	Hu, Sun & Liu[31]	Dividing disruption based on its source (customers, vehicles, roads)	No operational measure for disruption	Shipping distance addition increase and customer uncertainty level
24	Ivanov et al. [32]	Disruption not formally defined	No operational measure for disruption	Transportation cost.

Tabel 3. Transportations Planning Strategy.

Transportation Planning	Models	Reference
Strategic Planning Problems	Facility Location Problem	Daskin&Snyder [23], Ahmadi-Javid&Seddighi [29]
Tactical Planning Problems	Network Flow Planning	Ivanov et al. [32], Hu, Sun & Liu [31], Cui et al.[30], Ahmadi-Javid&Seddighi [29], Sun&Sconfeld [27], Udenta et.al [26], Chen&Miller-Hooks[25], Uddin&Hyunh [17], Ishfaq [15], Di, Lai&Zuddas [20], Huang, Hu&Zhang [13], Zilko, Kurowicka&Goverde [12], Gedik et al [10], Miller-Hooks, Zhang&Faturechi[18], Burgholzer et al[14], Fialkof et al. [22]Starita et al. [24]
Operational Planning Problems	Routing and Schedulling Problem	Jiang, Wang & Ding[28], Liu, Zheng & Zhang[21], Narayasnawami &Rangaraj[11], Pant, Barker & Landers[19], Azad et al. [16]

Modeling approach is implemented to identify and solve the problem based on the objectives function. According to the models objective, some previous papers solved the problems by using a variety of solution methodologies, such as optimization models, simulation models and heuristic/meta-heuristic model. Table 4 explain about method or approach used in the reviewed papers.

Tabel 4. Solution Methodology.

Solution Methodology	Reference
Optimization Approach/Exact	Narayasnawami &Rangaraj[11], Gedik et al[10], Zilko, Kurowicka&Goverde[12], Huang, Hu&Zhang[13], Ishfaq[15], Miller-Hooks, Zhang&Faturechi[18], Di, Lai&Zuddas[20], Liu, Zheng & Zhang[21], Ivanov et al.[32], Uddin&Hyunh[17], Azad et al.[16], Daskin&Snyder[23], Udenta et.al[26], Sun&Sconfeld[27], Cui et al.[30], Hu, Sun & Liu [31]
Simulation	Burgholzer et al [14]Pant, Barker & Landers [19]
Heuristic/Metaheuristic	Fialkof et al.[22], Starita et al.[24], Jiang, Wang & Ding[28], Ahmadi-Javid&Seddighi [29]
Combination	Chen&Miller-Hooks[25]

3. Conclusions and future research directions.

This paper reviews literatures concerning disruptions on multimodal transportation network, specifically for freight transportation. Disruptions can be divided into three levels: link, node, and network. Most of papers studied disruption on the link and node levels. There are only a small number of papers that studied disruptions on the network level. Beside disruption level, type of disruption also need to be studied further, especially type of disruption in the network (such as high level of natural disaster) [15]. This type of disruption may result in the supply network paralyzed totally.

Proper strategy need to be made to face each type of disruption. Steps done in transportation planning includes: risk mitigation (estimation of probable economic effect), preparedness (estimation of effect on deviation level of performance network), as well as stabilization and recovery (proposing effective, efficient, and flexible recovery strategy). Some strategies used in the above literatures are alternative route, mode, depot, and departure schedule determination. Some literatures also used buffer capacity and inventory, as well as location and allocation determination strategies. Coordination strategy needs to be developed further on this topic of disruption.

Performance measure used in most of literatures was the cost, while some papers used the time as their performance measure. There are many cost components that can be used as performance measure. Especially when cost analysis is combined with transportation network controlling model such as adaptation cost in order to be resilient or robust. On the other side, there is a huge opportunity to research this field with taking more than one performance measures into consideration. In this case, multi criteria decision making approach can be used. In addition, [3] said that the balance between efficiency and flexibility in supply network with disruptions still need to be developed since there were only a few papers did it. It is an interesting topic when facing uncertainty conditions.

Regarding the solution methodologies, there is a big opportunity to use another method that can solve the problems. According to [6], when facing a huge problems, a more sophisticated approaches should be employed to address the complex problems. Dealing with such complex problems may create computational issues and hence approaches to bring more efficient methodology should be made. Various techniques such as parallel computation and algorithms may promise significant improvements in solving huge problems.

References

[1] Pujawan N 2004 A Framework For Assesing And Coping With Supply Uncertainty And Complexity in *Proc. of the fifth Asia Pacific Industrial Engineering and Management Systems Conf.* p 3.2.1-11

[2] Pujawan N 2004 Assessing supply chain flexibility : a conceptual framework and case study *Int. J. Integr. Supply Manag* **1** (1) pp 79–97

[3] Angkiriwang R, Pujawan N and Santosa B 2014 Managing uncertainty through supply chain flexibility : reactive vs proactive approaches *Prod. Manuf. Res* **2** (1) pp 50–70

[4] Gilaninia S, Ganjinia H and Mahdikhanmahaleh B A 2013 Difference between Internal and External Supply Chain Risks On Its Performance *Singaporean J. Bussiness Econ. Manag. Stud.* **1** (8) pp 12–18

[5] Demir E, Huang Y, Scholts S and Van Woensel T 2015 A selected review on the negative externalities of the freight transportation : Modeling and pricing *Transp. Res. E* **77** pp 95–114

[6] Steadieseifi M, Dellaert N P, Nuijten W, Van Woensel T and Raoufi R 2014 Multimodal freight transportation planning : A literature review *Eur. J. Oper. Res* **233** pp 1–15

[7] Kim Y, Chen Y and Linderman K 2015 Supply network disruption and resilience : A network structural perspective *J. Oper. Manag.* **34** pp 43–59

[8] Ivanov D, Dolgui A, Sokolov B and Ivanova M 2016 Disruptions in in supply and recovery policies : state-of-the review Disruptions in recovery *IFAC-PapersOnLine* **49** 12 pp 1436–41

[9] Morlok E K and Chang D J 2004 Measuring capacity flexibility of a transportation system," *Transp. Res. Part A Policy Pract* **38** 6 pp 405–20

[10] Gedik R, Medal H, Rainwater C, Pohl E A and Mason S J Vulnerability assessment and re-routing of freight trains under disruptions : A coal supply chain network application *Transp. Res. E* **71** pp 45–57

[11] Narayanaswami S and Rangaraj N 2013 Modelling disruptions and resolving conflicts optimally in a railway schedule q *Comput. Ind. Eng.* **64** pp 469–81

[12] Zilko A A, Kurowicka D and Goverde R M P 2016 Modeling railway disruption lengths with Copula Bayesian Networks *Transp. Res. C.* **68** pp 350–68

[13] Huang M, Hu X and Zhang L A 2011 Decision Method for Disruption Management Problems in

Intermodal Freight Transport **2** pp 13–21
[14] W. Burgholzer W, Bauer G, Posset M and Jammernegg W 2013 Analysing the impact of disruptions in intermodal transport networks : A micro simulation-based model *Decis. Support Syst.* **54** pp 1580–86
[15] Ishfaq R 2012 Resilience through flexibility in transportation operations *Int. J. Logist. Res. Appl.* **15** 4 pp 215–29
[16] Azad N, Hassini E and Verma M 2016 Disruption risk management in railroad networks : An optimization-based methodology and a case study *Transp. Res. B* **85** pp 70–88
[17] Uddin M M and Huynh N 2016 Routing Model for Multicommodity Freight in an Intermodal Network Under Disruptions *Transp. Res. Rec. J. Transp. Res. Board* **2548** 9 pp 71–80
[18] Miller-hooks E, Zhang X and Faturechi R 2012 Measuring and maximizing resilience of freight transportation networks *Comput. Oper. Res.* **39** pp 1633–43
[19] Pant R, Barker K and Landers T L 2015 Dynamic impacts of commodity flow disruptions in inland waterway networks *Comput. Ind. Eng* **89** pp 137–49
[20] Di M, Lai M and Zuddas P 2013 Maritime repositioning of empty containers under uncertain port disruptions q *Comput. Ind. Eng.* **64** pp 827–37
[21] Liu C, Zheng L and Zhang C 2016 Behavior perception-based disruption models for berth allocation and quay crane assignment problems *Comput. Ind. Eng.* **97** pp 258–75
[22] Fialkoff M R, Omitaomu O A, Peterson S K and Tuttle M A 2017 Using geographic information science to evaluate legal restrictions on freight transportation routing in disruptive scenarios *Int. J. Crit. Infrastruct. Prot.* **17** pp 60–74
[23] Snyder L V and Daskin M S 2005 Reliability Models for Facility Location : The Expected Failure Cost Case Reliability Models for Facility Location : The Expected Failure Cost Case *Transp. Sci.* **39** 3 pp 400–16
[24] Starita S, Scaparra M P and Hanley J R O 2016 A dynamic model for road protection against flooding *J. Oper. Res. Soc.*
[25] Chen L and Miller-hooks E 2012 Resilience : An Indicator of Recovery Capability in Intermodal Freight Transport *Transp. sc.* **46** 1 pp 109–23
[26] Udenta F C, Jha M K, Mishra S and Maji A 2013 Srategies to Improve the Efficiency of a Multimodal Interdependent Transportation System in Disasters *Procedia - Soc. Behav. Sci.* **104** pp 805–14
[27] Sun Y and Schonfeld P 2016 Holding decisions for correlated vehicle arrivals at intermodal freight transfer terminals *Transp. Res. Part B* **90** pp 218–40
[28] Jiang L, Wang H and Ding B 2013 Disruption Management Recovery Model of Distribution Delay with Service Priority *Asian Soc. Sci.* **9** 2 pp 170–79
[29] Ahmadi-javid A and Seddighi A H 2013 A location-routing problem with disruption risk," *Transp. tion Res. Part E* **53** pp 63–82
[30] Cui J, Zhao M, Li X, Parsafard M and An S 2016 Reliable design of an integrated supply chain with expedited shipments under disruption risks *Transp. Res. Part E* **95** pp 143–63
[31] Hu X, Sun L and Liu L 2013 A PAM approach to handling disruptions in real-time vehicle routing problems Notes : The depot *Decis. Support Syst.* **54** pp 1380–93
[32] Ivanov D, Pavlov A, Pavlov D and Sokolov B 2016 Minimization of disruption-related return flows in the supply chain *Int. J. Prod. Econ.* **3** (12) pp 1–11

IOP Conf. Series: Materials Science and Engineering **337** (2018) 012044 doi:10.1088/1757-899X/337/1/012044

Initial sustainability assessment of tapioca starch production system in Lake Toba area

Asido Situmorang[1] and Yosef Manik[1*]
[1]Engineering Management Study Program, Faculty of Industrial Technology, Institut Teknologi Del, Laguboti 22381, North Sumatera, Indonesia

*yosef.manik@del.ac.id

Abstract. This study aims to explore to what extent the principles of sustainability have been applied in a tapioca industry located in Lake Toba area and to explore the aspects that open the opportunities for system improvement. In conducting such assessment, we adopted the life-cycle approach using Mass Flow Analysis methods that covers all cassava starch production processes from fresh cassava root till dry cassava starch. The inventory data were collected from the company, in the form of both production record and interviews. From data analysis the authors were able to present a linked flow that describes the production process of tapioca starch that quantifies into the functional unit of one pack marketable tapioca starch weighs 50 kg. In order to produce 50 kg of tapioca, 200 kg cassava root and 800 kg of water are required. This production efficiency translates to 25% yield. This system generates 40 kg of cassava peel, 60 kg of pulp and 850 kg of waste water. For starch drying 208.8 MJ of thermal energy is required in the form of heating fuel. The material flow analysis is employed for impact assessment. Several options in improving the operation are proposed includes utilization of pulp into more valuable co-products, integration of waste treatment plant to enable the use of water recycled from the extraction operation for the washing process, and to application of a waste water treatment system that produces biogas as a renewable energy, which reduces the consumption of fuel in dryer unit.

1. Introduction

Creating a balance between the dimensions of development such as economics, social and environment is a necessity. With the balance of these three pillars will be able to meet the current needs without compromising the fulfillment of the needs of future generations. Development that meets the needs of the present without compromising the ability of future generations to meet their own needs [1]. To achieve this balance requires the concept of sustainability. The term "sustainable development" was first introduced into the international policy debate by the World Conservation Strategy. It became established as a new global paradigm after "Our Common Future", the final report of the Brundtland Commission had been published and the preparatory work for its follow-up conference, and the United Nations Earth Summit 1992 had begun. This indicates that sustainability becomes a shared conversation / vision that will be applied almost in all countries in the world.

Lake Toba region is a natural ecosystem that provides strategic functions of the basic human needs and economic development. Recently, the Government of Indonesia is making every effort to accelerate the development of Lake Toba region into a main tourist destination in the country. Even though tourism has traditionally been considered the main driver for social and economic development for the region, the development of Lake Toba region has is also marked by the rise of industry sectors, particularly the agricultural processing industry. Tapioca industry is one of the agricultural processing industries sector located in the area of Lake Toba. In many studies, tapioca industries are known to be vulnerable to environmental impacts, particularly those related to social and economic aspects. As an important area, sustainability is only possible if the management is based on a friendly approach to environmental management and integrated. Implementation of sustainability principles in strategic policy and operational practice of tapioca industry in Lake Toba area is essential.

Embarking from this reality, the authors consider it important to conduct a sustainability assessment of the starch production system in Lake Toba Area. This study aims to understand the extent to which the principles of sustainability applied in operation management of tapioca industries in the area of Lake Toba that leads to disclose various aspects that have the opportunity for system improvement. This study is a part of a larger agenda in the research roadmap of the Faculty of Industrial Technology at Institut Teknologi Del, which contributes in policy design and development of Lake Toba Area.

2. Methodology

Various tools and methods for assessing and benchmarking environmental impacts of different product system have been developed, ranging from the internationally standard Life Cycle Assessment/LCA [2] up to the contemporary integrated systems modeling framework for Life Cycle Sustainability Assessment/ LCSA [3]. In this study we apply the Material Flow Analysis/MFA technique, which is a life-cycle approach to analyze flows and stock of a coherent group of substance in a flow model, as a tool in assessing the sustainability in particular the environmental impact.

2.1. Goal and Scope

The goal of this Material Flow Analysis (MFA) study is to assess the sustainability of tapioca production system through systematic analysis of energy and material flow in a “gate-to-gate” model. The inventory data were collected from primary data gathered from a tapioca starch industry located Toba Samosir Regency during the period between February and April 2017. The scope of this study covers all stages in the production of tapioca starch ranging from cassava root until packed tapioca starch.

2.2. Functional Unit

The performance of the tapioca production system model is evaluated for the functional unit of one pack marketable tapioca starch weighs 50 kg.

2.3. System Boundary

The system boundary of the tapioca production system models cover the entire unit processes (i.e., gate-to gate) that can be grouped as follow: sieving, washing, cutting, rasping, extraction, separation, peeling, drying, shifting, and packaging (Figure 1).

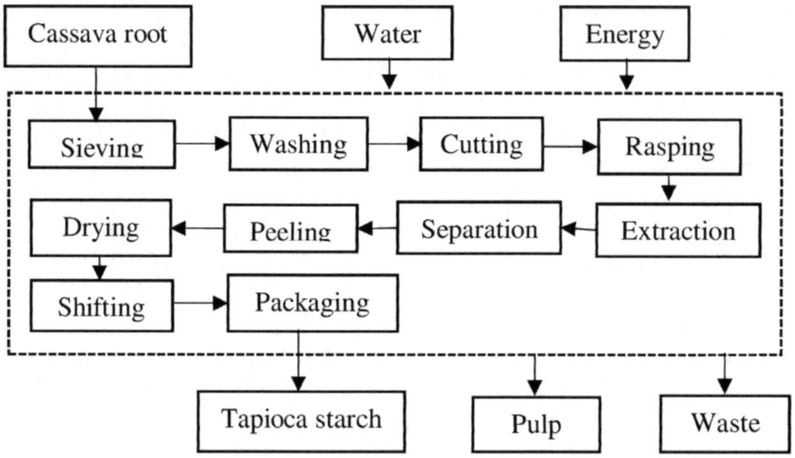

Figure 1. System boundary of the study

3. Data and Analysis

In this study, a comprehensive analysis was implemented covering all stages of tapioca starch production. The boundary of the system considered here includes the stages of the process directly related to the main product (tapioca starch). The flow model of the system is depicted in Figure 2. The following sections describe each step of the process from sieving to packaging.

Prepared cassava will be flown at loading ramp to dry sieve to be cleaned from peel, impurities in the form of soil and others. At this stage, the peel that produce of 20% of the total fresh cassava roots. The cassava is then cleaned using a washing machine. Washing machine uses constantly drained water. After that, the clean cassava will be brought to the Cutting machine through a conveyor that moves upward.

Cassava is cut into smaller pieces to facilitate the process of dissolution on rasping. Cassava that has been cut into smaller parts will be shredded on rasping machine. In this section, there is addition of water to facilitate the dissolution process and facilitate the flowing of yam porridge to the extractor.

The result of this grater will be extracted on the extractor machine so that the resulting starch is really thick. The ratio between starch and solid waste produced in the extraction process is 1: 1. Solid waste generated in the extraction process will be channeled to fiber press for reduced water content and solid waste processing will be easier. After passing the extractor, thick starch will be channeled to separator. This separator serves to increase the viscosity of starch in the liquid starch by reducing the water content contained in the liquid.

Viscous starch is turned into wet flour by using peeler. Peeler works by absorbing water. In the peeler, fine cloth is utilized to filter the starch that has been separated from the water. Wet flour is flown with steam generated by the boiler. After that, ambient air heated with superheated steam is then used to dry the wet powder. Dried flour still contains flour that has a texture that is not yet smooth and also there are still many lumps. Therefore, the dry flour will be fine sieved through a gap density up to 120 meshes. At the end, smooth flour selected from shifter will be packed of 50 kg weight. The final product has the moisture content of 12% with quality standard in accordance with SNI 01-3451-1994 [4].

The liquid waste generated from all stages of the process has the BOD and COD levels of around 5,000 mg/l and 10,000 mg/l. This waste is then channeled into a waste treatment facility, which technically consists of 8 stages of settling ponds covers an area of 1.5 ha. After the sedimentation process, the BOD and COD of the liquid waste are turned into the levels 150 mg/l and 300 mg/l, respectively.

Figure 2 depicts the linked-flow of the whole process step along with material and energy balance. Summary of the mass balance is given in Table 1.

IOP Conf. Series: Materials Science and Engineering **337** (2018) 012044 doi:10.1088/1757-899X/337/1/012044

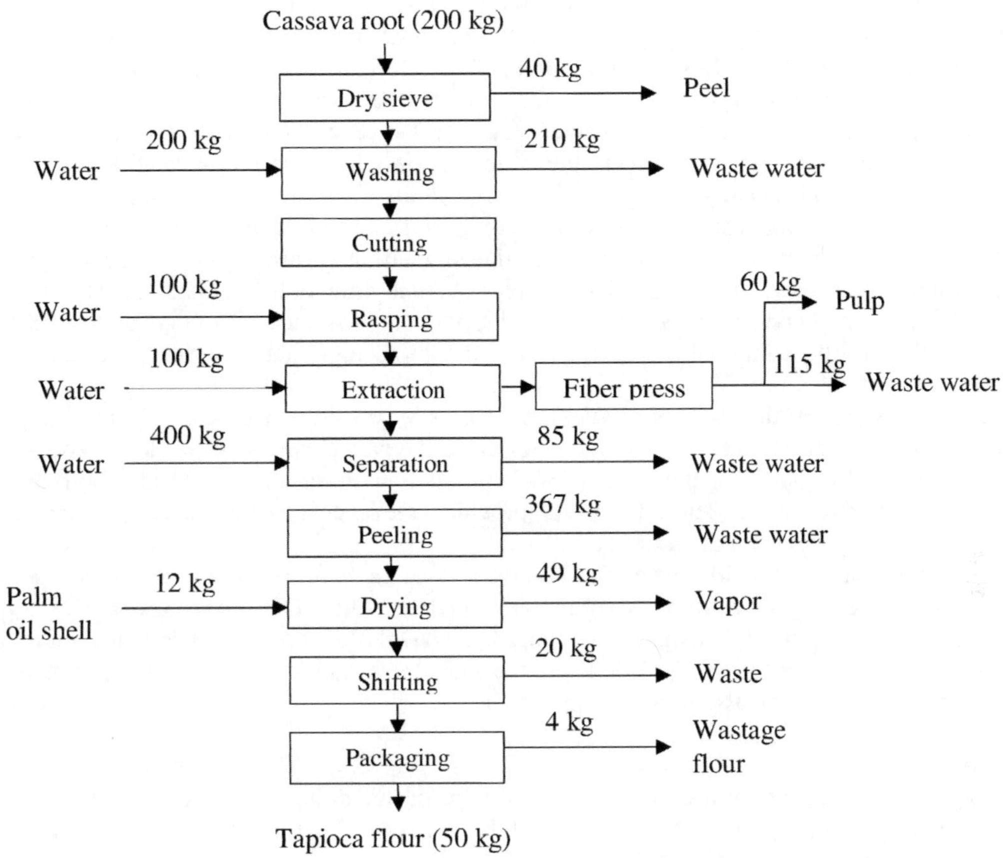

Figure 2. Linked-flow of the production model

4. Results and Discussion

Based on the inventory data explain in Figure. 2, mass and energy flow analysis were able to be summarized throughout the production of 50 kg of tapioca starch in the Table 1. For the drying process, 12 kg of oil palm shell is utilized as fuel for the boiler. Each kg of the shell is converted to 17.4 MJ / kg [5]. So that in producing 50 kg of tapioca spend energy equal to 208.8 MJ.

The productivity level of a tapioca company can be assessed from the obtained yield level. The results of the analysis show that companies in the Lake Toba region produce a yield percentage of 25%. Compared to the average yield of tapioca in Indonesia which is between 25% and 36% [6][7], the yield is considered low. The low yield is mainly attributed to the quality of raw material, which can be the factor of the variety of the cassava, the age of cassava at the time it is harvested, or conditions that leads to decrease in starch levels before entering the processing stage. [8].

Table 1. Mass balance of the production model

Inputs (kg)			**Outputs (kg)**		
1.	Cassava root	200	1.	Tapioca starch	50
2.	Water	800	2.	Cassava peel	40
			3.	Pulp	60
			4.	Waste Water	850

The company has been treating liquid waste through a sedimentation process prior to discharging it into the water stream to BOD level of 150 mg/l and COD 300 mg/l. Compared with the limits of Industrial Waste water that is BOD 200 mg/l and COD 400 mg/l [9], the company simply shows the quality of processed complies with the quality standard of waste water. This indicates that the processed liquid waste of the company is relatively safe. However, any failure in water treatment will be harmful to the ecosystem, considering the water stream where the post-treated waste is disposed empties directly to Lake Toba.

Several options in improving the system performance are proposed as follow. As seen on Table 1, out of 50 kg final product, 60 kg of pulp is co-produced. Currently, the pulp co-products are being sold to the community as animal feed at very low prices. Actually, the company can earn more revenue if they could create higher-value products out of the pulp, for instance as a food ingredient to make bread, substituting the commonly used flour into the basic ingredients of the cake, or other food varieties.

The use of water in the tapioca production process is considered excessive due to the lack of recycling system usage. There is opportunity to reduce the water use in the washing process. One of the possibilities is by increasing the speed of water into the washing section directly about the cassava root coming from dry sieve. This effort will provide kinetic energy that is able to accelerate the process of removing soil from the root peel.

In the processing of liquid waste, the company uses the sedimentation technique alone. This technique is not very effective in reducing the level of BOD and COD of the company. Waste water treatment can be improved by starting it with the coagulation process and then carried out the adsorption process with activated carbon. At the end, the water can be re-used primarily for the washing process, which can also minimize the water use.

5. Conclusion

This study has been able to assess some aspects on the sustainability of tapioca production system in Lake Toba area by applying is a life-cycle approach to analyze flows and stock of a coherent group of substance in a flow model. This study discloses a wide-range of shortcomings in the application of sustainability in the case study area, in particular those related to productivity (Profit) and environmental impact (Planet). However, rooms for improvement are also identified to help the stakeholders improve the system performance for cleaner production. Further research is needed to perform the impact assessment into a more detailed impact categories, such as global warming potential, eco-toxicity potential, etc. as well as to perform the assessment on the social aspects (People), which has not yet been incorporated in this study.

References

[1] World Commission on Environment and Development (WCED) 1987 *Our Common Future* (Oxford: Oxford University Press)

[2] ISO 14040:2006. *Environmental management—Life cycle assessment—Principles and framework*. 20 pages.

[3] Halog A and Manik Y 2016 Life Cycle Sustainability Assessments *Encyclopedia of Inorganic and Bioinorganic Chemistry* (John Wiley & Sons) pp 1–17

[4] SNI 1994 *Tepung Tapioka SNI 01-34511994-1994* (Jakarta: Badan Standarisasi Nasional)

[5] Kaválek M, Havrland B, Pecen J, Ivanova T and Hutla P 2013. Oil palm shell use as alternative biofuel *Agronomy Research* **11** pp 183–188

[6] Balitbang Pertanian 2008 *Teknologi Budidaya Ubi kayu*

[7] Howeler R 2006 *Cassava in Asia: Trends in Cassava Production, Processing and Marketing.* Workshop on Partnership in Modern Science to Develop a Strong Cassava Commercial Sector in Africa and Appropriate Varieties by 2020, Bellagio, Italy, May 2-6, 2006. 38 pages

[8] Wills R B H, Lee T H, Graham D, McGlason W B and Hall E G 2005 *Postharvest: An introduction to the Physiology and Handling of Fruit and Vegetables.* 2nd Ed. (AVI Publ. Co)

[9] Keputusan Menteri Lingkungan Hidup No. 5 Tahun 2014 *Batasan Air Limbah untuk Industri Kepmen LH No. KEP-51/MENLH/10/1995*

Valuating Indonesian upstream oil management scenario through system dynamics modelling

I Ketut Gunarta[1*] and F A Putri [1]
[1]Department of Industrial Engineering, Institut Teknologi Sepuluh Nopember, Kampus ITS Sukolilo-Surabaya 60111, Indonesia

[*]ik.gunarta@gmail.com

Abstract. Under the existing regulation in Constitution Number 22 Year 2001 (UU No 22 Tahun 2001), Production Sharing Contract (PSC) continues to be the scenario in conducting oil and gas upstream mining activities as the previous regulation (UU No. 8 Tahun 1971). Because of the high costs and risks in upstream mining activities, the contractors are dominated by foreign companies, meanwhile National Oil Company (NOC) doesn't act much. The domination of foreign contractor companies also warned Indonesia in several issues addressing to energy independence and energy security. Therefore, to achieve the goals of energy which is independence and security, there need to be a revision in upstream oil activities regulating scenario. The scenarios will be comparing the current scenario, which is PSC, with the "full concession" scenario for National Oil Company (NOC) in managing oil upstream mining activities. Both scenario will be modelled using System Dynamics methodology and assessed furthermore using financial valuation method of income approach. Under the 2 scenarios, the author will compare which scenario is better for upstream oil management in reaching the goals mentioned before and more profitable in financial aspect. From the simulation, it is gathered that concession scenario offers better option than PSC in reaching energy independence and energy security.

1. Introduction

The domination of oil, coal, and gas happened in Asia – Pacific region, including Indonesia, which still mainly rely on those resources. Oil dominates the usage of energy in Indonesia for around 32%, followed by coal for 23% and natural gas for 13% in 2014. Industrial sectors dominated the energy usage for 48%, followed by transportation sectors for 35%, household sectors for 11%, commercial sectors 4% and other sectors 2%.

All the energy needs are projected to be rising for the years ahead due to the growth of Indonesian Gross Domestic Product (GDP) which is composed by the energy user sectors. With the basic scenario that there will be 6% increase in GDP per year since 2014, energy needs will increase for 5.8 times in 2050. The needs of oil will increase up to 40.7% in 2050 due to the high usage especially in transportation sector. Meanwhile, the needs of natural gas will only increase up to 13.8% in 2050. The certain increase in energy needs, especially oil and gas, unfortunately not followed by Indonesia's ability in fully supplying both oil and gas. Indonesia has been facing the supply deficit.

Since 2008, Indonesia could only produce oil and gas for 977 thousand oil barrels/day and continues to decline until 786 thousand oil barrels/day in 2015 [1] . Indonesia's oil production declined for 3.9% annually since 2009. To fulfil the oil demand, Indonesia had to import both oil both in crude and refined forms, from several countries. In 2013, Indonesia imported 355.61 million barrels of crude oil, 237.41 million barrels refined oil, and 41.11 million barrels of gas. Indonesia, which became the 30th country with great oil reserves, even imported oil from South Korea and Taiwan which became the 94th country with great oil reserves [2]. With the increase in oil and gas demand, as the consequences, the amount of oil and gas import also increases from 122 million barrels in 2014 to 1.235 million barrels in 2050 [1]. Indonesia has turned to the net importer of oil and gas since 2004, even Indonesia used to be the net exporter of oil and became one of OPEC (Organization of Petroleum Exporter Countries) members.

The decrease in oil production caused by various factors. The upstream oil and gas mining activities is still concentrated on revitalizing the mature oil fields. Mature oil fields have no longer produce oil economically, even there's optimization effort on those fields using the water injection or Enhanced Oil Recovery (EOR). The most contributing factor is the lack of exploration activities of new oil and gas reserve in Indonesia. The new fact that 60% Indonesia's potential oil reserves located in Indonesian offshore deep-water and requires an advanced technology with high investment costs also caused the low rate of oil exploration and furthermore other upstream mining activities.

Since 1971, the upstream oil and gas mining activities are regulated by Constitution Number 8 Year 1971 (UU No 8 Tahun 1971) using Production Sharing Contract (PSC) scheme. Indonesian government act as the owner of all oil fields in Indonesia. National Oil Company (NOC) own the oil fields but doesn't do the upstream mining activities. The upstream mining activities are done by the contractors who signed the contract (Kontrak Kerja Sama/KKS) with SKK Migas as the regulator of oil and gas mining activities in Indonesia. Under the KKS, the ownership of oil fields held by Indonesian government from the beginning until the end of upstream mining activities. The production could only occur when the oil fields are proven economically-produced by the government. After production, there will be result sharing between the contractor and the government with specified percentage as stated in KKS. The KKS contractor must provide the whole initial investment costs for upstream mining activities but there is cost recovery scheme that will reimburse the "costs of oil" after the fields produce economically. When the fields can't produce economically, or even the exploration fails, the risks of loss will be owned by the contractors. Under the existing regulation in Constitution Number 22 Year 2001 (UU No 22 Tahun 2001), PSC continues to be the scheme in conducting oil and gas upstream mining activities. Because of the high costs and risks in upstream mining activities, the contractors are dominated by foreign companies, meanwhile NOC doesn't act much. The domination of foreign contractor companies also warned Indonesia in several issues addressing to energy independence, energy security, and energy equality.

Therefore, to achieve the goals of energy which is independence and security, there need to be a revision in upstream oil activities regulating scheme. The scenarios will be comparing the current scheme, which is PSC, with the "full concession" scenario for state-owned oil company in managing oil upstream mining activities. Both scenario will be modelled using System Dynamics methodology and assessed furthermore using financial valuation method of income approach.

Using the System Dynamics methodology and financial valuation, the author will compare which scenario is better for oil and gas management in reaching the goals mentioned before and more profitable in financial aspect. The chosen scenario will furthermore be the recommendation for the regulator in oil and gas management in Indonesia.

2. Theoretical Framework

This chapter contains the basic theory and method utilized in solving the formulated problems. The literature review consists of oil and gas management overview, oil and gas industry, system dynamics modelling, and financial analysis.

2.1. Oil and Gas Management

Oil and gas management in Indonesia have existed for more than 130 years, started since the first discovery of oil in North Sumatra in 1885 during Dutch colonialism era that makes the management of oil and gas industry done by Dutch companies. Since then Indonesia has been running various management model to accommodate oil and gas production and consumption for all Indonesian people. The history of oil and gas management is divided into 3 eras those are colonialism era, post-Independence era, and modern industrial era.

After the independence of Indonesia August 17, 1945 the government release several constitutions about oil and gas. The first constitution "UU No. 40 Tahun 1960" about oil and gas mining, was not only the previous concession system but also holding the sovereignty of oil and gas resources but also stating that the oil, gas and mineral located in Indonesian legal mining area are national wealth managed by country and oil and gas mining only undertaken by country and only conducted by national company. In order to support the constitution, the government established three national companies named Permina, Pertamin, and Permigas. Besides doing oil and gas mining activity, those companies also supervise the foreign oil companies' operation. As the adjustment to new regulation became tough, Pan – American Oil made a deal with the government in 1962 with the terms and condition more compatible with UU No. 40 Tahun 1960. This accord furthermore became the model for the three foreign oil companies in doing operational changes in Indonesia. Caltex, Stanvac, and Shell signed the accord in September 1963, named "Kontrak Karya", and became the new oil and gas management concept in Indonesia. As the foreign companies no longer holding the status as concession holder, the management is still held by them. In the implementation of this contract, there's almost no significant changes as the government role only limited to supervision and the foreign companies still doing operation as the previous system (concession). After G30S/PKI occurred in Indonesia in 1965, new regime ruled in Indonesia and led by Soeharto, changing the previous regime that was ruled by Soekarno. Permina then ruled by Dr. Ibnu Sutowo who criticized Kontrak Karya, pushing the new system where both government and contractor taking role in oil and gas management. The big foreign companies once again couldn't accept this new change and unready to release the management right as the terms and condition in PSC. This condition attracted smaller foreign oil company, Independence Indonesian American Oil Company (IIAPCO), to operate back in Indonesia in 1966. After 2 months negotiation, IIAPCO signed Production Sharing Contract (PSC) with Permina, and became the first PSC agreement in the history world oil and gas industry. Japan Petroleum Exploration Company (Japex), Refining Associates Ltd. Canada (REFICAN), Kyushu Oil Development Company Ltd., and Asamera Oil Indonesia Ltd. moreover following IIAPCO's footstep in signing PSC contract.

Behind the great achievement of PSC, the legality of PSC was still questioned due to the absence of regulation stating about PSC, since UU No. 44 Tahun 1960 only recognizing Kontrak Karya. Meanwhile, the other regulation, UU No. 1 Tahun 1967 about foreign investment (Penanaman Modal Asing/PMA), only regulating foreign investment about miningsector except oil and gas. Then in 1971, the government released UU No. 8 Tahun 1971 about Pertamina which become the legal foundation of PSC. PSC was stated in article 12 verse 1 and verse 2. The PSC development is divided into 3 phases as shown in table 1.

Table 1. Comparison between PSC Generations in Indonesia

	PSC First Gen	PSC Second Gen	PSC Third Gen
First Tranche Petroleum	None	None	20%
Cost Recovery Ceiling	40%	100% (no ceiling)	80% (due to FTP)
Investment Credit		20%	17% to 20%
DMO	DMO was defined as 25% of equity oil at 0.2$/barrel	25% of equity oil, full price for the first 60 months and 0.2 $/barrel there after	25% of equity oil, full price for the first 60 months and 10% of export price there after
Equity to be Split			
Government : Contractor	65%:35%	85%:15%	85%:15%
Oil : Gas	NA	70%:30% or 65%:35%	70%:30% or 65%:35%

2.2 Indonesia's Oil and Gas Reserves

Oil and gas industrial activities have produced various vocabularies related to reserves. Collaboration of several oil and gas institutions in the world (SPE, AAGP, WPC, SPEE, and SEG) releases oil and gas resource management system called Petroleum Resources Management System (PRMS) with the latest edition released in November, 2011. PRMS is the integrated system for basic classification and categorization of petroleum reserves and resources. PRMS divided each project to 3 main classes called Reserves (satisfied commerciality), Contingent Resources (satisfied sub-commerciality), and Prospective Resources (undiscovered). The illustration of PRMS is provided in figure 1. To fulfill the commerciality terms, the amount of project reserves can be estimated. Based on this classification, reserves are classified into 3 kinds, those are proved (1P), probable (2P), and possible or potential (3P). Where the uncertainty ranges from low estimation (1C), best estimation (2C), and highest estimation (3C).

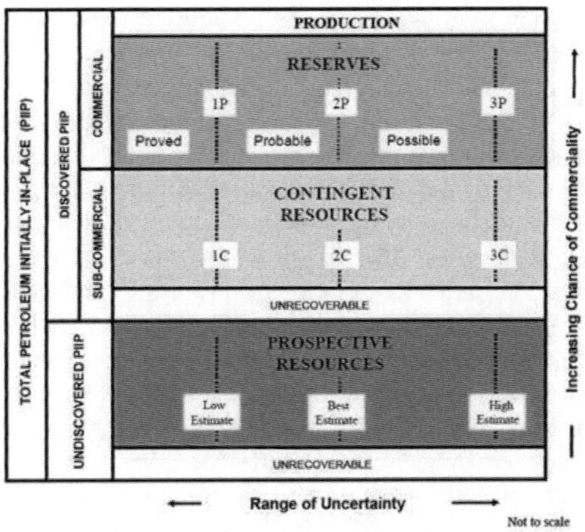

Figure 1. Reserve and Resource Classification Framework in PRMS 2011

Indonesian potential oil reserves have been decreasing slightly each year from 2010 to 2015. While the proven oil reserves declined gradually since 2010 to 2015. This phenomenon caused the almost stable total oil reserves from 2010 to 2015. For several years coming, the demand of oil and gas are projected to increase. The demand of oil will increase 3 times compared to 2014 (BPPT, 2016). Meanwhile, oil production is projected to decrease continuously until 52 million barrels in 2050.

Despite the rising demand of oil, the oil production showed conversely performance. Oil production has never exceeded oil consumption since 2010, which was only 1003 thousand barrels/day. Oil production continued to decrease until it reached 852 thousand barrels/day in 2014 with the decease rate of 5% per year (BPPT, 2016). Production raised slightly to 876 thousand barrels/day in 2015, yet it still couldn't fulfil the oil demand in 2015. As the consequence of declining production and demand of oil and gas, the government allowed importing oil and gas. Amount of oil import is projected to increase to 933 million barrels in 2050 (basic scenario) and 1,235 million barrels (high scenario).

2.3 System Dynamics Modelling

System dynamics has been known as the systemic thinking discipline in system science. This methodology was firstly introduced by Jay Forrester from Massachusetts Institute of Technology (MIT) in 1963 and has been improved continuously. System dynamics was used limitedly as industrial problem-solving tool in its early emergence. In its development, system dynamics has been used for analyzing and solving social, economic, managerial, political, and environmental issues which is complex, dynamic (fast-changing), and uncertain.

There are 3 common perspectives in analyzing system, those are black box system (input-output analysis and econometrics), grey box system (operation research), and white box system (soft-system methodology, viable system, hyper-game, and system dynamics). System dynamic offers comprehensive and integrated thinking which can simplify complexity without losing the essential object and also suitable in analyzing mechanisms, pattern, and system tendency based on analysis of system's complex, dynamic, and uncertain structure and behaviour. Studying system comprehensively doesn't mean that there's no boundary between the system and environment. Non-significantly influential variables will be the boundary in system analysis and causing the closed system.

A system which has causal loops can't be analyzed partially and this causes system dynamics better than partial analysis in analyzing system that has causal loops. Causal loop diagram illustrates the causal relationships between elements using inter-related arrows so it creates causal loop where the beginning of arrow means the cause and the end of arrow means the effect. Causal loop diagram is constructed with many elements involved in it. Therefore, before simulating the system, causal loop diagram is often used as the conceptual mapping of the system to simplify complexity. Causal loop diagram must be made as real as it could be to illustrate the nature of the studied system.

To understand the behavior of system, this research conducts using a model instead of doing trial-and-error research. Model is the representation of system in the real world by emphasizing the main elements of the system. Building a model could be cheaper and time-saving rather than doing trial-and-error study in the real system. Model can be made in conceptual form and simulation form. Conceptual model is a non-software specific description of the simulation model that is to be developed, describing the objectives, inputs, outputs, content, assumptions and simplifications of the model. Meanwhile, simulation model can be made through system dynamics software, such as Vensim, Powersim, and Stella © (iSee System). Stella © (iSee System) could build the simulation model visually using computer and could see the behavior of system quickly. Simulation conducted through procedures as follows: (1) Defining objective, scope and requirements (2) Collect and analyze system data (3) Build the model (4) Validate the Model (5) Conduct experiment (6) Result Model verification is the process to determine whether the simulation model correctly reflects the conceptual model. Testing is conducted to verify whether the simulation results absolutely predictable outcomes based on test data. Simulation models often generate slightly different result depending on configuration and input data. Verification is the process of debugging the model, seeking for the bug-free model.

Meanwhile model validation is the process of establishing confidence in the soundness and usefulness of a model [4]. Verification and validation must be conducted simultaneously throughout the model development process [4]. Several testing methods are required for validating the simulation model, including the usage of statistical testing techniques. Some of the testing methods by Forrester and Senge (1980) are Structure-Verification Testing, Parameter-Verification Testing, Boundary-Adequacy Testing, Sensitivity Testing, and Extreme Conditions Testing.

2.4 Financial Valuation

From the financial aspect, value means the sum of economic benefits expected in the future by the owner, where each benefit is discounted to the present value with the adequate discount rate [5]. Value of a company (Market Value of Invested Capital/MVIC) is the capital market value which is the sum of liability value and equity value where equity value comes from the value of equity plus the market added value. American Society of Appraisers divided valuation to 3 approaches, those are market approach, asset-based approach, and income approach. Income approach is used in this research.

Income approach is conducted based on anticipation principle where business interest defined as the sum of economic benefits generated by that business interest in the future. Company value is estimated by quantifying income flow generated by investments and converted to present value by opportunity cost of capital. Income approach is divided into Discounted Cash Flow (DCF) method and Capitalization method.

2.4.1 Discounted Cash Flow Method. Discounted cash flow uses projection of all economic benefits in the future such as free cash flows or other income variable and discounting each benefit to present value with a particular discount rate that reflects the cost of investment capital. DCF method uses time value of money theory. DCF method also based on principles that the value of investment determined by its ability in generating future cash flows, a new investment is valuable if that investment could give bigger return than the costs incurred for that investment, and the value of assets is present value of expected cash flows from that asset investment in the future. The basic formula used in DCF is shown below:

$$PV=\sum_{i=1}^{n}\frac{E_i}{(1+k)^i} \quad (1)$$

where PV = present value
E_i = economic income expected in period-i
k = discount rate
i = period in the future where the economic income will be received

Economic benefit meaning can be various depending on the valuation object and purpose of valuation. Generally, economic benefits are dividends, net profit, Earnings Before Interests and Taxes (EBIT), Earnings Before Interests Taxes Depreciation and Amortization (EBITDA), Net Income After Tax (NIAT), and net cash flow to equity. Net cash flow to equity's formula is shown below.

$$NCFe = NIAT + NCC - CAPEX - NCWC + NCLTD \quad (2)$$

Where NIAT = Net Income After Tax
NCC = Non-Cash Charges (Depreciation, Amortization, and Retained Tax)
CAPEX = Capital Expenditures
NCWC = Changes in Net Working Capital
NCLTD = Net Changes in Long-Term Debt

Discount rate defined as the return percentage to convert the amount of money that is expected in the future to present value. Discount rate can be in the form of Minimum Attractive Rate of Return (MARR) or Weighted Average Cost of Capital (WACC).

2.4.2 Direct Capitalization Method. In direct capitalization method, capitalization rate is more comprehensive method where the capitalization rate only changes one single income flow to present value.

3. Research Methodology

This chapter contains the methodology that is used in conducting the research in form of flowchart and also the explanation of methodology flowchart.

3.1. Data Collection and Processing Phase

Data collection is done paralelly with literature study. All data which is related to the research, such as oil and gas reserves, oil and gas consumption, and also oil and gas production are collected for supporting the research. In this phase, dynamic hypothesis is conducted to conceptually modelling the system using Causal Loop Diagram to map the problem and understanding the relation between stakeholders. Furthermore, the alternative scenarios for oil and gas management in Indonesia must be determined as comparison for the system improvement.

3.2. System Modelling Phase

The Causal Loop Diagram (CLD) that has been made in previous phase can't be simulated directly to produce the result. CLD elements must be transformed into stock, rate, and converter to convert the CLD into Stock and Flow Diagram. STELLA is one of the system dynamics modelling software. Stock and Flow Diagram furthermore needs to be simulated to gain the result.

3.3. Model Testing
Simulation model must be credible in order to represent the real system correctly. Every model used in simulation must be verified and validated to test the credibility of the model. Verification is the process to check whether the operational logic of the model compatible with the logic in flow diagram. Meanwhile validation is the process to determine whether the model is the accurate representation of the real world. Validation is done through 5 methods, those are model structure testing, boundary sufficiency testing, model parameter testing, mean comparison testing, and extreme condition testing.

3.4. Scenario Development and Financial Valuation
After verifying and validating the simulation model, as the improvement of the system, the better resulted scenario must be developed. Scenario will be developed through financial valuation that will assess the value of each scenario in achieving the goals of energy independence, energy security, and energy equity.

4. Simulation Design & Output Analysis
This chapter contains the simulation results for both before concession scenario and after concession scenario.

4.1 Dynamic Hypothesis
To understand the dynamics of the system, the historical data that related to the system are collected. This final project set time horizon for 20 years, which consists of 10 years historical data (since 2005 – 2015) and 10 years future projection (started from 2015 – 2025). The data then led to theory development, called dynamic hypothesis, to account for the problematic behavior of the system. Dynamic hypothesis consists of variable identification of the system and conceptual modelling using Causal-Loop Diagram to show the inter correlation of the system elements.

4.1.1 Variable Identification. All variables that are related to oil management system must be identified to understand the function and relation of each variable in affecting the work of the system. Oil management system, in this case, will be divided into 5 sub models to understand the role of the variables in the system. The sub models are Production, Consumption, Oil Market, Investments, and Income.

4.1.2 Causal-Loop Diagram. Causal-Loop Diagram is a tool for conceptual modelling used to illustrate the causal relationships between elements using inter-related arrows so it creates causal loop where the beginning of arrow means the cause and the end of arrow means the effect. Causal-loop diagram used in this research is provided in appendix A.

4.2 Stock and Flow Diagram (Main Model). Main model is used as a basic guidance in developing the Stock and Flow Diagram. The relation between the entities can be shown in the figure 2.

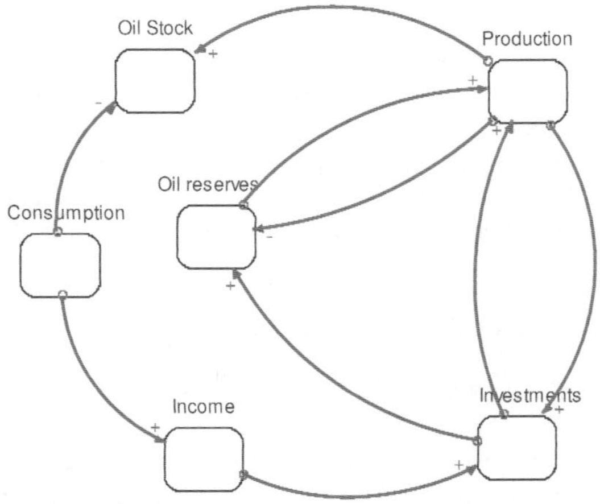

Figure 2. Main Model of Indonesian Oil Management

Production of oil will add the amount of oil stock available in the market. The more production will add more oil stock available, shown in positive relation of both entities. However, due to consumption, this activity will reduce the amount of oil stock. The more consumption occurs, the less oil stock available in the market. Then, when consumption occurs, the consumers will spend funding to consume oil. Consumption activity will generate income for oil producer (in this term NOC) and moreover for the government (due to the ownership of NOC by government, the income of NOC will be shared to government too). The more income that NOC and government have, the more investment in oil upstream mining activity will occur. When there are more investments, there will be more oil upstream mining activity such as exploration that will reveal new oil reserves both onshore and offshore. The more oil reserves available may increase oil production activity. More investments also led to more production activity due to the availability of fund to produce oil. Finally, the bigger oil production will result in bigger oil stock available for the market.

4.3 Verification and Validation

In the software, verification can be done by 2 processes, those are units checking and verify/repair model menu. Units checking is conducted to know whether all units used in the model are consistent or not. Units consistency is a crucial thing in simulation model, otherwise the simulation will not be able to run. After units checking, verification can be done through verify/repair model menu in the software to see whether there are any errors in the model and to see whether the total count of all entities in the model. From the verification process, there is no error found in the model. Therefore, it can be concluded that the model is already verified or it has reflected the real system.

Model validation is the process of establishing confidence in the soundness and usefulness of a model [4]. Validation aims to know whether the simulation result is appropriate with the modelled process. Model can be considered as good when its simulation result has relatively small mistakes.

Structure assessment testing is done through interview with the system owner which is NOC. All variables in each sub model has been assessed and considered to be valid qualitatively. From the parameter testing, according to the simulation output of sub model oil market, as the production rate declines (due to the decline of oil production), the amount of oil stock also declines as the consumption rate that consumes oil increases exponentially. The oil stock then fulfilled by crude oil import activity, therefore the amount could fulfil the demand. it is proven that the variables in the model are consistent with the logic of the variables.

4.4 Scenario Development

The difference between before concession and after concession scenario lies in the amount of potential oil reserves. In sub model production, amount of potential oil reserves is determined by the addition of new potential oil reserves explored from exploration activity. Potential oil found will determine the amount of proven oil and furthermore the oil production. In the 2nd scenario, there's an addition of new potential oil reserves that will be added when concession scenario is applied.

4.5 Simulation Result

The highlighted result of the simulation is summarized in the table below.

Table 2. Simulation Result of Existing Scenario and Concession Scenario

	Existing Scenario	Concession Scenario	Difference (Concession vs. Existing)
Potential Oil Found (barrels)	2.956.019.832,53	2.956.019.832,53	0%
Proven Oil (barrels)	614.022.158,89	19.646.632.184,62	3100%
Oil Stock (barrels)	61.629.763,42	1.671.463.169,02	2612%
Gross Income (USD)	2.894.996.887,11	92.629.642.736,88	3100%
Retained Earnings (USD)	45.109.669.504,23	226.913.830.076,09	403%
NPV (USD)	578.888.611.428,69	843.904.408.816,51	46%

After the simulation of both scenario, it can be interpreted that the concession scenario offers better option for upstream oil management in Indonesia because it resulted in the distinctive difference to achieve the goal of energy independence and energy security. Concession scenario will also strengthen the value of NOC to help NOC in doing upstream oil mining activity such as investment for finding new oil reserves.

5. Conclusion

The conclusions of this research are as follows:

1. Concession scenario offers the better result than the existing scenario. Addressing to energy independence goal, when NOC is given the concession, NOC could earn more proven oil due to the acquisition of oil wells. The amount of proven oil handled by NOC could up to 19.646.632.184,62 barrels, much higher than the existing scenario which is 614.022.158,89 barrels. There will be more oil wells managed by NOC, such as through the takeover of Mahakam Block, ONWJ Block, Sanga – Sanga Block (under discussion), and East Kalimantan Block (under discussion). When NOC is given more priority to manage oil fields, the more reserves and stock that NOC will handle and the more independent NOC is.
2. The acquisition of oil field will also help NOC in securing the amount of oil available in the market (oil stock). Before the concession scenario occurs, the amount of oil stock is 61.629.763,42 barrels. After implementing concession, the amount of oil stock will turn into 1.671.463.169,02 barrels. The higher amount of oil fields managed by NOC will lead to higher amount of production that will result in the higher amount of oil stock. Besides that, with concession scenario, NOC could minimize the amount of oil import because of its ability in fulfilling the demand. Therefore, the goal of energy security in fulfilling domestic oil demand can be accomplished.
3. Under the valuation method, the existing scenario has the value of US$ 578.888.611.428,6. Concession scenario offers the higher value which is US$ 775.492.486.128,98. With 34% difference of value amount between two scenarios, concession scenario offers higher value for NOC (where NOC represents the government). The higher value adds the power of NOC's capital and will add the ability of NOC in doing upstream oil mining operation both domestic and overseas and also the possibility of developing the non-conventional oil.

Appendices
Appendix A Causal Loop Diagram

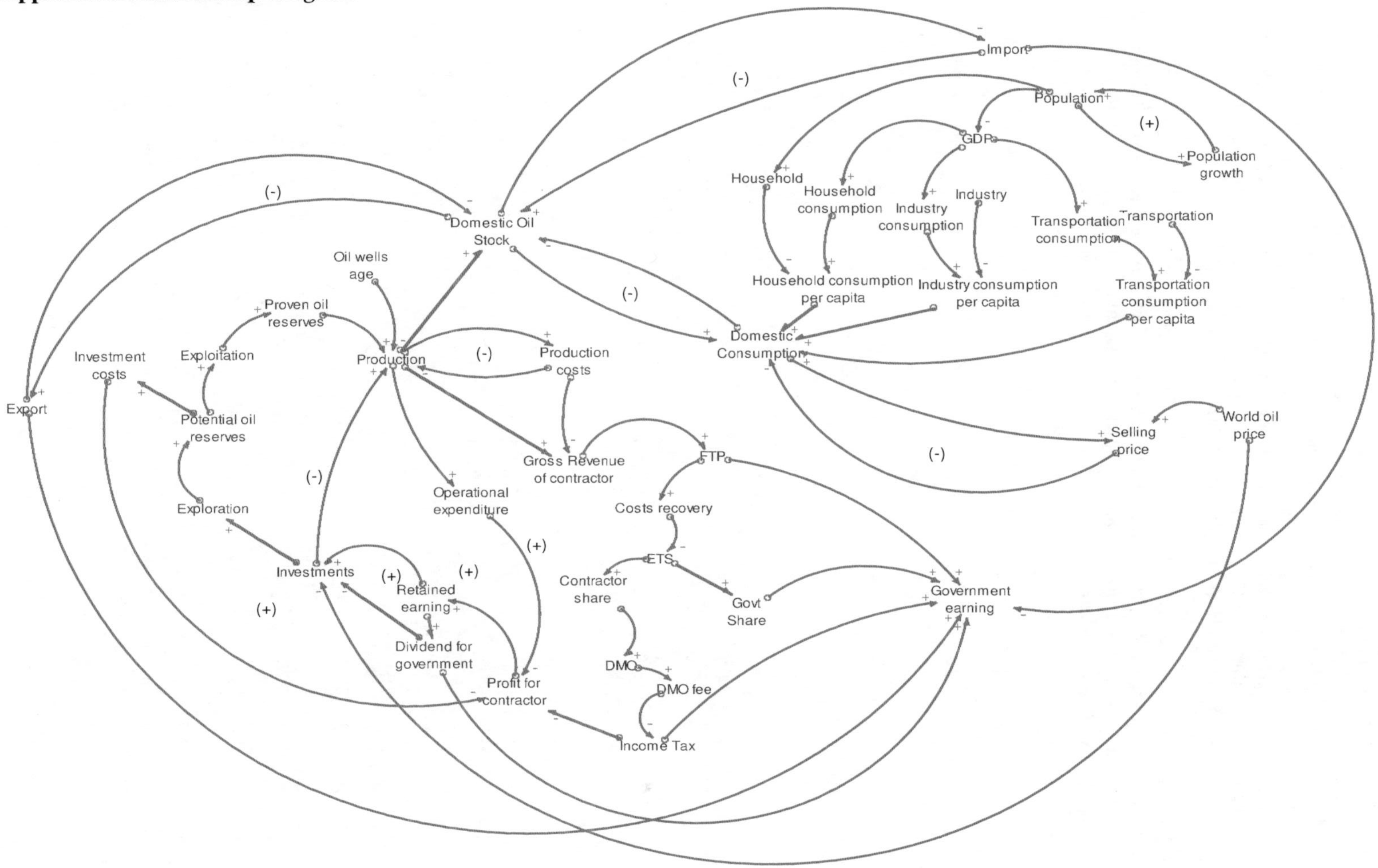

IOP Conf. Series: Materials Science and Engineering **337** (2018) 012045 doi:10.1088/1757-899X/337/1/012045

6. References

[1] SKK Migas 2015 Annual Report 2015 (Jakarta: s n)
[2] Indirasardjana P 2014 *Minyak untuk Presiden* (Jakarta: PT Gramedia)
[3] BPPT 2015 *Indonesia Energy Outlook 2016* (Jakarta: Center for Technology of Energy Resources and Chemical Industry BPPT)
[4] Richardson G P 2016 Model Validation Outlined by Forrester and Senge (New York: Rockefeller College of Public Affairs and Policy University at Albany - State University of New York)
[5] Prawoto A 2004 Penilaian Usaha (Jogjakarta: BPFE Jogjakarta)

Acknowledgments

Authors wishing to acknowledge assistance or encouragement from colleagues, special work by technical staff or financial support from organizations should do so in an unnumbered Acknowledgments section immediately following the last numbered section of the paper.

Policy impact of the Indonesian Central Bank certificate related on loan interest rate to the demand growth of property

B Wirjodirdjo [1*], **H Y Asjari**[2]
[1]Department of Industrial Engineering, Institut Teknologi Sepuluh Nopember, Kampus ITS Sukolilo-Surabaya 60111, Indonesia
[2]Telkom Indonesia, Jakarta, Indonesia

*budisantoso.wirjodirdjo@gmail.com, santoso@ie.its.ac.id

Abstract. The Indonesian economic indicators shown a positive progress in the last three years, Foreign exchange reserves position of the end of March 2017 stood at US $ 121.8 billion higher than the position of the end of 2015 amounted to US $ 105.9 billion of the end of 2015. This reserve would ensure the resilience and maintaining sustainable Indonesian economic growth in the future. Although Indonesia's foreign exchange is better, the structure of expenditure in the country is still less than ideal due the proportion of spending of consumer goods is far greater than the capital goods and tend to be unproductive spending. This needs to be regulated so that in the long term does not cause balance of payments deficit. Therefore, Indonesian Central Bank took a policy to raise interest rates for retail banks from 6% to 7.25% per annum gradually up to present. Policies relating to the interest rates on loans are intended to reduce the proportion of debt financing of consumer goods, however, these policies have implications to various economic sectors and one of those is property sector. A lot of research has been conducted related the impact of loan interest to the property sector but most of it is still in partial related to the ability the people to buy. However, this research has tried to see the implication of the macro Economic Policy of Indonesian Central Bank to the property sector as a systemic problem. This paper is going to present the study on the effects of these policies on the property sector, especially residence house. To obtain a comprehensive analysis and capture the relationship between interest rate policies and their impacts to the property sector, in this study the model developed and simulated using system dynamic methodology as an approach. Various scenarios are applied to the model to get an accurate information about how and when the effectiveness of the policy related to the property sector can be enforced. The result of this study can be delivered to the policy maker as feed-back to improving the policies related the interest rate.

1. Introduction

Property is a strategic and important sector as part of primary needs of human being after food and clothing. In addition, property sectors has a substantial multiplier effect to other sectors as well as supporting evolving industries. This has been the major challenges for Real Estate Indonesia (REI) considering backlog that happened in 2016 has reached 25 million units of residential needs. Time and efforts is needed in order to fulfil the residential needs [1]. Despite the major challenges ahead, property sectors business has been showing positive performance for the last three years for its attempt on fulfilling the residential needs. In 2015, the sectors donated around 28% to the economic growth. Indonesia's property development in 2017 is predicted to increase around 20% from its 2015 performance. Real estate is an interesting sector considering the need of property products keep increasing along with the economic development and population.

In May 2016, credits on flats and apartments grew about 87.4% (year on year), while back in January 2011, credit on this sector grew around 150% with relatively stable residential credit growth. Residential credit growth in 2016 was 16.96%, increased from 14.26% in May 2013 [2]. Currently, credit growth on flats and apartments is higher compare to residential (landed house).

The soaring property prices for the last few years and credit growth in Indonesia has been observed by Bank Indonesia. The central bank has been actively making efforts to anticipate crisis as happened in the United States back in 2008 considering residential property prices increased significantly during 1998-2006. Credit growth rate reach around 12.5% each year in the last eight years. Residential property prices index in 1998 is US 60 and increased to US 160 in 2006. The following year, residential property prices index was down to US 130 in 2008 and keep decreasing ever since. The crisis that was caused by subprime mortgage affected the price bubble in property sectors [3]. Property price bubble is a condition when the property price rose significantly in a short period and if ignored, the bubble bursts causing property prices fall followed by economic breakdown and might causing recession.

In order to address potential issues in the housing credits, the Government (in this case is Bank Indonesia) develop a policy or decree called Loan to Value ("LTV") where the potential properties buyers are expected to contribute down payment of 30% from the total credit, and banks are expected to provide maximum loan of 70%. This was intended to minimize bad debts and asymmetric information. As explained above, credit development in Indonesia is higher than what was expected, however, in the last few months, the property credits have been decreasing. In addition, the worsening of macro-economic condition (e.g. increase of BI Rate to 7.5%, expected inflation of 9.0%, and low economic growth) hinders the development of this sector.

This paper discusses property sector, where this sector relates to other sectors as a systemic reaction. Thus, to develop a banking policy related to the property sector, synchronization between the central bank (lending policy) and the Government (spatial policy) in the provision of housing, so it's not hampered and prudent.

This paper is trying to analyze and model the policies that were done by the central bank, which the Bank associated with the increase in loan interest rates and their impact on the property sector development plan. The methodology approach used is dynamic system methodology in the hope that the simulation result from the developed model can be used to estimate the prospects of the property sector in the future as well as to see the relation regarding the contribution of the property sector in maintaining economic growth.

2. Literature Review

Sertifikat Bank Indonesia (SBI) is the securities issued by Bank Indonesia in recognition of short term debt (1-3 months) with a discount/interest system. SBI is one of the mechanisms used by Bank Indonesia to maintain Rupiah stability. By selling SBIs, Bank Indonesia can absorb the surplus of outstanding primary money. The interest rate applicable to each sale of SBI is determined by market mechanism based on the auction system. Since early July 2005, BI has adopted the BI Rate mechanism (i.e. BI announces the target of SBI interest rate desired by BI for auction at certain period of time). BI Rate is used as a reference by market participants in following the auction.

The benchmark interest rate or BI Rate is a policy rate reflecting the monetary policy stance set by Bank Indonesia and announced to the public. In general, the function of the BI Rate is to regulate the inflation rate in Indonesia. Bank Indonesia will raise the BI Rate if future inflation is predicted to surpass the predetermined target, and on the contrary Bank Indonesia will lower the BI Rate if future inflation is predicted to be below the prescribed target [2].

Interest rates represent a future payment because there was a transfer of money in the past. Interest rates are divided into nominal interest rate and real interest rate. The nominal interest rate is the interest rate which already takes into account the rate of inflation. While the real interest rate is the interest rate that has not taken into account the inflation rate. The real interest rate is the nominal interest rate less the rate of inflation. Real interest rate that is generally used as a reference is a risk-free rate (or BI Rate).

There are several studies related to bank interest rate, for examples are Delis and Kouretas [4] conducted an empirical study of low interest rate tends to increase the banks' risks substantially. However, this could not be applied to all bank, it depends to the banks' characteristics in facing the interest rate impact on reduced asset risk from banks with higher equity capital.

There are other researches that emphasize more on dynamic behavior of a policy impact, such as Hwang et al. [5] who analyzed the effectiveness of mortgage rental policy by reducing the LTV and DTI ratio in 2008 in Korea. The policy is aimed to manage housing demand and stabilize house prices. Similar approaches in relation to dynamic behavior of housing market supply-demand was conducted by Kwoun et al. [6] which concludes that housing supply-demand is affected by macroeconomic conditions. By using system dynamics, the analysis states that this method is proper to be used since it explains the dynamic relation on unsold housing supplies.

Bouchouicha and Ftiti [7], by analyzing the dynamic interaction of real estate market in the United States of America and United Kingdom, also the macroeconomic environment using Dynamic Coherence Function (DFC). The paper provides a point of view that monetary policy is essential to avoid disruptions towards real estate market.

Contrary to research conducted by Ibicioglu and Kapusuzoglu [8] which analyzed the decisions impacts in interest rate policy of central banks on their decision towards investment in the Istanbul Stock Exchange (ISE). By using the impulse-response analysis, they found that the price movement at ISE was an effective market in monetary policy. Therefore, it is necessary to be cautious for any monetary policy that will have impacts on investment decisions especially on price considerations.

Research conducted by Ibicioglu and Kapusuzoglu [8] still leaves some implications of a policy that needs to be further explored and therefore Papadamou and Siriopoulos [9] examined the impacts of the MPC (Monetary Policy Committee) results which have risks on interest rates and life insurance companies in the UK. As a result, there are important implications for finding monetary authorities in aiding the stability development of financial industry through central bank policies.

Rubio and Carrasco-Gallego [10] analyzed the implications from macro-prudential and monetary policies on financial and welfare stabilities. With macro prudential regulations in the form of LTV which able to affects credit growth. Likewise, obtaining the optimal parameters of monetary and macro-prudential policies that are coordinated and not. From this it was found that when there is coordination, the stability of the economy and welfare is able to be maintained.

Based on the literature study above, it is learned that there is a research gap that has not been mentioned by previous researchers related to the complexity of property setbacks which existence is not independent but related to other sectors as systemic relations. Therefore, policies synchronization are fundamental in between the central banks in terms of financing policies with the government in terms of spatial policies, hence that the fulfillment of housing demands is not hampered and still retained prudential principles.

3. Model Conceptualization

Model conceptualization begins by identifying the interacting and affecting variables towards central bank policy system and government within the property sector development. To simplify the identification and modelling, diagrams of input-output and interacting variables are constructed. Then, casual loop diagram and stock and flow diagram are formed towards object of observation.

The identification of related variables that affects the system is collected including variables in the property sector, banking and macroeconomics. Next stage is to compose input-output diagram to describe input-output variable from systemic complications related to central bank and government policies. On the input-output diagram, variables are classified between controllable and uncontrollable input, as well as controllable and uncontrollable output, and the environment. Figure 1 shows the input-output diagram used for this paper.

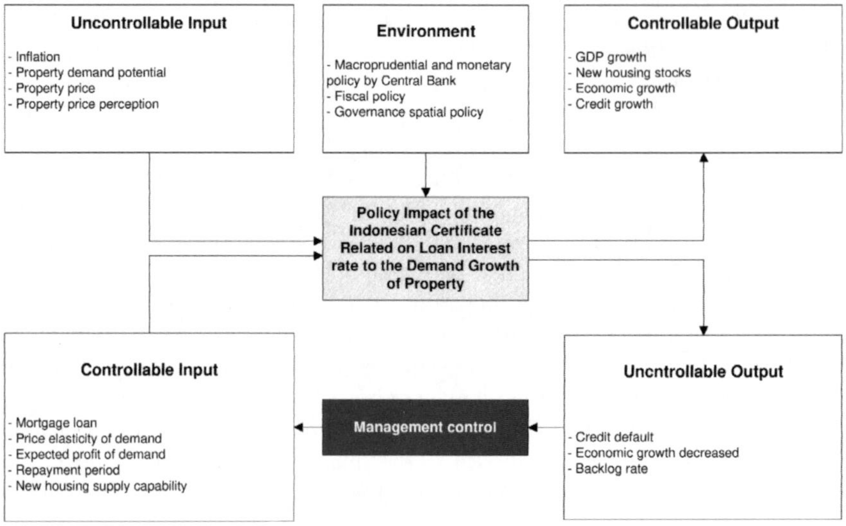

Figure 1. Input Output Diagram

The following stage is the construction of casual loop diagram in order to observe the systemic dependency between main variables on central banks and government policy complications in relation to the development of the property sector as shown in Figure 2.

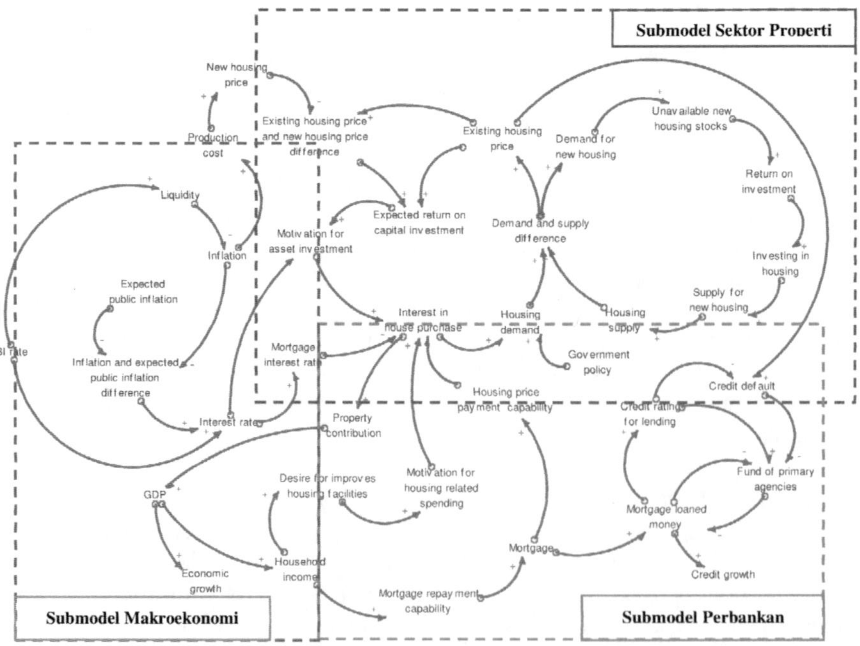

Figure 2. Causal Loop Diagram

Stock and flow diagram is constructed to illustrate the interaction between variables in accordance with logic structure on modeling software being used. The design of stock and flow diagram consider the research objective and must be able to illustrate the impact of policy instrument towards observed system.

The model is divided into 5 sub models called as a modules, each module represents the perspective of central bank and government policy sub-system within the property sector development. Those module are: 1) Supply Demand Landed House, 2) Apartment Supply Demand, 3) Macroeconomic, 4) Banking and 5) Credit default.

4. Policy Scenario Development

The policy scenarios to be taken under possible conditions can be controlled by stakeholders in handling the development of the property sector. In addition, the improvement scenario is determined based on the parameters that affect the performance of the system by testing the sensitivity or extreme test conditions that have been done on the model validation test. These policy scenario are based on the expected future conditions, where there is an increase and decrease in the value of the variables that will be used as scenarios.

- Scenario 1: Increase fraction of Central Bank's Reference Rate, this is policy rate reflecting the monetary policy stance set by Bank Indonesia and announced to the public.
- Scenario 2: Decrease of Maximum Proportion Granted for Mortgage by Banks in the form of allowable loan ratios for credit implementation. This policy is aimed at further improving the prudential aspects of banks in lending property and minimizing default. This scenario is to reduce the proportion of Loan to Value from initially 70% lowered to 50%.
- Scenario 3: Fraction of Land and Building Tax Fee, the aims of this policy to influence the aggregate demand side of an economy in the short run. The scenario is done by raising the tax fraction increased 0.5% from the existing condition.
- Scenario 4: Fraction of Residential Resettlement is intended to optimize land allocation for urban settlement. The scenario is to activate this variable and give contribution value of 10 % for 10 years
- Scenario 5: Proportion of Apartment and Housing Development, this is local government policy in managing land use by arranging the proportion of apartment and housing development to 52:48.
- Scenario 6: Combination of Scenarios, from each of the conditions described scenarios above, there are a combination that may occur between several conditions. It will be arranged for 3 scenarios: pessimistic scenario, moderate scenario and optimistic scenario (improvement scenarios). Table 1 shows the linkage between policy parameters and the expected future conditions.

Table 1. Relation between Parameter & Condition (state) of Policy Scenario

No	Parameter	Future Prospect		
		A. Pessimistic	B. Moderate	C. Optimistic
1	BI Rate	6,00	7,50	8,00
2	LTV	100%	70%	50%
3	Tax	0,25%	0,50%	1,00%
4	Resettlement	0	0	10%
5	Proportion (Apartment : Landed house)	40:60	50:50	60:40

5. Simulation Scenario Result

The pessimistic scenario is built on the following key parameter conditions: 1) the BI Rate fraction is reduced to 6.00% due to lower inflation, 2) the absence of LTV so that the borrower can borrow with 100% credit, 3) the tax rate is reduced to 0.25%, 4) the absence Resettlement, 5) the proportion of apartments: housing at 40:60 as a result of the community's desire for land ownership. It is shown that the pessimistic scenario has an impact on the supply demand, macroeconomic, banking and credit default aspect. In the supply-demand aspect, it appears that the more stocks for housing, the greater the land needed for development. In macroeconomic aspects, there is a decrease in economic growth and consistent increase in per capita income. In the banking aspect, trade-off between credit developments with credit liquidity occurs where an increase in credit development can reduce credit liquidity. Finally for the default credit sub-model¸ development probability occurs so that, this risk also interference with other variables that are affected by this variable.

The moderate scenario is built on the following key parameter conditions: 1) the BI fraction rate remains at 7.50%, 2) the LTC is 70%, 3) the fixed tax rate at 0.50%, 4) the absence of resettlement, 5) no proportion of the construction of the apartment with housing. It is shown that the moderate scenario has an impact on the supply-demand, macroeconomic, banking and credit default aspects. In the supply-demand aspect, it can be seen that low stock level for both apartment and housing, causing the required land is also low. However, from here it can be seen that there are still many unfulfilled requests. In macroeconomic aspects, there is a decrease in economic growth and consistent increase in per capita income. In the banking aspect, trade-off between credit developments with credit liquidity occurs where an increase in credit development can reduce credit liquidity. Finally for the default credit sub-model¸ development probability occurs so that, this risk also interference with other variables that are affected by this variable.

Optimistic scenarios are built on the following key parameters: 1) the fraction of the BI Rate rises to 8.00% due to increased inflation, 2) the LTV is 50%, 3) the tax rate is increased to 1.00%, 4) the resettlement, 5) the proportion of apartments: housing at 60:40 due to the efficiency of land use for the house. It is shown that the optimistic scenario has an impact on the supply demand, macroeconomic, banking and credit default aspects. In the aspect of supply demand, it is seen that the more stock for the apartment, thus causing the efficient land needed for development. The enlargement of stock figures needed for housing. This happens because the transfer to the apartment is done, so it takes encouragement to the community to switch to the apartment. In macroeconomic aspects, there is a decrease in economic growth and consistency in increased per capita income. In the banking aspect, trade-off between credit developments with credit liquidity occurs where an increase in credit development can reduce credit liquidity. Finally for the default credit sub-model¸ development probability occurs so that, this risk also interference with other variables that are affected by this variable. To compare quantitatively the three combinations of scenarios, see Figure 3 below and Figure 4.

Aspect	Parameter	Pessimistic	Moderate	Optimistic
Supply Demand	Stock of Apartemen (Unit)	(15,606)	(7,477)	14,245
	Stock of Landed House (Unit)	5,112	(8,323)	(18,284)
Supply Demand	Land Occupied (m2)	7.008.205,50	6.076.447,42	3.896.017,71
Macro Economic	Economic Growth (%)	5,68	5,64	5,64
	Income per Capita (Rp)	73.945.441,50	73.851.819,12	73.930.764,03
Banking	Credit Liquidity (Rp)	28.432.021.113,44	28.825.983.759,85	29.065.165.360,53
	Credit Development (Rp)	4.882.542.530,21	4.143.367.418,53	3.714.569.636,97
Credit Default	Probability of Credit Default (%)	5,91	6,59	6,36

Figure 3. Combination Scenario Simulation Result

Figure 4 can be seen that the dominance of excellence for all aspects is possessed by optimistic scenarios, where the optimistic scenario answers the objectives of the development of the property sector. Optimistic scenarios can reduce the rate of land use for homes, this is appropriate in supporting the needs of the government in limiting land use for buildings on the growing population. However, the stock of apartments is bigger than housing. It is considered beneficial in terms of efficiency but not for investment. What is needed is a form of social management in the transfer of residential use from housing to apartments. In addition, in terms of per capita income and economic growth, this scenario is superior and supportive in economic growth as well as for banking. In essence, with this optimistic scenario, the property sector can promote economic growth and support in terms of the fulfillment of homes by the community.

Fiscal policy has no significant effect on the development of the property sector. This can be seen from the trend of changes in the fraction of tax rates that cannot dominate against other scenarios on the response variable that has been set. But on the other hand, this fiscal policy contributes when combined with other variables in influencing response variables.

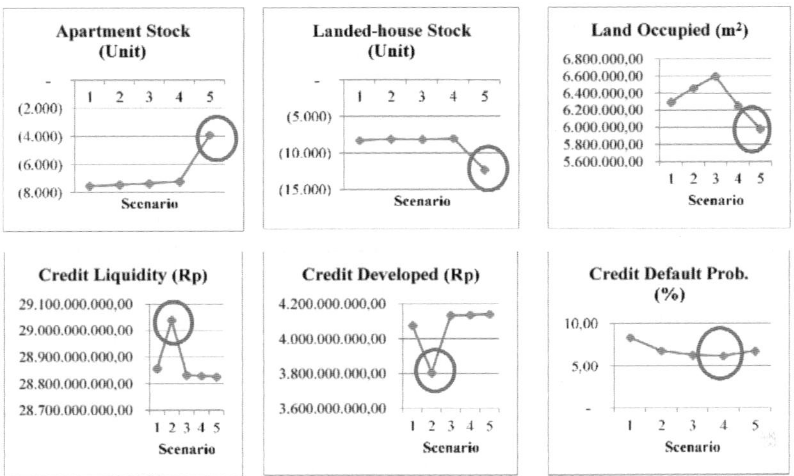

Figure 4. Combination Scenario Simulation Result

6. Conclusion

From the results of simulations and analysis that have been done, it can be concluded:

1. The policy of the central bank and the government has an impact on the development of the property sector. Policies that can be given by the central bank in the form of BI Rate determination as monetary policy, in order to manage macroeconomics, while LTV as macro prudential policy that provides limits on the amount of credit that can be done by the borrower. The policies that can be provided by the government in the form of tax rates as a fiscal policy, the implementation of resettlement and the proportion of provision of housing and apartment occupancy. The fiscal policy has no significant effect on the development of the property sector.
2. The policy giving policy partially has no significant effect and there is a trade off on the development of property sector. So it takes a combination of scenarios in the form of policy synchronization of the central bank and government in the business of fulfilling the house for the community. Combination scenario is arranged to get three kinds of scenario that is pessimistic scenario in the form of decrease parameter scenario, moderate scenario in the form of existing condition and optimistic scenario in the form of increase of reference parameter. All three combinations are selected as the greatest possibility that will happen in the future.

The scenario shows that the optimistic scenario answers the purpose of the development of the property sector. Optimistic scenario in the form of an increase in BI Rate of 8.00, a decrease in the proportion of credit (LTV) to 50%, increase of tax rate to 1.00%, resettlement by 10% for 10 years, and the proportion of apartment development: housing by 60:40. This can reduce the rate of land use for homes, this is appropriate in supporting the needs of the government in limiting land use for buildings on the growing population. In essence, with this optimistic scenario, the property sector can promote economic growth and support in terms of the fulfillment of housing by the community.

References

[1] Boediono 2013 *Boediono: Pemerintah-REI Kerjasama Cukupi Kebutuhan Perumahan* (M. F. Anugrah, Interview)

[2] Bank Indonesia 2013 *Perkembangan Properti Komersial* (Jakarta: Divisi Statistik Sektor Real Bank Indonesia)

[3] Ragimun 2012 *Overheating Sektor Properti di Indonesia.* (Jakarta: Policy Memo. Badan Kebijakan Fiskal, Kementerian Keuangan)

[4] Delis M. D and Kouretas G P 2010 Interest rates and bank risk-taking *J. Banking & Finance* **35** pp 840-55

[5] Hwang S, Park M and Lee H 2011 *Dynamic analysis of the effects of mortgage-lending policies in a real estate market* **57** pp 2106-20

[6] Kwoun M J, Lee S H, Kim J H and Kim J J 2011 Dynamic cycles of unsold new housing stocks, investment in housing and housing supply-demand *Mathematical and Computer Modelling:* **57** (Amsterdam: Elsevier) pp 2094-105

[7] Bouchouicha R and Ftiti Z 2012 Real estate markets and the macroeconomy: A dynamic coherence framework *Economic Modelling* **29** (Amsterdam: Elsevier) pp1820-9

[8] Ibicioglu M and Kapusuzoglu A 2012 An empirical analysis of impact of central bank policy interest rate on the decisions of share investors: evidence from Turkey *Procedia-Social and Behavioral Sciences* **62** pp 489-93

[9] Papadamou S and Siriopoulos C 2013 Interest rate risk and the creation of the Monetary Policy Committee: Evidence from banks' and life insurance companies' stocks in the UK *J. of Economics and Business* **71** (Amsterdam: Elsevier) pp 45-67

[10] Rubio M and Carrasco-Gallego J A 2014 Macroprudential and monetary policies: Implications for financial stability and welfare *J.of Banking & Finance* (Amsterdam: Elsevier)

Acknowledgments

First of all, I would like to extend my sincere gratitude to the conference steering comite who has given me an opportunity to share and to contribute a paper in this seminar. Secondly, I would like to thank Mr. Hasyim Asjari, Mr. Allain Wijanarko, Ms. Safira Mutiara and Mr. Theodore who have prepared the proposal up to the collecting data of this research. Thanks to their help and contributions, this research can be well conducted.

IOP Conf. Series: Materials Science and Engineering **337** (2018) 012047 doi:10.1088/1757-899X/337/1/012047

Model of urban water management towards water sensitive city: a literature review

D I Maftuhah[1*], M Anityasari[1], M Sholihah[1]
[1]Department of Industrial Engineering, Institut Teknologi Sepuluh Nopember, Kampus ITS Sukolilo-Surabaya 60111, Indonesia

*diesta@ie.its.ac.id

Abstract. Nowadays, many cities are facing with complex issues such as climate change, social, economic, culture, and environmental problems, especially urban water. In other words, the city has to struggle with the challenge to make sure its sustainability in all aspects. This research focuses on how to ensure the city sustainability and resilience on urban water management. Many research were not only conducted in urban water management, but also in sustainability itself. Moreover, water sustainability shifts from urban water management into water sensitive city. This transition needs comprehensive aspects such as social, institutional dynamics, technical innovation, and local contents. Some literatures about model of urban water management and the transition towards water sensitivity had been reviewed in this study. This study proposed discussion about model of urban water management and the transition towards water sensitive city. Research findings suggest that there are many different models developed in urban water management, but they are not comprehensive yet and only few studies discuss about the transition towards water sensitive and resilience city. The drawbacks of previous research can identify and fulfill the gap of this study. Therefore, the paper contributes a general framework for the urban water management modelling studies.

1. Introduction

Water is one of abundant primary natural resources in which has many benefits on it. This natural resource has many impacts for whole aspects in human life. Today, water becomes the main aspect in the world, especially in cities. Moreover, maintaining its sustainability is very complicated and complex. Social, culture, climate change, and other environmental problem has influenced to the water management itself. By considering the water management in cities, instead of national scale, many stakeholders have key role to manage and ensure water supply and demand. These stakeholders are connected to each other. The water management on cities is often considered as Urban Water Management (UWM). The paradigm of urban water management recently concerns on developing infrastructure in response to urban development that shifts to implement more sustainable [1] and comprehensive urban water management [2]. Sustainable urban water management helps socio economic developments without compromising the current and future water supply and needs [3]. Some cities are still struggling with the supply of important services, such as water and sanitation. Meanwhile, others are struggling to cope with many vulnerabilities created by the existing system of water management. Cities throughout the world are facing complex problem related to the effect of climate change, population growth, and hydrological variability, while ensuring sustainable water resources management and the protection of water environments [4]. Therefore, many countries especially developed and developing countries are trying to find guidance on solving the institutional challenges and the existing infrastructure to support a transition from urban water management into Water Sensitive City (WSC).

The concept of Water Sensitive City has developed as integrated vision of urban water management approach which meets not only the water needs, but also simultaneously give many benefits to increase the livability and resilience of a city [1]. This concept is also related to a new urban water governance paradigm regarding the use of decentralized water systems [5]. Water Sensitive City is based on the holistic point of view to manage the integrated water cycle in order to take care and increase the health of receiving waterways, mitigating flood risk, creating public spaces, as well as creating harvesting, cleaning and recycling water. It uses water management as a means of delivering better outcomes which are more broadly and able to deliver many critical objectives for the city liveability [6]. Eventually, water sensitive approach is developed with paradigms that water can contribute for connected, keen, and liveable communities. These paradigms of water management can be developed to make any decisions for stakeholders who have roles in water management system. It is important to breakdown all the key principles of water management in order to get resilient outcomes in all problems related to water. Modelling is one of approaches that can be a decisive tool to manage water management system considering systemic thinking of all relevant aspects in water management, especially urban water management and water sensitive city. Many research in modelling water management had been conducted. There are still few research that are mapping the literature review of modelling water management, including urban water system. Nevertheless, it is also hardly found research that are mapping the literature review in modelling water sensitive city. Therefore, this research reviews some research mapping not only in urban water system but also in water sensitive city, so that the research gap can be filled for scientific contributions.

2. Method proposed and literature review

Some previous research had been conducted in water management system, including sustainable urban water management (SUWM) towards specifically in water sensitive city (WSC). This paper reviewed some research related to the transition of urban water management towards the water sensitive city. Selection of some research related to water management depends on some criteria. This research was developed into two criteria, such as the concept of water management and classification of modelling approaches of water management, including urban water management and water sensitive city. After gathering data from some previous research, the next step is to conduct literature review. In reviewing literature, the research classifies the previous papers based on the criteria explained before such as the concept and the modelling approach of water management. Hence, from the classification of previous research, it can be found the research gap to be fulfilled. At the end, the discussion and recommendation about further research in water sensitive city specifically are included in this research.

2.1. Urban water management

Based on the information above, some concepts of urban water management, including sustainable urban water management had been widely researched in various aspects. The framework for cities transition through moving towards sustainable urban water conditions by considering ideological and technological contexts as a benchmarking tool and critical reassessment had been developed [1][7]. The complexity and solutions for eco-efficiency assessment of urban water systems are also explained [8]. Moreover, the interaction and understanding among governance were examined for sustainable urban water management [9][10]. In addition to that, some research reviewed the challenges and opportunities of transitioning to decentralized water management [11] and the transition scenarios to leapfrog for a sustainable urban water future [12]. Some research were also conducted in various aspects of urban water management, such as location aspect of urban storm water in tropics [13] and in coastal region [2] or city [14]. The sustainable utilization of urban water resources based on catastrophe theory was comprehensively assessed [15]. Moreover, some literatures of the adaptation of climate change and water resource management were also been reviewed [16].

Meanwhile, there were many research on modelling urban water management. A model based on system dynamics and agent based model was constructed to integrate modelling platform for simulating changes to socio-economics, technical and natural systems and for examining the effects of variety of water demand strategies on urban water system [17]. Moreover, The Dynamic Urban Water Simulation Model (DUWSiM) incorporated urban water balance concepts with the land use dynamics model MOLAND and the climate model LARS-WG to provide long term planning of urban water supply and demand with the effects of urbanization scenarios and climatic changes in urban water cycle[3]. In addition, dynamic simulation-optimization model was developed to analyze the adaptive management of urban water distribution system contamination threats [18]. The model only focused on drinking water network, not any other water types, such as review of the assessment of water demand modelling and management [19][20]. Furthermore, the agriculture water savings and the strategy for agricultural water management were provided and to be quantified and the optimization model for agricultural water transfer was developed to analyze the economic value of irrigation water [21]. Yet, the model considered was only for agricultural aspects, no other aspects considered. The research of urban water management were developed in other field such suburban area by using Life Cycle Assessment (LCA) methods [22][23]. They developed modelling tool to perform LCA of urban water system environmental insights to stakeholder's issues related to forecasting scenarios influencing the system such as water demand, scarcity, and technologies. However, it did not focus on forecasting scenarios linked to changes in wastewater treatment and associated water quality impacts. LCA were also applied in sustainable urban water resources management to assess the impacts of water utilization under uncertainty [24]. In the other hand, the assessment of water resources and water ecological carrying capacity by using integrated system dynamics model had been reviewed [25] and water management system in a green building had been modeled [26]. Furthermore, potential of hydropower as a co-benefit in balancing urban water portfolio and flood risk management by using system dynamics modelling were conducted. This research only focus on how to dealing with the water infrastructure to its benefit in balanced urban water portfolio and flood mitigation [27].

2.2. *Water Sensitive City*

To ensure the sustainability and resilience of city and its water system, it requires comprehensive aspects of technological, planning, economic, cultural and institutional dimensions considered in developing leapfrogging strategies and investment plans. The transition change processes are not easy to conduct. Technical innovation is not enough to be considered. Therefore, it is needed to understand the social and institutional dynamics that have important roles in attempt to move deeply rooted water management systems into new directions water sensitivity and resilience [1]. Monash researchers had conducted research within the EU FP7 PREPARED to develop the prototype of socio-technical model of Urban Water System, DAnCE4Water (Dynamic Adaptation for eNabling City Evolution for Water), which combines a Social Transitions Module [28] and can assess the joint effect of socio-technical factors and transitions to more sustainable urban water infrastructure. This research also included the UrbanBEATS model (Urban Biophysical Environments and Technologies Simulator) which made important breakthroughs on how to manage water infrastructure within a virtually generated urban setting [29]. In addition to that, DanCE4Water developed module of water supply and drainage networks within an evolving city in Australia [30]. Meanwhile, a strategic program was constructed to provide operational guidance for planners, designers, and decision-makers in planning and managing initiatives to facilitate sustainability transitions to Water Sensitive City [31].

In the Water Sensitive City, a hybrid mix of water systems and sources will operate at an interval of scales providing sustainable water services that will protect environmental quality and diversity, generation fairness (equity), and landscape amenity [32]. The implementation of water systems have to be in line with governance arrangements including strategies to enhance believe among stakeholders and to facilitate knowledge sharing for making decision. The collaboration and leadership are also essential for transitioning to Water Sensitive City [33].

The existing models were based on different objectives or have different methods/approaches. After reviewing literature from some previous research presented in table 1, it can be found that there are still few papers which considering high level planning and decision support tools to analyze the effects and the sustainability urban water management as well as water sensitive city. Moreover, it is hardly found some research in modelling the framework of Water Sensitive City comprehensively.

Table 1. Previous research on water management

No	Author	Year	Methodological Properties		Topics			Aspects		
			Analytical Approach	Model Approach	Urban Water Management	Sustainable Urban Water Management	Water Sensitive City	Water Infrastructure	Water Quality	Water Supply and Demand
1	Brown et al [1]	2009	x		x		x	x	x	x
2	Diaz et al. [2]	2016	x		x			x	x	x
3	Willuweit and O'ullivan [3]	2013		Simulation		x		x		x
4	Wong and Brown [4]	2009	x				x			
5	Brown et al . [6]	2008	x				x	x	x	x
6	Marlow et al. [7]	2013	x			x		x		
7	Stanchev and Ribarova [8]	2016	x		x			x		x
8	van de Meene et al.	2011	x			x		x		
9	Furlong et al. [10]	2016	x		x			x		
10	Chelleri et al. [11]	2015	x		x					
11	Poustie et al. [12]	2016	x		x					
12	Limand Lu [13]	2016	x			x		x	x	x
13	Li et al. [14]	2016		Non-linear regression model	x				x	
14	Chen et al. [15]	2016		Catasthrope assessment model	x					x
15	Olmstead [16]	2014	x		x					x
16	Baki et al [17]	2012		Simulation (System Dynamics and Agent-based Modelling)	x					x
17	Rasekh and Brumbelow [18]	2015		Dynamic simulation optimization model	x				x	
18	Cominola et al. [19]	2015		Multivariate analysis and behavioral modeling	x					x

Table 1. Previous research on water management (Continued)

No	Author	Year	Methodological Properties		Topics			Aspects		
			Analytical Approach	Model Approach	Urban Water Management	Sustainable Urban Water Management	Water Sensitive City	Water Infrastructure	Water Quality	Water Supply and Demand
19	Bosmann and Eser [20]	2016	x		x					x
20	Loubet et al [22]	2014	x		x			x		x
21	Loubet et al [23]	2016		Simulation and Life Cycle Assessment	x			x		x
22	Cai et al. [24]	2016		Optimization model and Life Cycle Assessment		x		x		x
23	Wang et al [25]	2014		System Dynamics Cellular Automata-Markov Model	x				x	x
24	Kalantzis et al. [26]	2016		Mental Model	x			x		
25	Sahinet et al. [27]]	2016		System Dynamics Model	x			x		
26	de Haan et al [28]	2013		Socio Technical model	x			x		x
27	Bachet et al [29]	2015	x			x		x		x
28	Urich et al [30]	2013		Simulation-Model(DAnCE4Water)	x			x		
29	Ferguson et al [31]	2013	x				x	x		x
30	Dobbie et al [32]	2016	x				x	x		
31	Floyd et al [33]	2014	x				x	x		

3. Solution model framework

Based on the previous research review, model framework of water management considering the transition urban water management towards water sensitive city was developed for constructing the conceptual model. The following Figure 1 represents the model framework of researched topic. The following framework is constructed by concluding variables from previous research and classified with the intersection between the concept of water management. Each parts of water framework have its intersection. According to the research so far, water infrastructure was developed within urban water management and water sensitive city. This aspect is considered due to the importance indicator to leapfrog from urban water management towards water sensitive city. Furthermore, water supply and demand was developed within sustainable urban water management. Many research found were conducted in the related aspect. Meanwhile, many research about water quality were developed within urban water management and sustainable urban water management.

The framework above is the first input data for modelling the urban water management towards water sensitive city. The next step is to construct the conceptual model by using causal loops diagram. Causal loop diagram is designed to model the logic of existing system based on the research framework before proceeding to the simulation model using simulation software. It represents causal relationship and loops between variables in the system. By using causal loop diagram, the expected model can be viewed comprehensively and it is known for simulating system dynamics. In other words, conceptualization of the model begins with first by identifying the variables that interact in the system related to urban water management as well as water sensitive city concept. Causal loops diagram is made to show the main variables that will be described in the model. With the existence of causal loops, it is able to understand the relationship among variables, and how far the influence of variables on system behavior and how the impacts of each variable to the others are.

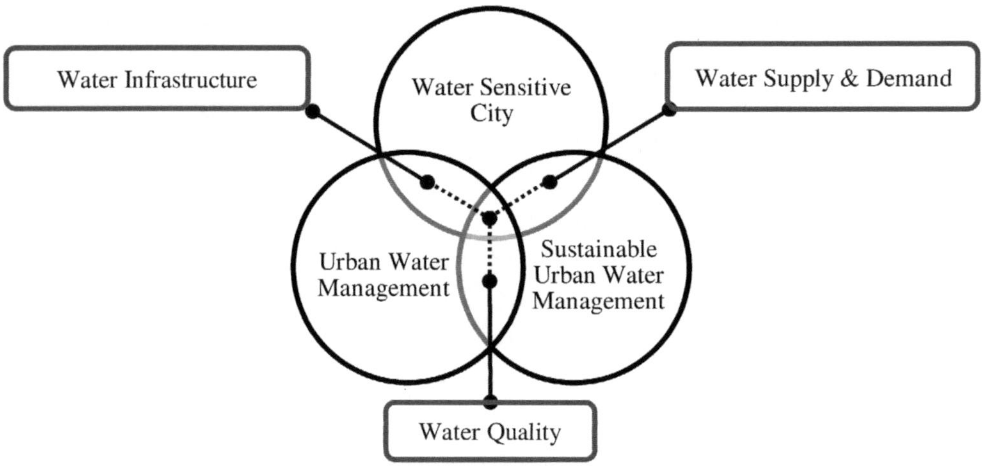

Figure 1. Integrated Research Framework.

4. Conclusions

Based on the data analysis, it can be concluded that there are many complex problems related to water management, including sustainable urban water system and water sensitive city. Besides that, there are few research conducted to identify and solve the problem as an integrated system in the city. All research so far about Water Management is still partial in specific aspect, such as the quality of water as well as water supply and demand separately. Furthermore, few research about water supply and demand in city related to transition in Water Sensitive City are hardly found. Although there was a research that developing integrated model of sustainable urban water system using system dynamics, yet the model did not involve how to leapfrog and to make transition into Water Sensitive City. Moreover, there is no research using modelling approach in the field of Water Sensitive City with comprehensive aspects. The conceptual model had been constructed and aimed to become important inputs for the future simulation model. This conceptual model will become an input in modelling stage to view the logic of the real system and the relationship among variables in urban water management and water sensitive city.

5. References

[1] Brown R R, Keath N, and Wong T H F 2009 Urban water management in cities: historical, current, and future regimes *J. of Water Science & Technology* (London: IWA Publishing) **59** pp 847-855

[2] Diaz P, Stanek P, Frantzeskaki N, and Yeh D 2016 Shifting paradigms, changing waters: transitioning to integrated urban water management in the coastal city of Dunedin, USA *J. of Sustainable Cities and Society*

[3] Willuweit L and O'Sullivan J J 2013 A decision support tool for sustainable planning of urban water systems: presenting the dynamic urban water simulation model *J. of Water Research* **47** pp 7206-7220

[4] Wong T H F and Brown R R 2009 The water sensitive city: principles for practice. *J. of Water Science & Technology* (London: IWA Publishing) **60** pp 673-682

[5] Yu C, Brown R, and Morison P 2012 Co-governing decentralized water systems: an analytical framework. *J. of Water Science & Technology*, **66** (12), 2731–2736

[6] Brown R R, Keath N, and Wong T H F 2008 Transitioning to water sensitive cities: historical, current, and future transition states *Proc. of 11th Int. Con. on Urban Drainage* (Scotland)

[7] Marlow D R, Moglia M, Cook S, and Beale DJ 2013 Towards sustainable urban water management: a critical reassessment *J. of Water Research* **47** pp 7150-7161

[8] Stanchev, P and Ribarova, I 2016 Complexity, assumptions and solutions for eco-efficiency assessment of urban water systems *J. of Cleaner Production*

[9] van de Meene S J, Brown R R, Farrelly M A 2011 Towards understanding governance for sustainable urban water management *J. of Global Environmental Change* **21** pp 1117-1127

[10] Furlong C, Gan K, and De Silva S 2016 Governance of integrated urban water management in Melbourne, Australia *J. of Utilities Policy*

[11] Chelleri L, Schuetze T, and Salvati L 2015 Integrating resilience with urban sustainability in neglected neighborhoods: challenges and opportunities of transitioning to decentralized water management in Mexico City *J. of Habitat International* **48** pp 122-130

[12] Poustie M S, Frantzeskaki N., and Brown R 2016 A transition scenario for leapfrogging to a sustainable urban water future in Port Vila, Vanuatu *J. of Technological Forecasting & Social Change* **105** pp 129-139

[13] Lim H S, and Lu X X 2016 Sustainable urban storm water management in the tropics: an evaluation of Singapore's ABC waters program *J. of Hydrology*

[14] Li Y, Li Y, and Wu W 2016 Threshold and resilience management of coupled urbanization and water environmental system in the rapidly changing coastal region *J. of Environmental Pollution* **208** 87-95

[15] Chen Y, Zhang S, Zhang Y, Qu Z, Song G, and Zhang J 2016 Comprehensive assessment and hierarchical management of the sustainable utilization of urban water resources based on catastrophe theory *J. of the Taiwan Institute of Chemical Engineers* **60** pp 430-437

[16] Olmstead S M 2014 Climate change adaptation and water resource management: a review of the literature *J. of Energy Economics* **46** pp 500-509

[17] Baki S, Koutiva I, and Makropoulos C 2012 A hybrid artificial intelligence modeling framework for the simulation of the complete, socio-technical, urban water system *Proc. of 2012 International Congress on Environmental Modelling and Software* (Leipzig)

[18] Rasekh A and Brumbelow K 2015 A dynamic simulation–optimization model for adaptive managementof urban water distribution system contamination threats *J. of Applied Soft Computing* **32** pp 59-71

[19] Cominola A, Giuliani M, Piga D, Castelletti A, and Rizzoli A E 2015 Benefits and challenges of using smart meters for advancing residential water demand modeling and management: a review. *J. of Environmental Modelling & Software* **72** pp 198-214

[20] Bosmann T and Eser E J 2016 Model-based assessment of demand-response measure - a comprehensive literature review *J. of Renewable and Sustainable Energy Reviews* **57** pp 1637-1656

[21] Zhang D and Guo P 2016 Integrated agriculture water management optimization model for water saving potential analysis *J. of Agricultural Water Management* (Elsevier) **170** pp 5-19

[22] Loubet P, Roux P, Loiseau E, and Bellon-Maurel V 2014 Life cycle assessments of urban water systems: A comparative analysis of selected peer-reviewed literature *J. of Water Research* **67** pp 187-202
[23] Loubet P, Roux P, Schneider L G, Maurel V B 2016 Life cycle assessment of forecasting scenarios for urban water management: a first implementation of the WaLA model on Paris suburban area *J. of Water Research* (Elsevier) **90** pp 128-140
[24] Cai Y, Yue W, Xu L, Yang Z, and Rong Q 2016 Sustainable urban water resources management considering life-cycle environmental impacts of water utilization under uncertainty *J. of Resources, Conservation, and Recycling* **108** pp 21-40
[25] Wang S, Xu L, Yang F, and Wang H 2014 Assessment of water ecological carrying capacity under the two policies in Tieling City on the basis of the integrated system dynamics model *J. of Science of the Total Environment* **472** pp 1070-1081
[26] Kalantzis A, Thatcher A, and Sheridan C 2016 Mental models of a water management system in a green building *J. of Applied Ergonomics*
[27] Sahin O, Stewart R, Giurco D, and Porter M 2016 Renewable hydropower generation as a co-benefit of balanced urban water portfolio management and flood risk mitigation *J. of Renewable and Sustainable Energy Reviews*
[28] de Haan F J, Ferguson B C, Deletic A, Brown R R 2013 a socio-technical model to explore urban water systems scenarios *Water Science & Technology* **68** 714–21
[29] Bach P M, McCarthy D T, and Deletic A 2015 Can we model the implementation of water sensitive urban design in evolving cities? *Water Science & Technology* **71** 149
[30] Urich C, Bach P, Sitzenfrei R, Kleidorfer M, McCarthy D T, Deletic A, Rauch W 2013 Modelling cities and water infrastructure dynamics *Proc. of the ICE-Engineering Sustainability* **166** 301–308
[31] Ferguson B C, Frantzeskaki N, and Brown R R 2013 A strategic program for transitioning to a water sensitive city *J. of Landscape and Urban Planning* (Elsevier) **117** pp 32-45
[32] Dobbie M F, Brown R R, Farrelly M A 2016 Risk governance in the water sensitive city: practitioner perspectives on ownership, management and trust *J. of Environmental Science & Policy* (Elsevier) **55** pp 218-227
[33] Floyd J, Iaquinto B L, Ison R, and Collins K 2014 Managing complexity in Australian urban water governance: transitioning sydney to a water sensitive city *J. of Futures* (Elsevier) **61** pp 1-12

Acknowledgment
The authors gratefully acknowledge some institutions in city level, province level, and even national level such as BAPPEKO, PDAM, BBWS, PDAB, PT Jasa Tirta I, Dinas PU-CKTR, Dinas PU-BMP, DKP, and so on for the support in conducting research on Water Sensitive City in Surabaya city.

IOP Conf. Series: Materials Science and Engineering **337** (2018) 012048 doi:10.1088/1757-899X/337/1/012048

Modelling impulsive factors for electronics and restaurant coupons' e-store display

P K Ariningsih[1*], M Nainggolan[1], and I A Sandy[1]
[1]Department of Industrial Engineering, Universitas Katolik Parahyangan, Bandung, Indonesia

*paulina.ariningsih@unpar.ac.id

Abstract. In many times, the increment of e-store visitors does not followed by sales increment. Most purchases through e-commerce are impulsive buying, however only small amount of study is available to understand impulsive factors of e-store display. This paper suggests a preliminary concept on understanding the impulsive factors in Electronics and Restaurant Coupons e-store display, which are two among few popular group products sold through e-commerce. By conducting literature study and survey, 31 attributes were identified as impulsive factors in electronics e-store display and 20 attributes were identified as impulsive factors for restaurant coupon e-store. The attributes were then grouped into comprehensive impulsive factors by factor analysis. Each group of impulsive attributes were generated into 3 factors. Accessibility Factors and Trust Factors appeared for each group products. The other factors are Internal Factors for electronics e-store and Marketing factors for restaurant coupons e-store. Structural Equation Model of the impulsive factors was developed for each type of e-store, which stated the covariance between Trust Factors and Accessibility Factors. Based on preliminary model, Internal Factor and Trust Factor are influencing impulsive buying in electronics store. Special factor for electronics e-store is Internal Factor, while for restaurant coupons e-store is Marketing Factor.

1. Introduction

Global retail and e-commerce retail condition had been improved significantly around Asia Pacific countries. In 2015, the growth had reached 28.4%, which can be considered as big number compared to North America (13%) and global (19.9%) [1]. The value of Asia Pacific e-commerce is predicted to be the greatest globally [2]. That numbers were triggered by the growth of credit card penetrations, wider product choices of global brands especially in China [3] and the increment of small business which are utilizing e-commerce transactions [4]. Among many popular products which are traded via e-commerce channels, Electronics and Restaurant Coupons are ranked in the tail of top 10 products [1]. Meanwhile, trends of e-commerce transactions are be persuaded by the consumer centric technology which means that user interface design of the store might be one of the most important factors to gain competitive advantages [3].

In traditional transactions, electronics and restaurant are usually to be bought after the product or service embedded in had been well-known by the consumer itself or recommended by other else. In electronics brick and mortal store, the consumer usually buys after they try the product to understand the quality of products. In restaurant, consumer may buy the product based on other people recommendations, however the returning transaction would also happen after the consumer feel satisfy with the products or services they've experienced by them self. Both electronics and restaurant brick and mortal store allow the consumer to feel the senses (visual, odour, tactical, taste, and audio) of the products which are very opposite with ecommerce store. Each page of ecommerce store shall be the medium or point of sales under very limited virtual user experiences.

In the meantime, as well as in the brick and mortar store, impulsive buying behaviour in e-commerce store cannot be ignored. This activity is triggered by the pleasure or recreational activity during browsing the store [5]. From retailer's perspective, the impulsive buying behaviour can be triggered by product layout and display [6]. Thus, user interface design of ecommerce store shall be a very carefully and strategically considered to improve ecommerce performances as point of sales [7].

Due to its value, potential, and research gap, it's important to understand the impact of each display attributes and factors toward impulsive buying strategies. The strategy may be used for the retailer to improve the sales of their potential products as well as to the shopaholic to understand the basic causal of their behaviour. Previous study had revealed ecommerce impulsive buying behaviour in online fashion store [8], which may lack of generalization for other type of products or business process. Therefore, it is important to explore the similar research to electronics e-store and restaurant coupon's e-store.

This paper presents exploration study to understand impact of web display attributes towards impulsive buying by 5 major chapters. Second chapter presents and discuss the generation of impulsive attributes and development of impulsive factors by factor analysis. Chapter three discuss the development of impulsive model by structural equation model. Reader may find the discussion of data processing as well as managerial implication. While in last chapter, reader may find conclusions and future potential research.

2. Impulsive Factors Development

The development of impulsive factors for each store is done through several steps, as detailed in Table 1. For both store, it began with the development of impulsive attributes and it finalized with factor analysis.

Impulsive attributes were developed by literature survey and also Critical Incident Techniques (CIT). Both methods were used to complement each other's. CIT is also able to make sure the attributes fit with consumer's point of view. Afterwards, each attributes found are then ensured their fitness to e-store web features. Potential attributes shall reflect the display web features that can be tested the impact to impulsive buying respond.

The process of attributes finding were done separately for each e-store type using two sample of pioneer of e-store in Indonesia. The samples are: Fave (formerly Groupon), Bhinneka.com. We elude the sampling from e-supermarkets such as Lazada, Alibaba, Amazon or MatahariMall to avoid bias that might happen during the application to e-store system.

Questionnaire was generated for each group of e-store based on the final attributes. Experienced respondents were filling up the questionnaire. Based on the questionnaire result, qualified potential attributes were grouped by factor analysis. Eigen value ≥ 1 is used to decide the number of factors to be developed. However, before factor analysis was conducted, a correlation test is performed to eliminate potential attributes that has low correlation coefficient (< 0.25) thus, final attributes are gained.

Table 1. Result of Impulsive Factor Development Process.

No.	Process	Parameter	Electronics E-Store	Restaurant Coupons E-store
1.	Identification Attributes from Literature Study	Number of literature studied	3 those are [5], [9],[10]	4 those are [12], [13], [9]
		Number of attributes found	11	8
2.	Identification Attributes from CIT	Category of respondent	• browse the electronics e-store at least once a week • in previous time impulsively buying at least once	• browse the restaurant coupons e-store at least once a week • in previous time impulsively buying at least once
		Number of sample	13	16
		Number of additional attributes found	4	5
3.	Fitting attributes	Number of qualified potential attributes	31	20
4.	Questionnaires	Respondent category	• browse the electronics e-store at least once a week • in previous time impulsively buying at least once	• browse the restaurant coupons e-store at least once a week • in previous time impulsively buying at least once
		Pearson correlation validity test	>0.25 (p-value = 0.05)	>0.35 (p-value = 0.05)
		Alpha-cronbach reliability test	0.837	0.919
		Number of sample	162	140
5.	Correlation test	Number of final attributes	17	14
6.	Factor Analysis	KMO Test	0.893	0.890
		Bartlett's Test of Sphericity significance	0.000	0.000
		MSA Score	>0.70	>0.70
		Number of factors developed	3	3
		Name of factors developed	• Accessibility Factor • Internal Factor • Trust Factor	• Buying Accessibility Factor • Marketing Factor • Trust Factor

2.1. Electronics Factor Results

Accessibility Factor contains attributes that allows consumer to understand and access the products easily including products' interesting characters and popularity. The accessibility attributes are: zoom feature (ZOOM), most selling products information (MSP), most viewed product information (MVP), newsletter page (LN), popular category menu (MKP), and advanced search features (LPS). **Internal Factor** assembles final attributes that contains basic feature of an e-store display profile. Those are: picture of products (TP), product specification information (PS), categorical menu (MENU), payment confirmation page (KP), tracking order page (LTO). **Trust Factor** includes all final attributes that convince the satisfaction after buying, i.e: inventory information (ISP), testimony (TESTIMONIAL), delivery lead time (LWP), product quality information (QP), e-store credibility (KREDWEB), and brand of products (MERK).

2.2. Restaurant Coupons Factor Results

As result of the factor analysis, 3 factors were generated, i.e.: Buying accessibility factor, Trust factor, and Marketing factor. Slightly different with Accessibility factors of electronics e-store, the **Buying Accessibility Factor** of restaurant coupons e-store defines all finals attributes which triggers or acting as catalysts of buying process, i.e.: instant voucher (IV), information of product expiry (VV), picture of product (TP), credit information (KK), payment method information (MP), delivery lead time (WP), and product variety (VV). In line with electronics e-store, **Trust Factor** corresponds with all final attributes which guarantee the satisfaction prior buying. Final attributes of this factor are: flexibility of redemption information (JLP), brand information (MERK), and expiry product information (KKP). **Marketing Factor** deals with all attributes that contains privilege of the product based on the marketing strategy, i.e.: percent discount information (PD), nominal discount information (ND), bundling information (B1G1), and gift information (GIFT).

3. Models Developments

Model of the research are referring to simplification of Stimulus – Organism – Respond model or SOR as used on [8] as seen in Figure 1. Because impulsive respond is spontaneous decision, organism is considered as black box [14], therefore it is assumed can be ignored. For each of e-store type, 3 scenarios were developed. Structural Equation Model (SEM) is able to recognise the interactions among attributes, between attributes and factor, factor and dependence variables. One of the SEM advantage is that this method can model multivariate variables (dependent or independent) [15]. Therefore, business practitioners can get a clear picture of their business by a SEM models.

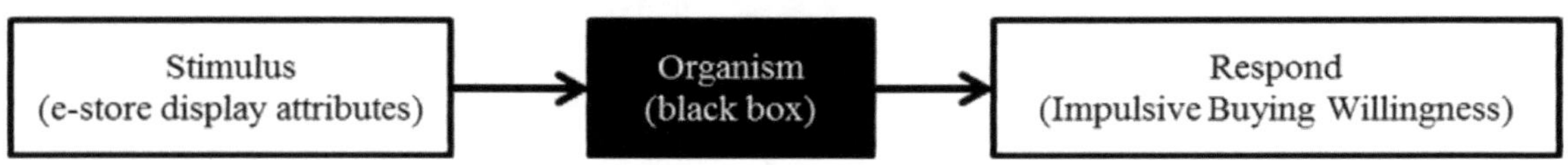

Figure 1. Modified SOR Framework.

Then, a scenario which has better model fit based on the SEM, is selected. The model fit parameters used are CFI, TLI, and RMSEA. CFI and TLI score should closer to 1, while RMSEA score should closer to 0. Detail result of each model for each e-store can be seen in Table 2, while detail of each scenario can be seen in sub section 3.1 and 3.2.

Table 2. Model Fit Parameter.

	Electronics E-store			Restaurant Coupons E-store		
	CFI	TLI	RMSEA	CFI	TLI	RMSEA
Scenario 1	0.661	0.564	0.124	0.722	0.669	0.138
Scenario 2	0.818	0.762	0.095	0.841	0.810	0.104
Scenario 3	0.819	0.765	0.095	0.845	0.815	0.103

3.1. Electronics E-store Model

First scenario was generated with assumption that all stimulus factors are correlated to Impulsive Buying Willingness without any covariance amongst each factor. Second scenario was assuming that consumer buying will be influenced by Trust Factor and Internal Factor; meanwhile Accessibility Factor will influence directly Trust Factor. While third scenario was generated by assuming that the

Accessibility Factor will influence Impulsive Buying Willingness indirectly through Trust Factor. In other hand, the Trust Factor is influenced by Accessibility Factors.

Scenario 3 has best model fit and shall be selected, because it has lowest RMSEA score (0.095), highest TLI score (0.765) and CFI score (0.819). Figure 2 shows the path diagram for scenario 3. As shown in the path diagram, Internal Factor is major influencer for Impulsive Buying Willingness with factor loading 1.07. Meanwhile Accessibility Factors and Trust Factors are empowering Internal Factors by 0.605 and 0.458 For Internal Factor, three major influencer attributes are Tracking Order Page (1.162), Categorical Menu (1.132), and Picture of Product (1.00). Three major influencer attributes for Trust Factor are Brand (1.224), Product Quality Information (1.139) and Information of Delivery Lead Time (1.105). Meanwhile, three major influencer attributes for Accessibility Factors are Most Selling Product Information (0.79), Most Viewed Product Information (0.73) and Popular Category Menu (0.7).

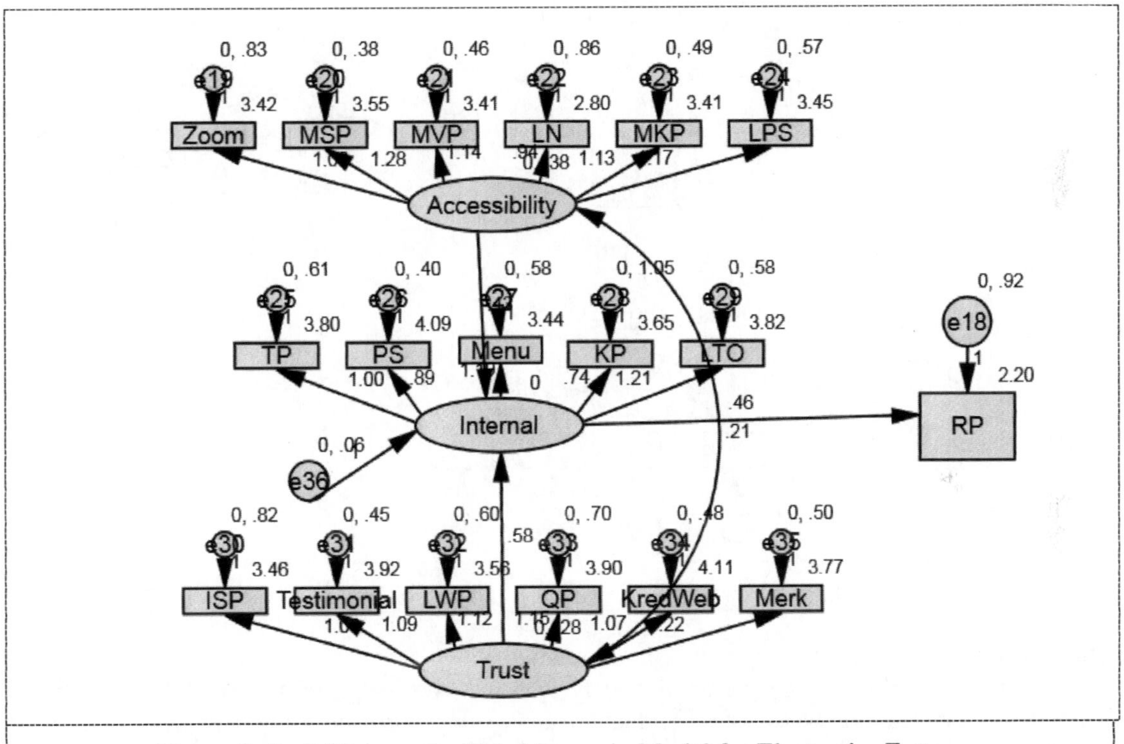

Figure 2. Path Diagram for Third Scenario Model for Electronics E-store.

3.2. Restaurant Coupon E-store Model

Similar with restaurant coupon e-store model in this type of e-store, three scenarios were developed to be tested. First scenario was generated with assumption that all stimulus factors are correlated to Impulsive Buying Willingness without any covariance amongst each factor. Second scenario assumes that only Trust Factor and Marketing Factor influencing impulsive buying. This scenario considers that the Accessibility Factor will influence Impulsive Buying Willingness indirectly through Trust Factor. Third scenario is assuming that Marketing Factor and Accessibility Factor are influencing respond, while Trust Factor is influencing Accessibility.

Here, scenario 3 also has best model fit, with lowest RMSEA score (0.103), highest TLI score (0.815), and highest CFI score (0.818). Figure 3 shows the path diagram for scenario 3. In Figure 3, Accessibility is shown to be the major influencer for Impulsive Buying Willingness with factor loading 0.65, whilst Marketing's factor loading is 0.09. Three major influencer attributes for Accessibility are Voucher Variety Information (1.275), Payment Method Information (1.22), and

Product Expiry Information (1.219). Three major influencer attributes for Marketing Factor are Gift Information (1.247), Number of Discount information (1.1) and Percent Discount (1.00). Meanwhile, three major influencer attributes for Trust Factors are Information of Flexibility Redemption (1.00).

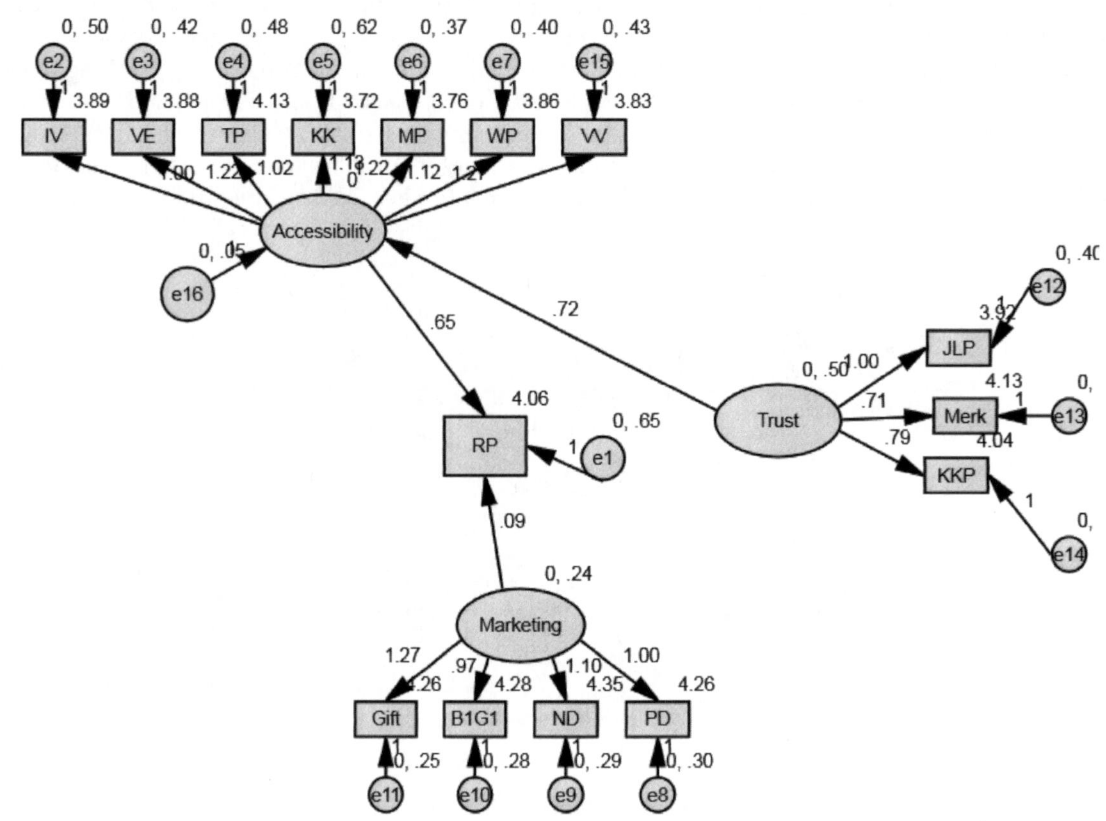

Figure 3. Path Diagram for Third Scenario Model for Restaurant Coupons E-store

4. Discussion

As might be seen on previous chapter, it is clearly seen that Accessibility factor and Trust factor are generated for both e-store types as those factors are related with the buying process in general. Recognition of product is one of the most important aspects on point of sales that sure the identification of product and produce direct stimuli to consumers' cognitive process. The easier a store creates appealing product identification, the easier a consumer is exposed with the buying stimuli.

It is also shown by the factor analysis; that there is different consumer behaviour for electronics e-store and restaurant coupons e-store. This difference is caused by the characteristics of satisfaction and experiences on using the products. Restaurant coupons e-store impulsive buying behaviour is triggered by the marketing factors while electronics e-store impulsive buying behaviour is influenced by the internal factors. Products of restaurant have shorter consuming time compared with the electronic products. Thus the expectation of satisfaction for electronics product is higher than restaurants'. To ensure the expectation of satisfaction span, consumers have willingness to sacrifice more for the product as long as the product itself guarantee to give the constant satisfaction throughout the consuming time. Therefore, in electronics e-store factors, internal factor is generated.

The third scenario of electronic store model was developed based on the actual buying characteristics of electronics product. In electronics product, it appears that during browsing activity, the basic feature of e-commerce store will be effective if consumer can access and understand the characteristics of product and trust to the satisfaction that could be given by buying the product.

Meanwhile, in restaurant coupons, the marketing factor is generated rather than internal factor. Based on the questionnaire respond, most restaurant coupons buying, a consumer can be distinguished into two major types: loyal brick and mortar store consumer and newbie consumer. The loyal brick and mortar store consumers have familiar with and have a clear expectation on quality of restaurant. Thus, whenever they find that desired benefit can be obtained with lower sacrificial, the trigger of pleasure getting from buying impulsively is higher. Meanwhile, for newbie consumers who are very opposite to the loyal consumer, a lower sacrificial or promotion may give lower risk on exploring the restaurant quality.

The proposed model, in other hand, is shown that the benefit of lower sacrificial only matter in small amount. Food, however, are basic and general commodity which easily being substituted. Therefore the accessibility can also be important factor in leading the buying decision. Consumer trust may influence impulsive behaviour indirectly via accessibility. The higher the trust, the higher accessibility factor influences impulsive buying behaviour.

Based on the impulsive factors and attributes, an e-store manager can measure the performance of their website toward impulsive buying. By measuring the performance, manager may improve the e-store display performance toward impulsive buying. The magnitude of each attributes gives the idea of prioritizing the improvement attributes. Further research may investigate whether this suggestion is effective to be applied in business process.

5. Conclusion

In this paper, impulsive attributes of an electronics and restaurant coupon e-store are identified. 16 attributes tested in influencing impulsive buying behaviour in electronics e-store, while 14 attributes shown their influence toward impulsive buying in restaurant coupon e-store. Accessibility factor and Trust factor are identified for both electronics and restaurant coupons. In electronics e-store, Internal factor is developed distinctively. Unique factor for restaurant coupons e-store is Marketing factor. The model suggested for each factors are showing that accessibility and trust are important factors to lead impulsive buying behaviour. However, different impact appears due to characteristics of product. In electronics e-store, both factors become indirect influencer for impulsive buying behaviour through internal factor. But in restaurant coupons e-store, accessibility factor have direct impact, and trust factor is indirect influencer through accessibility. This research is limited to particular e-store; therefore it is suggested to investigate generalization application of this research upon of e-mall or other product categories. It is also suggested to test this model on measuring and improving the e-store display performances.

6. References

[1] Ecommerce Foundation 2017 *Global B2C E-commerce Report 2016* https://www.ecommercewiki.org/wikis/www.ecommercewiki.org/images/5/56/Global_B2C_Ecommerce_Report_2016.pdf [access date: 5 May 2017]

[2] Anonim 2016 *4.eCommerce Market Size* https://www.export.gov/article?id=eCommerce-Market-Size [access date: 17 May 2017]

[3] Allen R 2016 *Ecommerce Trend to Watch 2016 [Infographics]* http://www.smartinsights.com/ecommerce/ecommerce-strategy/ecommerce-trends-watch-2016-infographic/ [access date : 5 May 2017]

[4] The CPA Australia 2017 *The CPA Australia Asia Pacific Small Business Survey 2016: General Report for Australia, China, Hong Kong, Indonesia, Malaysia, New Zealand, Singapore and Vietnam*

[5] Vojvodic K and Matic M 2014 *Challenges of E-Retailing: Impulsive Buying Behavior, Challenges for The Trade of Central and Southeast Europe* pp 155-171

[6] Amos C, Holmes G R and Keneseon W C 2014 A Meta-Analysis of Consumer Impulse Buying *Journal of Retailing and Consumer Services* **21** pp 86-97
[7] Ha S H 2002 Helping Online Customer Decide through Web Personalization *IEEE Intelligent System* November December 2002
[8] Ariningsih P A, Nainggolan M, Sandy I A and Widyasti D 2016 Understanding the Impulsiveness Effect of a Web Design on Online Fashion Store *Proceedings of the 2016 Int. Conference of Industrial Engineering and Operations Management* Kuala Lumpur, Malaysia pp 592 – 597
[9] Liu Y, Li H and Hu F 2013 Websites Atrributes In Urgin Online Impulse Purchase: An Empirical Investigation On Consumer Perceptions *Decision Support Systems* **5** Issue 3 pp 829-837
[10] Ozen H and Engizek N 2014 Shopping Online Without Thinking: Being Emotional Or Rational? *Asia Pacific Journal of Marketing and Logistics* ***26***(1) pp 78-93 doi:10.1108/APJML-06-2013-0066
[11] Banerjee S and Saha S 2013 Impulse Buying Behaviour In Retail Stores - Triggering The Senses *Asia Pacific Journal of Marketing &Management Review* **1** (2) pp 1-21
[12] Amos C, Homes G R and Kenneson W C 2013 A Meta-Analysis of Consumer Impulse Buying *Journal of Retailing and Consumer Service* pp 86-90
[13] Dawson S and Kim M 2009 External and Internal trigger cues of Impulse Buying Online *Direct Marketing: An Int. Journal* **3** Issue 1 pp 20-34
[14] Solomon M R 2007 *Consumer Behavior: Buying, Having, and Being* (7th ed.) Prentice Hall
[15] Hair Jr J F, Black W C, Babin B J and Anderson R E 2010 *Multivariate Data Analysis* 7th edition (USA: Pearson Prentice Hall)

Acknowledgments

Authors are expressing sincere gratitude for Antoni Salim, S.T and Jovin Wijaya, S.T for their best assistance in data collecting and preparing the data. Authors are also thanking to Research and Community Services Centre (LPPM) of Universitas Katolik Parahyangan for funding this research through Hibah Internal 2015.

Implementation of enterprise resource planning systems: a user perspective

E Reitsma[1*], P Hilletofth[1] and U Mukhtar[2]
[1]Department of Industrial Engineering and Management, School of Engineering, Jönköping University, Jönköping, Sweden
[2]Department of Management Sciences, GIFT University, Punjab, Pakistan

*ewout.reitsma@ju.se

Abstract. The purpose of this study is to evaluate critical success factors (CSFs) for the implementation of an enterprise resource planning (ERP) system from a user perspective. Users play a vital role when implementing an ERP system, but their perspective has been neglected in the literature. A better understanding of their perspective promises to contribute to the design of more effective ERP systems, its implementation, and management. In order to identify the user perspective, a survey was conducted within three Pakistani companies that recently have implemented a new ERP system. The questionnaire was developed based on thirteen CSFs deduced from literature. Based on each CSF's level of importance, they are ranked in order of importance and divided into three groups: most important, important and not important. Findings reveal that users believe that management should prioritize the following four CSFs when implementing an ERP system: education and training, strategic decision-making, communication, and business process alignment.

1. Introduction

Supply chain management is of major importance in most industries. In essence, it concerns collection and analysis of data to provide better basis for decision-making [1]. Information systems are needed to achieve this due to complex supply chain structures and processes [2]. One information system that has been central in operating and managing supply chains is the enterprise resource planning (ERP) system [3]. Despite ERP systems becoming a fundamental tool in many industries, the rate of failure in their implementation is high [4]. The reasons for failure have been widely discussed in the literature and research concentrates on identifying the conditions that are believed to increase the probability of success of an implementation project. These conditions are referred to as critical success factors (CSFs) and are mostly proposed from a generic perspective [5][6][7]. Understanding these CSFs and how they influence the project outcome decreases the risks of failure, and provide helpful guidance for organizations [8]. However, to be successful with an ERP system implementation, users play a vital role [9]. So far, very limited empirical research has been done with regard to the user perspective towards ERP system implementation [10].

The purpose of this paper is to evaluate CSFs for ERP system implementation from a user perspective. A better understanding of this perspective promises to contribute to the system design, its implementation, and management. Previous research has focused on one manufacturing company from Germany [11] and thus there is a need to extend the research to other industries and countries.

This research further investigates the user perspective towards CSFs for ERP system implementation in Pakistan. A survey was conducted within three companies from Pakistan that recently have implemented a new ERP system, which was developed based on CSFs deduced from literature [11]. The first case company (α) operates in the education industry, the second case company (β) operates in the manufacturing industry, and the third and final case company (γ) operates in the beverages industry.

2. Theoretical framework

Reitsma and Hilletofth [11] created a framework consisting of 13 CSFs for ERP system implementation. In addition, they investigated the user perspective towards the included CSFs and labelled them as: 'not important', 'important' or 'very important from a user perspective' (Table 1). Their empirical research was based on one manufacturing company from Germany [11] and thus need to be extended to other industries and countries.

Table 1. Framework of CSFs for ERP system implementation, based on [11].

CSF	Code	User label	Definition
Project team	CSF1	Most important	The project team needs to consist of the best people and has to include a project champion, employees from different functions and levels, and external consultants when ERP expertise is missing internally.
Top management involvement	CSF2	Not important	Top management should reinforce the commitment of all employees in the organization and create policies that determine and approve new organizational structures, roles, and responsibilities.
Strategic decision-making	CSF3	Most important	A well-defined business plan and vision should define how the organization operates behind the implementation effort and has to outline proposed strategic and tangible benefits, resources, costs, risks and timeline is critical.
Communication	CSF4	Important	Effective communication should be established at every organizational level and has to include the formal promotion of the project and its teams and advertisement of project progress.
Project management	CSF5	Important	Project management should include a clear definition of objectives, development of both a work and a resource plan has to focus on the identification of the equipment required to operate the system.
Vendor support	CSF6	Important	Vendor support should be established in the form of technical assistance, maintenance, and updates, which has to be facilitated by a committed vendor that oversees the entire implementation's life cycle.
Minimum customization	CSF7	Most important	Departments should not rearrange the chosen ERP system in order to prevent interdepartmental issues and should have access to the same data and system.
Organizational change management	CSF8	Not important	The organization should utilize change management techniques and tools that must be defined and evaluated with the best practices in the industry.
Business process alignment	CSF9	Important	A catalogue of best business processes should be selected and followed to stay on the right track and avoid conflicts with the procedural rigidity of an ERP system.
Software testing	CSF10	Most important	The organization should establish rigorous and sophisticated testing of the software in order to simplify the ERP system's implementation.
Performance measurement	CSF11	Most important	Performance measurements should be identified to manage expectations, keep track of all occurrences and to measure the achievements against the milestones and targets.
Education and training	CSF12	Most important	Education and sufficient training requires investment, promotes an effective and correct use of the ERP system, and should be provided for users from the beginning of the ERP system implementation project.
Technical possibilities	CSF13	Most important	All kinds of differences in ERP systems offered in the marketplace should be evaluated based on its own strategy, size, business field, its business processes, and its internal and external relationship structure.

3. Research methodology

The aim of this study is to evaluate CSFs for ERP system implementation from a user perspective. In the first step of the research, it was necessary to set the scene by presenting relevant research from the research field. This literature review is based on the study of Reitsma and Hilletofth [11], as they propose thirteen general CSFs for ERP system implementation and focussed on the user perspective (Table 1). In the second step of the research, a survey was conducted to identify the perspective of users towards the CSFs identified in the literature review. A questionnaire was developed based on the thirteen CSFs and each question contained a description of the corresponding CSF. The respondents were asked to grade each questionnaire item on a Likert-scale from one to five (1 = not important and 5 = very important). The questionnaire was distributed within three companies from Pakistan that recently had implemented a new ERP system (Table 2).

Table 2. Summary of demographic profile of respondents.

	Case company α	Case company β	Case company γ
Company location	Pakistan	Pakistan	Pakistan
Company type	Education	Manufacturing	Beverages
Company size in terms of employees	+-1200	+-600	+-2000
Implemented ERP system since	2015	2014	2012
ERP provider	Microsoft Dynamics	Microsoft Dynamics	SAP ERP
Number of ERP users	650	170	870
Number of distributed surveys	250	80	150
Response rate (%)	13	24	33

The collected data was analysed with GRA to extract the most important CSFs from a user perspective. The higher the GRA value, the more emphasis the respondents put on the particular CSF. GRA is an impact evaluation model, which measures the degree of similarity, or difference between two sequences and is based on the grade of relation [12]. When using GRA, a multi-criteria problem needs to be formulated by using a set of alternatives $(x_1, x_2, \dots, x_{13})$, in this study the CSFs, and criteria $(k_1, k_2, \dots, k_5)$, in this study the Likert-style rating scale. Each criterion is assigned weightings and a preference index (PI). If a higher number of respondents choosing the value of each k will have a positive impact on x, the PI is equal to one. Otherwise, it is equal to zero. Since k_1 and k_2 represent the Likert-style rating scales that indicates that users label the corresponding CSF is 'not important' or 'slightly important', their PI is equal to zero. The PI of k_3, which represents the median Likert-style rating scale outcome 'moderately important' is also equal to zero, as it not clearly indicates a positive outcome for x. k_4 and k_5 are equal to one, since they represent the Likert-style rating scales 'important' and 'very important', which clearly indicate a positive outcome for x. The next step is to formulate the following decision matrix D:

$$D = \begin{Bmatrix} x_1(k_1) & \dots & x_i(k_1) & \dots & x_m(k_1) \\ \dots & \dots & \dots & \dots & \dots \\ x_1(k_j) & \dots & x_i(k_j) & \dots & x_m(k_j) \\ \dots & \dots & \dots & \dots & \dots \\ x_1(k_n) & \dots & x_i(k_n) & \dots & x_m(k_n) \end{Bmatrix} \quad (1)$$

where:

$$PI_j = \begin{cases} 1, & \text{Increasing} \\ 0, & \text{Decreasing} \end{cases} \quad (2)$$

Afterwards, matrix D' needed to be computed:

$$D'=\begin{Bmatrix} x_1(k_1)' & \dots & x_i(k_1)' & \dots & x_m(k_1)' \\ \dots & \dots & \dots & \dots & \dots \\ x_1(k_j)' & \dots & x_i(k_j)' & \dots & x_m(k_j)' \\ \dots & \dots & \dots & \dots & \dots \\ x_1(k_n)' & \dots & x_i(k_n)' & \dots & x_m(k_n)' \end{Bmatrix} \quad (3)$$

where:

$$x_i(k_j)' = \frac{x_i(k_j) - \underset{\forall j}{\min}\left\{x_{i(k_j)}\right\}}{\underset{\forall j}{\max}\left\{x_{i(k_j)}\right\} - \underset{\forall j}{\min}\left\{x_{i(k_j)}\right\}} \quad (4)$$

Then, a pre-reference sequence $y_0=\left\{y_{0\,(k_j)};k=1, 2, 3, \dots, m\right\}$ was determined:

$$y_0(k_j)=\begin{cases} \underset{\forall i}{\min}\left\{x_{i(k_j)}\right\} & \text{if } PI_J=0 \\ \underset{\forall i}{\max}\left\{x_{i(k_j)}\right\} & \text{if } PI_J=1 \end{cases} \quad (5)$$

The reference sequence $y'_0=\left\{y_{0\,(k_j)}'\right\};k=1, 2, 3, \dots, m$ was computed by turning all decreasing criteria into opposite direction:

$$y_0(k_j)'=\begin{cases} 1-y_{0(k_j)} & \text{if } PI_J=0 \\ y_{0(k_j)} & \text{if } PI_J=1 \end{cases} \quad (6)$$

Finally, each sequence was compared with the reference sequence by calculating the grey relational coefficient: $y\left(y_0(k), x_i(k)\right) = \frac{\Delta \min + \zeta\Delta \max}{\Delta_{oj}(k) + \zeta\Delta \max}$. ζ serves as the equation's contrast control and the value of 0.5 is applied [12]. The mean of the grey relational coefficient was used as the grey relational grade: $y(y_0, x_i)=\frac{1}{n}\sum_{j=1}^{n} y\left(y_0(k),x_j(k)\right)$. This made it possible to evaluate the user perspective towards CSFs for ERP system implementation and rank them in order of importance (Table 3).

4. Findings and analysis

The GRA values listed in Table 3 reveal that users from one case company (α) label two CSFs as 'not important' for ERP system implementation, including minimum customization and performance measurement. All other CSFs have a GRA value above ζ 0.5, meaning they are regarded as 'important' by case company α, β, and γ. However, the GRA values also indicate that different emphasis was put on each CSF. Therefore, based on their GRA value, the CSFs can be ranked in order of importance and labelled as either 'most important', 'important', or 'not important'.

The findings reveal that there is a level of consensus and discrepancy between the users from the three included case companies, as they all label the same four CSFs for ERP system implementation as the 'most important'. The first CSF labelled as 'most important' is education and training (CSF12) which should therefore be prioritized by top management when implementing an ERP system according to users. This shows that users agree with Stedman [13], who stated that a well-prepared and educated user reduces the consultant hours needed, and increases the support of and commitment to the system. This also indicates that users also agree with Stedman's [13] statement that adequate training and the education level of the ERP system's users should be prioritized.

The second CSF labelled as 'most important' is strategic decision-making (CSF3), meaning that users from the three included case companies agree with Umble et al [14] that it is necessary to constitute a clear understanding of strategic goals for ERP system implementation and that key people throughout the organization have to create a clear, compelling vision of how the company should operate.

Table 3. Ranking of the CSFs in order of importance.

	Case company α			Case Company β			Case Company γ		
Rank	CSF	GRA	Label*	CSF	GRA	Label	CSF	GRA	Label
1	CSF3	.827	MI	CSF11	.844	MI	CSF12	.882	MI
2	CSF5	.807	MI	CSF10	.802	MI	CSF13	.832	MI
3	CSF1	.783	MI	CSF12	.784	MI	CSF11	.764	MI
4	CSF4	.727	MI	CSF4	.742	MI	CSF5	.680	MI
5	CSF6	.727	MI	CSF3	.690	MI	CSF3	.667	MI
6	CSF12	.695	MI	CSF7	.690	MI	CSF4	.620	MI
7	CSF9	.609	MI	CSF2	.667	MI	CSF2	.614	MI
8	CSF8	.535	I	CSF13	.667	MI	CSF8	.611	MI
9	CSF2	.528	I	CSF6	.667	MI	CSF9	.606	MI
10	CSF10	.514	I	CSF9	.660	MI	CSF7	.600	MI
11	CSF13	.513	I	CSF5	.637	I	CSF1	.590	I
12	CSF7	.442	NI	CSF1	.575	I	CSF10	.572	I
13	CSF11	.431	NI	CSF8	.562	I	CSF6	.551	I

*MI=most important, I=important and NI=not important

Users also agree with Gargeya and Brady [15], who propose that planning should be closely identified with maintaining scope during an implementation, since excessive costs and developmental delays are costly, and sometimes-fatal results of ineffective planning.

The third CSF labelled as 'most important' is communication (CSF4), meaning that users agree with Davenport [16] that interdepartmental communication is necessary across functional boundaries in ERP context to integrate business functions. In addition, this validates the work of Wee [17], who stated that an effective communication while implementing an ERP system is critical to the success of the entire project, meaning that expectations need to be communicated at every organizational level and that management of communication, education, and expectations throughout the organization is of high importance. Additionally, this indicates that users agree with Rosario [18], who proposed that their input should be managed in acquiring their requirements, comments, reactions and approval.

The fourth and final CSF labelled as 'most important' by users is business process alignment (CSF9). This indicates that users agree with Akkermans and van Helden [19] who stress the necessity of interdepartmental cooperation. Users also agree with Gargeya and Brady [15] that the reengineering of the processes results in a changed corporate culture and that these actions have to be managed on the human and operational level. Lastly, users recognize the statement of Bingi et al [20] who mention that whereas the adaptation increases the feature-function fit between the organization and the software, the resistance against the change decreases and less training is needed.

When comparing these findings with the study from Reitsma and Hilletofth [11], users from both studies label education and training as well as strategic decision-making as 'most important'. However, this study also differs as users from the included case companies label two additional CSFs as 'most important' (communication and business process alignment) and do not label any of the CSFs as 'not important'.

5. Concluding remarks

This study aimed to investigate the user perspective towards CSFs for ERP system implementation. A survey was distributed to three Pakistani companies to identify the user perspective towards the CSFs identified in a literature review. The findings revealed that users from only one case company (α) regard two CSFs as not important for ERP system implementation, including minimum customization and performance measurement. All other CSFs were labelled as important by case company α, β, and γ. However, the analysis also indicates that different emphasis was put on the different CSFs. Users from all three case companies label the following four CSFs as 'most important' for ERP system implementation: (1) education and training, (2) strategic decision-making, (3) communication, and (4) business process alignment. Education and training and strategic decision-making, were also labelled as 'most important' CSFs by users from a previous study conducted by Reitsma and Hilletofth [11].

The CSFs that are labelled as 'most important' should according to users be prioritized when implementing an ERP system. This paper is a step closer towards identification of the user perspective, which ultimately promises to contribute to the design of more effective enterprise systems, its implementation, and management. Since this study utilised a survey strategy to highlight the perspective of users towards ERP system implementation, the underlying reason for the outcome is not explained. If further research provides similar findings, it would be interesting to conduct an exploratory study.

6. References

[1] Hilletofth P and Lättilä L 2012 Agent based decision support in the supply chain context *Industrial Management & Data Systems* **112** (8) pp 1217-1235

[2] Hilletofth P 2008 Differentiated supply chain strategy: response to a fragmented and complex market *Licentiate thesis* Chalmers University of Technology, Gothenburg, Sweden

[3] Hilletofth P, Hilmola O-P and Ujvari S 2010 Teaching ERP in logistics curriculum: A case experience from Sweden *Int. Journal of Business Information Systems* **6** (3) pp 295-314

[4] Aloini D, Dulmin R and Mininno V 2007 Risk management in ERP project introduction: review of the literature *Information Management* **44** (6) pp 547-567

[5] Chang B, Kuo C, Wu C H and Tzeng G H 2014 Using fuzzy analytic network process to assess the risk in Enterprise Resource Planning system implementation *Applied Soft Computing* **28** pp 196-207

[6] Lech P 2016 Causes and remedies for the dominant risk factors in Enterprise System implementation projects: the consultants' perspective' *Springer Plus* **5** (238)

[7] Saade R G and Nijher H 2016 Critical success factors in enterprise resource planning implementation *Journal of Enterprise Information Management* **29** (1) pp 72-96

[8] Huang S M, Chang I C, Li S H and Lin M T 2004 Assessing risk in ERP projects: identify and prioritize the factors *Industrial Management & Data Systems* **104** (8) pp 681-688

[9] Davenport T 1998 Putting the enterprise into the Enterprise System *Harvard Business Review* **76** (4) pp121-132

[10] Young H K, Jane P, Boo Y C and Saumyendu G 2012 Understanding users' acceptance of Enterprise Resource Planning (ERP) system in project-based sectors *IEEE Transactions on Engineering Management* **59** (2)

[11] Reitsma E and Hilletofth P 2016 Enterprise resource planning system implementation: an end user perspective *7th Int. Conf. on Operations & Supply Chain Management* Phuket

[12] Deng J L 1986 The stability of the grey linear system *Int. Journal of Control* **43** (1) pp 313-320

[13] Stedman C 1999 *Survey: ERP costs more than measurable ROI Computerworld* **33** (6)

[14] Umble H, Haft R and Umble M 2003 Enterprise Resource Planning: implementations and critical success factors *European Journal of Operational Research* **146** pp 241-257

[15] Gargeya V B and Brady C 2005 Success and failure factors of adopting SAP in ERP system implementation *Business Process Management Journal* **11** (9)

[16] Davenport T 2000 *Mission critical: realizing the promise of enterprise systems* (Boston, Massachusetts: Harvard Business School Press)

[17] Wee S 2000 Juggling toward ERP success: keep key success factors high *ERP News*

[18] Rosario J G 2000 On the leading edge: critical success factors in ERP implementation projects *Business World (Philippines)* **27**

[19] Akkermans H and van Helden K 2002 Vicious and virtuous cycles in ERP implementation: a base study of interrelations between critical success factors *European Journal of Information Systems* **11** pp 35-46

[20] Bingi P, Sharma M K and Godla J 1999 Critical issues affecting an ERP implementation *Information Systems Management* **16** (2) pp 7-14

Augmented halal food traceability system: analysis and design using UML

Y V Usman[1,3*], A M Fauzi[1], T T Irawadi[1,2], T Djatna[1]
[1]Department of Agroindustrial Technology, Bogor Agricultural University, Bogor, Indonesia
[2]Department of Chemistry, Bogor Agricultural University, Bogor, Indonesia
[3]Department of Industrial Engineering, Pancasila University, Jakarta, Indonesia

*yulita@univpancasila.ac.id

Abstract. Augmented halal food traceability is expanding the range of halal traceability in food supply chain where currently only available for tracing from the source of raw material to the industrial warehouse or inbound logistic. The halal traceability system must be developed in the integrated form that includes inbound and outbound logistics. The objective of this study was to develop a reliable initial model of integrated traceability system of halal food supply chain. The method was based on unified modeling language (UML) such as use case, sequence, and business process diagram. A goal programming model was formulated considering two objective functions which include (1) minimization of risk of halal traceability failures happened potentially during outbound logistics activities and (2) maximization of quality of halal product information. The result indicates the supply of material is the most important point to be considered in minimizing the risk of failure of halal food traceability system whereas no risk observed in manufacturing and distribution.

1. Introduction

In term of supply chain management, logistic activities are divided into two main activities, i.e., inbound and outbound logistics. Inbound logistics involves supply and receiving of raw material, processing, packaging and dispatching activities, while outbound logistics activities consist of distribution activities. Inbound and outbound logistics combine within the field of supply-chain management. Those activities of logistic are illustrated in Figure 1.

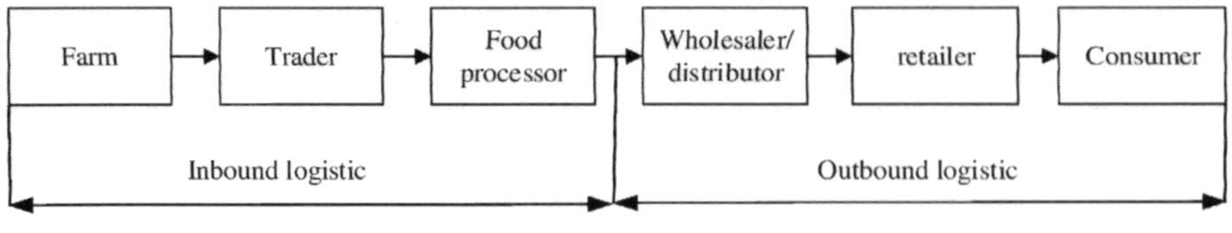

Figure 1. Logistic activities in food supply chain [1]

Halal traceability is one of critical success factors in halal supply chain management. Halal traceability means halal status should be maintained along the supply chain [2]. A fully integrated supply chain is needed to synchronize both inbound and outbound logistics in the halal traceability system to maintain product information and ensure the halal status from farm to fork. The traceability requires both forward and backward of product information (see Figure 2).

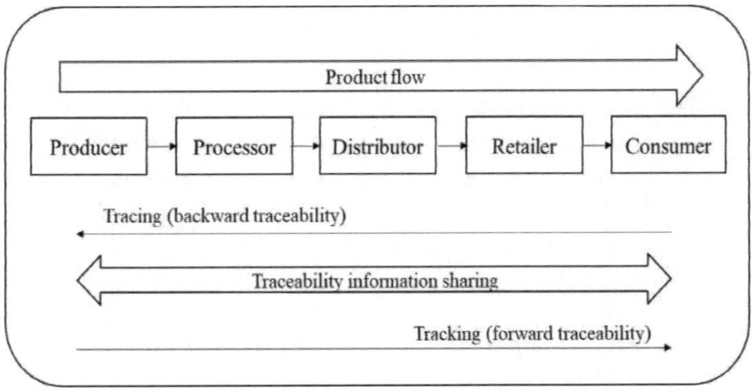

Figure 2. The concept of traceability in the food supply chain [3]

This study aims to identify and analyze the entities and the processes to develop halal food traceability system model by augmenting its system to cover the whole logistic activities through its supply chain. Based on the analysis, we design initial model to halal traceability system that encompasses inbound and outbound logistics. The model of this augmented halal food traceability system is expected to optimize the operational risk of the implementation of this system in food supply chain.

2. Methodology

The unified modeling language (UML) is used for analyzing and designing the model of this augmented halal food traceability system. As a methodology, standard diagrams of UML which relevant to this study are employed. They are use case, sequence and business process diagrams. Then the formulation for the risk of halal food traceability failure is analyzed by goal programming method. The step by step methodology is explained as follows.

2.1. Use case diagram

The step first of the system modeling is to identify the system's entities and then analyze the functional requirements of the system using use case diagram. This use case diagram represents the main actors and use-cases together with their relationships in the supply chain [4]. This study shows general group of actors and use cases to make the initial model become simple and understandable.

2.2. Sequence diagram

A sequence diagram visualizes the different processes or objects that exist simultaneously and the messages exchanged between those objects. The Sequence diagram is designed based on use case diagram.

2.3. Business process diagram

Business process diagram describes a system at the highest level. Business process diagram facilitates the graphical display from the control of the data flow among processes at each level of building a system. Business process diagram is also used to analyze the flow process that will be developed.

2.4. Optimization of the halal food traceability system model

The augmented halal food traceability system model is formulated by goal programming. Goal programming is the branch of multi-criteria decision making [5]. For this model, the formulation is considering two objective functions which include (1) minimization of risk of halal food traceability failures happened potentially during outbound logistics activities and (2) maximization of quality of halal food product information. The Goal programming model is given as below [5]:

$$Z_{min} = \sum_{k=1}^{m} (d_k^+ + d_k^-) \quad (1)$$

The d_k^+ and d_k^- are deviational variables that represent the level of achievement. It shows over and under achievement. Since goal programming for this study is the preemptive method where the level of importance of the variables is determined by the priority of the each variable [5].

$$Z_{min} = \sum_{k=1}^{m} P_k (d_k^+ + d_k^-) \quad (2)$$

$$Z_{min} = P_1^+ d_1^+ + P_1^- d_1^- + P_2^+ d_2^+ + P_2^- d_2^- \quad (3)$$

subject to:

$$\sum_{i=1}^{n} C_{ki} X_i + d_k^- - d_k^+ = Y_k, for\ k = 1, 2, \ldots, k \quad (4)$$

$$\sum_{i=1}^{n} a_{ij} X_i \leq or\ \geq t_j, for\ i = 1, 2, \ldots, n\ and\ j = 1, 2, \ldots, m \quad (5)$$

$$\forall d_k, \forall x_i \geq 0 \quad (6)$$

$$P_1 >>> P_2 >>> P_k \quad (7)$$

where:

k	=	the kth objective functions (k = 1, 2, ..., m)
X_i	=	the ith decision variable (i = 1, 2, 3.. n)
X_1	=	supplier
X_2	=	manufacturer
X_3	=	distributor/retailer. Because of distributor and retailer perform the same function, so in this study, we make it as just one variable to consider.
C_{ki}	=	coefficient of X_i on the kth objective function
Y_k	=	objective function (goal constraints)
a_{ij}	=	coefficient of constraint function
t_j	=	the amount of resources
P_k	=	the preemptive priority factors of the k goal
P_1	=	the preemptive priority factor to the risk of halal food traceability system failure
P_2	=	the preemptive priority factor to quality of halal food product information

3. Result and discussion

3.1. System entities and functional requirements

The augmented halal food traceability system was begun by identifying and analyzing system's entities. The entities that construct the system are input, process, output, stakeholder, resources, role, missions and object, and other related components (see Figure 3). The primary actors of the supply chain consisting of the supplier, manufacturer, distributor, and retailer [6]. Furthermore the stakeholder (actors) of the system and the functional activities, including their relationship, are shown in Figure 4. Based on system's entities and use case model was built the sequence diagram that represents information exchange among the actors and their processes in food supply chain. This sequence diagram can show the material and information flow both (see Figure 5).

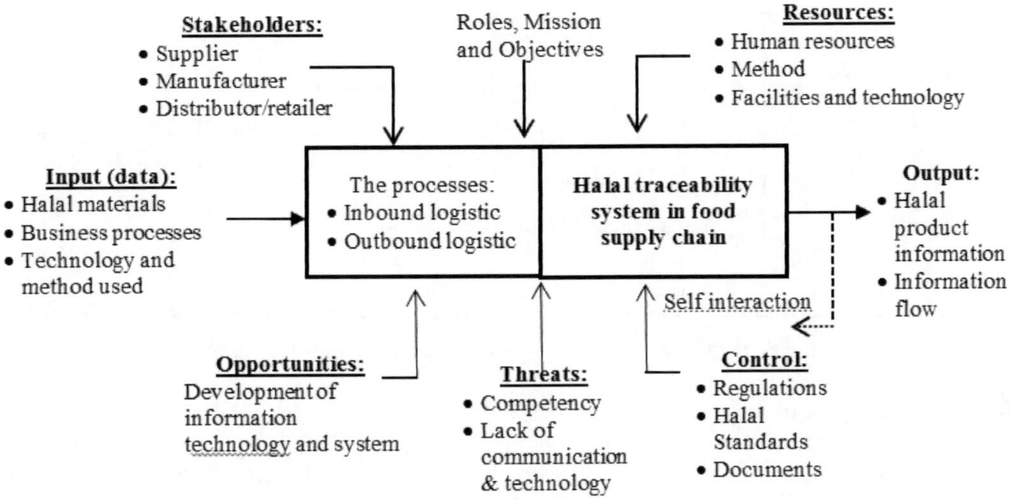

Figure 3. The entities of halal food traceability system

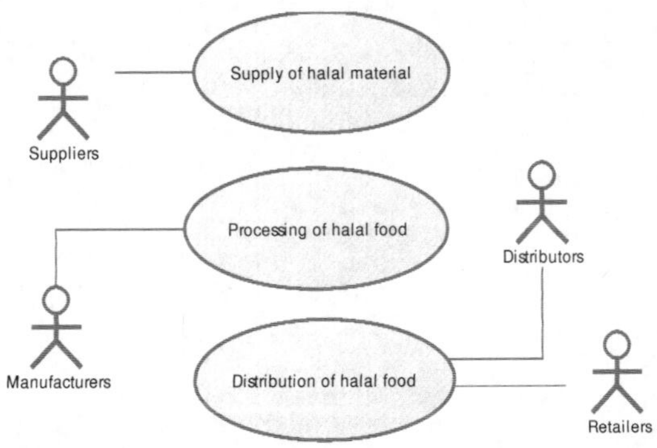

Figure 4. Use case diagram for halal food traceability system along logistic activities

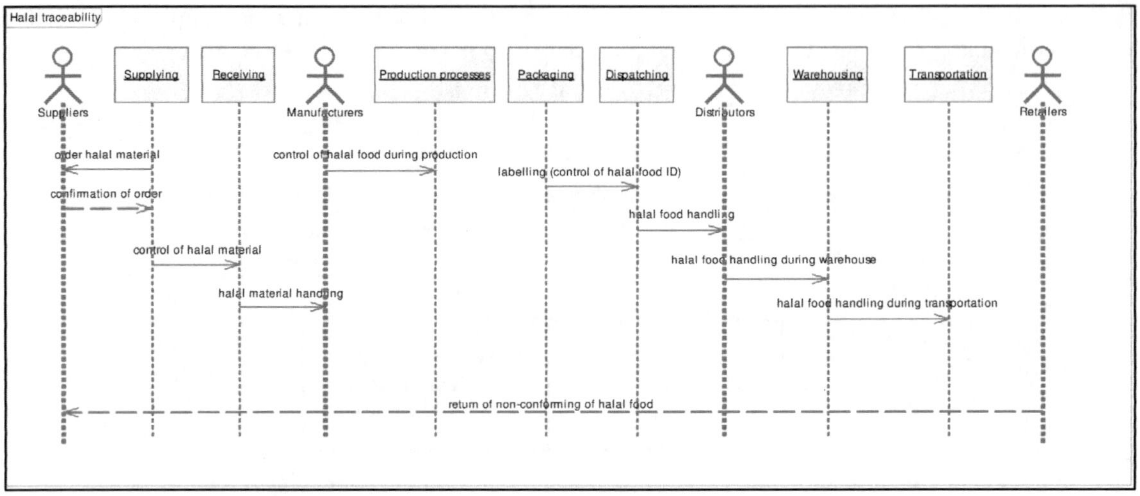

Figure 5. Sequence diagram for halal food traceability system

3.2. Business process model

The model of halal food traceability system is developed by using business process diagram. Business process model describes one or more linked procedures or activities which aim to achieve overall system goals. The result of analysis and modeling of business processes for halal traceability system in food supply chain is illustrated in Figure 6.

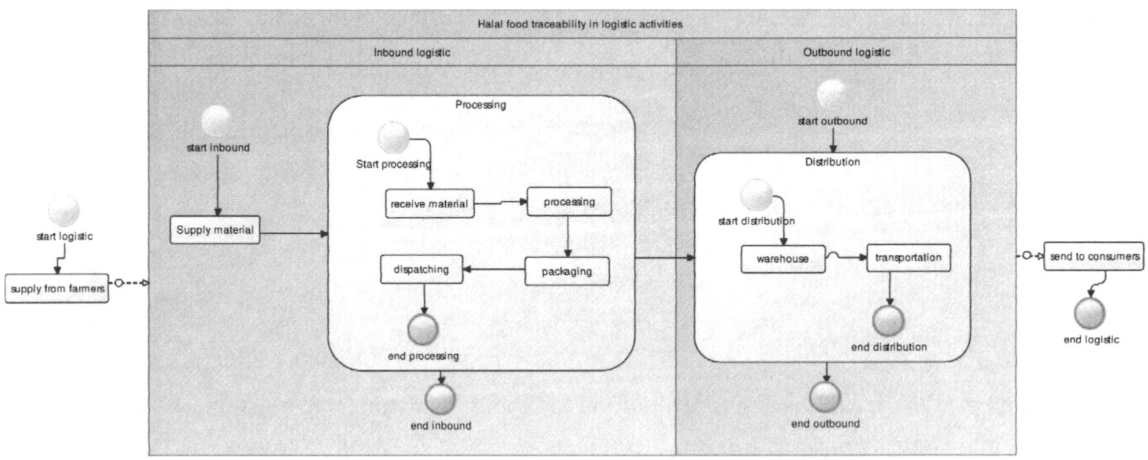

Figure 6. Business process model for halal food traceability system

3.3. Simulation of the halal food traceability system model

The simulation has been done to the optimization problem in halal food traceability model. The data that used in this case is generated from the random number in excel and use the scale of 1-5 (attribute number). Since the goal programming operates by the variable number then we have to do the normalization of the data. The goal programming problem can be solved with solver in excel [7]. Based on data processing obtained and the equations at section 2.4, then the goal programming formulated as follows:

- $0.47X_1 + 0.43X_2 + 0.50X_3 \leq 1$ risk of failures
- $0.43X_1 + 0.40X_2 + 0.46X_3 \geq 0$ quality of product information
- $X_1, X_2, X_3 \geq 0$

- $Z_{min} = P_1^+d_1^+ + 0*d_1^- + 0*d_2^+ + P_2^-d_2^-$ The objective function

subject to:

- $0.47X_1 + 0.43X_2 + 0.50X_3 - d_1^+ + d_1^- = 1$
- $0.43X_1 + 0.40X_2 + 0.46X_3 - d_2^+ + d_2^- = 0$
- $X_1, X_2, X_3, d_1^+, d_1^-, d_2^+, d_2^- \geq 0$

The solver result:

- $Z = 0; X_1 = 2.13; X_2 = 0; X_3 = 0; d_1^+ = 0; d_1^- = 0; d_2^+ = 0.91; d_2^- = 0$

From the simulation, the risk of failure halal food traceability system model is minimum or zero when the deviation of quality of product information is 0.91 and the value of supplier (X_1) is 2.13. The result of simulation also shows that the X_2 (manufacturer) and X_3 (distributor/retailer) does not work on the optimization of the problem as the values are zero. It shows that the supply of halal material influences the achievement of the objective. The halal material is not only limited to the product ingredient itself but also the material used for the packaging. Based on sensitivity analysis, the coefficient of X_1 (supplier) can be allowed to decrease. The optimal solution value will change as the objective function coefficient of X_1 is changing whereas the coefficient of the X_2 (manufacturer) and X_3 (distributor/retailer) remain the same.

Even though the simulation result showed that the critical point of halal traceability system in food supply chain is in the supply of material, the risk of failure may occur in manufacturing and distribution processes. For instance, in the fermentation process, the manufacturer may produce the non-halal product. In the distribution process, besides of the halal assurance of halal of packaging material, it also needs to pay attention in the handling of the product. It may happen contamination and mix of halal product and non-halal product cause of the sharing facilities or mislabelled. In the context of halal assurance, there is no tolerance for the risk of failure. For that reason, all the processes in halal food supply chain must be considered to assure the halalness of a product. The actors in every phase take their own responsibilities to do the processes and produce the product that complies with the Shariah law and provide the information of product during the processes it does.

4. Conclusion

The results obtained that the main actors (stakeholder) of the system which is divided into supplier, manufacturer and distributor/retailer. Related to the actors, the main functions of the system are supplying of halal material, processing, and distribution of halal food product. The information flow follows the material flow. The control of document and communication among the actors become important factors for the product information to be traced and tracked.

The optimization result showed that the risk of failure of halal traceability system in the food supply chain depends on the beginning of the process. The complete and accurate information from the supplier can decrease the risk of system failure. The optimized condition is achieved when the halalness of supplying of material can be assured. However, it must be considered the risks of producing non-halal products in the production process, the non-halal packaging material used, and possibly being contamination in the distribution process.

For the future work, the stakeholders and decision makers have to further analyze the critical processes throughout the food supply chain. The identification of resources may influence the optimization of the system. It is also important to find the better decision for the integrated halal food traceability system by using actual data.

References

[1] Dani S, Deep A 2010 Fragile food supply chains: Reacting to risks *Int J Logistics: Research and Applications* **13**(5) 395-410

[2] Ab Talib, M S, Hamid, A B A and Zulfakar M H 2015 Halal supply chain critical success factors: a literature review *J Islamic Marketing* ***6***(1) 44-71
[3] Bosona T, Gebresenbet G 2013 Food traceability as an integral part of logistics management in food and agricultural supply chain *Food Control* **33** 32-48
[4] Rhem A J 2006 *UML for Developing Knowledge Management Systems* (Ney York: Taylor & Francis Group)
[5] Schniederjans M J 1995 *Goal Programming: Methodology and Applications* (US: Springer)
[6] Dewi I A, Dania W A P and Wardani, B R K 2015 supply chain performance identification of horticulture product at cooperative brenjonk in Trawas, Mojokerto *Agriculture and Agricultural Science Procedia* **3** 163-8
[7] De Steiguer J E 2000 Applying excel solver to a watershed management goal-programming problem *USDA Forest Service Proc. RMRS-P-13* 325-9

An affective e-commerce design for SMEs product marketing based on kansei engineering

A N Habyba[1*], T Djatna[2], and E Anggraeni[2]
[1]Graduate School Program of Agro-Industrial Technology, Bogor Agricultural University, Bogor, Indonesia
[2]Department of Agro-Industrial Technology, Bogor Agricultural University, Bogor, Indonesia

*anik_bibib@apps.ipb.ac.id

Abstract. One of the SMEs problems in Indonesia in this information technology era is the inability to control the market. SMEs can use the e-commerce website to improve their competitiveness. It can be used as a marketing tool for SMEs to promote their products and expands the SMEs markets especially for the agroindustry SMEs where located in district area that still rely on local markets to sell their product. Some SMEs e-commerce websites have been developed in Indonesia but can not significantly increased the sales of SMEs product. Furthemore, the design of it is only able to meet the consumer need in function and usability. The development of e-commerce design should pay attention in high affective quality. This is because the affective responses effect the user's perception of cognitive quality, usability and ease of use of e-commerce. This study is aimed to make e-commerce that can meet the affective needs of users. The result of Kansei words selection and extraction using TF-IDF are four design concepts of e-commerce website. The formulation of new SMEs e-commerce website design is resulted from the integration of four design concepts and four design elements. The "Natural-Formal" concept has the greatest value than other concepts after QTT-1 analysis. This concept can implemented as the new SMEs e-commerce website design.

1. Introduction

Small and Medium Enterprises (SMEs) is one of the most important actors in Indonesia's development. In the last five years, agroindustry SMEs were able to contribute 48.85 percent to Gross Domestic Product (GDP) through tax payments [1].One of the problems of SMEs in Indonesia in the current era of information technology is the inability to attract the market. SMEs are not able to compete with other types of businesses that already adapt information technology in advance attract the market. This is the reason for the need of an information technology application for SMEs in order to encourage its growth in the current information technology era [2]. E-commerce can be used as a marketing tool for SMEs in promoting their products and expanding their product marketing network. E-commerce can be very helpful, especially agroindustry SMEs in the district area who still rely on local markets to sell their products. Using the e-commerce owners can increase revenue of the product and reducing operation and maintenance costs [3]. So the SMEs can improve their product competitiveness.

Nowadays there are some e-commerce sites that have been developed in Indonesia mostly to sell SMEs's product. For example krafie.com, usahadesa and kriya.com but their existing still can not significantly increased the selling of SMEs's product. These sites are a place for SMEs to sell their products online and are considered able to advance SMEs [4]. The competition between e-commerces in Indonesia was very tight, the SMEs's e-commerce have to compete with the bigger e-commerce websites which first succeed in Indonesia. The various offers of e-commerce have caused today's consumer have become further sophisticated then ever. The consumers desire product that match their own feelings of design [5]. [6] say initially e-commerce designers just focus on cognitive and usability function, whereas affective quality becomes more important than ever because the technology advances. Good affective quality of the e-commerce will affect its cognitive quality. So the developers of e-commerce have to meet the affective needs of e-commerce users. One method that can be used to determine the affective need of users is Kansei Engineering.

Kansei Engineering (KE) is a technology to assimilate human kansei to product design [7][8], in this paper the product is SMEs e-commerce website. The development of website design according to user concept will make the user have emotion with the website. In Indonesia, [9] using KE to design e-learning website interface and [10] using KE to evaluate the visual usability on mobile ecotourism applications. Our objective in this research are (1) to generate the new design concept of e-commerce website, (2) to identify the relevant e-commerce design element (3) to formulate the concept design of SMEs e-commerce website. The design could be used by SMEs e-commerce developers to make their emotionally appealing design better. This development would make user more satisfied and would increase sales of the product.

2. Research Methodology

This research frameworks are adopted based on [5], [6] and [11] to accomplish the research objectives. In this paper, SMEs e-commerce website has been chosen as product domain to develop. Furthermore, the research methodology is represented in Figure 1.

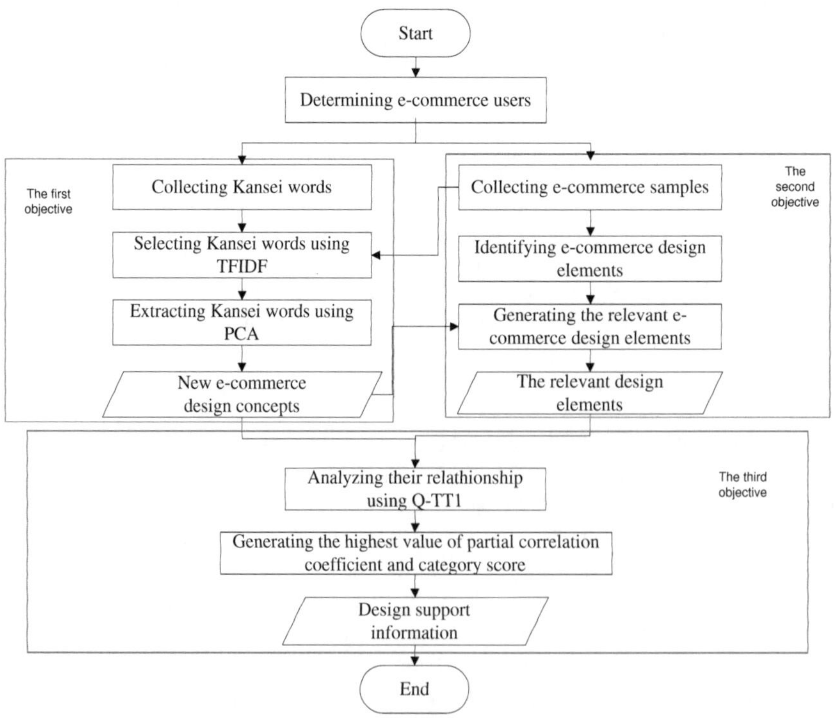

Figure 1. The methodology of the research

The research was began with the determination of the e-commerce users. The users of this e-commerce are consumer of agroindustry SMEs , owners and local goverment. Afterward, the KW were collected from literature research and then folowed by selecting and extracting them using Term Frequency-Inverse Document Frequency (TF-IDF) and Principal Component Analysis (PCA) method to achieve the first objective. From this, representative KW were chosen and were became the strategies of new e-commerce website design. For the second objective, design elements were identified for each sample collected and were evaluated according the new design concept. Finally, the new concept and design elements were synthesized using Quantification Theory Type I (QTT-1) method. In this step, once questionnaire were given to the users to evaluate e-commerce properties based on strategy design. R language software was used as a tool of data processing.

3. Result and Discussion

3.1 New Design Concept Identification

In this phase, firstly thirty Kansei words have been collected from literature research which illustrated SMEs e-commerce website. The KW that have been collected is represented in Table 1.

Table 1. The weight of Kansei words.

Number	Kansei Words	Weight	Number	Kansei Words	Weight
1	**Elegant**	**1.5**	16	Crowded	1.3
2	**Feminine**	**1.5**	17	Impressive	1.3
3	**Formal**	**1.6**	18	Calm	1.3
4	Warm	1.3	19	Bright	0.7
5	Beautiful	1.1	20	Unique	0.9
6	Informative	1.3	21	Lively	1.3
7	Childish	1.3	22	Comfortable	1.1
8	**Classic**	**1.5**	23	**Fresh**	**1.6**
9	Interesting	1.4	24	**Simple**	**1.6**
10	**Sweet**	**1.5**	25	**Natural**	**1.8**
11	Cool	1.4	26	Neat	0.9
12	**Creative**	**1.5**	27	Cute	1.3
13	Interesting	1	28	**Fun**	**1.6**
14	**Luxury**	**1.5**	29	**Futuristic**	**1.8**
15	Modern	1.3	30	Chic	1.3

These KW were evaluated using TF-IDF and researcher take 12 KW that have weight more than 1.5. They are elegant, feminine, formal, classic, sweet, creative, luxury, fresh, simple, natural, fun and futuristic. Thus, the result of evaluation has been analyzed using Principal Component Analysis (PCA) method in order to reduce the dimension of KW collected and followed by generating new concepts design. Principal components (PC) are computed based on evaluation of twelve Kansei words using semantic differential questionnaire. PCA method is used to decide which of principal component will be retained according to Kaiser's criterion proposed by Coghlan. The result of variances each principal component is shown in Table 2. Suhr says cumulative proportion of variance explained can be 70-80% based on Kaiser criterion too [12]. The result show that PC1 until PC4 have significant slope and they total proportion of variance 76%. These four PC were chosed to be developed as design concept of the SMEs e-commerce website.

Table 2. The value of variances and proportion of variance.

	Variance	Proportion of variances	Cumulative Proportion
PC1	**1.75**	**0.26**	**0.26**
PC2	**1.62**	**0.22**	**0.47**
PC3	**1.47**	**0.18**	**0.65**
PC4	**1.13**	**0.11**	**0.76**
PC5	1.07	0.10	0.86
PC6	0.86	0.06	0.92
PC7	0.82	0.06	0.97
PC8	0.56	0.03	0.99
PC9	0.13	0.001	1.00

3.2 Design elements

In order to answer the second research objective, nine e-commerces which popular in Indonesia have been collected. There are three SMEs e-commerce websites that have been collected for sample and others are popular e-commerces in Indonesia. The result of design element identification phase showed that there were 4 categories of design elements, it shown in Table 3. Each design element has different from types of its own, ranging from two until three. The next step is evaluating 4 pair of concepts based on 9 samples of e-commerces design. This evaluation using semantic differential questionaire with 1-7 point scale, where 1-7 represented standard look and the most attractive look, respectively. Table 4 shows the assesment result, as following:

Table 3. Design elements.

	Element design	1	2	3
X1	Body background color	white	grey	
X2	Typography	custom	standard	
X3	Page style	frame	table	none
X4	Dominant item	picture	ads	text

Table 4. Numerical data source for nine representative e-commerce website samples.

Sample	X1	X2	X3	X4	Natural - Formal	Simple - Complex	Classic - Modern	Standard - Fresh
1	1	1	2	1	3.6	5.1	4.9	5.9
2	2	1	2	3	3.9	4.2	3.6	3.8
3	2	1	1	2	4.2	4.7	3.7	4.3
4	1	2	2	3	4.8	3.7	4.6	4.0
5	1	1	1	2	5.4	5.0	4.9	5.6
6	2	1	3	2	3.9	5.6	4.0	4.8
7	2	1	1	2	4.3	5.3	4.9	4.6
8	1	1	3	2	4.3	3.3	3.3	3.6
9	1	2	1	1	4.8	3.2	4.0	3.3

3.3 New SMEs e-commerce website design formulation

The last step was to examine the relationship between four design elements (independent variables) and four pair of design concepts (dependent variable) using QTT-1 method. In this study calculation QTT1 are supported by R software, to generate R-square value, optimum category grade, and partial correlation coefficient (PCC). From R-square value shows that "Natural-Formal" concept has the highest value than other concepts; 0,91. So "Natural-Formal" concept will be used as a main image of SMEs e-commerce website design. The result of formulation design showed in Figure 2 and Table 5. The partial correlation coefficients (PCC) indicates the relationship between the four design elements and a pair of design concepts. The highest variable of PCC on "Natural-Formal" concept was "Page style" (X3 = 0.93), so this variable was great influence on customer perception

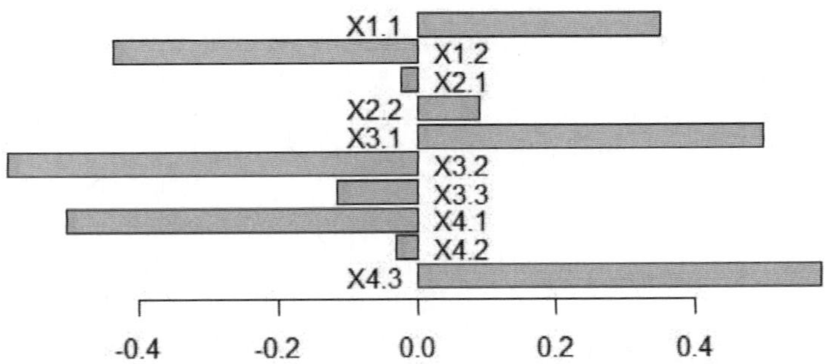

Figure 2. QTT1 bar graphic score category of natural-formal concept.

Table 5. Value of category grade and partial correlation coefficient each design elements.

Design Elements		Type		Category grade (from type grade)		Partial correlation coefficient (PCC)
				Natural	Formal	
X1	Body background color	X1.1	white		0.35	0.89
		X1.2	grey	-0.43		
X2	Logo typografy	X2.1	costum	-0.03		0.25
		X2.2	standard		0.09	
X3	Page style	X3.1	frame		0.5	0.93
		X3.2	table	-0.59		
		X3.3	none	-0.12		
X4	Dominant item	X4.1	picture	-0.5		0.88
		X4.2	ads	-0.03		
		X4.3	text		0.58	
						Constants = 4.4
						R square = 0.91
						R square = 0.65

4. Conclusion

This research focused on SMEs with agroindustry product to demonstrate how Kansei Engineering using QTT-1 Analysis was applied to assist e-commerce designers in decision making of the new SMEs e-commerce website design. The result of TF-IDF showed 12 KW have been selected, then they were extracted using PCA method and become 4 strategies of the new e-commerce design concept. They are "Natural-Formal", "Simple-Complex", "Classic-Modern" and "Standard-Fresh". Four concepts were analyzed with four design elements (body background color, logo typografy, page style and dominant item) of the e-commerce website using QTT-1 analysis. The result showed that "Natural-Formal" concept has the greatest value of R^2 than others. This design concept is recommended for a new SMEs e-commerce website design.

References

[1] Bank Indonesia 2015 Micro Small and Medium Enterprises Business Profile
[2] Sudaryanto R and Wijayanti R 2011 Strategy of SMEs Empowerment Facing Asean Free Market
[3] Khan A G 2016 Electronic Commerce : A Study on Benefits and Chalenges in an Emerging Economy Global *J. Management and Business Research: B Economics and E-commerce* **16** (01)
[4] Ramadhani S 2017 Five E-Commerces in Indonesia for SMEs More Forward. Available: http://goukm.id/*e-commerce*-marketplace-untuk-ukm/
[5] Lookman A M and Noor N M 2006 Kansei Engineering Concept in E-commerce website. *Proc. in Int. Conf. on Kansei Engineering and Intelligent Systems (pp. 117-124) KEIS 06*
[6] Lookman A M, Noor N M, and Nagamachi M 2008 Kansei Database System For Emotional Interface Design of E-Commerce Website. *Proc. in The Fourth International Cyberspace Conference on Ergonomics* (Cyberg 08)
[7] Nagamachi M 1999 Kansei Engineering: The Implication and Applications to Product Development. *Proc. in 1999 IEEE International Conference*
[8] Nagamachi M 2011 *Kansei/affective engineering* (CRC Press).
[9] Wicaksono F, Hadiana A and Asfi M 2016 Kansei Engineering Implementation on Web Based E-Learning Interface Design. *Proc. Nat. Conf. of Information Technology and Multimedia 2016* **4** (1)
[10] Aktiva R, Yani N, and Djatna T 2015 *A Design of Visual Usability for Ecotourism Mobile Aplication Using Kansei Engineering* (Master's thesis Bogor Agricultural University).
[11] Lookman A M and Noor N M 2008 Kansei Engineering : A Study on Perception of Online Clothing Websites. *Proc. of the 10th Int. Conf. on Quality Management and Operation Development 2008 (QMOD '07)*
[12] Suhr D D 2005 Principal Component Analysis vs Exploratory Factor Analysis. *Proc. SUGI 30*

IOP Conf. Series: Materials Science and Engineering **337** (2018) 012052 doi:10.1088/1757-899X/337/1/012052

The effect of emotional design and online customer review on customer repeat purchase intention in online stores

D S Dewi[1]*, A Sudiarno[1], H Saputra[1]***, and R S Dewi[1]**
[1]Department of Industrial Engineering, Institut Teknologi Sepuluh Nopember, Kampus ITS Sukolilo-Surabaya 60111, Indonesia

*dyahsd@gmail.com, **adithya.sudiarno@gmail.com, ***paraperantau@me.com

Abstract. The internet users in Indonesia has increased rapidly over the last decade. A survey conducted by Association of Internet Service Providers Indonesia shows that the internet users has reached 34.9% of total population in Indonesia. The increase of internet users has led to a shift in trading practice from conventional trade to online trade. It is predicted in the next years the number of online consumers in Indonesia will continue to increase, provide many opportunity for online business. The huge number of internet users is not necesarily followed by the high number of e-purchase. It is therefore become the interest of many researchers to investigate factors that influence the decision on online purchasing.This research proposes a model that assess the effect of emotional design and customer review to customer intention on e-repeat purchase. Online questionnaire is designed and is distributed randomly through google forms. There are 187 respondent filled the questionnaire from which only 162 respondents actually have experience in online purchase. These data are then processed by using statistical analysis. A model is developed by applying structural equation modeling (SEM) approach. This study revealed that customer reviews especially objective reviews has a significant effect toward repeat purchase. Whereas emotional design particularly visual attractiveness also shows a significant effect toward e-repeat purchase.

1. Introduction

The number of internet users in Indonesia has increased significantly. In 2005, for example, there were 16 million people used the internet which then increased to 88.1 million in 2014. A survey conducted by Association of Indonesian Internet Service Providers (also known as APJII- *Asosiasi Penyelenggara Jasa Internet Indonesia*) found that about 34.9% of total population in Indonesia used internet service [1]. Survey conducted by APJII shows that internet users in Indonesia is dominated by millenial generation, a generation who most of them used the internet intensively as a main support for their job. As the number of internet users in Indonesia increase rapidly, it has triggered a shifting from conventional transaction into online transaction. Online store begin to reduce the role of traditional market [2]. It is predicted that until 2020, the number of consumers who spend their money to buy a product via online store will continue to increase in Indonesia [3].

Currently, there are various online stores available in Indonesia. Some examples of these online stores which is mostly accessed in Indonesia are Tokopedia, Bukalapak, OLX, Blibli.com, FJB Kaskus, Lazada, Belanja.com, Indonesia eBay, Amazon Indonesia, MatahariMall.com, Elevania, and Zalora. While there are many options of online stores in Indonesia, a survey found that there are only 27%

people who actually shopped via online stores [1]. There are 3 highest reasons for this. First is due to 'do not know how', which represent a condition where the technology unable to fit particular user need as it not easily to be operated by the customer. Second reason is due to customer feeling of 'not secure', which lower the customer's intention to continue the transaction in online store. Third, is due to the price where online store sell higher price than conventional stores [4].

There are many factors that are believed to have affect on purchase intention. One of factors is Online Customer Review (OCR). OCR can be easily found in various online stores. Some research state that OCR can increase customer trust and purchase intention [5][[6]. OCR, or also known as customer reviews, testimonials, reviews, or electronic comments on an object or a subject in Internet, is a form of electronic Word of Mouth (eWOM) [5]. The presence of OCR in online store can help customer to get more description of the product or service, so that customers can set their expectation before they make purchase decision. Previous study has divided customers review into 12 dimensions where each dimension has its own indicators [7]. Another factor that can also affect the purchase intention is the design of online store site. In this context, emotional aspect of design need to be considered as it can indirectly effect customer's cognitive aspect during the purchasing process [8]. A repeat customer who make repeat purchase in the same online store is believed can provide five times more profit than new customers [10]. Because of this, it is very important for online business owner to understand how to attract existing customer to make repeat purchase. Curently, there are not much research on e-repeat purchase in regard to customer reviews and emotional design. This study, therefore, propose a model that can be used to analyze the effect of customer reviews and emotional design toward repeat purchase intention.

There are four primary objectives of this research which include; to investigate which indicator that contribute most to OCR; to find out which indicator that contribute most to emotional design; to find out the effect of OCR to repeat purchase intention on online store; and to analyze the effect of emotional design to e-repeat purchase intention. Latent variables used in this research include; customer reviews, emotional design, perceived ease of use, perceived usefulness, customer intention, and repeat purchase intention. These variables are chosen based on the results from previous studies. In this research, Structural Equation Model (SEM) is applied since the reseach use many attributes [5] [6][7][9][11].

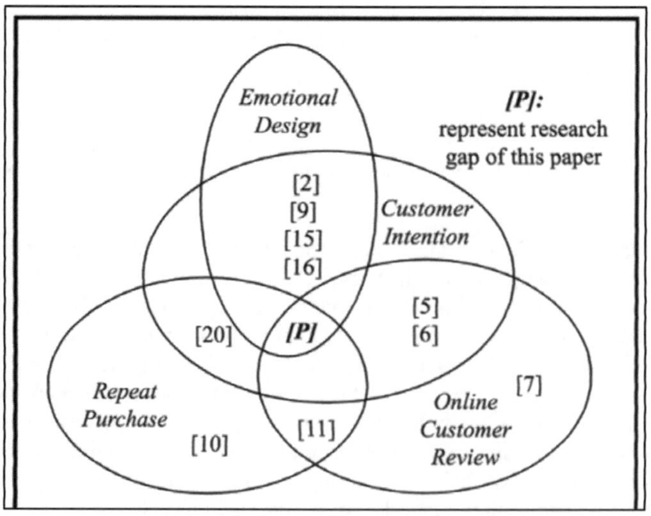

Figure 1. Diagram Venn of Research Position.

This research is intended to fulfill the gap in this research area. Research gaps were identified through a literature review. Current research position is illustrated in figure 1 where [P] represent position of this current research.

2. Literature review

Internet is an abbreviation of interconnection networking which simply can be defined as 'a global network of computer networks'. Among various benefits obtained from the usage of internet, it helps people to connect everywhere around the world as it now shifting a trend into online shopping. Internet site can be described as a collection of pages that display; information text data, picture, animation, voice, video, or a combination of these, whether they are static or dynamic.

2.1 Emotional design of online store

In designing an online store, visual appeal plays a key role in influencing customer's response and experience when interacting with an online store site [12]. It is important to consider emotional design on design process, as it has a significant effect on customer intention to make transactions [2][9][13][14][15][16]. Study of human computer science especially related to emotional design categorized emotional design into three types. First is visceral design that adapted from surrounding cultures, just like traditional art or kid toys, which its type can be easily recognized. Second is behavioral design that strongly related to function, understandability, usability, and physical feel of the product. Third is reflective design that related to a message about cultures, the meaning of the product or its function as well, which is able to trigger particular emotions just like feeling of proud or prestige from the product [8].

Online store is a dynamic mixture between technology, applications and business processes that connect corporate, consumer, and certain communities through; electronic transactions, trade, goods, services, and information made through the internet. Like other computer technology, the success of its implementation will depend on user acceptance. Technnology acceptance model (TAM) is an approach for predicting user acceptance attitude towards computer technology. Development of TAM begins by introducing four key variable which are perceived ease of use, perceived usefulness, behavioral intention to use, and actual system use [17]. This research propose a model that construted from two primary variables (perceived usefulness and perceived ease of use) which based on TAM, and additional variable that related to emotional design, namely, perceived visual appeal [9].

2.3 Online customer review

Online Customer Review (OCR) is a descriptive sentences or articles written by customer, mostly made based on users 's experiences from previous online purchase of product or service. OCR is a facility that allows customer to post their comments and opinions freely and easily about various products or services from online store. Some studies have proved that OCR can affect customer decision to purchase a product from online store [5][6]. OCR can be divided into several dimensions for examples; online review structure and style, positive online reviews, negative online reviews, helpfulness rating, number of online reviews, argument quality, author information, online review title, moderate online reviews, and online review shortness [7].

2.4 Repeat purchase intention

Repeat purchase intention is a purchase from the same customer more than three times [11]. Many studies are trying to understand what detemine customer purchase intention in purchasing decision. There are five stage of buying process; recognition of the need, search of information, evaluation of alternatives, buying decision, and post-purchase behavior [18]. Customer's satisfaction and trust as result of buying experience through these overall processes will affect the repeat purchase intention.

3. Research Methods

The current research is conducting through several steps. Detailed research steps can be seen in figure 2.

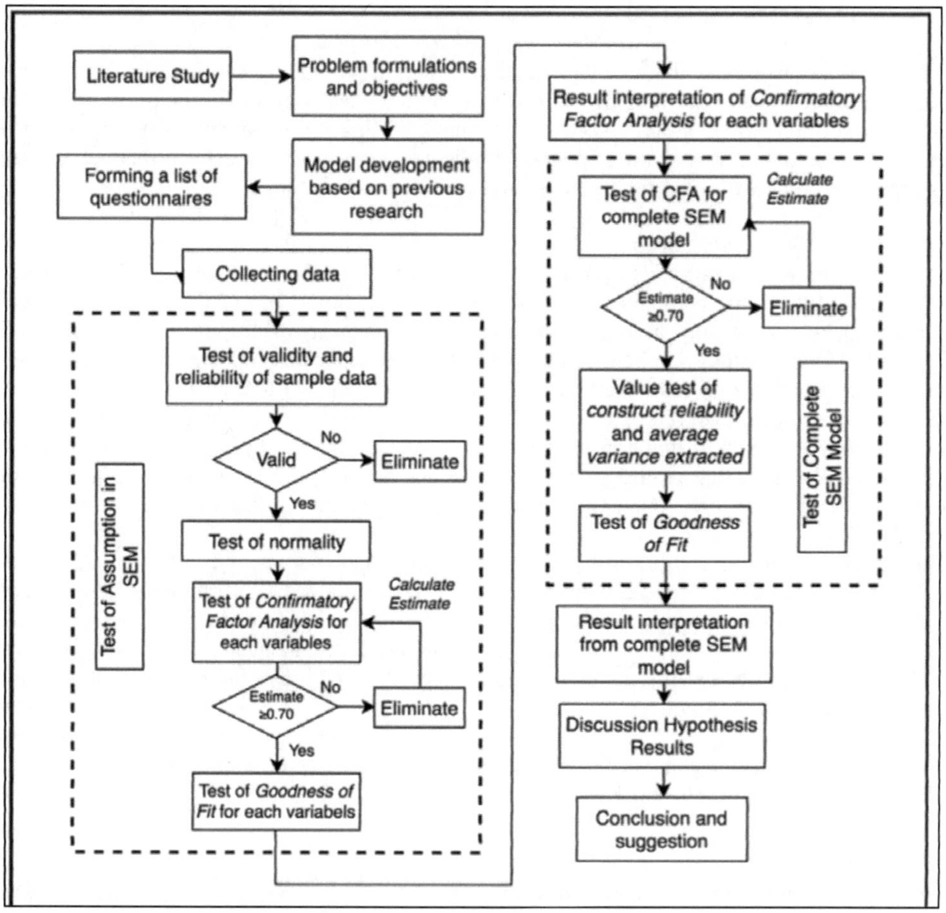

Figure 2. Flowchart of Reseearch .

Questionnaires are designed and shared via social media as a method for collecting data. There were 29 main questions in this research. Questionnaires is designed to use 6-point of likert scale; where score 1 represent 'Strongly Disagree' while score of 6 represent 'Strongly Agree'. By using Google Forms, respondents can easily access questionnaire via shared link on social media such as Facebook, Instagram, Twitter, and Path. There are 187 respondents participated in this survey. However, only 162 data were used as it more relevant for the research.

Several statistical tests was conducted to check the quality of questionnaires as a survey instrument. There are two group of statistical tests; test of assumption and test of complete model. In testing the assumption firstly, the validity test was conducted to measure the validity of questionnaires. The validity test evaluates whether all questions are relevant to the purpose of survey. The test is necessary since the the quality of data can be affected by the survey instrument [19]. Criteria on determining the validity of the questionnaires are; if *r-count* > *r-table* then the questionnaires are valid, but if *r-count* < *r-table* then the questionnaires are not valid. After validity test, next step is to conduct reliability test. The purpose of reliability test is to check the consistency of the result when measuring the same symptoms [19]. Criteria in determining whether the questionnaires are reliable or not is; if it is positive or *r-alpha* > *r-alpha* then the questionnaires are reliable, but if negative or *r-alpha* < *r-alpha* then the questionnaires are not reliable. After reliability test, the normality test is conducted to check whether the data distributed as normal distribution. The tests were run by using SPSS19. Next steps are confirmatory factor analysis and goodness of fit..

After the serial assumption tests, the complete model then is tested. The model analyzed by using Structural Equation Modeling (SEM) with software AMOS22. This process was performed to elucidate the relationship between variables that used in the proposed model. SEM is a set of statistical techniques that allow the testing of a series of relationships simultaneously. The underlying reason for the use of SEM is because of its ability to estimate the relationship between multiple connected variables.

4. Results and discussion

4.1 Proposed model

Variables used in this research are based on research by [7][9][17][9][20]. These include customer review, perceived visual appeal, perceived ease of use, perceived usefulness, customer intention, and repeat purchase. There are eleven hypotheses formulated in this study:

Hypothesis 1 (H1): Perceived usefulness affect customer intention
Hypothesis 2 (H2): Perceived ease of use affect perceived usefulness
Hypothesis 3 (H3): Perceived visual appeal affect perceived ease of use
Hypothesis 4 (H4): Perceived visual appeal affect perceived usefulness
Hypothesis 5 (H5): Customer review affect perceived usefulness
Hypothesis 6 (H6): Customer intention affect repeat purchase
Hypothesis 7 (H7): Perceived visual appeal affect repeat purchase
Hypothesis 8 (H8): Customer review affect repeat purchase

Figure 3 presents the proposed model. Each connected lines related to each hyphothesis.

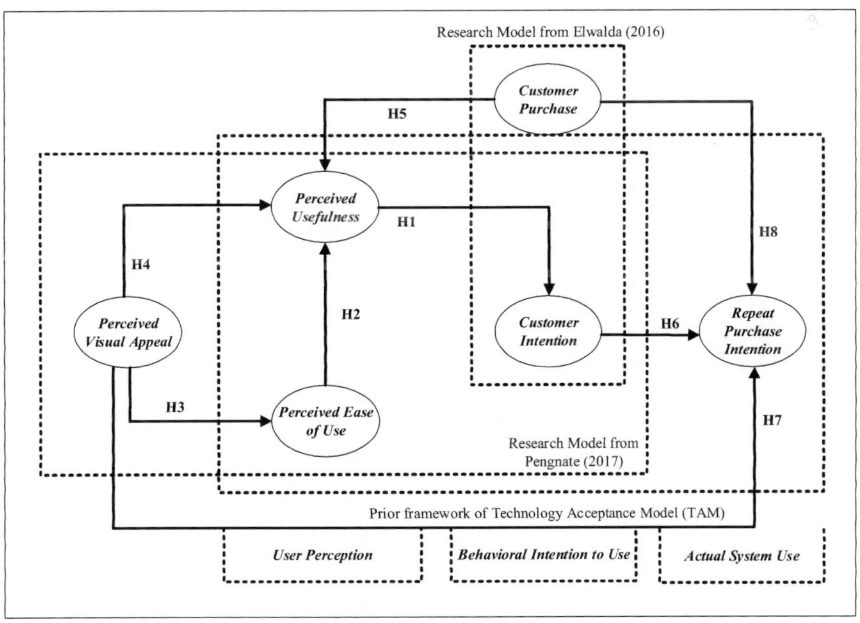

Figure 3. Proposed model

4.2 Validity, reliability, and normality test results
As mentioned previosly, the 162 data were tested for their validity, reliability, and normality. The validity test present that values of each *r-count* ranged from 0.360 to the highest at 0.746. Since the results shows higher value than *r-table* which is ≥ 0.153, mean that every questions were valid. Whereas the reliability test on each variable shows that all the question were reliable since value of each *alpha cronbach have* a higher value than *alpha cronbach's ≥ 0.8*. Moreover, the normality test based on scatterplot and KMO – Bartlett's Test shows each plot were scattered in a line to right upside of the graph, meaning that data were normal. The test result in KMO value 0.952 with bartlett's value of 0.000 shows that a high sensitivity of test.

4.3 Confirmatory factor analysis and goodness of fit result
Construct was made by using shaping tools which connect each indicators to main variable. There are six variables used in this research, each constructs were made based on these variables. From the test with AMOS 22, indicators for each variable that have estimate value ≥0.70 can be accepted, otherwise should be eliminated. The indicators that meet this requirement are: 'objective reviews' with estimate value at 0.765; 'visually attractive' with estimate value at 0.869; 'performance improved' with estimate value at 0.788; 'easy to learn' with estimate value at 0.823; 'ordering from online store' with estimate value at 0.881; and 'desire to revisit' with estimate value at 0.886.

4.4 SEM model
After Confirmatory Factor Analysis (CFA), the complete SEM model is established. The overall construct with 6 variables and their own indicators can be seen in figure 4.

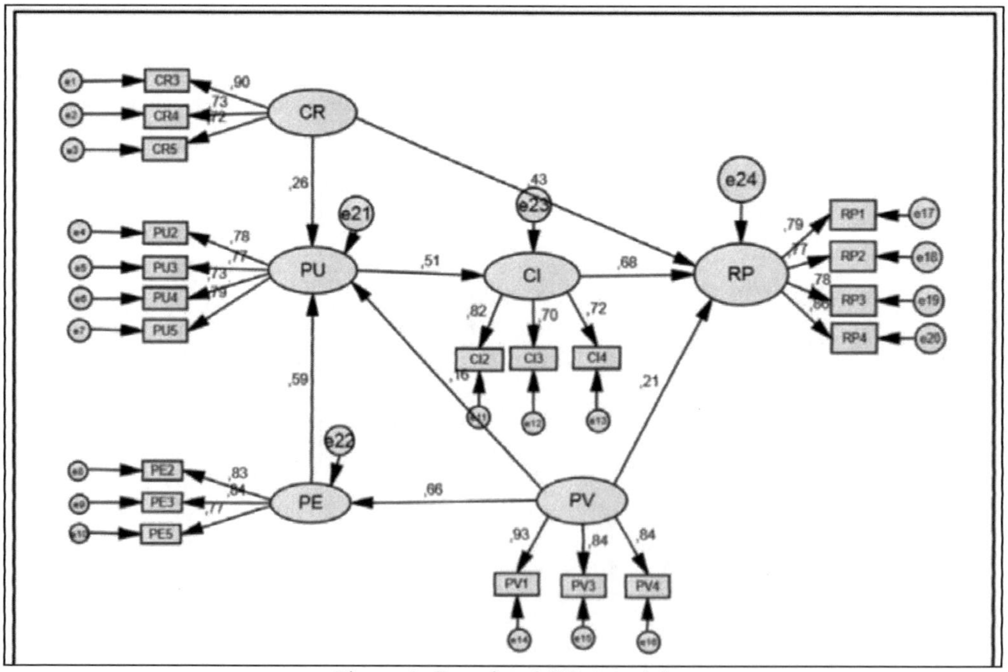

Figure 4. Completed SEM model

Statistical tests are conducted to measure the effect between variables and their level of significance. The effect between variables can be seen from their estimate value. The greater the value of the relationship between variables means the stronger effect between variables. On the SEM model there are requirements for parameter estimation where construct reliability should have value ≥0.70 and average variance extracted should have value of ≥0.50 [21].

The test result shows the value of each indicator and variables met the requirements, as every construct reliability are ≥0.70 and average variance extracted are ≥0.50 as demonstrated in table 1 [22].

Table 1. Construct reliability and average variance extracted of each variables.

Variable	CR	AVE	Variable	CR	AVE
Customer Review	0.831	0.625	Perceived Ease of Use	0.852	0.590
Perceived Visual Appeal	0.902	0.754	Customer Intention	0.791	0.560
Perceived Usefulness	0.854	0.661	Repeat Purchase	0.875	0.636

As the complete model met the reliability and validity requirements, the goodness of fit test now can be executed.The result is depicted on table 2.

Table 2. Goodness of fit from complete model

Goodness of fit index	Cut of Value	Value	Decision	Reference
CMIN/DF	≤3.0	2.300	Acceptable Fit	Ferdinand, 2000
RMSEA	0.05 – 0.10	0.090	Moderate	Byrne, 2001
CFI	≥0.80	0.881	Permissible	Byrne, 2001
GFI	0 - 1	0.819	Model Fit	Ferdinand, 2000

Based on the result of the goodness of fit, the conclusions are:

a. CMIN / DF (the minimum sample discrepancy function / degree of freedom) value of the proposed model is 2,300. It means acceptable fit [22].
b. RMSEA (root mean square error of approximation) is 0.090. It means moderate criteria, where the value range from 0.05 to 0.10 [23].
c. CFI (comparative fit index) is 0.881. It means permissible, with the category of ≥ 0.80 [23]
d. GFI (goodness of fit index) is 0.819. Although the recommended value is greater than 0.90 [22], but if the values close to 1. It can considered as fit. In this model the value of GFI were close to 1 so it is fit.

From AMOS22, the value of t-value and P-value is obtained. This becomes the based of decision whether the result are strongly significant or not. Table 2 depicts hypothesis decision for the proposed model. It shows only H4 was rejected as its t-value ≤ 1.96. P values with *** (three stars symbol) represent a condition where the P values almost reach 1.000 which mean that the hypothesis is strongly significant, while for H4 its P value is only 0.096.

Table 3. Result of proposed hypothesis analysis.

	Hypothesis	t-value ≥1.96	P	Decision
H1	*Perceived Usefulness* can affect *Customer Intention*	2.936	***	Strongly Significant
H2	*Perceived Ease of Use* can affect *Perceived Usefulness*	4.732	***	Strongly Significant
H3	*Perceived Visual Appeal* can affect *Perceived Ease of Use*	3.020	***	Strongly Significant
H4	*Perceived Visual Appeal* can affect *Perceived Usefulness*	1.530	0.096	Not Significant
H5	*Customer Review* can affect *Perceived Usefulness*	9.257	***	Strongly Significant

H6	*Customer Intention* can affect *Repeat Purchase* Intention	8.281	***	Strongly Significant
H7	*Visual Appeal* can affect *Repeat Purchase* intention	1.961	***	Strongly Significant
H8	*Customer Review* can affect *Repeat Purchase* intention	3.809	***	Strongly Significant

5. Conclusion

Conclusions has been made in relation to the four main objectives. First, this research found that the strongest indicator that affect customer review (CR) is objective review (CR3). Second, reseach also indicate that the indicator that affect most percieved visual appeal (PV) is visual attractiveness (PV1). Third conclusion is, the high quality of objective review can greatly affect e-repeat purchase. Finally, it is important to considering the emotional design aspect on designing online store since visual attractiveness can significantly affect customer to make a repeat purchase.

6. Future research

For further research, it issuggested to explore other variables that might affect e-repeat purchase. Current construct could be improved by considering variables, such as, satisfaction and trust.

7. References

[1] Asosiasi Penyelenggara Jasa Internet Indonesia (APJII) 2014 *Profil Pengguna Internet Indonesia.* Perpustakaan Nasional RI: Katalog Dalam Terbitan

[2] Gunarsih T 2012 *Pengaruh desain web dan kualitas informasi terhadap minat pembelian pada toko online*. Universitas Teknologi Yogyakarta. Yogyakarta

[3] Godjali H, Wong T T, Marino R, Sleigh A and Davarzani L 2012 *Ready for Indonesia's Digital Future?* (Jakarta: Accenture Indonesia)

[4] Deloitte Consumer Insights 2015 Capturing Indonesia's Latent Markets *Consumer Business*

[5] Minnema A, Bijmolt T H A, Gensler S and Wiesel T 2012 To keep or not to keep: effect of online customer review on product returns *Journal of Retailing* **92** pp 253-267

[6] Elwalda A, Lu K and Ali M 2016 Perceived derived attributes of online customer reviews *Computer in Human Behaviour* **56** 306-319

[7] Gottschalk S A and Mafael A 2017 Cutting throught the online review jungle – the investigating selective eWOM processing *Journal of Interactive Marketing* **27** pp 89-104

[8] Norman D A 2004 *Emotional Design: Why We Love (or Hate) Everyday Things* (First Ed.). New Year. Basic Books

[9] Pengnate S F and Sarathy R 2017 An experimental investigation of the influence on website emotional design feature on trust in unfamiliar online vendors *Computers in Human Behavior* **67** pp 49-60

[10] Chiu C M, Hsu M H, Lai H and Chang C M 2012 Re-examining the influence of trust on online repeat purchase intention: the moderating role of habit and its antecedents. *Decision Support Systems* **53** pp 835-845

[11] Moriuchi E and Takahashi I 2016 Satisfaction trust and loyalty of repeat purchase customer within the Japanese online supermarket trade. *Australian Marketing Journal* **24** pp 146-156

[12] Nielsen J and Loranger H 2006 *Prioritizing web usability* (CA: New Riders Press)

[13] Kukar-Kinney M 2009 The relationship between consumer's tendencies to buy compulsively and their motivaitons to shop and buy on the internet *Journal of Retailing* pp 298-307

[14] Wang Y J 2011 Aesthetics and the online shopping environment: understanding consumer responses *Journal of Retailing* **87** pp 46-58

[15] Hasan B 2016 Perceived irritation in online shopping: the impact of website design characteristics *Computer in Human Behaviour* **52** pp 224-230

[16] Gunawan I H 2016 *Pengaruh e-store design quality pada pilihan konsumen (studi eksperimen online shop: tokopedia, matahari, dan jualo)* Universitas Kristen Maranatha. Bandung

[17] Davis F D 1989 *Perceived usefulness, perceived ease of use, and user acceptance of information technology.*

[18] Kotler P 2003. *Marketing Management* (New Jersey: Prentice Hall)
[19] Hair J , Black W C, Babin B J and Anderson R E 2009 *Multivariate Data Analysis: Seventh Edition.*
[20] Hsu M H, Chang C N and Chuang L W 2015 Understanding the determinants of online repeat purchase intention and moderating role of habit: the case of online group-buying in Taiwan *International Journal of Information Management* **35** pp 45-56
[21] Ghozali I 2013 *Structural Equational Modeling*
[22] Ferdinand A 2000 *Structural Equation Modelling dalam penelitian manajemen* Penerbit Universitas Diponegoro. Semarang.
[23] Byrne B M 2001 *Structural Equation Modeling with AMOS: Basic Concepts, Applications, and Programming* Lawrence Erlbaum Associates Publishers. London.

IOP Conf. Series: Materials Science and Engineering **337** (2018) 012053 doi:10.1088/1757-899X/337/1/012053

Usability analysis of indoor map application in a shopping centre

R S Dewi[1]* and R K Hadi
[1]Department of Industrial Engineering, Institut Teknologi Sepuluh Nopember Kampus ITS Sukolilo-Surabaya 60111, Indonesia

*ratna.sari.dewi80@gmail.com

Abstract. Although indoor navigation is still new in Indonesia, its future development is very promising. Similar to the outdoor one, the indoor navigation technology provides several important functions to support route and landmark findings. Furthermore, there is also a need that indoor navigation can support the public safety especially during disaster evacuation process in a building. It is a common that the indoor navigation technologies are built as applications where users can access this technology using their smartphones, tablets, or personal computers. Therefore, a usability analysis is important to ensure the indoor navigation applications can be operated by users with highest functionality. Among several indoor map applications which were available in the market, this study chose to analyse indoor Google Maps due to its availability and popularity in Indonesia. The experiments to test indoor Google Maps was conducted in one of the biggest shopping centre building in Surabaya, Indonesia. The usability was measured by employing System Usability Scale (SUS) questionnaire. The result showed that the SUS score of indoor Google Maps was below the average score of other cellular applications to indicate the users still had high difficulty in operating and learning the features of indoor Google Maps.

1. Introduction

A mobile navigation application (app) is a software which provides cellular users with various navigational functions and features [1]. Such an example, by using input from the users, some navigation apps are able to determine the optimum route to travel from users' position to another specific location in a geographic region [1]. The mobile app can be used for both outdoor and indoor navigations. An outdoor mobile navigation app is supported by the technology which combines cellular signals and Global Positioning System (GPS) signals for determining the location of a mobile phone user [2]. In case of indoor mobile navigation, since GPS signals cannot penetrate most solid objects such as buildings, mountains, and water [1][2], IPS (Indoor Positioning System) technologies are being developed by companies like Google, Microsoft and Nokia to offer pinpoint location accuracy even while users are inside a building [3]. IPS provides users with several indoor navigational functions. According to a research conducted by IndoorAtlas at 2016 [4] on 298 respondents, two most desirable functions of IPS were the navigation or finding the route (42%) and searching for points/locations of interest (31%). Twenty eight percent (28%) of the participants also mentioned that there was a need for a function of IPS which can help public security and anticipating a disaster [4]

In order to provide optimum benefits and functionalities for its users, an indoor mobile navigation app should have acceptable usability level. According to ISO 9241-11[5], usability is the extent to which a system, product, or service can be used by specific users to achieve specific goals with effectiveness, efficiency and satisfaction in a specified context of use. Furthermore, the usability level of an app will influences the user performance in operating it. For indoor navigation applications, the performance indicators could be the speed and accuracy of searching the points of interest or guiding users to the desired destination.

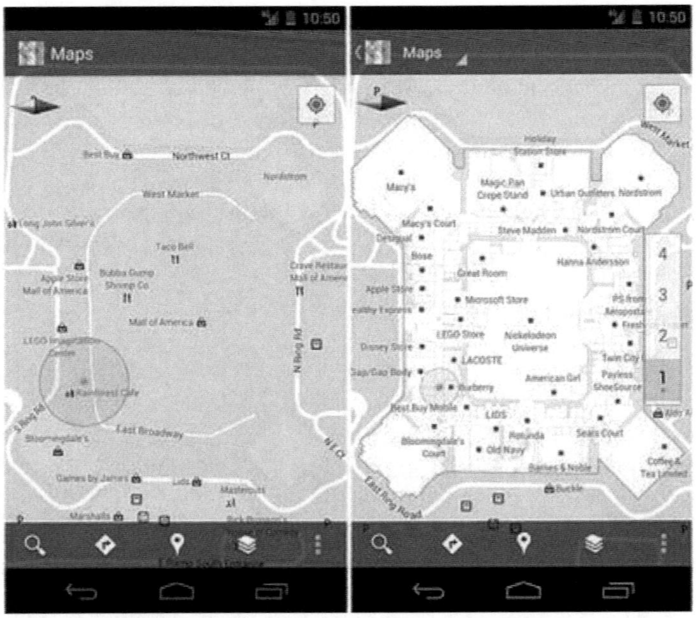

Figure 1. Indoor Google Maps interface.

Given the benefits of indoor navigation apps and considering that they are recently introduced in Indonesia, a usability analysis of this type of applications for Indonesian population is needed. Among several indoor map applications which were available in the market, this study chose to analyse the usability indoor Google Maps due to its availability and popularity in Indonesia. Indoor Google Maps is an indoor navigation product embedded on the Google Maps app which was firstly introduced in Google Maps version 6.0 for Android [6]. In the app, the details of the floor plans will appear automatically when a user is viewing or enlarging the map of a building which is equipped by indoor map data (see Fig. 1). Furthermore, indoor Google Maps uses Wi-Fi technology to perform indoor positioning [7]. In Indonesia, the indoor Google Maps app is available in around 60 buildings across 13 cities in Indonesia, ranging from shopping centres, airports, museums and universities [8].

2. Methodology

2.1. Object of the study and participants

This study was conducted at Tunjungan Plaza Shopping Mall, one of the biggest shopping centres in Surabaya Indonesia. The shopping centre was selected due to its high spatial complexity which was suitable for testing the usability of indoor Google Maps. Twenty two participants (11 males and 11 females) aged between 20 to 22 years took part in this study. All the participants were regular users of android smartphones and based on the preliminary interview, all of them had limited knowledge on the layout or floor plans of Tunjungan Plaza Shopping Mall.

2.2. Experiment tasks and apparatus

Each of the participants was required to do four experiment tasks as follows:

- *Task #1*: open Google Maps app and find indoor Google Maps feature.
- *Task #2*: search a location of a clothing store, namely "Coach" in the indoor Google Maps and follow the route suggested to arrive at the clothing store.
- *Task #3*: find public facility signs/icons in the Tunjungan Plaza's indoor Google Maps.

- *Task #4*: Go back to your initial location before you conduct Task #2.

In conducting all the four experiment tasks, each participant used a same smartphone which was Asus Zenfone 5 A500CG. During the study, the smartphone was constantly connected to the internet.

2.3. Usability testing

After each of participants completed the 4 experiment tasks, he/she was required to fill in the System Usability Scalability (SUS) questionnaire to measure his/her perceived satisfaction on the indoor Google Maps which indicated the usability of the app. SUS was one of most popular standardized usability questionnaires [9][10]. Regarding Sauro and Lewis [10] the questionnaire was employed by 43% of recorded usability studies.

The SUS questionnaire consists of 10 items, half of them worded positively and the rest worded negatively, listed as follow [10]:

1. I think that I would like to use this system frequently.
2. I found the system unnecessarily complex.
3. I thought the system was easy to use.
4. I think that I would need the support of a technical person to be able to use this system.
5. I found the various functions in this system were well integrated.
6. I thought there was too much inconsistency in this system.
7. I would imagine that most people would learn to use this system very quickly.
8. I found the system very cumbersome to use.
9. I felt very confident using the system.
10. I needed to learn a lot of things before I could get going with this system

A factor analysis conducted by Lewis and Sauro [11] on the ten items resulted on a two-factor solution. Item 1, 2, 3, 5, 6, 7, 8, and 9 were on one factor namely “usable” and items 4 and 10 on the other hand were grouped in one factor called “learnable” [10]. In this current study, the order of the items was presented randomly to each participant to minimize the sequence effect.

Each item of the SUS questionnaire was presented to the participants as 5-point scales numbered from 1 (anchored with “Strongly disagree”) to 5 (anchored with “Strongly agree”) [12] .If a participant fails to respond to an item, the centre of the rating scale (i.e., 3) was assigned on it [12] . On the next step, score contribution of each item was determined [12]. For positively-worded items (1, 3, 5, 7 and 9), the score contribution is the scale position minus 1, while for negatively-worded items (2, 4, 6, 8 and 10), it is 5 minus the scale position [12]. The overall SUS score was calculated by multiplying the sum of the item score contributions by 2.5, accordingly, SUS scores range from 0 to 100 in 2.5-point increments [12]

Previous researches such as Sauro and Lewis [10] and Bangor et al. [13] on the psychometric properties of the SUS provided norms [12] to justify whether a SUS score for a specific product is on acceptable level. A survey on 324 usability studies conducted by Sauro & Lewis [10] showed that the average of SUS scores is 62.1 for overall, 59.44 for usable scale and 72.72 for learnability scale. Regarding Bangor et al. [13], the SUS score below 50 is considered as unacceptable. Furthermore, Bangor et al. [14] developed a grading scale in which similar to a grading system in schools. SUS scores below 60 were an “F,” between 60 and 69 were a “D,” between 70 and 79 were a “C,” between 80 and 89 were a “B,” and 90 and above were an “A” [14].

3. Result

Table 1 presented the SUS score of indoor Google Maps gathered in this study. The overall, usable, and learnable SUS scores were 44.09, 44.32, and 43.18, respectively. Compared to the lower limit, mean, and upper limit of SUS scores from Sauro and Lewis [10], the SUS scores of indoor Google Maps were still below the mean score (see Fig.2). The further analysis conducted by mapping to the SUS score of indoor Google maps with the SUS norm provided Bangor, et al. [13][14] As presented at Fig.3, SUS score of indoor Google maps was considered as unacceptable. The overall SUS score of 44.09 was equivalent to the “F” in letter grade and “OK”on adjective ratings.

IOP Conf. Series: Materials Science and Engineering **337** (2018) 012053 doi:10.1088/1757-899X/337/1/012053

Table 1. Indoor Google Maps' SUS Scores.

	Question	Strongly Disagree	1	2	3	4	5	Strongly Agree	Score
1	I think that I would like to use this system frequently.		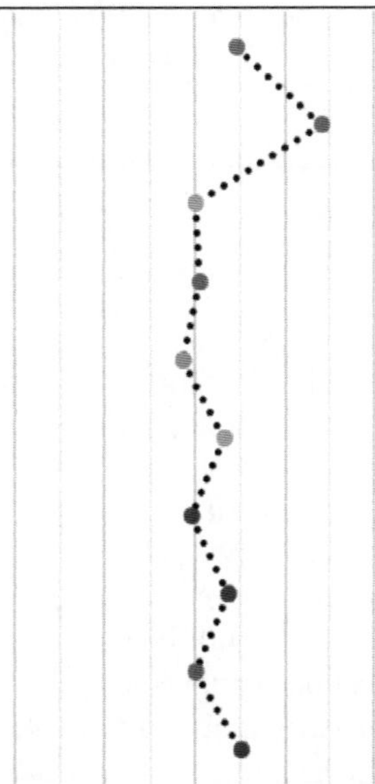						3.45
2	I found the system unnecessarily complex.								4.41
3	I thought the system was easy to use.								3.00
4	I think that I would need the support of a technical person to be able to use this system.								3.05
5	I found the various functions in this system were well integrated.								2.86
6	I thought there was too much inconsistency in this system.								3.32
7	I would imagine that most people would learn to use this system very quickly.								2.95
8	I found the system very cumbersome to use.								3.36
9	I felt very confident using the system.								3.00
10	I needed to learn a lot of things before I could get going with this system.								3.50

Subscale	Items	SUS Score
Overall	All	44.09
Usable	1-3, 5-9	44.32
Learnability	4 & 10	43.18

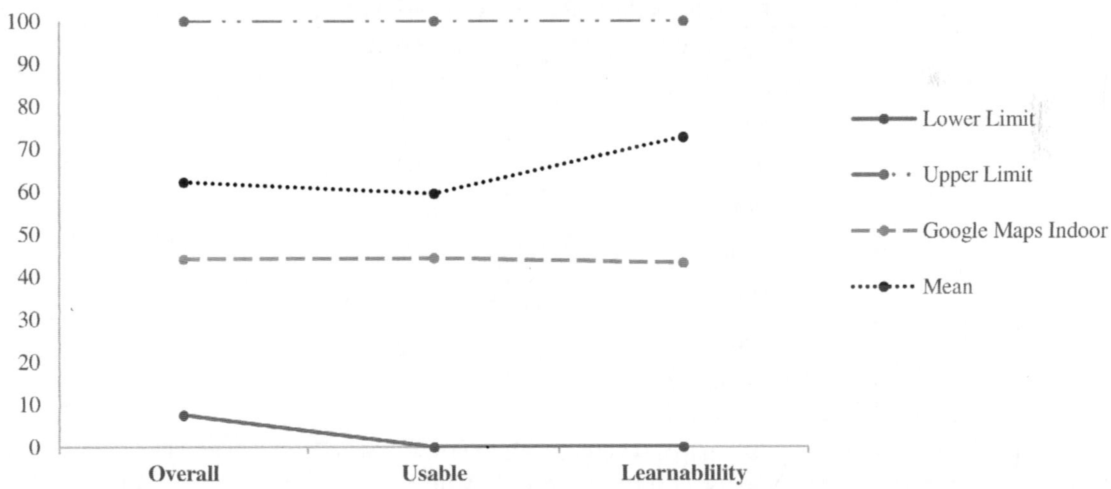

Figure 2. The overall, usable, and learnable SUS Scores of Indoor Google Maps compared to lower limit, mean, and upper limit from Sauro & Lewis [10].

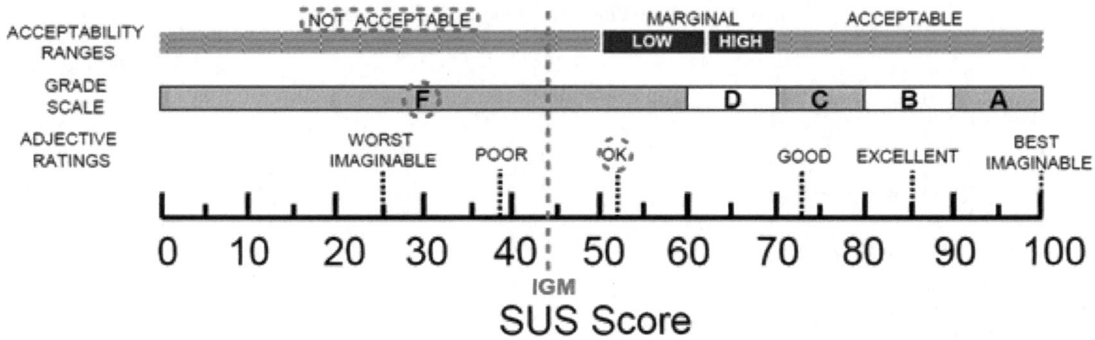

Figure 3. The overall SUS score of indoor Google Maps (IGM) on acceptability range, school grading scale and adjective ratings proposed by Bangor et al. [13][14].

4. Discussion and conclusion

In general, the relatively low SUS score of indoor Google Maps indicated the need of further improvement of the application. Although participants were likely agree that they will need the application frequently in the future (item 1), they found that the indoor Google Maps was unnecessary complex (item 2). Most participants also thought that they need to learn a lot of things before they could operate the app (item 10) which made them to think that they may need to get support of a technical person (item (4). Furthermore, the app was also considered as cumbersome to use (item 8) by most of participants. Table 2 also presented the SUS score of indoor Google Maps was below several well-known everyday products', such as Excel, GPS, Power Point Presentation (PPT), Word, and iPhone [12]. Moreover, the SUS score of indoor Google Maps was way below two other applications from Google (i.e., Gmail and Google search).

Table 2. SUS Scores of everyday products.

Everyday products	SUS score
Indoor Google Maps	44.09
Excel	56.5
GPS	70.8
PPT	74.6
Word	76.2
iPhone	78.5
Gmail	83.5
Google search	93.4

Furthermore, the age range of participants recruited in this study was 20-22 years old or grouped as millennial generation. It is commonly known that people with this age range are very familiar with digital media. Thus, if they still get difficulty in operating the indoor Google Maps, there is a chance that older people from previous generations (i.e., generation X, Y, baby boomers) will face more issues in operating indoor Google map. It is a note for the app developer to improve the quality of the interface so that it will accelerate the learning process as well as considering several learning limitations of older population.

From the observation during the participants conducted the four experiment tasks, the difficulties might rise from several issues such as inaccurate user positioning, wrong route recommendation, poor information arrangement, hidden place/business descriptions, and ineffective markers in the maps. Future study is needed to formulate the critical design criteria for an indoor map application as well as recommendations to improve its usability level. The future study should also incorporate other usability evaluation methods and recruiting larger number of participants with more diverse educational levels and/or age groups.

5. References

[1] McDonough W 2000 *United States of America Patent No. 6092076*
[2] Sullivan M 2012 *A brief history of GPS* (http://www.pcworld.com/article/2000276/a-brief-history-of-gps.html)
[3] Fredrick H 2017 *Why Doesn't GPS Work Inside a Building?* (http://techin.oureverydaylife.com/doesnt-gps-work-inside-building-18659.html)
[4] IndoorAtlas 2016 *The Rise of Indoor Positioning, A 2016 Global Research Report On The Indoor Positioning Market*
[5] Int. Organization for Standardization 1998 ISO 9241-11:1998(en) *Ergonomic requirements for office work with visual display terminals (VDTs) — Part 11: Guidance on usability*
[6] Ohazama C 2011 *Go indoors with Google Maps 6.0 for Android* (http://googlemobile.blogspot.co.id/2011/11/go-indoors-with-google-maps-60-for.html)
[7] Chi Z 2014 *Towards A Large Scale Indoor Localization Service with Crowdsensing Indoor Map Generation* (Nanyang: Nanyang Technological University)
[8] Deliusno 2015 *Google Maps Sediakan Peta Dalam Gedung di Indonesia* (http://tekno.kompas.com/read/2015/12/09/10130037/Google.Maps.Sediakan.Peta.Dalam.Gedung.di.Indonesia)
[9] Nunnally J 1978 *Psychometric Theory* (New York: McGraw-Hill)
[10] Sauro J and Lewis J 2009 *Correlations among prototypical usability metrics: Evidence for the construct of usability* (Boston: ACM) p 1609-18
[11] Lewis J R and Sauro J 2009 *The factor structure of the System Usability Scale. In: Kurosu, M. (Ed.), Human Centered Design, HCII 2009* Springer-Verlag, Heidelberg, Germany pp 94–103.
[12] Sauro J and Lewis J R 2016 *Quantifying the user experience: Practical statistics for user research.* Morgan Kaufman n.
[13] Bangor A, Kortum P T and Miller J T 2008 A*n empirical evaluation of the system usability scale. Intl. Journal of Human–Computer Interaction* **24**(6) pp 574-94
[14] Bangor A, Kortum P T and Miller J T 2009 *Determining what individual SUS scores mean: adding an adjective rating scale. J. Usability* Stud. **4**(3) pp 114–23

IOP Conf. Series: Materials Science and Engineering **337** (2018) 012054 doi:10.1088/1757-899X/337/1/012054

QC operator's nonneutral posture against musculoskeletal disorder's (MSDs) risks

F Kautsar [1*], D Gustopo[2] and F Achmadi[2]
[1]Department of Industrial Engineering, University of Merdeka, Malang, Indonesia
[2]Department of Industrial Engineering, National Institute of Technology, Malang, Indonesia

*fuad.kautsar@unmer.ac.id

Abstract. Musculoskeletal disorders refer to a gamut of inflammatory and degenerative disorders aggravated largely by the performance of work. It is the major cause of pain, disability, absenteeism and reduced productivity among workers worldwide. Although it is not fatal, MSDs have the potential to develop into serious injuries in the musculoskeletal system if ignored. QC operators work in nonneutral body posture. This cross-sectional study was condusted in order to investigate correlation between risk assessment results of QEC and body posture calculation of mannequin pro. Statistical analysis was condusted using SPSS version 16.0. Validity test, Reliability test and Regression analysis were conducted to compare the risk assessment output of applied method and nonneutral body posture simulation. All of QEC's indicator classified as valid and reliable. The result of simple regression anlysis are back (0.326<4.32), shoulder/arm (8.489>4.32), wrist/hand (4.86 >4.32) and neck (1.298 <4.32). Result of this study shows that there is an influence between nonneutral body posture of the QC operator during work with risk of musculoskeletal disorders. The potential risk of musculoskeletal disorders is in the shoulder/arm and wrist/hand of the QC operator, whereas the back and neck are not affected.

1. Introduction

Recently, the measurement of the Musculoskeletal disorders (MSDs) risk is generally not managed by the company, even MSDs are the most common complaints of labours, stated by [1]. The 2014 Labor Force Survey (LFS) Report published by [2] showed that there were 526,000 cases of MSDs in the UK, it almost three times higher than reported MSD's cases in 2013 (184,000 cases) MSDs are defined by [3] as the injuries or pain in the human musculoskeletal system when the muscles receive static loads and for a long periods. MSDs occur due to some risk factors, such as nonneutral working postures, static load of muscles, high workload, repetitive work, stated by **Error! Reference source not found.**. Musculoskeletal disorders (MSDs) refer to a gamut of inflammatory and degenerative disorders initiated or aggravated largely by the performance of work or associated work settings **Error! Reference source not found.**. It is the major cause of pain, disability, absenteeism and reduced productivity among workers worldwide **Error! Reference source not found.**. Although not fatal, MSDs have the potential to develop into serious injuries in the musculoskeletal system if ignored **Error! Reference source not found.**. Most MSDs are accumulative disorders resulting from protracted exposure to loads of varied intensity at work [6]. Risk factors often cited for musculoskeletal disorders in the workplace include rapid work pace and repetitive motion, forceful exertions, non-neutral body postures, and vibration [7]

Nonneutral body postures will increase the labour's workload that can be the potential causes of MSDs on certain body parts, stated by [8]. According to the results of MSD's research that has been conducted by **Error! Reference source not found.** back is body part with the highest risk of MSDs.

In this study QC operator's body posture at Widatra Bhakti ltd showed that the QC operators work in nonneutral body posture. It was supported by Mannequin pro simulation, the results are as follows : wrist = 0 Nm, elbow = 3 Nm and 2 Nm, shoulder = 4 Nm and 9 Nm, hip = 9 Nm and 10 Nm, Knee = 22 Nm and 21 Nm, Ankle = 23 Nm and 22 Nm, back = 26 Nm and 2 Nm Neck, these indicates there is an imbalance of labour torque

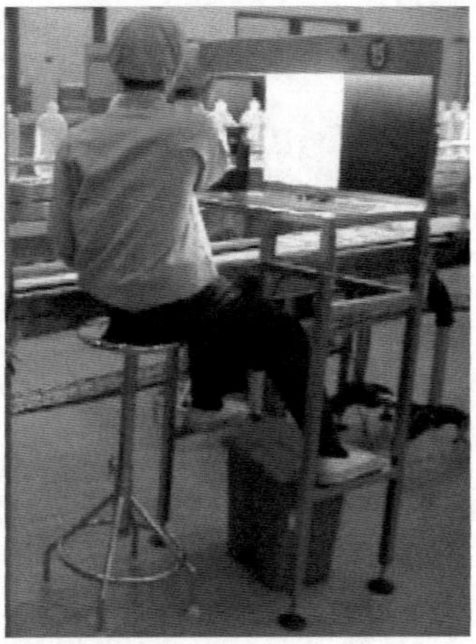

Figure 1. QC Operator when taking product at conveyor.

In this study, the researcher adopted a case study analysis method to confirm the influence of nonneutral working posture that is represented by torque load and MSDs risk. Case study analysis was conducted at Widatra Bhakti ltd. This study required collaboration Widatra Bhakti stakeholder to find the best solution for labour' safety and health.

2. Musculoskeletal Disorder

Musculoskeletal disorders (MSDs) can be defined as the injuries or pain in the human musculoskeletal system when the muscles receive static loads and for a long period. According to [3] musculoskeletal disorders usually affect the back, neck, shoulders and upper limbs, but lower limbs can also be affected. MSDs cover any damage or disorder of the joints or other tissues. Reference **Error! Reference source not found.** stated that the health problems range from minor aches and pains to more serious medical conditions requiring time off or medical treatment. In more chronic cases, MSDs can even lead to disability.

According to **Error! Reference source not found.** most work related MSDs develop over time. There is usually no single cause of MSDs; various factors often work in combination. Physical causes and organisational risk factors include: Handling loads, especially when bending and twisting, repetitive or forceful movements, awkward and static postures, vibration, poor lighting or cold working environments and prolonged sitting or standing in the same position. There is growing evidence linking MSDs with psychosocial risk factors (especially when combined with physical risks), including high demand of work or low autonomy and low job satisfaction

Quick Exposure Check is a tool developed by Li and Buckle [9] and enhanced by David et al. [10]. QEC is applied to assess level of exposure to ergonomic risks. One of the unique features of QEC is that the observed worker should rate the weights handled, time spent on observed task, level of hand force, visual demands, application of vibrating tools and difficulties to sustain with the work as well as the stressfulness of the work.

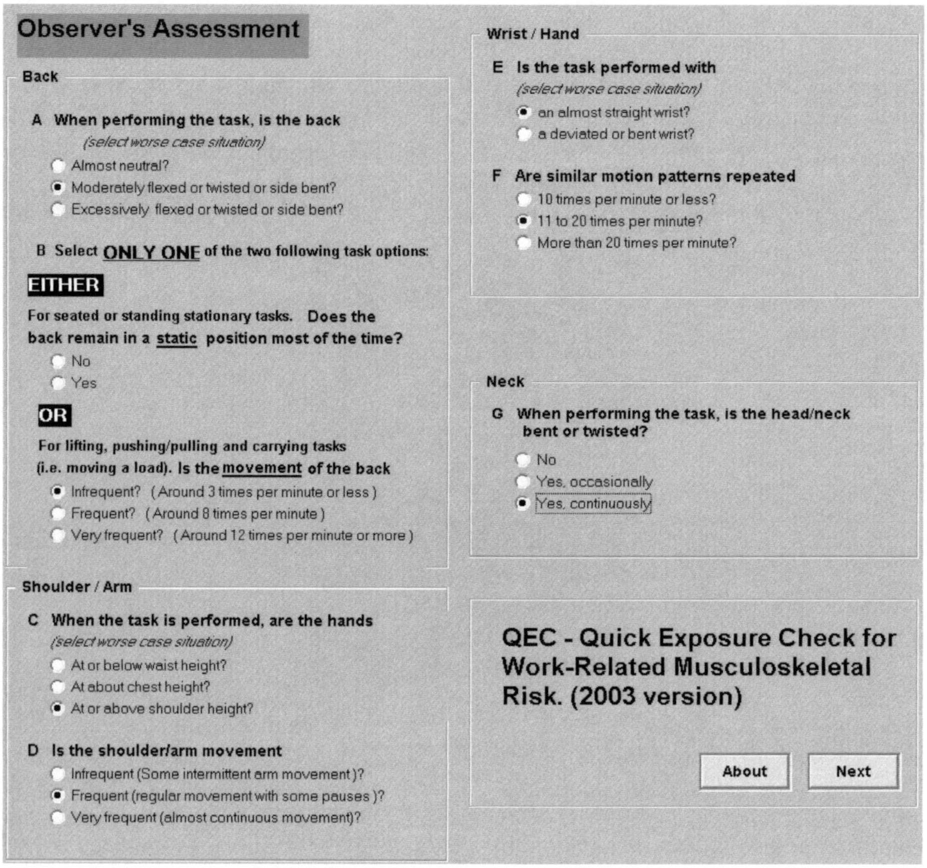

Figure 2. Observer's Assessment of Quick Exposure Check .

3. Research Methodology

This study uses Quick Exposure Checklist (QEC). This study focused on MSDs risk on four indicator of 24 QC operators body as study's object that included back, shoulder/arm, wrist/ hand and neck. This cross-sectional study was condusted in order to investigate correlation between risk assessment results of QEC (Dependent variable) and body posture calculation of mannequin pro (Independent variable) at a single point in time. Observer asked the workers about their tasks while completing the observer portion of the tool after that observer asked the workers to answer the worker assessment and observer scored the assessment.

QC operator's body postures were simulated by Mannequin pro. The simulation results measured its contribution to the variable work posture to determine which factor or torque load in which body part most dominant influence the operator work posture

Statistical analysis was conducted using SPSS 16.0 student version [11]. Validity test, reliability test and gegression analysis were done to compare the risk assessment output of applied method and nonneutral body posture simulation.

4. Result and Analysis

a. Validity and Reliability test of Quick Exposure Check

The results of validity and reliability test are presented in Table 1. According to this table, all of QEC's indicator classified as valid (The Corrected item total correlation (R) is higher than 0.05) and reliable (The Cronbach's Alpha is higher than 0.60)

Table 1 . Validity and Reliability of Quick Exposure Check.

Exposure	R		Result
Back			
When performing the task, is the back	0.612	> 0.05	Valid
For manual handling tasks only, is the movement of the back	0.484	> 0.05	Valid
Other tasks : are the tasks performed in static posture most of the time?	0.187	> 0.05	Valid
Shoulder/arm			
Is the task performed	0.207	> 0.05	Valid
Is the arm movement repeated	0.546	> 0.05	Valid
Wrist/hand			
Is the task performed	0.401	> 0.05	Valid
Is the task performed with similar repeated motion patterns	0.595	> 0.05	Valid
Neck			
When performing the task, is the head/ neck bent or twisted excessively	0.506	> 0.05	Valid
Cronbach's Alpha	0.705	> 0.06	Reliable

b. Regression Analysis

- Back postures with MSD's risk at Back

The result of simple regression analysis is presented in Table 2. According to this table, the F – Theory is higher than F-Empiric. It can be concluded that there is no correlation between nonneutral back posture and MSD's risk at the back of the QC operators (H_0 accepted)

Table 2. Regression Analysis result.

F – Empiric	F – Theory	Result
0.326	4.32	0.326<4.32

- Shoulder/arm postures with MSD's risk at Shoulder/arm

The result of simple regression analysis is presented in Table 3. According to this table, the F – Theory is lower than F-Empiric. It can be concluded that there is correlation between nonneutral shoulder/arm body posture and MSD's risk at the shoulder/arm of the QC operators (H_0 rejected)

Table 3. Regression Analysis result.

F – Empiric	F – Theory	Result
8.489	4.32	8.489 > 4.32

- Wrist/hand postures with MSD's risk at Wrist/hand

The result of simple regression analysis is presented in Table 4. According to this table, the F – Theory is lower than F-Empiric. It can be concluded that there is correlation between nonneutral wrist/hand body posture and MSD's risk at the wrist/hand of the QC operators (H_0 rejected).

Table 4. Regression Analysis result.

F – Empiric	F – Theory	Result
4.86	4.32	4.86 > 4.32

- Neck postures with MSD's risk at Neck

The result of simple regression analysis is presented in Table 5. According to this table, the F – Theory is lower than F-Empiric. It can be concluded that there is no correlation between nonneutral wrist/hand body posture and MSD's risk at the wrist/hand of the QC operators (H_0 accepted)

Table 5. Regression Analysis result.

F – Empiric	F – Theory	Result
1.298	4.32	1.298 .32

5. Analysis

Based on the empirical data obtained by the results of this study, there is an influence between non-neutral body posture of the QC operator during work with risk of musculoskeletal disorders. The potential risk of musculoskeletal disorders is in the shoulder/arm and wrist/hand of the QC operator, whereas the back and neck are not affected by the body posture.

The operator's movement of shoulder and wrist illustrate the high risk of MSDs by their moving shoulders and repetitive hands flexing during work. This condition occurs because the shoulder of the operator keep moving and the wrist always bent every time they picking up the product from conveyor.

On the other hand, the QEC's statement of moving/bending/twisting back and neck does not occur to the QC operators. This happens because the back of the operator does not bend and does not in a static condition and neither does the neck. Since the neck of the operator does not often bend or twisted and does not cause fatigue while working, in which the height of the work bench can be adjusted to the posture of each operator thus minimize potential risk of musculoskeletal disorders.

The dimensions of musculoskeletal disorders (MSDs) based on the concept of Quick Exposure Check (QEC) that explained by **Error! Reference source not found.** consist of 4 potential risk dimensions (back, shoulder/arm, wrist/hand and neck). It was found that most of the research's objects had the potential risk of MSDs on high scales in all four dimensions of potential risk with the order from the largest to the smallest was the back, neck, shoulders and wrist.

The results in this study indicated that the posture of the QC operator during work significantly influence the potential risk of MSDs, especially on the shoulder and wrist operator. The results of shoulders and wrists suggest that these dimensions potentially lead to MSDs in QC operators supported by observations where the constantly moving shoulders and wrists are always bent due to unsuitable conveyor location. This result is in accordance with the statement results of the study by [13] and **Error! Reference source not found.**. This indicates that working wih nonneutral / awkward posture will increase the risk of MSDs on shoulder and wrist.

6. Discussions

Case study analysis is the method that allows us to accomodate users need in a research. This will be helpful on reseacrh becasue the opinions revelaed are based on the real case in the field. In order to find the representative case study, the case must be chosen carefully and, it is quite challenging sometimes. The future research could be developed not only medication product company, but can be expanded to the other company type so that the research will produce a more general result. The number of objects can be added that the criteria of the study and the resulting robust criteria. In a subsequent study, it is expected that the management in safety and health in the company be considered, so the result becomes more representative when it is used in real life event.

7. Conclusions

By using the integration of mannequin pro and QEC analysis, this research developed MSDs risk indicator that can be used by company to assess their worker's safety and health. Refer to **Error! Reference source not found.** four parts of the body (i.e. back, shoulder/arm, wrist/hand and neck) that can be affected by MSDs. This study concluded that there is a significant effect of non-neutral body posture on the risk of MSDs especially at shoulder/arm and wrist/hand. It can occur because their shoulder/arm is constantly moving and their wrist/hand are flexed repeatedly during work. On the other hand, back and neck are constantly in a neutral position.

8. References

[1] European Agency for Safety and Health 2010 *Work European Risk Observatory Report* (Luxembourg) ISSN 1830-5946
[2] Health and Safety Executive 2014 *National Statistics* (Great Britain)
[3] Grandjean E 1993 *Fitting the Task to The Man* 4th edition (London: Taylor & Francis)
[4] Park B C, Cheong H K, Kim E A and Kim S G 2010 Risk factors of work-related upper extremity musculoskeletal disorders in male shipyard workers: structural equation model *analysis Safety and health at work* **1**(2) pp 124-33
[5] Jellad A, Lajili H, Boudokhane S, Migaou H, Maatallah S and Frih Z B 2013 Musculoskeletal disorders among Tunisian hospital staff: Prevalence and risk factors *The Egyptian Rheumatologist* **35**(2) pp 59-63
[6] Motamedzade M, Ashuri M R, Golmohammadi R and Mahjub H 2011 Comparison of ergonomic risk assessment outputs from rapid entire body assessment and quick exposure check in an engine oil company *Journal of research in health sciences* **11**(1) pp 26-32
[7] Collins R M, Van Rensburg D J and Patricios J S 2011 Common work-related musculoskeletal strains and injuries: CPD article *South African Family Practice* **53**(3) pp 240-6.
[8] Kumar V K, Kumar S P and Baliga M R 2013 Prevalence of work-related musculoskeletal complaints among dentists in India: a national cross-sectional survey *Indian Journal of Dental Research* **24**(4) pp 428-438
[9] Li G and Buckle P 2009 *Evaluating change in exposure to risk for musculoskeletal disorders: A practical tool* (Suffolk: HSE Books)
[10] David G, Woods V, Buckle P and Stubbs D 2003 Further development of the Quick exposure Check (QEC) In *Ergonomics in the Digital Age* The XVth Triennial Congress of the International Ergonomics Association
[11] Santoso S 2010 *Panduan Lengkap Menguasai SPSS 16* (Jakarta: PT. Elex Media Komputindo)
[12] Bulduk E Ö, Bulduk S, Süren T and Ovalı F 2014 Assessing exposure to risk factors for work-related musculoskeletal disorders using Quick Exposure Check (QEC) in taxi drivers *International Journal of Industrial Ergonomics* **44**(6) pp 817-20
[13] Farooq M and Khan A A 2014 Effects of shoulder rotation combined with elbow flexion on discomfort and EMG activity of ECRB muscle *International Journal of Industrial Ergonomics* **44**(6) pp 882-91

Effective production planning for purchased part under long lead time and uncertain demand: MRP Vs demand-driven MRP

M J Shofa[1*], A O Moeis[2] and N Restiana[3]
[1]Department of Industrial Engineering, University of Serang Raya, Indonesia
[2]Department of Industrial Engineering, University of Indonesia, Indonesia
[3]PT Krakatau Steel Tbk, Indonesia

*m.j.shofa@gmail.com

Abstract. MRP as a production planning system is appropriate for the deterministic environment. Unfortunately, most production systems such as customer demands are stochastic, so that MRP is inappropriate at the time. Demand-Driven MRP (DDMRP) is new approach for production planning system dealing with demand uncertainty. The objective of this paper is to compare the MRP and DDMRP for purchased part under long lead time and uncertain demand in terms of average inventory levels. The evaluation is conducted through a discrete event simulation with the long lead time and uncertain demand scenarios. The next step is evaluating the performance of DDMRP by comparing the inventory level of DDMRP with MRP. As result, DDMRP is more effective production planning than MRP in terms of average inventory levels.

Keywords: Demand-driven MRP, long lead time, MRP, uncertain demand

1. Introduction

Nowadays the business environment is in the stochastic system with fluctuating demand, long lead time, inaccurate forecast, huge variety of product, and complex network impact on production planning and inventory control [1], meanwhile MRP as production control system is appropriate for the deterministic environment [2]. Therefore, MRP has many deficiencies to face current business environment.

In stochastic condition, some researches proposed modified MRP to anticipate uncertain demand and lead time to get good service level goals [3][4][6][7]. DDMRP method is workable under long lead time and uncertainties environment with variablity of demand and supply. Nevertheless, there are few of researches elaborate the benefit of DDMRP. It is indicated that there is opportunity to compare MRP and DDMRP. This study fills this gap by means of evaluating the performance of DDMRP and MRP for purchased part under long lead time and uncertain demand in terms of average inventory level in the industrial world.

2. Literature review

MRP was first popularized by Joe Orlicky's in 1975, in figure 1 depicts the history of MRP from time to time, begin with MRP, closed-loop MRP, manufacturing resources planning (MRP II), advanced planning and scheduling systems (APS), and the enterprise resources planning (ERP). There are 5 inputs in MRP: Master Production Schedule (MPS), Bill of Material (BOM), item master, orders, requirements a lot of items that are needed and MRP combine them [7][8].

MRP is appropriate under deterministic environment. Poorly, most production systems are stochastic. Therefore, the deterministic assumptions of MRP are often too restrictive. So that, under demand uncertainty MRP combined with lot sizing [9], safety stock and safety lead time [10], while under lead time uncertainty, MRP's approached with markov process and newboys model [11], brunch and cut algorithm [12], and periodic order quantity [13]. Under Inaccurate forecast, MRP combined with properly forecasting models [14] and under long lead time MRP combinde with lot size and and planned lead time setting [15].

Finally, traditional MRP is not relevant with the current environment. Wherein companies that have experienced in MRP got chronic problems and the risk of high variation, overstock, and shortage in supply planning and customer demand. Such conditions, impact on three main factors, inventory performance, service level performance, high expedite related expenses and waste **Error! Reference source not found.**.

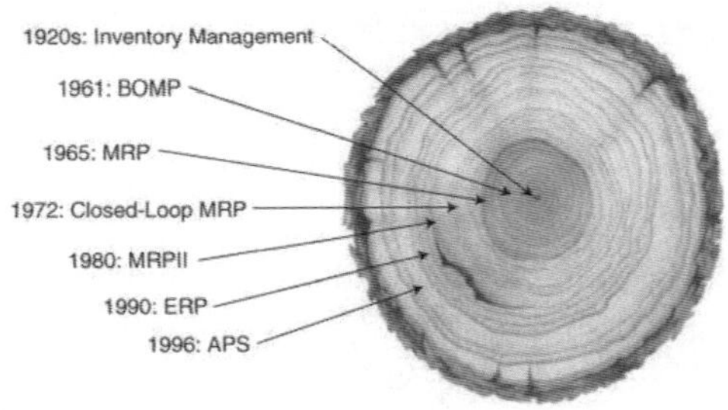

Figure 1. Plannning tool evaluation [1].

To become an agile company, which consists of four distinctive competencies: cost, quality, dependability, and flexibility, companies are required to produce their products at low cost, high-quality products and services, short lead time and varied volume, and also improve the value of customer through customization. Demand-Driven MRP (DDMRP) approach deals with those challenges. DDMRP can adopt a stochastic environment to become an agile company.

DDMRP was introduced by Carol Ptak and Chad Smith in 2011 in Orlicky's Material Requirement Planning 3rd Edition book. DDMRP built from Orlicky's MRP that still relevant coupled with innovation in compressing lead time that refers to the demand-driven world. DDMRP moves from the world of push and promote to position and pull [16].

Figure 2 depicts that DDMRP consists of a core of MRP, distribution requirements planning (DRP) logic, theory of constraint (TOC), lean principles and innovations. From those, DDMRP develope more realistic product planning and inventory control method [1].

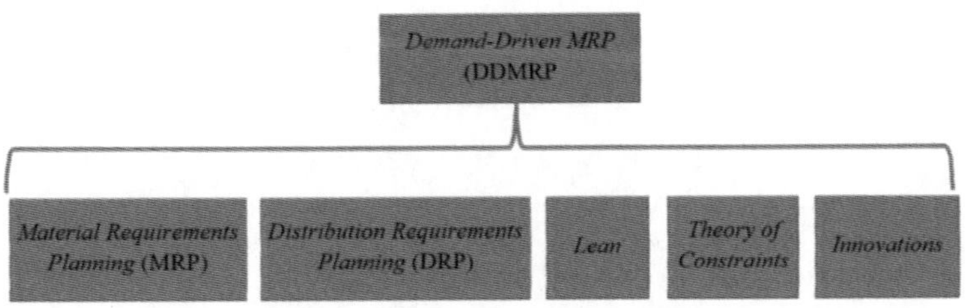

Figure 2. Demand-driven MRP.

DDMRP represents a change from the conventional idea of MRP. In DDMRP, demand is not defined by the statement "what we can and will build" but the statement "what we can and will sell [17]. See the figure 3, there are five steps of DDMRP: strategic inventory positioning, buffer profiles and levels, dynamic adjustment, demand-driven planning, visible and collaborative execution. These steps is deviden into modelling or remodelling environment, plan, and execution.

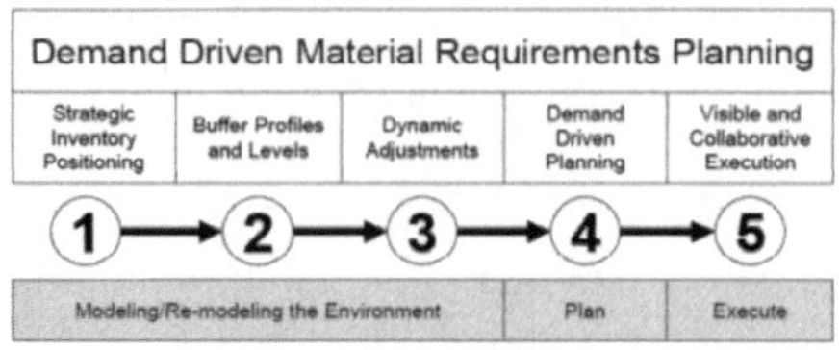

Figure 3. Five components demand-driven MRP.

Strategic inventory positioning is the first step in DDMRP. It consider where inventory should be placed. The six positioning factors are used to determine the initial positioning strategy such as customer tolerance time, market potential lead time, demand variability, supply variability, inventory leverage and flexibility, and the critical operation protection. In strategic positioning is introduced a realistic lead time called actively syncrhonized replenishment lead time (ASR lead time) -core concept behind DDMRP-. It is defined by the longest unprotected or unbuffered sequence in the BOM for a particular parent.

In DDMRP, buffer profiles and levels provide the appropriate buffer position and strategic replenishment to reduced variability from both demand and supply. It revised from the conventional inventory management in term of the seasonal level inventory into three color-coded zones that comprise the total buffer. These zones are green, yellow, and red zone. Green represents an inventory position that requires no action, yellow represents a part has entered its rebuild or replenishment. Red represents a part that is in jeopardy my require special attention [1].

Figure 4a shows the zone stratification in buffer stock is segmented in green (G), yellow (Y) and red (R) zone, and figure 4b shows the inventory effective-Ineffective curve with overlaid color-coded zones and displasys the meaning of each colored zone. This color coding system will be used in both planning and execution priority management.

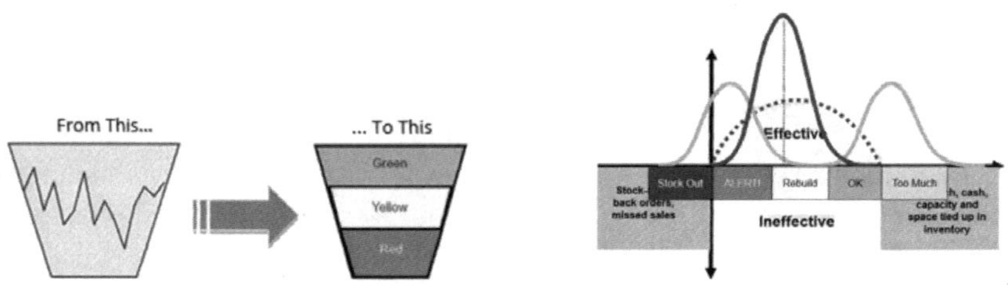

(a) (b)

Figure 4. (a) Moving to zone stratification in a stock buffer (b) Effective/Ineffective with buffer zone [1].

To calculate each buffer profiles (G, Y, and R), the average daily usage / ADU over the percentage of lead time is taken. It is calculated follow some formulations, wherein:

Green Zone = ADU (Average daily usage) * Lead time * GZ Lead time (1)
Yellow Zone = ADU * Lead (2)
Red Zone base = ADU * Lead time * RZ Lead (3)
Red Zone safety = Variablity factor * Red zone base (4)

and, buffer zone is calculated:

Top of Red = Red Zone (5)
Top of Yellow = Top Of Red + Yellow Zone (6)
Top of green (TOG) = red zone base + red zone safety + yellow green + green zone (7)

Dynamic adjusment is buffer profile adjust. Adjusting buffer pofile is used to adapt company production planning to a dynamic environment. There are three type of adjusment: recalculated adjusment, planned adjustment, and manual adjustment.

Demand driven planning is supply generation. It is based what the available stock equation place the part. Open order is needed when a part stocked are available in yellow zone (rebuild zone), wherein:

Available stock = on hand + on order – demand (pas due, due today, qualified spike) (8)

To be able to define threat of spike to the inventory, order quantity spike threshold (OST) should be defined, OST is measured from the percentage of red zone, for this percentage Ptak and Chad provides guidance with 50 percent of the red zone and OST is used for all stocked items [1]. As for the time horizon becomes the criterion of order spike is time where the environment can react rationally to changes that occur. Horizon spike order is usually defined in the lead time of the relevant part. And finally, visible and collaborative execution. It is challenge priority by due date. Priority by due date gives the real day-to-day inventory and materials priorities.

3. Research Metodology

3.1 Flow of The Research

Figure 5 displays flow of the research. In order to compare the effectiveness of MRP and DDMRP, simulation run for 180 days. The indicator that will see is in term of average inventory level. A case study is conducted on manufacturing company in Indonesia. This study focused on Piston-1 with 45 days of lead time. Input data is taken in the period January-December 2015.

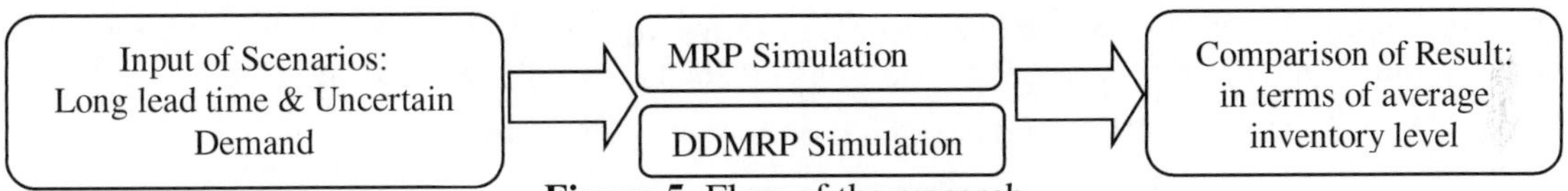

Figure 5. Flow of the research .

3.2 Input Parameters

This research investigated on purchased part Piston-1 with lead time is 45 days (table 1) and demand forecasts for one year is on table 2.

Table 1. Lead time and part type.

Lead time (day)	Type
45	Purchsed

Table 2. Annual Demand Forecast.

Month	1	2	3	4	5	6	7	8	9	10	11	12
Demand Forecast	4700	51480	49500	43200	36360	44040	37020	38880	39840	41040	37020	6564

Scenarios were given as shown in table 3 with the actual demand is generated based on random factor number to make stochastic environment, simulation run on two systems: MRP and DDMRP for 180 days.

Table 3. Demand variability (daily).

Week	Demand (daily)						
1	0	5505	1992	0	1992	5505	0
2	0	2283	0	0	0	5440	0
3	2817	1878	5505	0	5505	0	0
4	1992	5780	2817	5505	0	240	5700
5	5452	0	2817	0	2067	2067	0
6	2817	5440	2067	5440	0	0	0
7	240	2817	5505	0	2058	240	2283
8	0	2283	1905	0	1905	1905	1905
9	0	240	1992	2283	0	0	5452
10	0	2067	0	2817	240	2058	0
11	0	2817	0	5440	3900	2067	2817
12	0	0	0	2817	0	0	5505
13	2283	3900	0	0	0	2283	0
14	0	0	5505	3900	0	1878	0

Week	Demand (daily)						
15	2067	0	3900	5780	2817	0	0
16	0	5480	0	1992	2817	5780	1905
17	2817	240	0	1992	0	5480	0
18	2058	5780	0	3900	240	0	0
19	5780	5452	2283	0	0	0	5440
20	0	2817	0	1992	5505	5780	0
21	5700	2058	0	0	0	2817	0
22	0	5480	0	5505	0	0	5505
23	1992	3900	1905	0	0	2058	0
24	0	0	240	2817	1878	0	0
25	240	2817	0	2283	0	0	0
26	0	0	0	0	5480	0	0

4. Result and discussion

4.1 Production planning using MRP

From the input parameter: demand forecasts, lead time data, and demand variability, MRP and DDMRP is generated. In MRP implementation find that the inventory level is ranged 31,541-162-195 pcs/day and the average level inventory is 106852 pcs/day, as shown in figure 6. The inventory conditions is on instable condition in the end, if any spike conditions the inventory will be stockout.

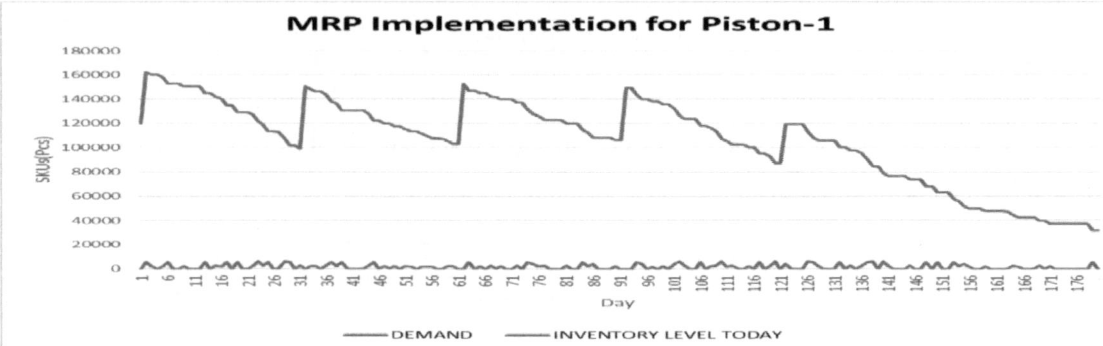

Figure 6. Demand and inventory condition of piston-1 using MRP method.

4.2 *Production planning using DDMRP*

DDMRP method is simulated. Firstly, remodelling with strategic inventory positioning, buffer profiles and levels and dynamic adjustment before generating the production planning system. Using strategic replenish position the lead time for part Piston-1 is decoupled from the supplier lead time.

Table 3a shows DDMRP Parameter based on remodelling of DDMRP method. ADU is calculated from the forecasts and Green Zone (GZ) lead time and Red Zone (RZ) lead time is stated based on data converted in percentage. Variability factor and initial state inventory is based on actual data. From the ADU and ASR lead time, buffer profiles can now be investigated as displayed table 3b.

Then, simulation using DDMRP is run. In this simulation find the inventory level is ranged 80,304-120,000 pcs/day and the average inventory level is 95,284 pcs/day. Figure 7 shows the inventory level is on stable condition, it can be preventive for spike in demand.

Table 3. a) DDMRP parameter and b) Buffer profile and levels of piston-1.

(a)

Item	Piston-1
ADU (unit)	1313
Lead time (day)	45
GZ lead time	30%
RZ lead time	30%
Variablity Factor	50%
Initial state Inv (unit)	12000

(b)

Item	Piston-1
Green	17726
Yellow	59085
Red Base	17726
Red Safety	8863
Top of Green	103399
Top of Yellow	85673
Top of Red	26588

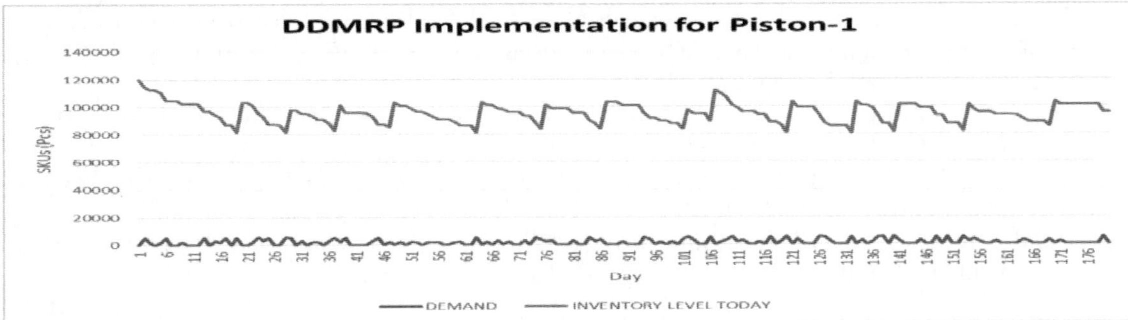

Figure 7. Demand and inventory condition of parent part SA-12, SA-22, and SA-02 using DDMRP .

5. Conclusion

Base on the purposeof this study is to evalute the performance of DDMRP and MRP in terms of average inventory level, DDMRP improve the inventory level from 106,852 pcs/day to 95,284 pcs/day (11% reduced) and make inventory on the stable condition, So that DDMRP is more effective production planning than MRP.

This research is just a prototyping of how to realize the production planning with DDMRP and MRP. Further research can compare DDMRP with other production planning method.

6. References

[1] Ptak C and Smith C 2011 *Orlicky's material requirements planning 3rd ed* (New York: Mc Graw Hill)

[2] Suresh V and Sivasubramanium R 2010 The implementation of material requirement planning techniques in crane hook assembly *Int. Journal of Applied Engineering Research*, **vol. 5**, no. 8, p. 1371–1379

[3] Xie J, Zhao X and Lee T S 2003 Freezing the master production schedule under single resource constraint and demand uncertainty *Int. Journal Production Economics*, **vol. 83**, pp. 65-84.

[4] Koh S C L and Saad S M 2003 MRP-controlled manufacturing environment disturbed by uncertainty *Robotics and Computer Integrated Manufacturing*, **vol. 19**, no. 1-2, p. 157–171.

[5] Dolgui and Prodhon C 2007 Supply planning under uncertainties in MRP environment: a state of the art *Annual Reviews in Control*, **vol. 31**, pp. 269-279.

[6] Niu J and Dartnall J 2008 Application of fuzzy-MRP in fast moving consumer goods manufacturing industry, in *Proc of the 2008 Winter Simulation Conference*.

[7] Gaspersz V 2004 *Production planning and inventory control; berdasarkan pendekatan sistem terintegrasi MRP II dan JIT menuju manufacturing 21*. (Jakarta: PT Gramedia Pustaka Utama)

[8] Jacobs F and Chase R 2011 *Operations and supply chain management third edition* (China: McGraw-Hill).

[9] Callarman T E and Hamrin R S 1983 A comparison of dynamic lot sizing rules for use in a single stage MRP system with demand uncertainty *Int. Journal of Operations and Production Management,* **vol. 4**, pp. 39-48.

[10] Van Kampen T M, Van Donk D P and Van Der Zee D J 2011 Safety stock or safety lead time: coping with unrealibility in demand and supply Int. *Journal of Production Research*, **vol. 4**, pp. 15 -24.

[11] Ould-Louly M S and Dolgui A 2004 The MPS parameterization under lead time uncertainty *Int. Journal of Production Economics*, **vol. 90** , p. 369–376.

[12] Ould-Louly M S and Dolgui A 2008. An approach for the MRP parameterization under lead time uncertainty: brunch and cut algorithm. in *Proceedings of the 17th World Congress The Int. Federation of Automatic Control* (2008) (IFAC, 2008) , **vol. 17**, p. 128.

[13] Louly M A, Dolgui A and Al-Ahmari A 2012 Optimal MRP offsetting for assembly systems with stochastic lead times: POQ policy and service level constraint *J Intell Manuf,* **vol. 23** , p. 2485–2495.

[14] Yenradee P, Pinnoi A and Charoenthavornying A 2001 Demand forecasting and production planning for highly seasonal demand situations: case study of a pressure container factory *Science Asia*, **vol. 27** , pp. 271-278.

[15] Enns S T 2001 MRP performance effects due to lot size and planned lead time settings *Int. Journal of Production Research*, **vol. 39**, no. 3, pp. 461-480.

[16] Ptak C and Smith C 2011 *Standing of the shoulders of giant to see the future of formal planning, Demand Driven Institute*, LLC.

[17] Smith C and Ptak C 2013 *Demand driven MRP and master production scheduling (MPS),* Demand Driven Institute.

IOP Conf. Series: Materials Science and Engineering **337** (2018) 012056 doi:10.1088/1757-899X/337/1/012056

Redefining the business process of Department of Food Security and Agriculture in Government of Surabaya City

H Cahyono[1], N A Wessiani[2]
[1] Department of Business Management, Universitas Internasional Semen Indonesia, Gresik, Indonesia
[2] Department of Industrial Engineering, Institut Teknologi Sepuluh Nopember, Kampus ITS Sukolilo-Surabaya 60111, Indonesia

*hdicahyono@gmail.com

Abstract. Government of Indonesia has been launched the bureaucratic reform program since 2010. One of the action is conducted restructuring organization in all city governments. Department of Food Security and Agriculture in Government of Surabaya City is the result of merger from two Department, namely Bureau of Food Security and Department of Agriculture. This merger makes Department of Food Security and Agriculture to redefine their business process. The new business process is needed to be defined in order to align the new structure with the long term strategic planning of Surabaya City Government. This research aims to redefine the business process of Department of Food Security and Agriculture in Government of Surabaya City. The CIMOSA model is adopted for identifying the activities in the business process. The new business process is important for the department to allocate their resource, mainly the human resource and as the main input for the department to build their standard operating procedure.

Introduction

Government of Indonesia has been launched the bureaucratic reform program since 2010. The grand design of bureaucratic reform 2010 – 2025 is stated in the Regulation of President of Republic Indonesia number 81 year 2010. The main reason that underlies the bureaucratic reform program is the public demand for improving the quality of public services and capacity building and accountability of bureaucratic performance. At the end in the year 2025, this bureaucratic reform program is expected to attain its main goal for realizing good corporate governance with a professional bureaucracy, high integrity, and good public services [1]

The bureaucratic reform program defines eight areas of changes, namely Organization, Governance, Regulations, Human Resource, Audit, Accountability, Public Service, and the Mindset and Culture Set of Government Employees[1]. Each of areas of changes has its main goal to be attained. The result expected from those eight areas of changes could be shown in table 1 below.

Table 1. The Result Expected from Eight Areas of Changes in the Bureaucratic Reform Program

No	Areas of Changes	Results
1	Organization	Organization that has right function and right sizing
2	Governance	The system, process, and procedure that is clear, effective, efficient, measureable, and align with the good governance principles
3	Regulations	Regulations that are more orderly, non-overlapping, and conducive
4	Human Resource	Human resource that has high integrity, neutral, competence, capable, and free from corruption, collusion, and nepotism
5	Audit	Improve the government administration to be clean and free from corruption, collusion, and nepotism
6	Accountability	Improve the capacity and accountability of bureaucratic performance
7	Public Service	Excellence service that meet society needs and demands
8	Mindset and Culture Set of Government Employees	Bureaucratic that has high integrity and performance

As shown in table 1, the organization area of changes is expected to have an organization that has right function and right sizing. The action taken for attaining this goal is conducting restructuring organization in all city government. Act of Republic Indonesia number 23 year 2014 has governed the restructuring organization of local governments [2].

Department of Food Security and Agriculture is one result of restructuring organization that was conducted in Government of Surabaya City. This department is a merger from two different departments, namely Bureau of Food Security and Department of Agriculture. This merger process has an impact on the redefinition of existing business processes within the department. The redefinition of business process will be very important for the department to allocate and manage their resources, mainly their human resource, in order to align with the long term of strategic planning of Surabaya City government.

This paper aims to redefine the business process of Department of Food Security and Agriculture. Many models of business process have been introduced by several experts. This research will adopt the CIMOSA model to redefine the business process.

1. CIMOSA

CIMOSA is stand for Computer Integrated Manufacturing Open System Architecture that was develop by the ESPRIT Consortium AMICE (Kosanke, Vernadat, & Zelm, 1996). CIMOSA is one of enterprise modelling framework that also has been widely used in many sectors. As an enterprise modelling framework, CIMOSA considers the enterprise as[3]:

- a set of communicating concurrent processes governing the execution of elementary actions called *functional operations*
- a finite set of agents, called *functional entities,* executing the functional operations required by business processes and processing *enterprise objects.*

Adopting CIMOSA for modelling the business process in organization can ignore the type of organization. CIMOSA provides generic model for defining the business process. CIMOSA is divided into several levels. The detailed the business process that should be mapped, then it will need more detail level of CIMOSA. The zero level of CIMOSA states that every organization whatever the type of organization consists of three main processes which are manage process, core process, and support process. Manage process includes set direction, set strategy, and direct the business. Core process includes develop product, get order, and fulfill order. Support process can consist of human resource

IOP Conf. Series: Materials Science and Engineering **337** (2018) 012056 doi:10.1088/1757-899X/337/1/012056

process, information technology process, maintenance process, etc. Manage and support process will directly support the core process[4].

2. Results

Redefining the business process of Department of Food Security and Agriculture requires several steps. The first step is reviewing the historical business process that occurred in two initial department before they merger. Based on the first step, this research could capture the critical process that exist in each department before they merger. The next step is analysing the factors that underlie the merger process. And the final step is identifying the critical processes that should be occurred in the new department through aligning with the long term strategic planning of Government of Surabaya City. The data collection methods that are used in this research involve brainstorming, focus group discussion, and depth interview. Adopting the CIMOSA model, the core process as part of business process in Department of Food Security and Agriculture of Government of Surabaya City can be shown in table 2 below.

Table 2. Core Process Mapping of Department of Food Security and Agriculture

Level 0	Level 1	Level 2	Level 3	Level 4	Level 5	Level 6
Business Process Office of Food Security and Agriculture	Core Process	Coaching	Coaching the Agriculture Section	Determination of Intervention Parties	Defining of the intervention goals	
					Implementation of Training of Utilization of Appropriate Technology for Cultivation of Food Crops and Horticulture	
				Assistance / Development utilization of appropriate technology for cultivation of food crops and horticulture	Defining of the intervention goals	
					Implementation Assistance / Development Utilization of Appropriate Technology for Cultivation of Food Crops and Horticulture	
			Coaching the Food Section	Counseling, Supervision and Inspection of Food Safety to Customers and business communities	Counseling, Supervision and Inspection of Fresh Foodstuffs	Outreach to Customers and business agent
						Fresh Food Supervision
						Supervision of Fresh Food and Fresh Food Inspection
				Socialization of food reserves management and prevention of food insecurity	SKPG Analysis and Mapping of Food Prone Areas	
					Socialization of Food Reserves Management and	Socialization of Food Reserves Management

Level 0	Level 1	Level 2	Level 3	Level 4	Level 5	Level 6
					Prevention of Food Insecurity	Socialization Prevention of Food Insecurity
				Local Foodstuff Processed Food Training	Determination of Intervention goals and Implementation of Foodstuff Training	
			Coaching the Farming Section	Supervision and inspection of animal products	Supervision of Animal Products	
					Inspection of Animal Origin Products	
				Technical socialization of slaughtering of qurban hygienic and halal animals	Determination of intervention goals	
					Implementation of Technical Socialization of Slaughtering of Hygienic and Halal Sacred Animals	
				Prevention and control of animal diseases and zoonosis	Animal clinic service	
					Maintenance of animal and livestock health	
					Implementation of the bird flu vaccination	
					Treatment of worms in livestock farming	
				Training and mentoring appropriate technology livestock farming	Training of appropriate technology for cultivating livestock	
					Assisting appropriate technology for cultivating farming	
				Increased HR agricultural extension	Development of agricultural extension workers	
					Institutional development of the target group	
				Organizing appropriate technology exhibition of agricultural products	Facilitate the creation of innovative products / agricultural products	
			Coaching Fisheries and Marine Section	Guidance, supervision, and training of aquaculture	Supervision of aquaculture	
					Training of aquaculture fishery	

Level 0	Level 1	Level 2	Level 3	Level 4	Level 5	Level 6
				Coaching, supervision, and training Fishing	Surveillance of fisheries	
					Training of fisheries	
		Development	Developing agriculture section	Provision of plant, materials and breeding support facilities and infrastructures production food crops and horticulture	Provision of Seeds	
					Seed Development	
					Provision of Supporting Facilities for Breeding	
					Procurement of Facilities and Infrastructure for Cultivation of Food Crops and Horticulture	
					Maintenance of Facilities and Infrastructure for Cultivation of Food Crops and Horticulture	
				Development and maintenance of coastal protected areas Surabaya beach and urban agricultural center	Development of Coastal Protected Area Surabaya Beach	
					Development of Urban Agricultural Center	
					Maintenance of Coastal Protected Areas Surabaya Beach	
					Maintenance of Urban Agricultural Center	
					Planting Protective and Productive Plants	
				Development of agricultural promotion	Participation in Exhibition	
					Implementation of Farmers Market	
					Implementation of Agricultural Contest	
			Developing food section	Development of food diversification	Preparation of food pattern score Expectations	
					Preparation of Food Balance Sheet	
				Organizing food diversification activities	Competition of food diversity	
					Organizing Movement Event Populating Fish Eating	

Level 0	Level 1	Level 2	Level 3	Level 4	Level 5	Level 6
			Developing farming section	Procurement of training facilities and infrastructure and cultivation of farming	Development of infrastructure facilities of people's livestock centers	
					Maintenance of infrastructure facilities of livestock farms	
				Procurement of training facilities and infrastructure and cultivation of farming	Procurement of livestock breeding training facilities	
					Procurement of aquaculture farms	
			Developing Fisheries and Marine section	Development and procurement of facilities and infrastructure of aquaculture fishery	Development of fishery infrastructure facilities	
					Maintenance of fishery facilities and infrastructure	
				Procurement of Fishing Facilities and Infrastructure	Procurement of fishery facilities and infrastructure	
					Procurement of fishery facilities and infrastructure	
		Monitoring	Monitoring agriculture section	Monitoring of Facility Utilization, Facility and Agricultural Infrastructure	Monitoring Utilization Facilities Facility and Infrastructure Food Crops and Horticulture	
					Monitoring Application of Appropriate Agricultural Technology Applications	
			Monitoring Food section	Food Availability Monitoring	Food Needs Survey at Home Level	
					Food Availability Survey	
			Monitoring Farming and Counseling section	Monitoring of the utilization of facilities of infrastructure and application of appropriate technology application for farming	Monitoring the utilization of facilities and infrastructure of farming	
					Monitoring application of appropriate technology applications for farming	
			Monitoring Fisheries and Marine section	Monitoring of the utilization of facilities for infrastructure and application of	Monitoring the utilization of facilities and infrastructure of aquaculture fishery	

Level 0	Level 1	Level 2	Level 3	Level 4	Level 5	Level 6
				appropriate technology applications, aquaculture and marine aquaculture	Monitoring of the utilization of facilities and infrastructure of catch fisheries Monitoring application of appropriate technology applications for aquaculture and marine aquaculture	

3. Conclusion

Redefining the business process in the Department of Food Security and Agriculture of Government of Surabaya City is very important since this department has experienced a merger process and restructuring organization. As the result of redefining the business process, mainly there are three main process in level 0 of CIMOSA that are manage process, core process, and support process. Core processes of Department of Food Security and Agriculture consists of three main core processes, that are coaching, developing, and monitoring process. Each of this sub main process also has its own activities and sub activities that already define in level 3, 4, 5, and 6 of CIMOSA in the Department of Food Security and Agriculture of Government of Surabaya City. This new business process could be used by the department for allocating and managing their resources to attain their strategic goals.

4. Reference

[1] Goverment of Republic Indonesia 2010 *Regulation of President of Republic Indonesia number 81 year 2010 about Grand Design of Bureaucratic Reform Program year 2010 - 2025.* Indonesia: Goverment of Republic Indonesia

[2] Government of Republic Indonesia. 2014 *Act of Republic Indonesia number 23 year 2014 about Local Government.* Indonesia: Government of Republic Indonesia

[3] Kosanke K, Vernadat F B, and Zelm M (1996) CIMOSA process model for enterprise modelling *7 international conference on the Design of Information Infrastructure Systems for Manufacturing DIISM '96 Eindhoven the Netherlands 15–18 September 1996* (pp. 59-68). Eindhoven: Springer Boston MA

[4] Maull R S, Childe S, Bennet J P, Weaver A M and Smart P A 1995 *Different Types of Manufacturing Process and IDEFO Models Describing Standard Business Processes* University of Plymouth Plymouth: University of Plymouth

IOP Conf. Series: Materials Science and Engineering **337** (2018) 012057 doi:10.1088/1757-899X/337/1/012057

Investment feasibility tracking: the importance of measuring and tracking the success level of the project during commercialization phase

Y A Saputra[1*], V E D Setyaningtyas[1], E Latiffianti[1], S H Wijaya[1] and O S A Ladamay[2]

[1]Department of Industrial Engineering, Institut Teknologi Sepuluh Nopember, Kampus ITS Sukolilo-Surabaya 60111, Indonesia

[2]Balai Besar Pelaksanaan Jalan Nasional Wilayah VIII Jatim – Bali, Ministry of Public Works and Housing of Indonesia

*yandrian@ie.its.ac.id

Abstract. Measuring project success level is a challenging activity. This area of works has attracted many researchers to look deeper into the method of measurement, success factor identification, risk management, and many others relevant topics. However, the project management scope is limited until the project handover stage. After a project handover, the control of a project management changes from Project Management Team to the project owner/commercialization team. From an investor's point of view, the success of a project delivery needs to be followed by the success of commercialization phase. This paper aims to present an approach on how we track and measure the progress and success level of a project investment in the commercialization phase. This is an interesting topic which probably often being forgotten in many practical case. Our proposed concept modify Freeman and Beale concept by estimating the variance between the Planned Net Present Value / Annual Worth (as it is in the Feasibility Study Document) and the Actual Net Present Value / Annual Worth (until the point time of evaluation). The gap will lead us to the next analysis and give us some important information, especially exposing whether our project investment performs better than the planning or underperformed. Some corrective actions can be suggested based on the provided information. Practical cases to exercise the concept is also provided and discussed; one case in a property sector in the middle of commercialization phase, and another case in a Power Plant investment approaching the end of commercialization phase.

1. Introduction

The Project Management Institute defined Project Life Cycle consist of project start – organizing and preparing - carrying out the work - and closing the project [1]. However, after project closing, in many construction/manufacturing project, it will be continued with commercialization phase. Munss and Bjeirmi tried to extent the scope of project after the closing/handover [2]. They tried to highligt which is the project scope actually not only in the short term until the project deliverable is ready for use. From the investor point of view, it just the beginning. A good Project Management team who responsible with the development or construction project need to be followed by a good commercialization team to achieve different kind of objectives. It's also important part of project and need to be considered (even it's not become the part of Project Management scope).

Jugdev and Muller stated that the definition of project success nowadays has changed from the PMI definition about Project Life Cycle to the entire Project and product life cycle / commercialization stage [3]. Both part have the sampe importance for investor / owner point if view. We are often heard this statement: "The operaration was succeed but the patiet died". In other hand, sometimes a project even it was categorized as failure project, it successed during the commercialization (Sydney Opera House, took 15 years to build and 14 times overbudget).

Now, we focus on commercialization stage. How we can tracking and monitoring the project success level during the commercialization stage? In construction stage, it's normally evaluated with Earned Value Analysis (EVA), Earned Schedule, or Earned Duration Management (EDM) [4]. For success probability in the construction planning considering Quality – Cost – Time we known the concept of Project Reliability [5]. For commercialization stage, Freeman and Beale suggested the Net Present Value from Discounted Cash Flow of Feasibility Study compared with the actual as the basis mechanism to track and monitor project success during commercialization stage [6].

Even from theoritical/research point of view it's already defined, the project success measurement during commercialization stage seems to be forgotten especially in Indonesian case. In many public infrastructure case, from public road until the toll road, it was completed with a comprehensive feasibility study (FS) document as the basis of judgement go or not go the project. In toll road case, the FS document will be the basis consideration to determine the fare. However, in practical, we never found the process of tracking and monitoring the actual progress and comparing with the planning document. Similar case in some industry, e.g.: power plant, port, or some manufacturers.

The question about the project success during commercialization we found is raised up by the Senior Managers or CEO in case of the asset/project/product reaching near end of their life time, e.g.:

- Do we achieve the target?
- When the actual Break Even Point / Payback period is occurred?
- Can we shut down the asset next year and replace with the new one?

Although it can be easily answered since the concept is already well defined, we found there are some obstacles when we want to evaluate this issue when the asset reaching near end of their life time. The several obstacles, i.e.:

- Do we keep the Feasibility Study document (from 15 -30 years ago?)
- Do we have good record data about the actual cost, revenue, and investment spending during the commercialization years?
- Currency exchange may have an effect (in term cost, revenue, or investment). Indonesian industry experienced hitted by currency exhange fluctuation during 1997 – 1998.
- Some company maybe doing revaluation with their asset during the commercialization period.
- Too late for corrective action if the commercialization performance is worst than the plan.

We named the project success measurement in commercialization stage as Investment Feasibility Tracking. We want to complement the method based on Freeman and Beale idea from investor point of view. We also want to add another alternative graphical method, the Annual Worth (AW) method, as basis of evaluation. The graph will consist of 3 annual worth components of feasibility analysis. i.e.: Annualized Revenue, Annualized Cost, and Annualized Investment.

2. Investment Feasibility Tracking Concept

Freeman and Beale define the project success during commercialization (they named it as success of a venture) can, therefore, be measured objectively via a DCF technique called Net Present Value (NPV) [5]. It can be used to show past and prognose of future performance during commercialization stage. The budgeted/planned NPV as FS Document is compared with the actual NPV. The concept of NPV is discussed in many Engineering Economics Handbook [7]. We try to complement Freeman and Beale concept with following concern.

1. *Period of Evaluation*

Our first concern, do we need to track the deviation every year or we wait until the project reaching near end of their lifetime? It's similar with the question when the right time to evaluate a long term planning in strategic management [8]. The answer is everytime when one from two conditions occurred, e.g.: the assumption is changed or the deviation (actual vs planning) is significat different.
About the time of evaluation, in our opinion, the 1st revision can be revisited in the end of hand over stage (from construction to commercialization). Some deviation during construction project and assumption updated may need during the period. Next revision is conducted as needed, depend on 2 requisite above.

2. *Basis document of evaluation*

For basis of evaluation, in our opinion, it's not necessary to be always compared with FS Document. For the first time revision, the FS Document (the number zero document) will be revised by the 1st revision document. Anytime we revaluate the actual and future planning again, the last document will replace the previous one. Comparation each update with FS Document may not be relevant since 15-30 years time horizon is a long time, with a high potential of deviation.

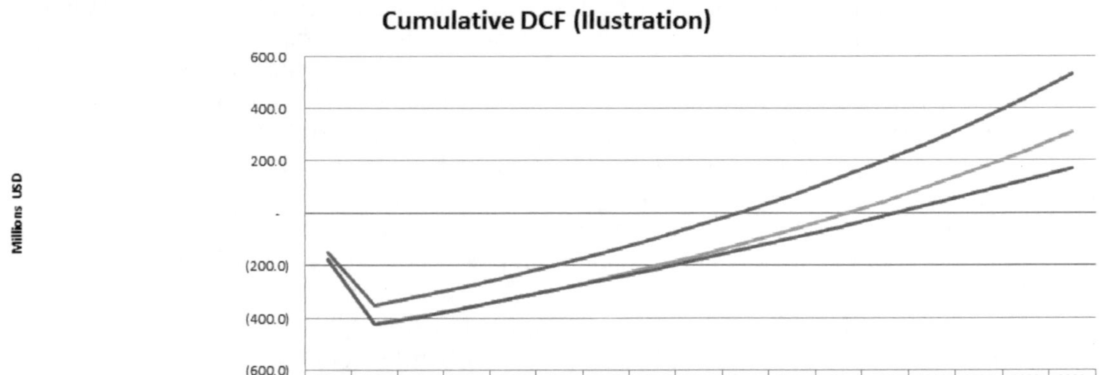

	2014	2015	2016	2017	2018	2019	2020	2021	2022	2023	2024	2025	2026	2027	2028	2029	2030
Cum DCF (USD), Rev 0 - FS	(150.0)	(350.0)	(315.0)	(277.6)	(237.5)	(194.6)	(148.7)	(99.6)	(47.1)	9.1	69.2	133.6	202.4	276.1	354.9	439.3	529.5
Cum DCF (USD), Rev 1 - beginning 2016	(175.0)	(425.0)	(393.5)	(360.1)	(324.7)	(287.2)	(247.4)	(205.3)	(160.6)	(113.2)	(63.0)	(9.8)	46.6	106.4	169.8	237.0	308.2
Cum DCF (USD), Rev 2 - FS, beginnning 2021	(175.0)	(425.0)	(396.5)	(361.4)	(325.8)	(289.7)	(252.6)	(214.6)	(175.6)	(135.6)	(94.6)	(52.6)	(9.6)	34.4	79.4	125.4	172.4

Figure1. Investment Feasibility Tracking, revision 0, 1, and 2.

For instance, figure 1 indicates the illustration of our concern. In the beginning (year 0 = 2014), as FS Document the project owner expected NPV (in specific discount rate) USD 529.5 million by the end 2030 with 2 years construction period. The Break Even Point is expected in 2023. However, after the end of construction period (2016), the construction cost overrun. Market is lower than expected in 2014 and growth rate of business 1% lower. The 1st revision revised the Project NPV into USD 308.2 million and BEP 3 years longer (2026). After 4 years commercialization running, the actual net cash flow actually higher than expected, but slowing in the last 2 years due to many new competitor completing their construction process. Management need to revise future growth assumption. The 2nd revision revised the project NPV into USD 172.4 million and BEP 1 year longer (2027).

3. *Graphical Method of Annual Worth as an alternative*

Graphical Method of Annual Worth is an alternative to show Investment Feasibility Tracking. The concept of Annual Worth actually equivalent with the Present Worth with different time point of view (Annuity vs present) [7]. The ilustration concept is showed in Figure 2.

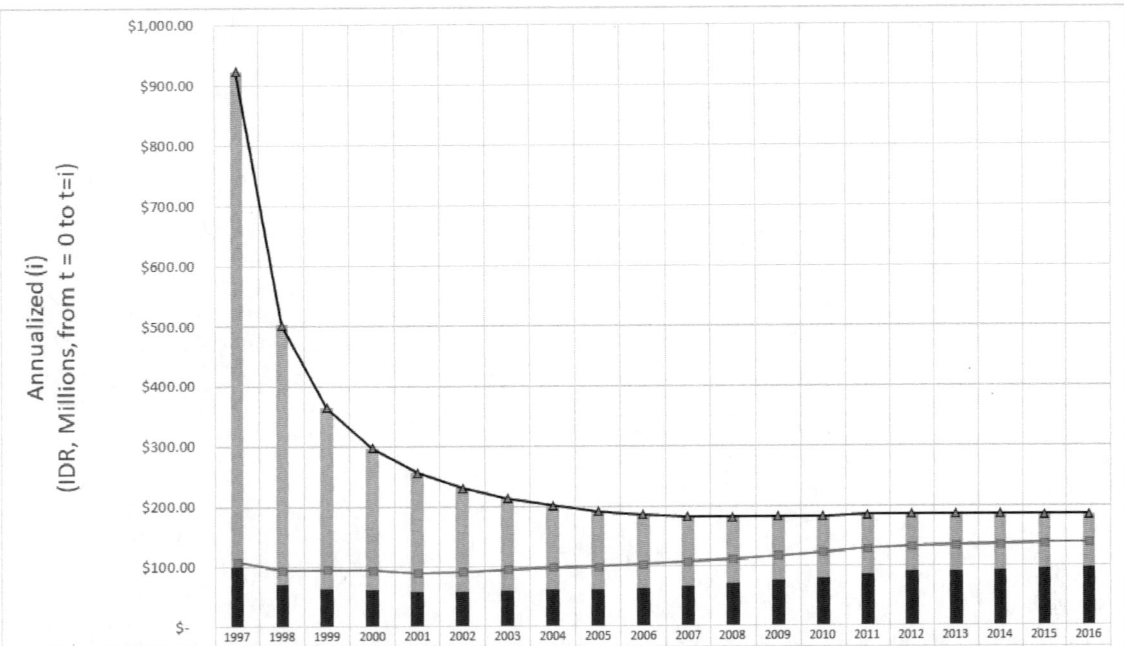

	1997	1998	1999	2000	2001	2002	2003	2004	2005	2006
Annualized Investment Cost	$823.1	$431.1	$300.8	$236.0	$197.3	$171.8	$153.7	$140.2	$129.9	$122.0
Annualized O & M Cost	$99.76	$70.85	$63.84	$61.11	$59.08	$58.80	$59.83	$60.93	$61.70	$64.18
Annualized Revenue	$108.5	$93.72	$95.05	$94.04	$89.52	$91.16	$95.42	$98.74	$99.59	$103.3
Annualized Total Cost	$922.8	$502.0	$364.7	$297.1	$256.4	$230.6	$213.5	$201.2	$191.6	$186.2

	2007	2008	2009	2010	2011	2012	2013	2014	2015	2016
Annualized Investment Cost	$115.4	$110.0	$105.6	$101.9	$98.83	$96.39	$94.57	$92.58	$90.84	$89.26
Annualized O & M Cost	$66.94	$71.13	$76.32	$80.66	$85.94	$90.09	$91.05	$93.34	$94.74	$96.12
Annualized Revenue	$107.3	$111.8	$117.2	$122.7	$128.6	$132.7	$134.0	$136.3	$137.5	$139.1
Annualized Total Cost	$182.4	$181.2	$181.9	$182.5	$184.7	$186.4	$185.6	$185.9	$185.5	$185.3

Figure 2. The Ilustration concept of Annual Worth

We divide the series into 3 parts, Annualized Investment Cost, Annualized Revenue, and Annualized O & M Cost (minus depreciation). Each year data of annualized is indicating the value of annual worth from the beginning of the project (after project handover, t = 0, 1997) until t = the period year of evaluation. For instance, in 2005, annualized revenue was IDR 599.59 Million; it's indicating the annual worth of revenue since 1997 to 2005. By the end of 2016, Annualized Revenue (from 1997 – 2016, IDR 139.1 million) is lower than annualized total cost (from 1997 – 2016, IDR 185.3 millions). It means the project not reached the Break Even Point yet.

The graphical method of Annual Worth provides slightly different approach comparing to the Net Present Value Method. It shown 3 parameters simultaneously (annualized investment, operation & maintenance cost, and revenue), giving more comprehensive analysis, i.e.:

- Does our EBITDA margin have enough amount of money to cover investment capital cost of recovery?
- We can track the gap between Annualized Revenue and Annualized Total Cost in every year of commercialization.
- Does our growth of revenue can deal with the growth of Operation & Maintenance Cost?

3. Practical Assessment

In this section, we would like to discuss the practical assessment the concept of Investment Feasibility Tracking in 2 project commercialization evaluation in Indonesia. Some discussion and finding are provided.

1. *10th years Hotel*

X Hotel was developed in 2005 and last year is the 10th year of their operation. The deviation between FS Document and 10th years operation is showed in figure 3. From year to year operation, if management not aware with Investment Feasibility Tracking issue, it will lead them to the wrong information. Every year, the net cash flow is positive. It seems to be good. However, since 3rd year of

IOP Conf. Series: Materials Science and Engineering **337** (2018) 012057 doi:10.1088/1757-899X/337/1/012057

operation, the difference between actual and FS went wider. Unfortunately, the comparison is occurred in 10th years of operation, a little bit too late for corrective action.

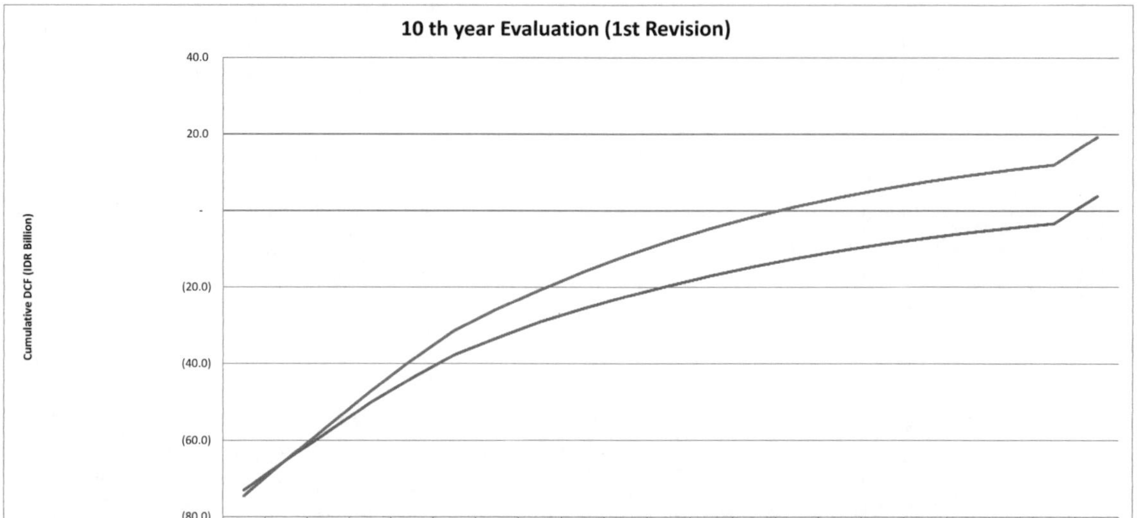

	1	2	3	4	5	6	7	8	9	10	11	12	13	14	15	16	17	18	19	20	21
Cum. DCF (i = 15%, FS)	(74.7)	(65.3)	(56.3)	(47.4)	(39.1)	(31.4)	(25.8)	(20.6)	(16.0)	(11.8)	(8.0)	(4.6)	(1.5)	1.1	3.5	5.6	7.5	9.2	10.7	12.0	19.3
Cum DCF (i = 15%, t = 10th year)	(73.1)	(65.3)	(57.8)	(50.3)	(43.8)	(37.7)	(33.3)	(29.1)	(25.6)	(22.4)	(19.5)	(16.9)	(14.6)	(12.4)	(10.5)	(8.7)	(7.2)	(5.8)	(4.5)	(3.4)	3.8

Figure 3. Investment Feasibility Tracking of a Hotel in 10th year operation (1st revision)

Net Present Value deviation between FS and actual condition in 10 years operation almost reached IDR 10.4 Billion. The early warning detection actually can be detected since 3rd and 4th year operation. However, since there is no evaluation, the gap become wider and little bit too late to handle. Management didn't realize the situation since the hotel could provide net cash flow around IDR 10-12 Billion/year with 3-4% growth/year and saving around IDR 1 Billion in the construction process.

If the management choose to continue the current business, with current condition without any Capital Investment to rehab or renew the facilities, the gap become worst. The business actually still has negative cumulative DCF, until the end of period. By the end of 20th year, it assume the hotel will be sold IDR 100 Billion (change the cum DCF into positive). From our point of view, rather than continuing the bleeding and waiting an investor buy the asset in the end of 20th year, it's better if there is any possibility to divest the business anytime within next 10 years. Minimum valuation required to reach positive cum DCF (NPV) is IDR 79 Billion.

2. *Power Plant (at the end of the lifetime)*

We can't show any data or figure related with this practical assessment due to confidentiality agreement with the company. However we want to share the interesting behaviour we found when we deal with the problem. Let say Y Power plant was built in 1994 with an investment (for instance) USD 100 million. If we converted the value to IDR, it's around IDR +/- 250 Billion in that day. However, Indonesia had suffered by monetery crisis in 1997 and the IDR to USD exchange rate rose until 6x. The Power Plant Valuation in the end of 2000 reach almost IDR 1200 Billion. The company renewed their asset valuation in balance sheet based on this information. After 20 years of their operation, the CEO asked a question can we shut down the plant right now?

To answer this question, we need to apply Investment Feasibility Tracking concept. Based on asset acquisition (in IDR) in 1994 and track all O & M cost, additional investment, and revenue during the operation period, we found the break even point has reached in 2006. However, if we change the currency basis from IDR to USD, we found that the Power Plant not reached the break even point yet (by the time of evaluation). It still need to be operated for more few years. Similar conclusion we found when we tried to change the 0 year basis from 1994 to 2000 (due to revaluation issue) and adjust all revenue and cost during 1994 – 1999 to 2000 basis (a matter of discounted cash flow problem). We found the power plant not reached the break even point yet.

From the contradictive result above, we suggest the power plant company to refer the 2nd (based on USD) or 3rd result (revaluation result). Nothing wrong with the 1st result, but in investment point of view, even we already reached break even point in 1st result, however, it's impossible we can replace the current asset with a similar one (due to different currency exchange issue, the price of new asset in USD probably not significant different, but in IDR does). The electricity tariff need to be renegotiated to meet the break even point target within next few year or if it impossible, the company need to have innovation program to reduce the operation and maintenance cost.

4. Conclusions

Based on the discussion and practical above, here some expected benefit from the Investment Feasibility Tracking implementation, i.e.:

- Early warning for management, remind the false pattern related with the project, related with 3 aspects simultaneously, investment, revenue, and cost.
- Fast risk mitigation / corrective action to response the deviation.
- Understanding the current position of Financial Performance in term of Feasibility Performance. In will give importance information what the impact of our current year performance to the overall project. It will provide comprehensive information for owner to set up future target to chase the project performance target.
- Maintain the data, keep update with current condition information, and may provide valid suggestion for future FS Document.

We also need to highlight the tracking (revision) process is conducted as requirement, but it's better when we found the actual deviation is higher or there is some wrong assumption. Beware with currency issue in investment and operation and revaluation asset.

5. References

[1] Project Management Institute 1987 *Project Management Body of Knowledge (PMBOK)* Project Management Institute
[2] Munns A K and Bjeirmi B F 1996 *The role of project management in achieving project success International journal of project management* 14(2) pp 81-87
[3] Jugdev K and Müller R 2005 *A retrospective look at our evolving understanding of project success* Project Management Institute
[4] Khamooshi H and Golafshani H 2014 *EDM: Earned Duration Management a new approach to schedule performance management and measurement* International Journal of Project Management 32(6) pp 1019-1041
[5] Saputra Y A and Ladamay O S A 2011 *Project reliability: Probability of a project meets its Quality-Cost-Time target under uncertainty* International Journal of Electronic Business Management 9(3) p 220
[6] Freeman M and Beale P 1992 March *Measuring project success* Project Management Institute
[7] Thuesen H G, Fabrycky W J and Thuesen G J 1989 *Engineering economy* Englewood Cliffs NJ: Prentice-Hall
[8] David F R 2011 *Strategic management: Concepts and cases* Peaeson/Prentice Hall

System analysis for technology transfer readiness assessment of horticultural postharvest

M Hayuningtyas [1] and T Djatna[2]
[1]Graduate Program of Agro-industrial Technology, School of Graduate and Post Graduate Studies, Bogor Agricultural University, Indonesia
[2]Department of Agro-industrial Technology, Bogor Agricultural University, Indonesia

*maulida_tia2016@apps.ipb.ac.id

Abstract. Availability of postharvest technology is becoming abundant, but only a few technologies are applicable and useful to a wider community purposes. Based on this problem it requires a significant readiness level of transfer technology approach. This system is reliable to access readiness a technology with level, from 1-9 and to minimize time of transfer technology in every level, time required technology from the selection process can be minimum. Problem was solved by using Relief method to determine ranking by weighting feasible criteria on postharvest technology in each level and PERT (Program Evaluation Review Technique) to schedule. The results from ranking process of post-harvest technology in the field of horticulture is able to pass level 7. That, technology can be developed to increase into pilot scale and minimize time required for technological readiness on PERT with optimistic time of 7,9 years. Readiness level 9 shows that technology has been tested on the actual conditions also tied with estimated production price compared to competitors. This system can be used to determine readiness of technology innovation that is derived from agricultural raw materials and passes certain stages.

1. Introduction

System are a form of integrated entity or element for reach objective[1]. System analysis is set of concepts, approaches, methodologies and tools for complex design. System complexity have a two things, there are dinamic and desain complexity. Analysis system begins by analyzing needs that show relationships between integrated entities, communities in achieving goals by building effective communication on each element. Technology transfer is process of transfer of knowledge through a technology, which is vertical on results of laboratory to application in community. A technology requires integrated readiness to be able to be applied by the community. Technology Readiness Level (TRL) is a measure of technological readiness level that is defined as an indicator showing how ready or mature a technology can be applied and adopted by a user / potential user [2]. Tekno-Meter is a tool that calculates TRL applied by the industry or for tangible products issued by Kemenristekdikti on 9 stages.

In the last four years, horticulture commodity in Indonesia have 35-40% of losses on farm to off farm, 60% of 40 occures in postharvest handling [3]. Postharvest handling like GHP (*Good Handling Practices*) must doing on procedure, can be reduce the postharvest losses. GHP need more postharvest technology to save and reduce number of horticulture losses more 10-15% from 40%.

The objective of this work are assess of technology a readiness from each level, 1-9 on TRL and to minimize time of transfer technology in every level.

2. Related Work

TRL is a systematic measurement system that supports the assessment of the maturity or readiness of a particular technology and the ratio of maturity or readiness between different technology types [4]. These stages have parameters and questions provided to assess the readiness of a technology to be commercialized [5]. Commercialization effort is closely linked to understanding consumer behavior. Consumer is no longer choosing the product of the function alone, but the combination of components or parameters are measured and unmeasured [6].

New approaches are necessary to overcome some traditional barriers between researchers and industry for implementation of innovative technology [7]. The technological readiness provides pillar measures the agility with which an economy adopts existing technologies to enhance the productivity of its industries, with specific emphasis on its capacity to fully leverage information and communication technologies (ICT) in daily activities and production processes for increased efficiency and competitiveness [8]. PERT is targetted not just for scheduling type of project such as construction, computer programming, planning and maintainance project, but the method can be used to research and development product or technology, including TRL implementation scheduling effectively [9].

3. Problem formulation

3.1 Current condition

Various research on postharvest horticulture technology with many aims to facilitate and accommodate the problems that occur in society as the end user. Sometimes the results of research or innovation technology is not socialized to the community well and massive, so less useful. Gap or distance between innovation results with user acceptance of the technology must be calculated and connected to the maximum, so the technology can be used. Technology undergoes a bottleneck at the dissemination stage to the wider community that requires new technology, or most of the technology fails in the process.

Based on that problem, so there needs to be technical parameters preparation of technology can be applied. How a technology can compete and be ready to pass the assessment phase so that it can be applied to the community / industry as a user. Furthermore, it is important to measure the readiness of these technologies so that they can be mass produced and used by farmers as primary consumers and industries as producers of products.

3.2 System analysis technology transfer readiness assesment

Currently the use of the Tekno-Meter counting device can only assess most of the resulting technology, it is necessary to create parameters relevant to the product or technology characteristics that will be generated with levels from 1-9 so that technological readiness can be measured. This tool have 3 stages, first level 1-3 is basic technology research where parameters on that level describe about basic technology. Level 4-6 is level of technology development, parameters describe about how technology upgrade the production scale or pilot scale. The last, level 7- 9 is system/subsystem development parameters describe about technology can be exist on real industry.

Every technology produced should be attempted to meet the criteria or parameters that exist within the tool or Teknometer. Technological assessment if it stops at a certain level, should be attempted to improve the technology according to recommended criteria/parameters. It will be repeated if the technology has not been able to pass through stage 9. The parameters in the tool can be conditioned to maximum so that fulfillment of the criteria is not too long.

4. Methodology

4.1 Identification System
System identification as a needs analysis of technology transfer readiness assesment system which consist of system boundary/scoping. Beside that have objectives, stakeholder, resources, parameter on each level, time required for technology transfer. These element being analyzed in relation with stakeholder activities along this system. The data and information collection used in this system from the parameter on Teknometer, like experimental stage, kind of machinery, simulation, prototype, etc.

4.2Business Process Model and Notation (BPMN)
Relation and connectivity between processes being depict literally in graphics to give a detail analysis of the system. A system analysis for technology transfer readiness assesment based on knowledge and information by expert interview, which is stakeholder or actor involved on this system. Stakeholder this is researcher, industry and Institute for Agricultural Technology Transfer (IATT), and with literature like technology information from books and document reports. Data and information like data of excellent technology, technology publication, feasibility technology and etc. The information obtained will be processed and displayed in BPMN 2.0.

4.3 A system analysis for technology transfer readiness assessment
Each technology generated should be assessment with certain parameters that are structured to saw the competitiveness of the technology, so it is certain that technology can be adapted by industry. The assessment to technology of each parameter and each level will be the input data to be processed in the approximate method approach. Other than that the time required by the technology from laboratory to be ready adopted requires optimum time. The optimum assesment of this system analysis is solved by Relief and PERT method:

4.3.1 Relief. RELIEF is a feature selection algorithm used in binary classification (generalisable to polynomial classification by decomposition into a number of binary problems) [10]. Relief method serves to determine the weight of each criteria that will affect the level or level specified and can show the interests of each criteria. Data for criteria or parameters which calculate the weight from case about postharvest technology such as, characteristic of technology, performance of the elements technology, operation components can be improve fresh red chili shelf life.
Assesment for this technology based on basic and development research, with answer kind of parameters to presented part of the level. For example, on existing condition level 1 have 3 parameters. Relief give a ranking with the weight on each parameters, and minimalist parameter can representated same conditon about the technology. Each level (N) have a number of parameter (n), on each level results minimum have 80 point from 100 than can be continue to next step.

4.3.2 PERT. Program Evaluation and Review Technique is a common method used in scheduling a complex system, so that the whole system can run with the maximum. PERT technique is a method that aims to reduce as much as possible delay, or disruption of production, and coordinate various parts of a job in a way through and accelerated the completion of the project [11]. PERT is analyzed using time data from each level of technological readiness level, so that the maximum time used in the implementation of a technology can or can be applied and useful in the community. Can be ascertained that a technology can be done from laboratory research to the community with a certain time in accordance with the technological specifications submitted [12].
Data collection for PERT method is from researcher whose have the technology to give a change and suggestion about assesment that new technology. They feel the stage, time for basic technology like research laboratory, kind of equipment, and environment study. They make their a shelf assesment to know the technology to be adopted by the industry.

IOP Conf. Series: Materials Science and Engineering **337** (2018) 012058 doi:10.1088/1757-899X/337/1/012058

$$Te = \frac{T_0 + 4T_m + T_p}{6} \quad ((1)$$

$$V = \frac{[T_p - T_0]^2}{[6]} \quad ((2)$$

Remarks :
To = optimistic duration
Tm = most likely duration
Tp = pessimistic duration

5. Result and Discussion

5.1 Identification System
A requirement analysis of system analysis for technology transfer readiness assesment of horticultural postharvest involve the availability of consumer need, kind of postharvest technology, integrated element and stakeholder which consist of reseacher, consumer, industry and technology provider. This system analysis can produce the technology adopted by industry and have a beneficial to society. An identification system are shown in Fig. 1 :

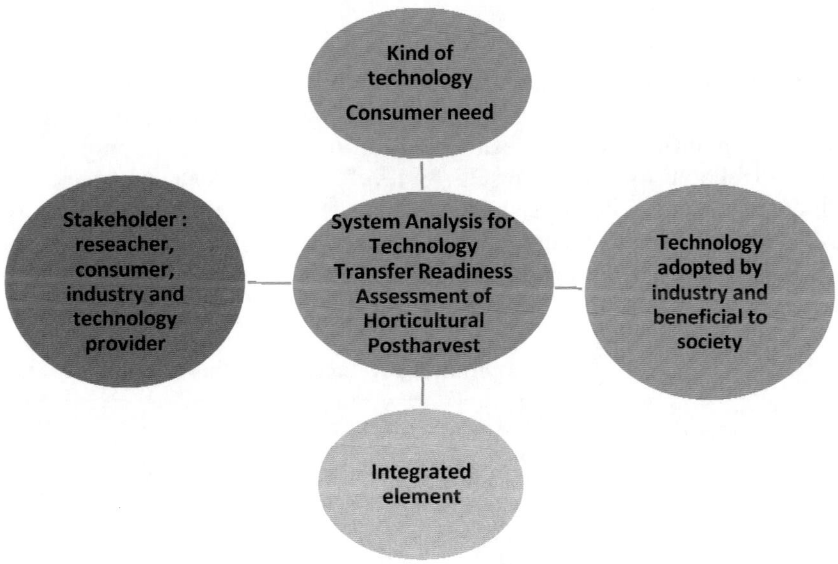

Figure 1. Identification of a system analysis for technology transfer readiness assesment

5.2 Business Process Model and Notation Analysis (BPMN)
Analysis of a system analysis for technology transfer readiness assesment of horticultural postharvest is based on kind and number of postharvest technology, availability technology of horticultural product in industry, and process on technology provider. The analysis process use BPMN are presented in Fig. 2:

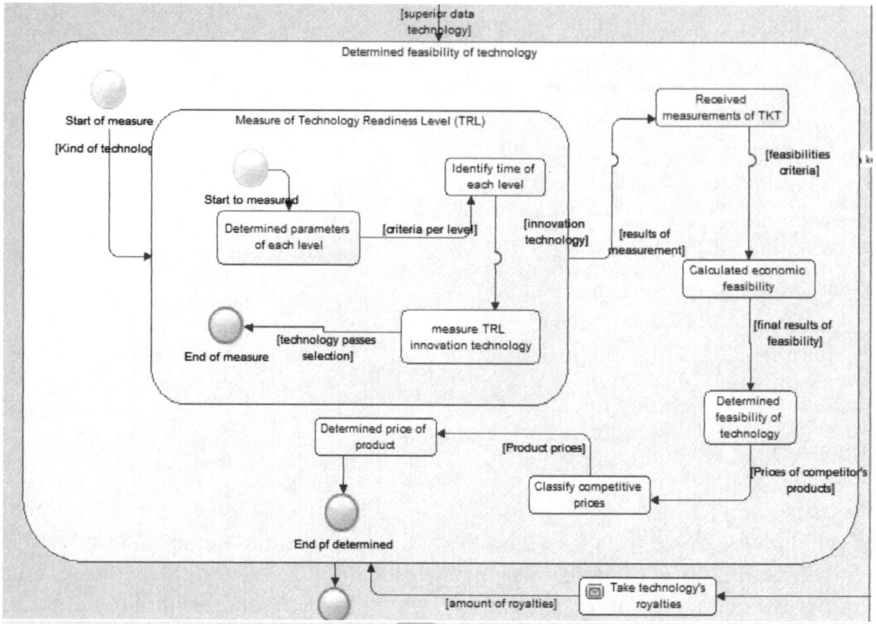

Figure 2. Analysis process of system technology transfer readiness assesment of horticultural postharvest in BPMN.

5.3 Results

Each level has certain parameters to select and rate a technology until it reaches the highest level. It is shown with parameters that have been modified and the result using the relief method on the effectiveness of the parameters used. For example, level 1 on the level of technology readiness consists of 3 parameters and relief method have a results for level 1, just have one parameter. The highest weight of 0.92 which emphasizes the literature or basic principles of the technology to be developed. Total parameter level 1 until 9 have a 76 parameters to get verification about technology which want to be transfered to industry. With relief method can be optimum the parameters, from 76 parameter to 49 parameters.

Table 1. Results optimum parameter on Relief Method.

Level	Existing	Relief Method	Score
1	3	1	100
2	12	5	95
3	9	8	95
4	8	4	90
5	8	6	86
6	6	2	80
7	13	8	62
8	9	7	53
9	8	8	22
	76	49	

Determination of levels in transfer technology is adjusted to the characteristics of the technology to be implemented. Levels are divided into 9 levels with each predefined criterion, the conditions of each statement in each level is 100 to proceed to the next level. A value less than 80 can not proceed to the next level. Critical factors that affect the readiness of technology can be known by the weight of use is in a certain level. For this problem, critical parameter have on level 7, because the weight as posistive high point and on level 7, product must be transfer to system development. If the stage is not passed then the technology has no prospect to be commercialized

Table 2. Results PERT Method for time required of technology transfer.

Level	Existing	Relief	Optimistic time	Pesimistic Time	Expected time	Most like time
1	3	1	2	5	3	2
2	12	5	8	15	9	8
3	9	8	16	23	17	16
4	8	4	8	14	9	8
5	8	6	17	23	18	17
6	6	2	6	11	7	6
7	13	8	13	21	15	14
8	9	7	12	20	13	12
9	8	8	13	18	14	13
		Month	**95**	**150**	**105**	**96**
		Year	**7,9**	**12,5**	**8,7**	**8,0**

The result shown in Table 2 is that with scheduling using the PERT method, it is evident that if the judging process is conducted sequentially from level 1 to 9 then the time required is divided into optimistic time, pessimistic time and predicted time in the assessment or transfer process technology. With an optimistic time, technology transfer from laboratory preparation to the technology can be properly adopted and applied as a whole to 7.9 years. So, researcher or innovator must be sure and design the hole system of technology and the technology can be exist for long time.

6. Conclusion and Future Work

Technology Readiness Level (TRL) is an assessment prepared to select the readiness of a technology to be generated and can be widely used by the community. Each level on TRL have a number of parameters, that statement accordingly with every level means and efectiveness of technology. Parameter / statement every level must be optimum, from 76 statement just on 49 statement for whole assesment system with the weight method for every statement. Based on score at each level, postharvest technology of horticulture is only up to level 6 in its development, and can not proceed to the next level if it has not fulfilled the desired parameters. Technology transfer time from basic technology to operational condition on industry on selection used PERT : optimistic time : 7,9 years, pesimistic time : 12,5 years, expected time : 8,7 years. This level of technological readiness can be utilized to create an ideal system for all technologies with different properties. In addition, linkages between levels and between statements can explain in detail the assessment is optimum.

References

[1] Eriyatno 2012 Ilmu Sistem (Surabaya : Guna Widya)
[2] Wikipedia Transfer technology
[3] FAO 2013 Postharvest losses for horticulture commodity and product.
[4] Anonymous 2012 Technology Transfer Assesment Readiness Level Kementerian Riset dan Teknologi dan Pendidikan Tinggi
[5] Graettinger C P, Garcia-Miller S, Siviy J, Van Syckle P J and Schenk R J 2002 Using the Technology Readiness Levels Scale to Support Technology Management in the DoD's ATD/STO Environments (A Findings and Recommendations Report Conducted for Army CECOM) Special Report
[6] Munichputranto F and Djatna T 2013 An Ontology Design for Commercialization Through Technology Transfer Based on Digital Business Ecosystem *Proc. of Industrial Engineering and Service Science*

[7] Morrissey M T and Almonacid S 2005 Rethinking technology transfer *Journal of Food Engineering* **67** 135–145.
[8] Ghasemi R, Petroudi S H H, Mahbanooei B and Kiasari Z M 2013 Relationship between "Infrastructure" and "Technological Readiness" based on Global Competitiveness Report: a Guidance for Developing Countries *Int Conf on Electronic Commerce and Economy*
[9] Hayun A 2005 Perencanaan dan Pengendalian Proyek dengan Metode PERT-CPM *Journal The WINNERS* **6** (2) 155-174
[10] Kira K and Rendell L 1992 The Feature Selection Problem: Traditional Methods and a New Algorithm *AAAI-92 Proceedings*
[11] Charles K A and Levin R I 1972 Perentjanaan dan pengawasan dengan PERT dan CPM (Jakarta: Bhatara)
[12] Hajdua M and Bokor O 2016 Sensitivity analysis in PERT networks: Does activity duration distribution matter *Automation in Construction* **65** 1–8.

A Study on Course Management System Implementation in Indonesian Higher Education Institutions

Y A Saputra[1*], M L Singgih[1], E Latiffianti[1], E Suryani[2], and Mudjahidin[2]
[1]Department of Industrial Engineering, Institut Teknologi Sepuluh Nopember, Kampus ITS Sukolilo-Surabaya 60111, Indonesia
[2]Department of Information System, Institut Teknologi Sepuluh Nopember, Kampus ITS Sukolilo-Surabaya 60111, Indonesia

*yandrian@ie.its.ac.id

Abstract. Information technology development nowadays has brought new colors in the higher education learning process. In Indonesia, the current trend showed a higher use of CMS to support the existing conventional learning method in the classrooms. This paper attempts to understand the characteristics of CMS implementation based on a survey at several higher education institutions in Indonesia. There were 9 selected higher education institutions observed in this study. The objectives were to find out the CMS implementation in terms of: 1) the management of CMS implementation, 2) the evaluation, 3) originality of materials, platform, and feature; and 4) participation level. The result showed that the use of CMS in these institutions, in general, was to support the classroom conventional learning method by providing a repository of lecture notes and communication forum/media outside the classroom. The management task mostly was taken care by a specific unit. A Moodle (freeware) was found as a typical platform in use, and none of the institutions chose to use paid platform i.e. Blackboard. The accessibility of CMS used was kept closed for limited group of people due to high cost of material originality assurance. Observation also found that there was not much attempts in evaluating the success of CMS implementation in each institution, whereas the success measurement was limited to the users' satisfaction level. The majority of institutions claimed a good internal participation level (with lecturers and students as the main users), but in general we found that lecturer participation in most institutions were low or even very low.

1. Introduction

Information and communication technology (ICT) development nowadays has led to many improvements in the education learning process. Technology development on Web 2.0 that brought up E-Learning 2.0 in 2004 has made the use of ICT in learning process became higher. The development of E-learning in form of Course Management System (CMS) has been a symbol of innovation adopted and implemented in many higher educations in the world [1].

Although the implementation has been going on for many years, pros and cons on CMS were still highly discussed in many research. Several cons on CMS highlighted that this system was originally implemented as a distance learning method, but in the development it has changed its purpose and act as a supporting media for in-classroom learning as well as a communication media for lecturer and students outside the classroom [2][3]. Thus, the investment purpose of the CMS development became questionable, as the use of the system itself is more for the in-campus learning rather than distance learning. However, regardless which role the CMS played, researchers also noted several benefits of

CMS implementation. Some suggested that CMS was a set of technology to support and leverage learning process [4], while other saw CMS as a simple tool to help lecturers/teachers with the least web design knowledge creating a website to support courses they teach [5]. CMS can be defined as a comprehensive web-based set of tools, either static or interactive, that support any aspects of learning process, including preparation, presentation, communication, participation, and interaction [6]. On top of that, based on the case studies in several higher educations in South Ontario, Canada; it is reported that students and lecturers found CMS very positive for learning process although the usage was limited to the basic function of CMS (as repository) [7].

In spite of the fact that pros and cons are still debatable, many higher education's thought of CMS in the long term as strategic and it would bring a better learning process. Massachusetts Institute of Technology (MIT) was a pioneer in the use of CMS as knowledge dissemination medium in learning process, not only for internal users but also for the external through a mechanism so called Open Courseware (www.ocw.mit.edu) developed in September 2002. This courseware was strategically beneficial for the internal stakeholders while at the same time it also put an idea in the external users' mind about MIT contribution for academicians in the entire part of the world. The fact has made the management of MIT to consistently focus on the attempt of periodic evaluation of the CMS implementation. Peer comparison with several US universities on CMS, Course Material Life Cycle, and the involved cost was an example of one of many attempts the MIT made to improve the implementation of CMS [8] . At about the same time, Universities and Colleges Information System Association (UCISA) UK also put a significant consideration about CMS implementation as one of ICT components to be used in the higher education learning process next to Managed Learning Environment (MLE) [9]. Started in 2001, UCISA has surveyed the implementation of Technology Enhance Learning (TEL) for its members in England, Wales, Scotland, and Ireland.

With the rapid development of ICT, CMS has become a new part of education learning process in the Indonesian higher educations. Large number of universities, both public and private, have developed and implemented CMS with high variation in terms of investment level, complexity level, and operational standards. Some higher educations developed its own platform; e.g. Universitas Gajah Mada (UGM) with E-Lisa (www.elisa.ugm.ac.id) and Amikom Jogja with http://elearning.amikom.ac.id; while the others developed their system using open source platforms; e.g. Institut Technology Sepuluh Nopember (ITS) with its www.share.its.ac.id, Universitas Indonesia with its www.scele.ui.ac.id, and Institut Teknologi Bandung (ITB) with www.kuliah.itb.ac.id. While the implementation of CMS is still in the development stage, some higher educations that have implemented CMS; such as UGM, Universitas Gunadarma, ITB, and Universitas Indonesia; try to disseminate the contents in their CMS as an Open Courseware as that used by MIT.

At the macro level, E-Learning Readiness of Indonesia was ranked 53 out of 60 countries observed [10]. This result reported in 2003 would change a lot in the nowadays measurement, especially because the CMS development in Indonesia has grown rapidly these recent years. In November 2013, Webometrics reported this growth by placing 11 Indonesian universities (higher educations) within 15 top universities in the South East Asia in term of openness.

Earlier research about CMS implementation in Indonesia was mainly limited at micro level; in general it evaluated the success of CMS implementation in a specific institution. This early research included a study that evaluated the acceptance of CMS in the *Universitas Pendidikan Indonesia* using the Technology Acceptance Model (TAM) approach [11] and an evaluation of E-Learning Maturity Model implementation in a certain university using an approach adopted from eMM 2.0 of New Zealand Standard [12]. This was few research that reported the evaluation of CMS in Indonesia at macro level as those available publications [8][9][13]. Therefore, this paper attempts to explore the profile of macro characteristic of CMS implementation in Indonesia based on the survey of public and private higher education institutions in Indonesia.

2. Methodology

This study used a survey as the main approach. The selection and the number of respondents in this survey were mainly defined with a consideration of the following factors: 1) the availability of information from the previous research, 2) respondent willingness to share the required information, and

3) limitation on time and resources. Questioners were designed to at least accommodate the early mapping on important information that is necessary to explain the profile of CMS implementation in the selected higher education. The survey was then conducted using the questioner form, through discussion session, and direct interview with the information source.
With the approach, finally there were 9 higher education institutions data to be used in this study as shown in Table 1.

Table 1. List of higher education institutions (universities) surveyed.

No	Institution Name	CMS Address	Rank Webometrics (2013)	
			South East Asia	Indonesia
1	Institut Teknologi Sepuluh Nopember (ITS)	www.share.its.ac.id	48	13
2	Institut Pertanian Bogor	www.lms.ipb.ac.id	34	7
3	Universitas Diponegoro	www.kulon.undip.ac.id	40	10
4	Universitas Gunadarma	www.elearning.gunadarma.ac.id; ocw.gunadarma.ac.id	28	5
5	Universitas Airlangga	www.aula.unair.ac.id	38	9
6	Universitas Paramadina	www.elearning.paramadina.ac.id	900	93
7	Universitas Kristen Petra	www.lentera.petra.ac.id	37	8
8	Universitas Surabaya	www.elearning.ubaya.ac.id	157	45
9	UIN Syarif Hidayatullah	http://ais.uinjkt.ac.id	829	75

The questioner sheet was divided into 4 aspects with 19 questions in total (5, 7, 4, and 3 for each). Table 2 explains each aspects and the questions.

Table 2. List of questions, the objective, and the reference of each question used in the survey.

No	Aspect	Objective	References
I.	**The management or the operation handling of the CMS**		
a.	Management structure	To find out whether it is outsourced or in-house. For an in-house, is it centralized or decentralized within the school, faculty, or department? For the centralized, what unit is assigned to manage the operation?	Question 3.2.a in Walker et al. [9] also used in the strategic section in Ingraham et al. [13]. Adopted by adding the characteristic of general practice found in Indonesia.
b.	Access restriction	Important to understand whether a CMS is accessible for everyone or closed only for internal use. The question also tries to identify the opportunity of *Open Courseware* development.	Added in this survey.
c.	Reasons for CMS implementation	To investigate the reasons behind the implementation of CMS.	The question was used in Landon et al. [8] to understand the strategic motive and the importance of CMS for the university. This question was also discussed in Walker et al. [9]
d.	Amount of investment cost and its components	To find out the amount of investment an institution willing to pay, and the components of the cost (including the major component).	Initial investment in the US based on Landon et al. [8] varied, ranged from US$ 23K to US$ 500K.
e.	Operating cost and its major component	To identify the operating cost and its major components.	Landon et al. [8] showed that operational and maintenance cost ranged from US$ 275K to US$ 1.3M. License fee for Blackboard was estimated at US$ 100K per year and for WebCT was about US$ 20-25K
II.	**The choice of platform and the management of contents and features**		
a.	Existing platform used	To understand the variation of platform used in each institution.	Based on Landon et al. [8] the platform used was dominated by Blackboard, while Walker et al. [9] suggested the major use of Moodle followed by the Blackboard.
b.	Reasons for platform selection	To understand the reasons for platform selection.	Based on Walker et al. [9] the main reasons were cost and reliability.

No	Aspect	Objective	References
c.	Evaluation or plan on platform change	To find out whether the institutions perform periodic evaluations on the use of existing platform. Who would perform the evaluation? Is there any plan to change the platform and why?	Based Walker et al. [9] a periodic evaluation was commonly found. Platform change was also commonly found in universities.
d.	Available features	To identify what features are currently available, and the possibility of Self-Developed features.	Question 3.16 in Walker et al. [9] The same question was also found in Landon et al. [8]
e.	Plan on feature development	To find out whether the institutions have a plan to develop their own features.	Landon et al. [8] confirmed that there were several plans in feature development among the US universities.
f.	Originality of contents and teaching material	The access restriction was suspected to be influenced by the originality of teaching materials. The institutions would make the access open when they can assure the originality.	The MIT Open Courseware required an intensive teaching material development with a great support from staffs Landon et al. [8]
g.	Strategy on content development and originality assurance	To find out whether there is a certain strategy to improve the content development and the material originality. If it is so, what it is and how much it would cost.	An incentive of US$ 3000 was awarded for each course to participate in the MIT OpenCourseWare. It required about US$ 5.1 Million, only for the development of contents Landon et al. [8]
III		**Participation Level**	
a.	Participation level of lecturers	The management record on lecturers participation level in the implementation of CMS	Participation level is normally measured in term of course number and the number of students involved[13]. It was reported that there was an aggregate estimation of 69% student enrollment recorded as active users in the US Landon et al. [8]. The strategy to increase the participation level was discussed in section 15a in Walker et al. [9]
b.	Strategy to raise the participation level of lecturers	The use of specific strategy to raise the participation level of lecturers	
c.	Participation level of students	The management record on students participation level in the implementation of CMS	
d.	Strategy to raise the participation level of students	The use of specific strategy to raise the participation level of students	
IV		**Evaluation**	
a.	Availability of evaluation on implementation success	To check whether there is an evaluation method developed in this case. Does it follow a certain assessment standard or the local standard based on the institutions' policy?	54 out of 98 universities in the UK performed an evaluation to check the impact of CMS implementation to the students' learning experience Walker et al. [9]; while 34 out of 98 universities performed an evaluation on the impact of CMS implementation to the pedagogic practice.
b.	The benefit of evaluation	The benefits identified by the institutions from the implementation of CMS based on the last evaluation	Question 3.25 in Walker et al. [9]
c.	Obstacles of CMS implementation	The obstacle identified by the institutions in the implementation of CMS based on the last evaluation	Additional question designed in this survey to complete information related to the benefits of CMS implementation.

3. Results

Based on the survey, we found several unique facts regarding the implementation of CMS in Indonesia that differed from those reported in the US and UK. Explanation on each aspect is provided in subsection 3.1 to subsection 3.6.

3.1. In-sourcing

The UCISA showed that 20 out of 98 universities in the UK employed a third party to host the CMS [9]. This indicates that CMS was no longer seen as something to be managed in house. In the other hand, the survey in this study indicated that the institutions generally managed CMS in-house. Most of the managements were centralized under a specific unit.

The process of CMS development varied. Some used vendors to help the development, while the rests developed their own CMS. We also found that CMS maintenance service providers were not commonly available in Indonesia. It is clearly different than what was found in the UK where the free or paid platform of CMS have already had several partners to provide the service that included not only platform development, but also the management and the maintenance of CMS.

3.2. Moodle Domination

There was only 1 out of 9 university that did not use Moodle as the platform of CMS. This university chose to use a Joomla-based platform instead of Moodle. There was also one institution used the mixed of Moodle and Joomla, while the rest 7 institutions used Moodle.

The reason for choosing Moodle was mainly due to the fact that it was an open source. That way the institutions could minimize the investment and operational cost, but at the same time this option forced the institutions to develop their own CMS. However, most respondents chose to develop their own CMS rather than to outsource.

The domination of Moodle was also caused by the period of CMS development in these institutions that were happened at nearly the same time. As the pioneer gained a success on the implementation, the followers then took the same step by choosing the same platform. This factor is a stronger influence than the previously mentioned, as we found that most institutions to choose the Moodle did not perform a comprehensive study on it. Extending the sample outside the survey, we can see that several public institutions in Indonesia, in majority used Moodle as their platform; e.g. Institut Teknologi Bandung (ITB), Universitas Indonesia (UI), and Universitas Brawijaya.

The Moodle domination is actually the same phenomenon reported in [9]. It was reported that Moodle has made Blackboard move to the second position in the platform domination in CMS; 58% of the universities in the survey have been reported to use Moodle. However, Moodle domination in Indonesia was far stronger.

Another notable finding is that most of higher education in Indonesia based on this survey, used a single platform of CMS, whereas in the US and the UK, it was very common for a university to have more than one CMS. For example MIT had Stellar and Sloan Space, California Berkeley used B-Space, Blackboard, and WebCT, Stanford University employed CourseWork, WebCT, Blackboard, and CCNet [8]. Additionally, a periodic evaluation on platform, including the needs to add more features and the needs of change, was commonly performed by the universities in the UK and US. The similar phenomenon was not found in Indonesia. Most institutions did not perform an evaluation. Thus, we can say that the basic reason for Moodle domination is merely influenced by the cost efficiency and trend instead of the result of comprehensive evaluation.

3.3. Investment, operations, and maintenance cost

The practice of CMS in the US at first was for a knowledge repository. By the development, it changed its function to be a media of knowledge dissemination for public. This change of course required a relatively high cost. In general, CMS implementation cost can be classified as follows [8]:

1. Cost of investment, operational and maintenance; this includes periodic maintenance, operational routines, human resource, troubleshooter, license fee, etc.
2. Cost of CMS development for students with disability
3. Cost of personnel to support the implementation of CMS
4. Cost of socializing and educating the users (students and lecturers)
5. Cost of material development and other supporting material

Investment cost for materials development in general was relatively not high. High cost in this particular item was only found in Stanford where it reached US$ 1 million, while the rest was commonly found in the range of US$ 20K up to few hundred thousand.

The biggest cost component for CMS implementation in US universities was found at the fifth. MIT spent about US$ 5.4 Million per year to develop the OpenCourseWare, while the cost was varied in the other universities with the smallest cost as low as below US$ 200K per year [8]. As for this survey, it is rather difficult to classify the cost in term of components suggested in [8]. To simplify, we divided the cost in two big parts: 1) investment cost, and 2) operations and maintenance cost (non-investment cost). Details on both costs based on the survey result are shown in Table 3.

Table 3. Comparison on investment and operation-maintenance cost of CMS implementation across the surveyed institutions.

Institutions Name	Investment cost	Highest expense	Annual O&M cost	Highest operational cost
IPB	19k – 37k USD	Software and hardware	19k – 37k USD	Course material development
Universitas Gunadarma	> 75k USD	Network infrastructure	> 75k USD	Incentive for content& material development
UIN Syarif Hidayatullah Jakarta	1k – 4k USD	Hardware, server, and network infrastructure	4k – 7.5k USD	Hardware
Universitas Diponegoro	< 1000 USD	Teaching/course material development and socialization	< 1000 USD	Managing the material, server maintenance, spam handling
Paramadina	3.5k – 7.5k USD	Server	3.5k – 7.5k USD	Data maintenance
Universitas Kristen Petra	3.5k – 7.5k USD	CPU Camel Application	< 1000 USD	Server sharing
UBAYA	<1000 USD	Server	< 1000 USD	Maintenance
ITS	19k – 37k USD	Hardware, software, and network infrastructure	19k – 37k USD	Data maintenance, server, and teaching material development
Universitas Airlangga	37k – 74k USD	Preparation and Lab. Building Construction	19k – 37k USD	Training (education the users)

In general, it was clear that the majority of the institutions in Indonesia spent an investment cost of less than 7500 USD. Only a few of the institutions were found spending more than 7500 USD. Universitas Gunadarma spent the biggest investment among all, with the cost of more than 75.000 USD that spent mostly in the development of network infrastructure. In term of operations and maintenance cost, once again Universitas Gunadarma spent the highest cost among respondents, with the value of more than 75.000 USD in an annual basis. The amount of the cost this university has spent can be seen as a serious commitment to be the leading university in the implementation of CMS in Indonesia. The biggest contribution in the O&M cost of Universitas Gunadarma was given by the cost of developing teaching material/contents development that were renewed in a yearly basis.

To sum up, the survey showed that the aggregate expense paid by the Indonesian institutions for CMS implementation is not as much as that found by the universities in the US or UK. It is probably one of the factors affecting the quality of features and contents of the CMS.

3.4. Internal evaluation

The survey result indicates that in average, the institutions did not retain an integrated application to perform an evaluation on the success level of CMS implementation. Only 5 out of 9 have developed evaluation tools although they were limited to the user's satisfaction.

Evaluation in the success of CMS implementation should not be merely a user satisfaction survey. It should have covered the assessment on the level of technology acceptance, the quality of teaching material, platform handling/management, organization, and other relevant assessments. The phenomenon of insufficient or the absence of evaluation, may also imply that it was not the culture of the organization. In term of concept, we suggest that evaluation on CMS implementation to be performed by applying the principals of E-Learning Maturity Model (eMM) 2.2 that was developed by the Ministry of Education of New Zealand [14] eMM 2.2 developed an evaluation system based on 5 dimensions of evaluations: organization, evaluation, learning, development, and support. Assessment can be performed by asking questions on each dimension with a scoring for each question. This mechanism should be relatively easy to adopt. It is also considerably good reference for self-assessment or internal audit related to the evaluation of the implementation of CMS in the university.

Poor evaluation practice found in the survey may also indicate that investment spent on CMS these days can be merely for a reason to be a part the current trending. In a better maturity level, these institutions/universities should see CMS as a necessity. Having said that, a mechanism of periodic evaluation would help the institution to formulate a strategy for continuous development.

3.5. Development of course material

The development of teaching material as the main content in a CMS has gain significant attentions in the universities. This success of this activity required a specific strategy. It includes but not limited to giving an incentives for developing a teaching material, educating the users to better formulate a teaching material, and optimizing the use of technology for better learning experience. In a long term, teaching materials can be disseminated broadly as an attempt to contribute and develop the community (known as Open Courseware).

In Indonesia, the strategy of teaching material development was not a main focus. Most institutions surveyed decently wished that the lecturers would participate to the implementation of CMS. Only a few institutions used an active strategy to encourage the lecturers to develop teaching material to be shared in the CMS as an Open Course. Universitas Gunadarma, ITS, and IPB were some to put the strategy in practice. Those three institutions have provided an amount of funding to make better quality of teaching material available in the CMS.

3.6. Other Findings

Apart from the above aspects discussed, the survey also highlighted the reasons behind the implementation of CMS in the surveyed institutions as follows:

- To strengthen the learning method and to provide variation in learning process
- To lift up the institution's rank
- To provide communication media/forum outside the classroom and outside the learning hours
- To support submission for assignment and online examination
- To facilitate teaching material update
- To improve the academic atmosphere
- To manage knowledge in teaching
- It is a trend in a learning method development

This study also observed the obstacles commonly found in implementing CMS as follow:

- Slow internet connection. It can be influenced by the price of connection or the availability and capability of the network offered.
- Cultural problem and conventional mindset that CMS is not an urgent support made CMS less important to some groups of people.
- Literacy problems on the usage of CMS, from both sides (lecturers and students) can be one of many reasons for low-level participations in the implementation of CMS.
- The availability of funding.

Table 4 summarizes the results in four surveyed aspects as explained in this section.

Table 4. Results summary.

No	Aspect	Findings
I	**The management of CMS implementation**	
a	Management structure	All of the respondents use an in-house management. None of them outsourced the management to the third party.
b	Access restriction	Most institutions restricted access only for registered internal users (closed access). There was only one institution that opened the access for both internal and external users. However, it also limited few contents to internal users only for license/patent reasons, standard compliance reasons, and other.

No	Aspect	Findings
c	Reasons for CMS implementation	– To give the students new experience in learning method variation and technology application. – For easier management of teaching material (repository) and to provide communication forum for students outside the classroom. – In a long term, it could lift up the institutions Webometrics rank (especially when it is made open).
d	Amount of investment cost and its components	See Table 3. The biggest expense on investment was commonly in the component of hardware and application development. There was only one institution where the most expense was on network infrastructure. In most institutions there was no License Fee because none of them use paid platform or feature. The investment cost of CMS in Indonesia was ranged roughly from 1000 to 75K USD, far below the investment of CMS in the US that ranged from 23K to 1 million US$.
e	Operating cost and its major component	Annual rough estimation (excluding cost of staffs): • 2 institutions spent less than 1000 USD • 1 institution spent 1000 to 7500 USD • 2 institutions spent 3500 to 75k USD • 3 institutions spent 19k – 37k USD • 1 institution spent more than 75k USD – See Table 3 for more details on operation cost. – The result confirmed the implementation of Open Course Ware and funding support to enrich teaching material as the biggest component other than maintaining hardware and software.
II	**The choice of platform and the management of contents and features**	
a	Existing platform used	9 institutions used Moodle as the platform, 2 of them combined the Moodle with Joomla. It confirmed those reported in [9], a relatively high migration to Moodle. Furthermore, it also confirmed the absence of License Fee because Moodle was free.
b	Reasons for platform selection	Most institutions explained the ease of use as the main reason, very few mentioned about doing a comparison between platforms. Moodle was dominating the selection, and the more institutions used Moodle, the more followers decided to use Moodle too. Another reason for choosing Moodle is because it is free. The finding showed that an in depth analysis was not something to define the selection of platform. The most important targets were low cost and that CMS was in place.
c	Evaluation or plan on platform change	Most of institutions mentioned that there was no evaluation method, nor did it plan to change the platform. Except for Universitas Gunadarma that planned to develop a videoconference based platform and web mobile. Although the implementation duration can be considered as long enough (3-6 years), doing the evaluation did not seem to be a culture.
d	Available features	The features used were commonly the standard available one. Only two universities took the advantages of Single Sign On (SSO). In this SSO, there is an integrated login access between CMS and other information system facilities available. The majority of users were at the basic level in term of feature usage. Most of the institutions used CMS only for a repository of teaching material and assignment submissions.
e	Plan on feature development	There was only one institution planned on exploring the multimedia contents. It was possibly do to the effect of poor evaluation method in general.
f	Originality of contents and teaching material	Most institution claimed that the originality of material was a high priority. However, observations showed that the level of originality mostly was around 50-75%, one institution was at 25-50%, and another claimed that it never did a specific measurement. Assuring the originality of material was considerably difficult to perform, unless there is a commitment from the institution managements to do so.
g	Strategy on content development and originality assurance	The content development strategy was one important factor to consider. Three institutions have been applying "Funding Grant" to encourage the submission of teaching material to be disseminated in the CMS. The rest has not considered the strategy. In Universitas Paramadina, Guidance for prevention of Academic violation was used to encourage the development of original teaching material.

No	Aspect	Findings
III	**Participation Level**	
a	Participation level of lecturers	Participation level of lecturers was considerably low. Most institutions reported that the participation level was between 25% and 50%. Higher level of participations (>50%) were found in three institutions.
b	Strategy to raise the participation level of lecturers	The answers on this question varied. The strategy included socialization, training (educating), responsibility assignment, etc.
c	Participation level of students	Most institutions were at 75%-100%. There were 2 institutions stated that they never did any measurement in the past.
d	Strategy to raise the participation level of students	Training (educating), socialization, coaching, and adding the number of courses available in the Moodle.
IV	**Evaluation**	
a.	Availability of evaluation on implementation success	The evaluation practice was minimum. Some of the institutions performed it but limited to the users satisfaction assessment. None of the institutions applied a comprehensive evaluation like eMM 2.2. It was also found that none of them has officially evaluated the impact of CMS implementation on the students.
b.	The benefit of evaluation	Because evaluation was not performed, it was hard to define the benefit for the institutions.
c.	Obstacles of CMS implementation	– Slow internet connection – Cultural problem and conventional mindset that CMS is not an urgent support made CMS less important to some groups of people. – Literacy problems on the usage of CMS, from both sides (lecturers and students) can be one of many reasons for low-level participations in the implementation of CMS. – The availability of operational funding support (when available, it is normally small).

4. Conclusion

A survey has been performed to 9 higher education institutions in Indonesia. It aimed to understand the characteristics of CMS implementation in terms of: 1) the handling of the operation and its mechanism, 2) the evaluation, 3) originality of material, platform, and feature; and 4) participation level.

In most surveyed institutions, we found that the purpose of implementing the CMS was mainly to support conventional learning where CMS functioned as a material repository and communication forum outside the classroom. This finding confirmed what has been suggested by [7]. This study also found that the management of CMS in most institutions was handled by a specific unit/office/department within the institution organization and none of the institutions outsourced the management of CMS. The majority of platform used was Moodle (freeware) and none of the institutions used paid platform like Blackboard. Most developed CMS was designed to be closed access due to high cost in assuring the material originality.

In term of cost, there was a significant difference between the amount of money spent for investment and operations and maintenance cost in the implementation of CMS in Indonesia compared to those in the US and the UK. This relatively small amount of fund made the Indonesian institutions hard to improve the quality of teaching materials and contents. Insufficient or even absence of attempts in the evaluation of the CMS implementation was another problem. The existing evaluation was simply to measure the satisfaction of the users. The majority of the institutions claimed a good level of students and lecturers participation, but this study found that lecturer participation level was relatively low.

Among 9 institutions surveyed, Universitas Gunadarma can be seen as the best one in implementing CMS, the year in which the implementation started, the expense it has spent, and the commitment of the management were evidence. In the other hand, the implementation of CMS in most institutions was at the early stage. At this stage, most institutions have tried to adopt and implement, but they were still dealing with many problems found in the development, management, and organization. If the institution management considered CMS simply as a support facility, then current phenomenon of the implementation can be seen as sufficient. But if the management would like to see it from different perspective, for example as a strategic tool to improve the effectiveness and efficiency of learning process, knowledge dissemination for community development, or a must-have system to lift up the institutions rank, then there should be plenty of space for improvement, and it can be started from the investment.

A potential future research in this topic would be best conducted by comparing the maturity level of the CMS implementation between institutions. The use of assessment concepts such as 2.2 [14] would be very interesting to discuss considering the finding regarding the evaluation practice in the institutions. Furthermore, in reference to the TEL survey (2012) conducted in every other year [9], the questioners in this survey can be improved to make it appropriate for periodic survey with a bigger sample size for mapping the implementation of CMS in Indonesia.

5. References

[1] Coates H 2005 Leveraging LMSs to enhance campus based student engagement *Educause Quarterly* **28**(1) pp 66-68

[2] West R E, Waddoups G, Kennedy M M, and Graham C R 2006 Understanding the experiences of instructors as they adopt a course management system *Educational Technology Research and Development* **55**(1) pp 1-26

[3] Morgan G 2003 Faculty use of course management systems *Educause Center for Applied Research.* Retrieved 9 September 2006 from http://www.educause.edu/

[4] Carmean C and Haefner J 2002 Mind over matter: Transforming course management systems into effective learning environments *Educause Review* **37**(6) **pp** 27-34

[5] Collis B and Boer W D 2004 Teachers as learners: Embedded tools for implementing a CMS *TechTrends* **48**(6) pp 7-12

[6] Malikowski S R, Thompson M E, and Theis J G 2007 A model for research into course management systems: Bridging technology and learning theory *Journal of Educational Computing Research* **36**(2) pp 149-173

[7] Lopes V, 2006 Course Management Systems and Campus-Based Learning *CSSHE Professional File No 29* Canadian Society for the Study of Higher Education

[8] Landon B, Henderson P, and Russel P 2006 MIT Peer Comparison on Course/Learning Management System *Course Material Life Cycle and Related Cost*, Masschusets Institute of Technology, USA

[9] Walker R, Voce J, and Ahmed J 2012 *Survey of Technology Enhanced Learning for higher education in the UK*, Universities and Colleges Information Systems Association (UCISA) United Kingdom

[10] Economist Intellegence Unit, 2003 The 2003 e-learning readiness rankings: A white paper from the Economist Intelligence Unit *Economist Intellegence Unit & IBM Corporation*

[11] Munir 2010 Penggunaan Learning Management System (LMS) di Perguruan Tinggi: Studi Kasus Universitas Pendidikan Indonesia *Cakrawala Pendidikan* No 1 pp 109-119

[12] Usaheni T 2013 Evaluasi tingkat kematangan e-learning berdasarkan E-learning maturity model (EMM) Di Universitas Riau *Thesis Chief Information Officer, Universitas Negeri Padang*

[13] Ingraham B, Watson B, McDowell L, Brockett A, and Fitzpatrick S, 2002 Evaluating and implementing learning environments: A United Kingdom experience *Educational Technology Review* [Online serial] ***10***(2) 28-51

[14] Marshal 2006 *E-Learning Maturity Model 2.2: Process Assessment Workbook* New Zealand Ministry of Education Tertiary E–learning Research Fund

Acknowledgements

This research is funded by DP2M DIKTI, The Ministry of National Education, Republic of Indonesia, under the funding scheme of *Penelitian Desentralisasi Unggulan Perguruan Tinggi Dana APBN Tahun Anggaran 2013 No. 013674.279/IT2.7/PN.08.01/2013.*

Readiness Assessment Towards Smart Manufacturing System for Tuna Processing Industry in Indonesia

D Anggrahini[1,3], N Kurniati[1,3], P D Karningsih[1,3], S M Parenreng[2], N Syahroni[3,4]
[1]Department of Industrial Engineering, Institut Teknologi Sepuluh Nopember - Kampus ITS Sukolilo-Surabaya 60111, Indonesia.
[2]Department of Industrial Engineering, Hasanuddin University, Makassar, Indonesia.
[3]Marine Research Centre, Institut Teknologi Sepuluh Nopember, Kampus ITS Sukolilo-Surabaya 60111, Indonesia.
[4]Ocean Engineering Department, Institut Teknologi Sepuluh Nopember, Kampus ITS Sukolilo-Surabaya, 60111, Indonesia

*dewanti@ie.its.ac.id

Abstract. Marine product processing is one of the top priority clusters in the national development. Tuna, as a kind of deep ocean fishes, has the highest number of production that significantly increased throughout the years. Indonesia government encourages tuna processing industry, which are mostly dominated by small to medium enterprises, to grow continuously. Nowadays, manufacturers are facing substantial challenges in adopting modern system and technology that will lead a significant improvement through the internet of things (IoT). A smart factory transform integrated manufacturing process, in a high speed processing to respond customer needs. It has some positive impacts, such as increasing productivity, reducing set up time, shortening marketing and other support activities, hence the process is being more flexible and efficient. To implement smart manufacturing system, factories should know the readiness at any level of them, technology capability and strategy appropriateness. This exploratory study aims to identify the criterias, and develop an assessment tools to measure the level towards smart factory.

1. Introduction

Indonesia, a maritime nations with 3.257.483 km^2 of ocean areas, has a great potentials of fish and other sea resources [1].As stated in Government Road Map, and Indonesian Chamber of Commerce and Industry report, marine product processing is one of the top priority clusters in the national development. Since the number of export commodities are increasing to 300% per year [2], they also provide bigger opportunity of working and poverty reduction. Based on the data in Statistics Indonesia [3], the volume of capture fisheries production reached 4.812.235 tonnes in 2009, and increased to 5.779.990 tonnes in 2014. Tuna, including Skipjack Tuna and Eastern Little Tuna are two commodities which can be categorised as high economic value and high demand product. It was exported to many countries, such as Japan, Singapore, China, Malaysia, Taiwan, Hongkong and Australia. To expand tuna's export market, Indonesia needs some structured improvement throughout its supply chain, from upstream to downstream sectors. An important way that should be implemented is how to apply technological innovation.

Modern manufacturing industries are facing big challenges that caused by current business environment, economic, social, and technology development. Manufacturers have to improve their capability in managing the whole agile and responsive value chain [4]. They need the integration of both virtual and physical structure which can easily adopt and communicate since innovation design phase, production, and distribution [5]. Since it can be more efficient, the application of cyber-physical system (CPS), known as smart manufacturing system or smart factory, is considered to support towards industry 4.0 (I4.0). The activities data in shop floor will be processed faster, to assist company in predicting and failure anticipating. Moreover, they can use the data for determining the number of products, doing customisation, material handling, and decision making. Thus, it will develop production flexibility. Furthermore, company competitiveness will be steadily improved.

Many researches on I4.0 were conducted in recent years. There is a contemporary study of smart technology implementation in 2014 [6]. An industry in Germany built a virtual representation of I4.0 and digital factory repository [7].A technical integration framework was also deployed from top to shop floor of I4.0 for small and medium enterprises [8]. Zezulka [9] introduced RAMI (reference architecture model industry) 4.0 as a modern manufacturing system, which the components represent important things in I4.0 applications. RAMI 4.0, pyramid 4.0 and component model of I4.0 were developed by Gesellschaft [10], Seliger [11]reviewed about I4.0 regarding to current issues on research and practical application, including macro and micro perspective framework, trend analysis, and development expectation, and some improvement opportunities in the field of sustainable manufacturing.

Refers to the research of [12] and [13] , it was known that transforming to I4.0 needs an integration of digital and intelligence in a value chain, and it requires multi-level synergy both vertical and horizontal hierarchy. This is linked to the study by [4] , which stated that most of manufacturers that have been evoluting to be smart factories got serious problems. The less readiness of some company stakeholders, limitation on technology capability, and inappropriate strategy are the major factors. So as [4] conducted a reseach on developing a model to measure maturity level of manufacturing industry which has been implementing smart factory concept. There are nine dimensions used in it, such as products, customers, operations, technology, strategy, leadership, governance, culture, and people. But there is still limited study focusing on assessing readiness level towards smart factory for a developing countries. On the other hand, the needs of performance improvement in marine product processing industry being highly demanded in this country. Thus, this research objectives is how to develop a readiness assessment towards smart manufacturing system for tuna processing industry in Indonesia. The findings of this exploratory study will extend literature on smart manufacturing system and I4.0 area, especially for tuna processing industry, and will provide insights for practitioners to face the challenges of modern industry in its relevant fields.

2. Methodology

Firstly, literature review and field analysis were two important steps in this study, to identify main problem, determine the research objectives, also domain complexity and gap analysis. However, the lack of information and insights about smart manufacturing system in Indonesia become a problem. Critical analysis on previous research has been the most fundamentals. The next stage was develop the model, by doing benchmarking to several maturity level measurements. Thirdly, determining and validating the assessment criterias, then developing research tools was the following stage. The questionnaire as a measurement tools, were constructed through discussion with the experts. Before it had been spreaded to some respondents, who came from several companies related to sea food processing industry, the tools been validated through expert judgements. Then, regarding to the questionnaires, the data had been analysed, to state what criterias are the most important to measure it. As the last stage, the conclusion would be explained for expanding theoritical framework of industry 4.0, smart manufacturing system, smart factory, and readiness assessment, also for practical implementation in tuna processing industry. The detail steps was pointed out in figure 1.

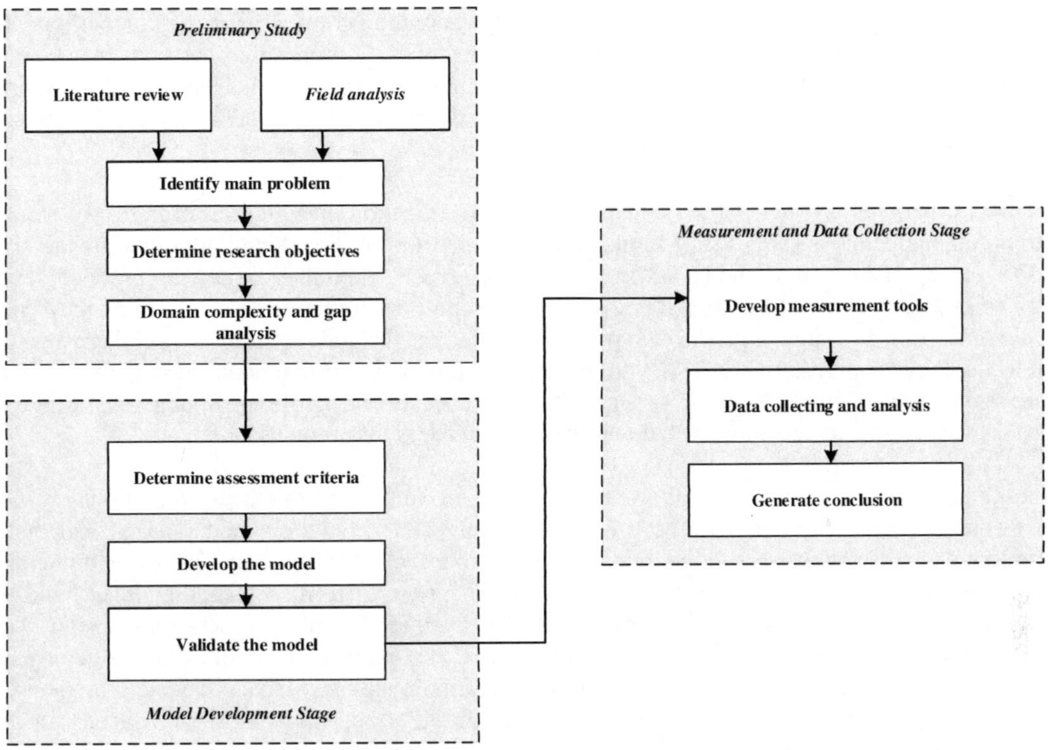

Figure 1. Research Flow Chart.

3. Result and Analysis

This study is an exploratory study, which reviewed, analysed and adjusted previous findings to develop new model on assessing the industry readiness towards smart manufacturing system. Refers to Schumacher [4] research, which developed a maturity assessment model using nine dimensions, this study find the same dimensions as the criteria. Since Indonesian companies consists of various scales, and dominated by small to medium enterprises (SMEs), there are some further adjustment to the following items in the measurement tools. Strategy dimension stands for assessing how the company develop its long-term objectives, the resources availability, business model adjustment in term of preparing a huge challenge. Most of the SMEs in Indonesia, did not have any structured vision and mission. They run the business by using the owner or the board of director (BOD) goals. Indeed, some of them already had the physical evidence of it. Thus, in this study defined the strategy dimension is how tuna processing industry has its specific plan, proper scheduling on resources and machines, to achieve its ideas.

The second dimension is leadership. It measures company's leader readiness, as the frontline executor, they should be a visionary with good spirit to transform. Unfortunately, some of them had never learned about I4.0. In some local tuna industries, the leaders follow the most common production process regardless the limitation of their technology. However, they also found some difficulties in managing the system. Various employee's characteristics, lack of knowledge and experiences also affected a lot. Through some discussion, this study deployed some questions to measure the eagerness of the leader, to adopt new concepts in modern manufacturing.

On transforming towards smart factory, customers have strong bond with the products. Customers have an important role by getting involved in accessing the detail of products and services. Several years ago, domestic enterprises applied push market system. They manufactured based on the existing capability, and let the customers bought it. However, in the recent years, the business environment gets very dynamic. By respecting to many equal players at the same field, they should provide high value of products. Market might choose what product, when and where they got it, among the competitors. This current competitive, encourage most of them to realise that sharing information between customer and manufacturer is important. Customers will decide and communicate what they need, and manufacturers try to fulfil it beyond the limitation. Since most of the companies are SMEs, they rarely got some problems on providing an open-access system.

The next criteria is product, which consists of individualisation and digitalisation of products. In industry 4.0 manufacturers should start thinking about a smart product, which integrated to their smart operations. Most of local tuna industries provide homogeneous products, otherwise produce several variants with no differences to others. They still focused on how to get order as much as they could. Then, the industries will run the production process and deliver the goods. Furthermore, there are some lack of technology capabilities. The SMEs often failed on applying improvement funding, as they stated the limitation of their revenue streams. One big difference between the criteria in this study and Schumacher [4] is how to accommodate the limitation on product customization.

Other dimensions are operations, culture, people, governance, and technology. Operations is used to assess the level of decentralization, the use of modelling and simulation, and internal stakeholder integration along the business activities. For tuna processing industry, that classified as manufacturing industry, operations will be the most important part to be transformed. On the other hand, the involvement of customer, during the process would be demanded in future operations design. Thus, several questions to measure interaction linkage on internal and external stakeholders were deployed as well. Another aspect is the work culture of the companies. Employees may have difference competences, but the most important is how they share information, insights and knowledge. I4.0 needs an open integrated database that can be accessed easily. The proper internet connection, modern ICT competences, utilization of mobile devices and machine-to-machine communication are required. Transforming to smart manufacturing system, companies should develop labour regulations, and the suitability of technological standards. Since there are consists of millions intellectual property, companies must protect it to avoid cyber security.

Figure 2. Research Variables.

Regarding to the group discussions and literature, this study found that the readiness of development team becomes moderate important. Company should have the establishment of a working team to develop and maintain the smart manufacturing system. The team consist of cross-functional people who have specific responsibility to implement it. They will be working consistently as the agent of change to accelerate enterprise transformation. Thus, this variable was included, to be deployed to measurement tools. Ten variables were defined, then being drawn in Figure 2.

Several Indonesian experts were involved in this study, and determined the weighting factor of each variable. For Tuna processing company as well known as the object, classified as manufacturing industry, has operations as the highest weight among all variables. A proper operation is required towards good manufacturing system. Technology becomes another important factors as well. Since all the databases in I4.0 is connected by the internet, so the need of advanced cyber-physical configurations. People, as the following dimensions, got weighted score as high as strategy, leadership, customer involvement, and products. Then development team, culture and governance are three dimensions with smaller weighted score. The detail of it as shown in Table 1.

Table 1. Variables Weighting Factor.

	Variable	Weight
A	**Strategy**	**0.1**
B	**Leadership**	**0.1**
C	**Customers**	**0.1**
D	**Products**	**0.1**
E	**Operations**	**0.2**
F	**Culture**	**0.05**
G	**People**	**0.1**
H	**Governance**	**0.05**
I	**Technology**	**0.15**
J	**Development Team**	**0.05**

The ten variables then well deployed into 65 (sixty-five) questions and been validated to 43 (forty-three) questions in questionnaire as the tools to assess the readiness of the companies. The questionnaire was validated by two experts. They are academics and researcher. At the next step, it had been distributed to 15 respondents, who represented some tuna processing industries, most in Java Island and three of them came from South Sulawesi.

Based on the analysis, this study found that there are 10 (ten) variables that can be used to assess the readiness of Tuna processing industry in Indonesia towards smart manufacturing system, such as strategy, leadership, customers, products, operations, culture, people, governance, technology and development team. This study also found that the existing companies, which mostly categorised as small to medium industry, do not ready to face the modern manufacturing system. However, some companies in Java Island have further maturity in strategic planning, people and work culture. They should improve more in technology and customer also product customization. On the other hand, company in Makassar has less preparation to transform to smart manufacturing system. Thus, it needs high efforts to develop the predetermined-system as well.

References

[1] Yuniasri R, Karningsih P D and Anggrahini D 2016 Quality Risk Mitigation Strategy in Baby Anchovy Supply Chain, Case Study PT Insan Citraprima Sejahtera (ITS: Final Project)

[2] Akbar B 2014 *Seafood Products Becoming Indonesia Priority Commodity* (Available online at http://beritadaerah.co.id/2014/11/12/pemerintah-jadikan-produk-laut-sebagai-komoditas-ekspor-unggulan/ accessed at August 2015)

[3] BPS 2015 Fisheries Production Based on Subsectors (tonnes) 1993-2013 (Available online at http://bps.go.id/linkTabelStatis/view/id/1711/ accessed at August 2015)

[4] Schumacher A, Erol S and Sihn W 2016 A Maturity Model for Assessing Industry 4.0 Readiness and Maturity of Manufacturing Enterprises *Proc. CIRP* **52** pp 161-166

[5] Gligor D M and Holcomb M C 2012 Understanding the Role of Logistics Capabilities in Achieving Supply Chain Agility : A Systematic Literature Review *Supply Chain Management* **17** (4) pp 438-53

[6] Radziwon A, Bilberg A, Bogers M and Madsen E S 2014 The Smart Factory: Exploring Adaptive and Flexible Manufacturing Solutions *Proc. Engineering* **69** pp. 1184-90

IOP Conf. Series: Materials Science and Engineering **337** (2018) 012060 doi:10.1088/1757-899X/337/1/012060

[7] Ergebnisbericht der Platform Industrie 4.0 2015 Platform Industrie 4.0 : Umsetzunsstrategie Industrie 4.0 (Berlin)

[8] Clemens F and Feldmuller D 2015 Industry 4.0 Learning Factory for Regional SMEs *Proc. CIRP* **32** pp 88-91

[9] Zezulka F, Marcon P, Vesely I and Sajdl O 2016 .Industry 4.0 – An introduction in the Phenomenon *IFAC Papers Online* **49** (25) pp 008-012

[10] VDI/VDE Gesellschaft Mess und Automatis-ierungstechnik 2016 Statusreport : Industrie 4.0- Technical Assets Basic Terminology Concepts, Life Cycles and Administration Models

[11] Stock T and Seliger G 2016 Opportunities of Sustainable Manufacturing in Industry 4.0 *Proc CIRP* **40** pp 536-41

[12] Zhou J 2013 Digitalization and Intelligentization of Manufacturing Industry *Advanced Manufacturing Journal* **1** (1) pp. 1-7

[13] Lanza G, Nyhuis P, Ansari S M, Kuprat T and Liebrecht C 2015 Befahigungs und Einfuhrungsstrategien fur Industrie 4.0 (Hannover)

Acknowledgement

This research is centralised in Marine Research Centre / Pusat Studi Kelautan of Sepuluh Nopember Institute of Technology (ITS), and funded by the marine research centre (sub major in LPPM – ITS research area) for a year. Furthermore, the model development and data collecting stages is supported by some experts come from academics and industrial practitioners.

IOP Conf. Series: Materials Science and Engineering **337** (2018) 012061 doi:10.1088/1757-899X/337/1/012061

The integration between Business Model Canvas and Manufacturing System Design

Y Prasetyawan[1*], N Maulida[1] and M R Lutvitasari[1]
[1] Department of Industrial Engineering, Institut Teknologi Sepuluh Nopember, Kampus ITS Sukolilo-Surabaya 60111, Indonesia

*yudhaprase@gmail.com

Abstract. Business Model Canvas (BMC) is an increasingly popular business design tool especially for a start-up business and new business player. In general, BMC seeks a balance between effective working patterns with suppliers, good relation with customers and ability to understand and manage internal resources. This balance will expedite the implementation of Manufacturing System Design (MSD). The existing use of BMC and MSD is frequently applied separately at various business levels. BMC business plan is primarily to have engagement with customers and explore potential revenue to increase profits, while MSD primarily aims to meet production targets with available resources. The purpose of this research is to provide a roadmap to align BMC and MSD. A series of simple mathematical (modified) and integration models are created to connect BMC and MSD. Several results in various industries (new, developed and mature) are presented and used as examples of implementation.

1. Introduction

The competition of manufactured products is no longer on the track of achieving the best quality, the greatest quantity, the shortest production time and the similarity of products in one variant - but more to achieve the best level of flexibility. Flexibility is triggered by the dynamics of customer wants and needs with the choices from the manufacturers. Innovation is the key to reach flexibility, so that the formerly production equipment based on mass production should be transformed into dynamic production. Some concepts such as flexible manufacturing system, mass customization, and also additive manufacturing are offered to accommodate innovation based on the dynamic production system. Dynamic production systems, among others, provide the choice of the amount produced, the sequence of production processes and the variation of the product. This option is not an easy task because the risk of cost and the level production equipment requirement have a significant contribution [1].

BMC is widely used for the start-up business, as a tool that can show business entities to manage and information and for employees and business partners to understand [2][3]. BMC consists of 9 components that are connected. First, BMC provides an overview to identify costs and revenues for analysing the potential profit. Second, BMC provides an overview of opportunity identification and customer relationship management - including what values that need to be considered. Third, BMC provides an overview of supplier needs that can be combined with internal resources management and development. BMC can be used as the reference for new business development or changed business orientation for companies that are already operating [4]. In the requirement of innovation in responding to customer demand flexibility, the BMC needs to be rewritten accordingly with the changes that able to answer the challenges [5]. Changes are not limited to offered values for customers or known as offer driven, but also on other parts of BMC as a business consequence. It is also driven by changes in infrastructure and partnerships (resource factor), changes in customer conditions (customer factor), changes in profit potential (finance factor), or a combination of multiple factors (multiple epicentres factor). BMC as a strategy guide is still limited to be used as a static guide for an early description of a business [6].

In the manufacturing industry, there is no product can be sold without production activities. The dynamics that can be generated by various factors in the BMC need to be answered with a good supplier's reliability and good internal production management (in some cases an internal and external combination - outsourcing) [7]. In an existing production system, various performance parameters can be measured easily in real time with the use of sensors. Problems arise when there are new products with variations in size, process or other specifications [8][9][10]. MSD is used to prepare a production system capability in responding to such changes [11]. Analysis and approximate estimation can be made with complex or simulated models [12][13]. For companies, the time to model and simulate is not equivalent with the need for rapid response to market. Furthermore, MSD needs to be used for internal synchronization of management with the support of reliable supply chain management capabilities [14][15][16][17][18]. A simple, straightforward MSD approach to the main parameters is an important thing for companies to do. Thus, the expected output of MSD is the basic layout design and the big picture mapping description.

Thus the integration between MSD and BMC is absolutely necessary especially in the dynamics of the competition [19][20][21][22][23]. Companies need a simple approach to accelerate response, without necessarily involving simulations with difficult models. The simplest approach should remain within the main BMC framework to provide an overview of the full business model and MSD framework to perform an effective and efficient production process. BMC remains aimed at customer satisfaction that leads to increased revenue, while MSD remains aimed at aligning due date from customers, manufacturing lead time and availability of raw materials from suppliers.

In this work, several sequential mathematical formulations are delivered with a simple BMC and MSD integration approach. The mathematical and integration formulations are further presented in different implementations for the industry in new, developing and mature levels - or generally understood as product life cycles. Hence, the proposed idea can be used at various business levels.

2. Method

In showing the potential profit of a business, BMC offers the concept of balance in the cost structure and revenue model. MSD, in this case, can be used to estimate the basic cost structure (cost of goods manufactured). In the start-up level, the cost is calculated in detail, while at the mature level the calculation is produced in percentage. Profit is the main goal of a business where the amount of revenue and the small cost will show efficiency. On the other hand, production management capabilities can be seen in supplier relationships, business processes, and resources - compared to potential customer management. MSD is an activity that can be used to assess the ability of BMC components in activities and resources.

The first step is to determine the scope of MSD. MSD is limited in nine components. Explanation of components accompanied by modification of the mathematical model becomes simpler. Modifications are made with a basic approach to the manufacturing principle that the number of inputs of production equals the number of production outputs. The main indicator is the degree of flexibility (dof), can also be used as an indicator of innovation (degree of innovation). The notation used in this simple mathematical model is as follows.

n	: the number of variant	Th_i	: handling time product i*
Q_i	: the demand of product i	PC	: production capacity
TQ	: total number of demand	AT	: available time*
Tc_i	: cycle time product i*	UT	: unavailable time*
Td_i	: delivery time product i*	TPT	: total production time*

Rp	: production rate	W	: weighted factor
U	: utilization	WIP	: number of work in progress
A	: availability	N_{days}	: number of operating days within a year
NF	: number of facilities	N_{act}	: number of activities

*All variable that contain time, should be determined in hour unit

2.1. Customer Demand or Product Variety (MSD1)

The variation of demand will be similar as product variety, where higher the number will result in higher the flexibility to meet the current trend in customer demand. Further analysis and extended capability is required where n_{demand} is greater than $n_{capability}$.

The number of demand can be described as follow:

$$TQ = \sum_{i=1}^{n} Q_i \tag{1}$$

The required processing time is

$$TPT = \sum_{i=1}^{n} Q_i . Tc_i \tag{2}$$

Then the degree of flexibility MSD1 is

$$dof_{MSD1} = \frac{n_{capability} - n_{demand}}{n_{demand}} \tag{3}$$

2.2. Production Capacity (MSD2)

Production capacity is normally calculated by dividing the total available time by the time needed to produce the whole product [24]. These parameters can be calculated in units of days, weeks, months or years. Production capacity may vary starting with the number of products ordered in a number of demands for a particular product, with no demand for other products. Or, by giving the same amount in each product variation, so it can use the average cycle time. The ratio between planned production capacity and installed capacity reflects the level of system flexibility.

$$PC = \frac{AT}{\frac{\sum_{i=1}^{n} Tc_i}{n}} \tag{4}$$

Then the degree of flexibility MSD2 is

$$dof_{MSD2} = \frac{PC_{planned}}{PC_{installed}} \tag{5}$$

2.3. Production Rate (MSD3)

This parameter shows the amount that can be produced per unit of time [24], so it is the opposite of Tc_i. The greater the production rate indicates an increasingly flexible manufacturing system, as it can produce larger quantities with a variety of products.

$$R_P = \frac{1}{\frac{\sum_{i=1}^{n} Tc_i}{n}} \tag{6}$$

Then the degree of flexibility MSD3 is

$$dof_{MSD3} = \frac{R_p . TPT}{PC} \tag{7}$$

2.4. Utilization (MSD4)

In general, this parameter is calculated by the use of overall manufacturing system facilities to produce output [24]. The utility level can also be used as a measure of flexibility which means it can maximize the use of facilities to produce various products.

$$U = \frac{TPT}{AT} \tag{8}$$

Then the degree of flexibility MSD4 is

$$dof_{MSD4} = U \tag{9}$$

2.5. Availability (MSD5)

The availability level is affected by the layout of the manufacturing system facility. Availability is a description of the reliability of a facility that depends on the structure of the series or parallel. Unavailability can be caused by machine malfunction, setup, or anything else [24]. When simplified, the production process layout can be divided into several stages in accordance with the stages of the process. The proportion of series and parallel will determine the degree of flexibility, where a large number of parallel facilities will increase flexibility.

$$A = (AT_{serial} - UT_{serial}) + (AT_{parallel} - UT_{parallel}) \quad (10)$$

Then the degree of flexibility MSD5 is

$$dof_{MSD5} = \frac{NF_{parallel}}{NF_{series}+NF_{parallel}} \quad (11)$$

2.6. Manufacturing Lead Time (MSD6)

This parameter is used to measure the time required for a product on the production floor [24]. The same number can be used as input for big picture mapping. The setup time is the time taken to perform a series of activities after the production order is approved until the production process begins. While the delivery time is a series of activities after the production process is completed until the product is ready to be sent to the customer. The measure of flexibility used is the proportion of cycle time to setup time and delivery time.

$$MLT_i = Tsu_i + Tc_i + Td_i \quad (12)$$

Then the degree of flexibility MSD6 is

$$dof_{MSD6} = \frac{\sum_{i=1}^{n} \frac{Tc_i}{MLT_i}}{n} \quad (13)$$

2.7. Work in Progress (MSD7)

Minimizing the number of work in progress will increase flexibility because in different time periods it does not require additional setup time.

$$WIP = (AT_{daily} - TPT_{daily}) \text{ x } N_{days} \quad (14)$$

Then the degree of flexibility MSD7 is

$$dof_{MSD7} = \frac{\sum_{i=1}^{n} QFG_i.TC_i}{AT_{daily}} \quad (15)$$

Where QFG_i is number of finished good i within an operating day.

2.8. Basic Layout Design (MSD8)

The basic layout will determine the flow of the production process, the product variation gives a longer travel time. The flexibility of the manufacturing system is measured by the total travel time of each product as short as possible. An example of using a basic layout can be seen in Figure 1. Then the degree of flexibility MSD8 is

$$dof_{MSD8} = \frac{\sum_{i=1}^{n} \frac{Tc_i - Th_i}{Tc_i}}{n} \quad (16)$$

2.9. Big Picture Mapping (MSD9)

The approach of manufacturing patterns in a big picture mapping can be simplified in the form as shown in Figure 2. The flexibility of these parameters will be measured using lean manufacturing principles, the proportion of value added activities to the total number of activities [25]. Then the degree of flexibility MSD9 is

$$dof_{MSD9} = \frac{Nact_{VA}}{Nact_{VA}+Nact_{NVA}+Nact_{NNVA}} \quad (17)$$

Where,

VA : value added
NVA : non value added
NNVA : necessary but non value added.

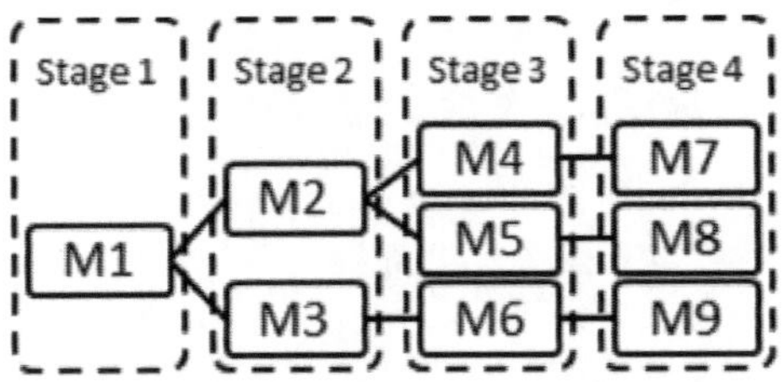

Figure 1. A typical basic layout design

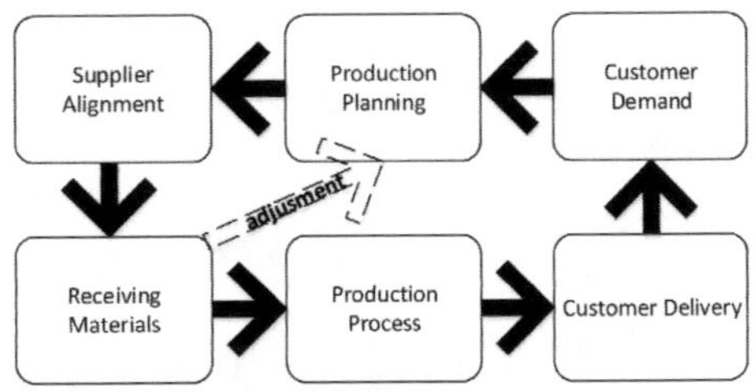

Figure 2. A typical big picture mapping

The second step is about determining of how to integrate MSD elements into the following BMC element.

1. Key Partners (BMC1)
2. Activities (BMC2)
3. Resources (BMC3)
4. Value Proposition (BMC4)
5. Customer Relationship (BMC5)
6. Channels (BMC6)
7. Customer Segments (BMC7)
8. Cost Structure (BMC8)
9. Revenue Model (BMC9)

The integration is to connect all elements of MSD into each BMC element, through a weighted factor. The weight of the contribution of the MSD element in BMC can be defined by the user of the model, through the use of the W_{ij} variable where i is the variable number of the MSD and j is the BMC variable number. MSD1 (customer demand) will be able to have a significant influence on BMC1 (key partner), with the more variation of the desired product the supplier's flexibility will also increase. A good deal with suppliers will determine the availability of raw materials. Contribution weights can vary at each level of the company and also between companies at the same level.

$$dof_{BMC_j} = \sum_{i=1}^{n} W_{ij} . dof_{MSD_i} \quad (18)$$

The third step is to provide a simple calculation example of the integration model. The calculation begins by assigning a weight to the W_{ij} parameter and performing the calculation for each dof value of the MSD element. The flexibility value of the BMC element (dof_{BMC}) will be obtained by multiplying the weights by the flexibility value of the MSD element according to the formula.

3. Result and Discussion

Numerical experiments were conducted on several companies with start-up, developing and mature level. Where the difference calculation will lie in the quantity and variation of demand. A new company has a small amount of product variation. In recent developments, product innovation is required and product variation is a challenge for new companies. Companies in the developed level show a greater number of product variations. While companies with mature levels have the most variety of products, with better production and innovation levels.

The results indicate that BMF elements degree of flexibility will also depend on the company's experience. Experience determines which MSD elements affect a particular BMC element. Determination of the right weights will result in the accuracy of the strategy result. The weighted value also shows the company's efforts to strengthen certain elements that are expected to deliver good results. Further, the value of MSD elements such as the production rate, availability and etc. can also be used as a preliminary description of the company competitiveness.

The model can be used to measure the company's innovation capability through the description of BMC elements. When companies want to increase business flexibility then the elements of MSD that support BMC elements should be expedited. BMC is no longer merely as a business strategy guide containing descriptions, but it can be estimated the value of business capability achievement in the current period.

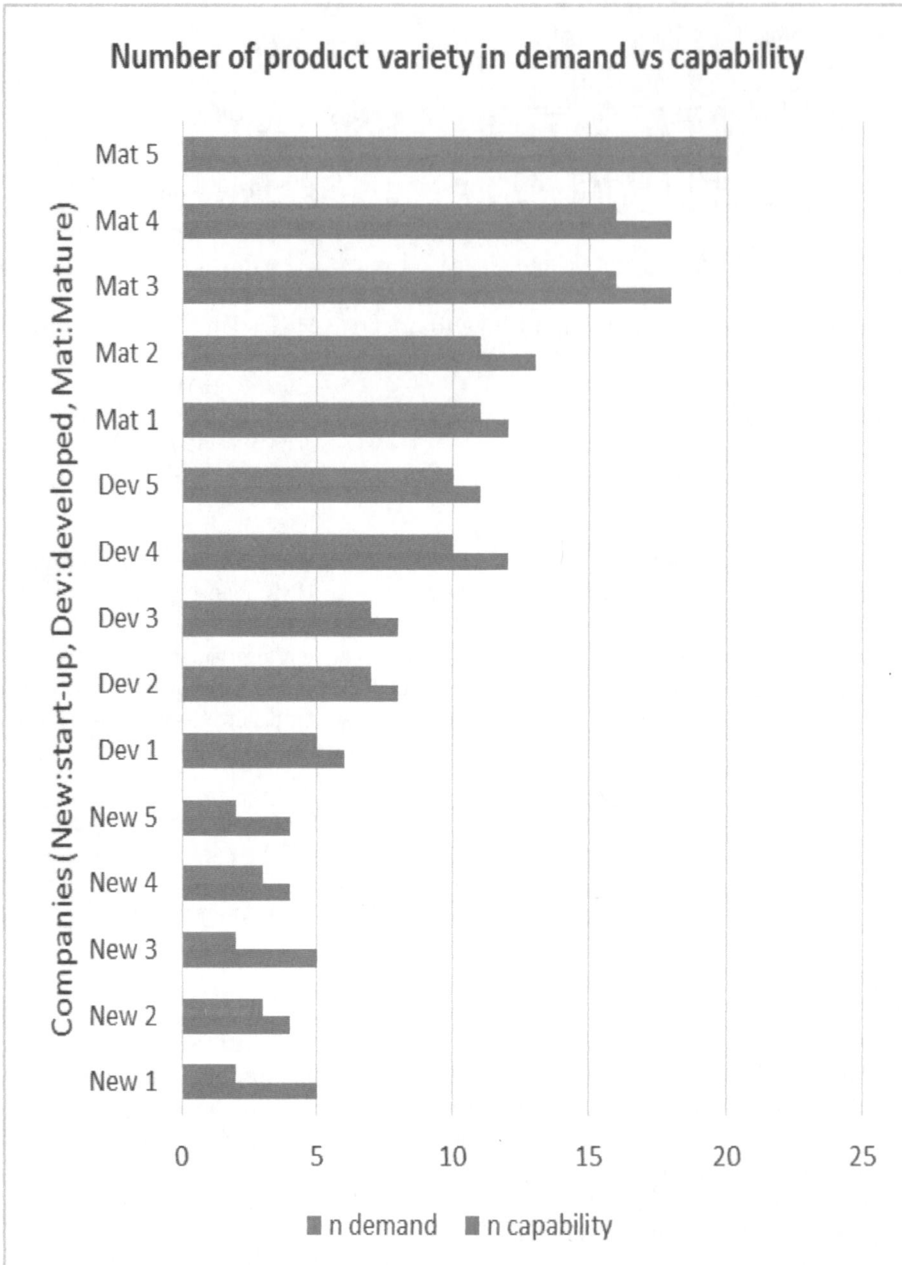

Figure 3. Degree of flexibility MSD1 input – demand and capability comparison

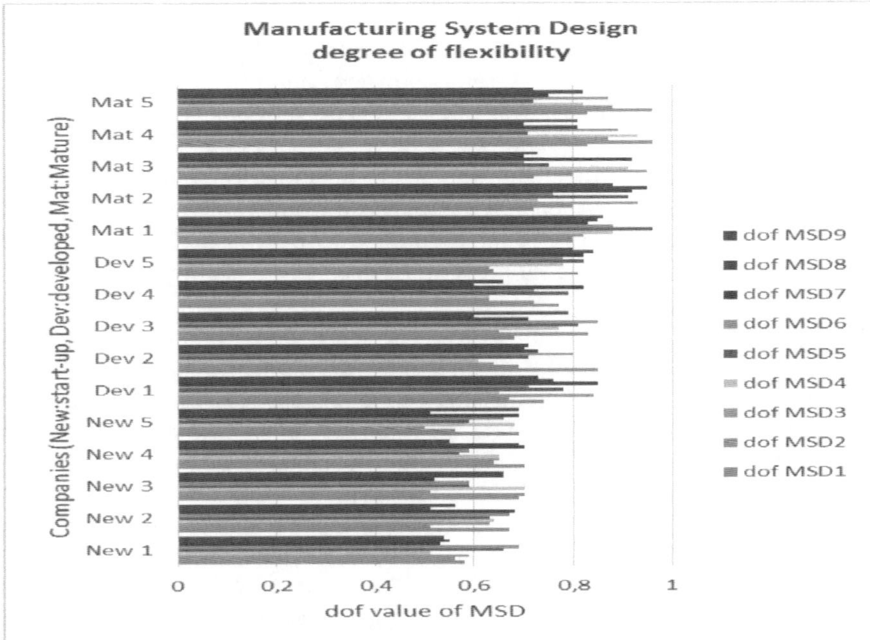

Figure 4. Degree of flexibility MSD elements

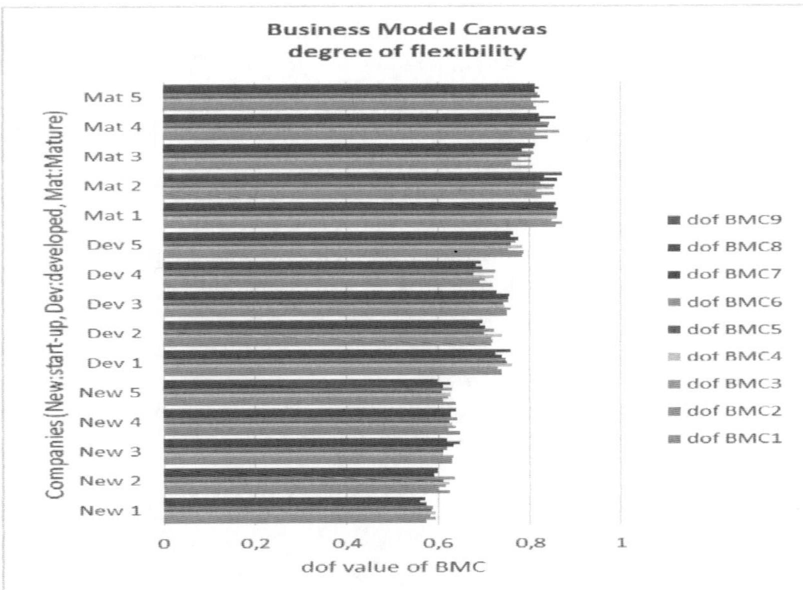

Figure 5. Degree of flexibility BMC elements

4. Conclusion

BMC can be used as a means to describe the corporate strategy that prioritizes flexibility, while MSD is a technical design to meet demand. A simple mathematical formula for integrating MSD into BMC has been described with the main degree of flexibility indicator to answer to the current growing customer demand. Numerical experiment in several corporates level has been transposed into varying degree of flexibility values. This simple integration model can then be used to determine the company's ability to meet demand.

5. References

[1] Osterwalder A and Pigneur Y 2010 *A Handbook for visionaries, Game Changers, and Challengers* In: Business Model Generation p 288
[2] Stefaan S and Richard B 2014 Analysis of Business Model *Journal of Competitiveness* 6(4) pp. 19-40
[3] Lindgren P 2012 Business Model Innovation Leadership: How Do SME's Strategically Lead Business Model Innovation? *Int. Journal of Business and Management* 7(14) pp 53-66
[4] Rodiguez M J, Martiez N M, Martiez J and Pirez A W 2014 Business Models in the Smart Grid: Challenges, Opportunities and Proposals for Prosumer Profitability *Energies* Vol **7** pp 6142-6171
[5] Fielt E 2013 Conceptualising Business Models: Definitions, Frameworks and Classifications *Journal of Business Models* Vol 1(1) pp 85-105
[6] Ching H Y and Fauvel C 2013 Criticisms, Variation and Experiences with Business Model Canvas *Int. Journal of Small Business and Enterpreneurship Research* Vol 1(4) pp 18-29
[7] Matt D 2008 Template Based Production System Design *Journal of Manufacturing Technology Management* Vol 19(7) pp 783-797
[8] Hager T, Wafik H and Faouzi M 2017 Manufacturing System Design Based on Axiomatic Design: Case of Assembly Line *Journal of Industrial Engineering and Management* Vol 10(1) pp 111-139
[9] Noble J S and Tanchoco J 1993 Design Justification of Manufacturing System – A Review *The Int. Journal of Flexible Manufacturing System* Vol 5 pp 5-25
[10] Tariq A 2012 A Comprehensive Production Oriented Technique for Cellular Manufacturing System Design *Journal Engineering and Applied Science* Vol 31(1) pp 61-69
[11] Prasetyawan Y, Agustiani E and Jumayla S 2017 Aligning Business Strategy of Incubator Center and Tenants *AIP Conf. Proc.* AIP Publishing p 30021
[12] Tahar R B M and Adham A A J 2010 Design and Analysis of Automobile Manufacturing System Based On Simulation Model *Modern Applied Science* Vol 4(7) pp 130-134
[13] Negahban A and Smith J S 2014 Simulation for Manufacturing System Design and Operation: Literature Review and Analysis *Journal of Manufacturing System* Vol 33(2) pp 241-261
[14] Cochran D S, Hendricks S, Barnes J and Bi Z 2016 Extension Of Manufacturing System Design Decomposition To Implement Manufacturing System That Are Sustainable *Journal of Manufacturing Science and Engineering* Vol 138(10) pp 1-10
[15] Rosyidi C N, Fatmawati A and Jauhari W A 2016 An Integrated Optimization Model for Product Design and Production Allocation in a Make to Order Manufacturing System *Int. Journal of Technology* Issue 5 pp 819-830
[16] Houshyar A N 2014 Literature Review on Dynamic Cellular Manufacturing System *Material Science and Engineering* Vol 58 pp 1-6
[17] Jovanovic J R, Milanovic D and Djukic R 2014 Manufacturing Cycle Time Analysis and Scheduling To Optimize Its Duration *Journal of Mechanical Engineering* Vol 60(7-8) pp 512-524
[18] Liu J P, Luo Z B, Chu L K and Chen Y L 2004 Manufacturing System Design with Optimal Diagnosability *Int. Journal of Production* Vol 42(9) pp 1695-1714
[19] Dudin M N 2015 The Innovative Business Model Canvas in the System of Effective Budgeting *Asian Social Science* Vol 11(7) pp 290-296
[20] Bonazzi F L Z and Zilber M A 2014 Innovation and Business Model: A Case Study Integration of Innovation Funnel and Business Model Canvas *Organizational Strategy and Behaviour*, Vol 16(53) pp 616-637
[21] Freiling J 2015 Business Model Innovation - A Concept between Organizational Renewal and Industry Transformation *Journal of Enterpreneurship, Management and Innovation* Vol 11(1) pp 3-10
[22] Eppler M J, Hoffmann F and Brescian S 2011 New Business Models through Collaborative Idea Generation *Int. Journal of Innovation Management* Vol 15(6) pp 1323-1341

[23] Joyce A and Paquin R L 2016 The Triple Layered Business Model Canvas: A Tool to Design More Sustainable Business Models. *Journal of Cleaner Production* Vol 135 pp 1474-1486
[24] Groover M P 2008. *Automation, Production Systems, and Computer Integrated Manufacturing* New Jersey: Pearson Education Inc
[25] Lasa I S, Vila R D C and Uriarte A G 2009 Pacemaker, Bottleneck And Order Decoupling Point In Lean Production Systems *Int. Journal of Industrial Engineering* Vol 16(4) pp 293-304

Model of relationship between personal factors and Occupational Health and Safety (OHS) management toward unsafe actions: a case study

M D P Syamtinningrum[1*], S G Partiwi[1], D S Dewi[1***]**
[1]Department of Industrial Engineering, Institut Teknologi Sepuluh Nopember, Kampus ITS Sukolilo-Surabaya 60111, Indonesia

*maeka.dps@yahoo.com,**srigunani@ie.its.ac.id,***dyah@ie.its.ac.id

Abstract. One indicator of a good company is when a safe business environment can be well maintained. In this work environment, the number of industrial accidents is minimum. Industrial accidents are the incidents that occurred in the workplace, especially in industrial area. Industrial accidents are generally caused by two main reasons, unsafe actions & unsafe conditions. Some research indicates that unsafe actions significantly affect the incidence in the workplace. Unsafe action is a failure to follow the proper procedures and requirements, which is led into accidents. From several previous studies it can be concluded that personal factors & OHS management are two most influential factors that affect unsafe actions. However, their relationship in influencing unsafe actions is not fully understood. Based on this reason the authors want to investigate the effect of personal factors and OHS management toward unsafe actions to workers. For this purpose, a company is selected as a case study. In this research, analyses were done by using univariate test, bivariate correlation and linear regression. The results of this study proves that two indicators of personal factors (i.e. knowledge of OHS & OHS training) and OHS management have significant effect on unsafe actions but in negative direction, while two indicators of personal factors (i.e. workload & fatigue) have positive direction of effect on unsafe actions. In addition, this research has developed a mathematical model that can be used to calculate and predict the value of unsafe actions performed by the worker. By using this model, the company will able to take preventive actions toward unsafe actions to reduce workers accidents.

1. Introduction

Data from Department of Labor and Transmigration of Indonesia show that until 2012 no fewer than six employees passed away (in average) per day due to workplace injuries. This number is considered high compared to European countries where only two employees passed away (in average) per day due to the same cause. Industrial injuries according to Petersen [1] is an incident of work injuries that occurred in the workplace, especially in industrial environments. industrial injuries are generally caused by two main factors namely unsafe actions and unsafe conditions. Unsafe action is human failure to follow the requirements and procedures of work properly, which causes the accidents. For example, not wearing Personal Protective Equipment (PPE) at workplace, intentionally violating safety regulations, working rashly, etc. While unsafe condition is unsafe and high risk conditions for the workers. There are two types of unsafe conditions in terms of when they occurred in task sequence ; unsafe condition which exists before commencement of a task and unsafe condition which progresses after commencement of a task [2].

Some research indicate that human factors play important role in the occurance of workplace injuries. Most of these studies state that 80% - 85% of workplace injuries are caused by unsafe actions due to worker negligence. Similarly, some research in Indonesia, for example, by [3], [4], [5], and [6] also found that the number of injuries due to unsafe actions become the major cause of workplace injuries. Based on these previous research, it is therefore interesting to know what factor actually drive the unsafe actions and how to reduce these unsafe actions in order to decrease the workplace injuries.

Previous research conducted by Ismail and Hashim [7] stated that Occupational Health and Safety (OHS) management is one of factors which dominantly affects unsafe actions. Whereas other research conducted by Javaid et al., Fam et al., Hidayat et al. [8], [9], [10] found that personal factors are factors that dominantly affects unsafe actions. These two factors may relate each others in affecting worker unsafe actions, however, the relationship between personal factors and OHS management towards unsafe actions has not been fully studied. The authors, therefore, are interested to know how the relationship between OHS management and personal factors can effect worker unsafe actions. Based on this, the main aim of this research is to identify the influence of personal factors and OHS management towards unsafe actions on the workers. The research is conducted at company X as a case study. Company X is a handicraft company in Yogjakarta which established in 1986. This study created a mathematical model that describes the relationship between personal factorsand OHS management and towards unsafe actions. The model can be used as a measurement tool to prevent worker unsafe actions. By applying the mathematical model, the company can determine how often the employees performing unsafe actions so that the company can prepare actions to prevent unsafe actions.

2. Methodology

After studies on literature, a preliminary study have been conducted in company X from August to September, 2016. The preliminary study was conducted through direct observation equiped with questionnaires. Based on the preliminary study, a model is then developed. The first step in developing the model is to determine the main variables of the model. The variables were identified from previous research. After variables are identified, the main variables which becomes independent variables is then selected. These main variables are variables that have significant influence on unsafe actions at company X. After this process indicators for each main variables are determined. The indicators is determined based on previous research combining with the results from company 's interview. Following this step, a quesionnaire is designes to include these indicators in order to get data that required in developing the mathematical model. The data obtained from questionnire were processed by using statistical tests with SPSS software. After the model is developed, the model is then verified and validated.The validation process is conducted by using paired sample t-test method. Paired sample t-test was used to determine whether there are differences in the average of two samples which paired or connected. When the results of the analysis indicate probability value or sig. (2 tailed) is greater than 0.05, this means no significant difference between the Y computation model and Y results of questionnaire, hence it can be concluded that the model can be used as a predictive model [11].

3. Model development

Workplace injuries can be caused by many factors. These include human factors, environment factors and interaction between human and supportive work equipment (figure 1). This study focuses only on unsafe actions which part of human factor.

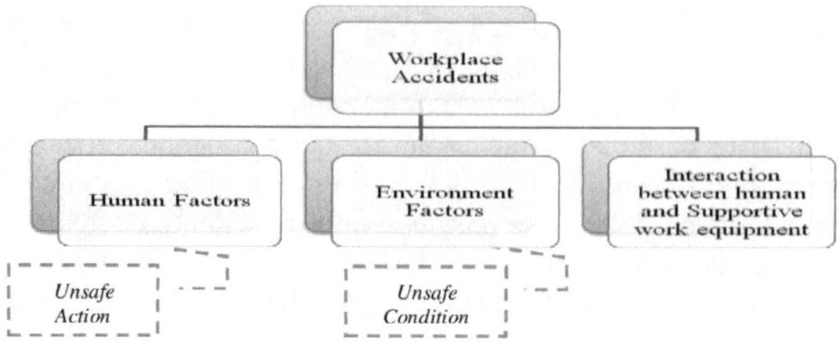

Figure 1. General causes of workplace injuries.

The research model of the unsafe actions which causes workplace injuries is created and presented in figure 2. The proposed reseach model is a closed system (not influenced by external aspects of other systems).

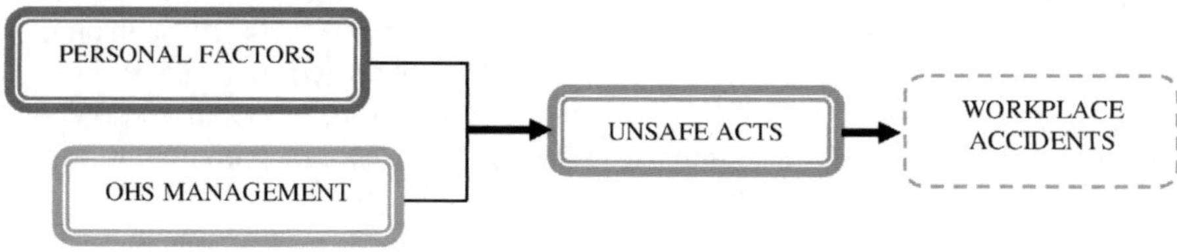

Figure 2. Research model.

3.1 *Parameter model*

The variables and indicators for the proposed model are identified by using direct interview and the result of previous studies. These variables and indicators can be seen in table 1.

Table 1. Variables that used in modeling

Variables	Indicators	Source
Personal Factor	Knowledge of OHS	[12][13]
	OHS training	[2]
	Workload	[14]NASA TLX *(Task Load Index)*
	Fatigue	[2] *Subjective Self Rating Test, Industrial Fatique Research Committe* (IFRC)
OHS Management	Commitment and policies of OHS	[13], primary data taken by using Focus Group Discussion with experts
	OHS Planning	[13]primary data taken by using Focus Group Discussion with experts
	OHS Implementation	[13]primary data taken by using Focus Group Discussion with experts
	OHS Checking and Corrective Action	[13]primary data taken by using Focus Group Discussion with experts
	OHS Review	[13]primary data taken by using Focus Group Discussion with experts
Unsafe Actions	Work SOP Violations	[14]Observation sheet of unsafe acts of the company
	Work Attitude	[14]Observation sheet of unsafe acts of the company

There were nine hypotheses formulated based on the result of previous studies. These hypotheses are described as follows:

- H1 : Knowledge of OHS has a negative influence on unsafe actions.
- H2 : Training OHS has a negative influence on unsafe actions.
- H3 : Workload has a positive influence on unsafe actions.
- H4 : Fatigue has a positive influence on unsafe actions.
- H5 : Commitment and policy of OHS management has a negative influence on unsafe actions.
- H6 : Management planning OHS has a negative influence on unsafe actions.
- H7 : Implementation of OHS management has a negative influence on unsafe actions.
- H8 : Inspection and corrective action on OHS management has a negative influence on unsafe actions.
- H9 : Review on OHS management has a negative influence on unsafe actions.

These hypotheses are tested by using multiple linear regression.

3.2 Conceptual Model

There are three models proposed in this research. The conceptual model can be formulated in the following formula:

Model 1 is the description of the relationship indicators of personal factors towards unsafe actions.

Y = f (personal factors towards unsafe actions)

$$Y = af + bf_1Xf_1 + bf_2Xf_2 + bf_3Xf_3 + bf_4Xf_4 \quad (1)$$

Information :

Y :	Unsafe actions	**bf_4 :**	regression coefficient value of fatigue
af :	intercept value of the equation personal factors towards unsafe actions	**Xf_1 :**	Knowledge of OHS
bf_1 :	regression coefficient value of knowledge of OHS	**Xf_2 :**	OHS Training
bf_2 :	regression coefficient value of OHS training	**Xf_3 :**	Workload
bf_3 :	regression coefficient value of workload	**Xf_4 :**	Fatigue

Model 2 is a description of the relationship of indicators OHS management towards unsafe actions.

Y = f (OHS management towards unsafe actions)

$$Y = am + bm_1Xm_1 + bm_2Xm_2 + bm_3Xm_3 + bm_4Xm_4 + bm_5Xm_5 \quad (2)$$

Information :

Y :	unsafe actions	**bm_5 :**	regression coefficient value of review of OHS
am :	intercept value of OHS management equation	**Xm_1 :**	commitment and policy of OHS
bm_1 :	regression coefficient value of commitment and policy of OHS	**Xm_2 :**	OHS planning
bm_2 :	regression coefficient value of OHS planning	**Xm_3 :**	implementation of OHS
bm_3 :	regression coefficient value of implementation of OHS	**Xm_4 :**	checking and corrective action of OHS
bm_4 :	regression coefficient value of OHS checking and corrective action	**Xm_5 :**	review of OHS

Model 3 is a description of the relationship indicator of personal factors and indicators of OHS management towards unsafe actions simultaneously.

Y = f (personal factor and OHS management towards unsafe actions)

$$Y = afm + bf_1Xf_1 + bf_2Xf_2 + bf_3Xf_3 + bf_4Xf_4 + bm_1Xm_1 + bm_2Xm_2 + bm_3Xm_3 + bm_4Xm_4 + bm_5Xm_5 \quad (3)$$

Information

Y : unsafe actions

Afm : intercept value of indicators of personal factors and OHS management towards unsafe actions

The rest is the same with information in model 1 & model 2.

In more detail, it can be explained that Y in the model is the value of unsafe actions, the prediction results from the value of X, while X is the value of personal factors and OHS management. Personal factors consist of knowledge of OHS, OHS training, workload and fatigue. While OHS management consists of commitment and policy of OHS, planning, implementation, checking and corrective action and review. Predicted value in the form of Y is a quantity that indicates the value of how often the employees performing the unsafe action at the company in accordance with the value scale of questionnaire phase 2. The higher the value of Y, the more frequent unsafe actions performed by a worker at the company. Meanwhile, the smaller the Y value, the lower unsafe actions performed by a worker at the company.

4. Data processing

Classic assumption test is a requirement in the application of multiple linear regression. In this research, the classical assumption tests that used are the normality test, multicollinearity test and heteroskedasticity test. The test results can be seen as follows

Normality test

The results of assumption normality test Kolmogorov-Smirnov multivariat from the data which collected from questionnaire obtained sig > 0.05 so it failed to reject H0. It can be concluded that the data distribution in this research is normal.

Multicollinearity test

Multicollinearity test is used to know whether or not the deviation of classical assumption of multicollinearity. The deviation of classical assumption of multicollinearity is the existence of linear relationship between independent variables in the regression model. Multicollinearity test was conducted by determining tolerance value and VIF. If the tolerance value > 0.1 and VIF <10 then there is no multicollinearity in the regression equation. The test results can be seen in the results of the calculation as follows: **a)** On Personal Factor variable, the value of tolerance 0.984> 0.1 and VIF 1.017 <10. **b)** In variable OHS Management, tolerance value of 0.962> 0.1 and 1.023 VIF <10. All variables of the research are qualified, so it can be concluded that the regression model did not experience multicollinearity and the research is asserted as ideal.

Heteroskedasticity test

Heteroskedasticity test is used to know whether or not there is deviation of classical assumption of heteroskedasticity.The deviation of classical assumption of heteroskedasticity is existence of variant inequality of residual for all observation in regression model. The test results of heteroskedasticity using SPSS were scatterplot graph shows that the absence of clear patterns or dots scattered randomly and spread both above and below the number 0 on the Y axis. This indicates that there is no heteroskedasticity in regression models, so a regression model can be used to predict magnitude of the value of unsafe actions based on variables of personal factors and OHS management.

5. Result and analysis

From the data processing by bivariate analysis test for two indicators of personal factors such as knowledge of OHS and training OHS also all indicators of OHS Management have a negative influence on unsafe actions. This is consistent with the hypothesis H1, H2, H5, H6, H7, H8, and H9. As for the two other indicators of personal factors such as workload and fatigue have a positive influence on the unsafe actions. The negative influence means that the indicator has the higher value, the smaller the chance occurrence of unsafe actions, while a positive effect means that the higher value of the indicator, the higher the chance occurrence of unsafe actions.

From the analysis of multiple linear regression obtained three models are as follows

Model 1

Using SPSS software, it can be formulated the following equation:

$$Y = 2.944 - 0.179Xf_1 - 0.025Xf_2 + 0.087Xf_3 + 0.109Xf_4 \tag{4}$$

It can be seen that the equation model 1 the highest coefficient value is equal to -0.179 in Knowledge of OHS. From the results of the validity of the paired samples t-test showed that the value sig (2-tailed) 0.999 greater than 0.05. It is proved that this model can be used as a predictive model of how often workers perform unsafe actions at the company. The company can use this model when the company wants to see the effect of personal factors only towards the unsafe acts on the workers. Yet this model is not complete because only consider the personal factors without involving the OHS management indicators.

Model 2

Using SPSS software, it can be formulated the following equation:

$$Y = 4.302 - 0.025Xm_1 - 0.119Xm_2 - 0.036Xm_3 - 0.19Xm_4 - 0.067Xm_5 \tag{5}$$

The highest coefficient value or the highest contribution rate is equal to -0.190 in checks and a corrective action of OHS. From the results of the paired samples test showed that the value sig (2-tailed) 0.994 greater than 0.05. It is proved that this model can be used as a predictive model of how often workers perform unsafe actions at the company. The company can use this model when the company wants to see the effect of OHS management only towards the unsafe acts on the workers. Yet this model is not complete because only consider the OHS management without involving the personal factors indicators.

Model 3

Using SPSS software, it can be formulated the following equation:

$$Y = 4.496 - 0.194Xf_1 - 0.036Xf_2 + 0.096Xf_3 + 0.214Xf_4 - 0.105Xm_1 - 0.122Xm_2 - 0.024Xm_3 - 0.266Xm_4 - 0.071Xm_5 \tag{6}$$

The highest coefficient value or the highest contribution rate is equal to -0.266 in the check and corrective action of OHS. From the results of the paired samples test showed that the value sig (2-tailed) 0.994 greater than 0.05. It is proved that this model can be used as a predictive model of how often workers perform unsafe actions at the company. The company can predict how often the worker in doing unsafe action with the overall influence of personal factors and OSH management indicators, so this model resulting in more detailed model considerations

These three models were used to test the influence of overall indicators of personal factors and OHS management towards the overall indicators of unsafe actions. From the results of paired samples t-test, all three proposed models can be used as a predictive models.

6. Conclusion and future research

The conclusions of this study are described as follows:

First, personal factors in the form of knowledge of OHS and OHS training have influence to unsafe action but in different direction (negative), while the workload and fatigue have influence in the same direction (positive) towards unsafe actions. Moreover, OHS management in form of commitment and policy of OHS, OHS planning, OHS implementation, OHS monitoring and checking, as well as, the corrective actions have influence towards unsafe actions but not in the same direction (negative).

Second is about when to use the proposed models. If the company wants to see the influence of only personal factors towards unsafe actions, the company can use model 1. This model, however, is incomplete, because it only considers the personal factors without involving the OHS management factors. Model 2 can be used when the company wants to see the effect of only OHS management towards unsafe actions. But similar to model 1, this model is incomplete because it only considers OHS management factors without involving personal factors. Model 3, on the other hand, is the most complete model where the company can used this model to predict how often the workers performs the unsafe actions as influence of both factors, personal factors and OHS management factors, so this model results in more detailed information.

Third, this research shows that all three models can be used as the first step in minimizing and preventing workplace injuries that relate to unsafe actions. Based on the results of the validation, model 1 is a model which is closest to the Y results of questionnaire.

This study only considers the influence of personal factors and OHS management towards unsafe actions. Thus, for future research, this research can be explored further by considering other factors such as equipment design, work environment, employment, and psychological work environment. Additionally, the model need to be tested in different company to gain better understanding on these relationship towards unsaved actions.

7. References

[1] Hosseinian S S and Torghabeh Z J 2012 Major Theories of Construction Accident Causation Models: A Literature Review *International Journal of Advances in Engineering and Technology* Vol. **4** Issue 2 pp 53-66.

[2] Sanders M S and McCormick E J 1993 *Human Factors in Engineering and Design* McGraw-Hill Inc New York.

[3] Maanaiya I 2005 *Faktor-faktor yang Berhubungan dengan Tindakan Tidak Aman* (Unsafe Act/ Substandart Practice) *pada Pekerja di bagian Press PT. YIMM* (Jakarta: Program Magister Jurusan Teknik Industri Universitas Indonesia.

[4] Bessie V A 2006 *Faktor-faktor yang Mempengaruhi Tindakan Tidak Aman pada Pekerja Teknisi di Press & Welding Division dan Assy & Painting Division PT. Toyota Motor Manufacturing Indonesia Karawang Plant* (Jakarta: Jurusan Teknik Industri Universitas Indonesia)

[5] Gandhatama C and Halim Y 2009 *Evaluasi Unsafe Act dan Unsafe Condition pada suatu Proyek* (Surabaya: Program Magister Jurusan Teknik Industri Universitas Kristen Petra)

[6] Haqi 2013 *Analisis Penyebab Unsafe Action dengan Pendekatan Human Factors Analysis and Classifications System (HFACS), Studi pada Pekerja Proyek Pembangunan Hotel PT. AXA* (Surabaya: Jurusan Teknik Industri Universitas Surabaya)

[7] Ismail F and Hashim E 2012 Steps for the Behavioural Based Safety: A Case Study Approach *International Journal of Engineering and Technology* Vol **4** chapter 5 pp 594-597.

[8] Javaid M U, Isha A S N, Ghazali Z and Langove N 2016 Psychosocial Stressors in Relation to Unsafe Acts Inter. *Review of Management and Marketing* Vol. **6** chapter 4 (special issue) pp

IOP Conf. Series: Materials Science and Engineering **337** (2018) 012062 doi:10.1088/1757-899X/337/1/012062

108-113.

[9] Fam I M, Kianfar A, and Mahmoudi S 2010 Evaluation of Relationship between Job Stress and Unsafe Acts with Occupational Accident Rates in a Vehicle Manufacturing in Iran *International Journal Of Occupational Hygiene* Vol. **20** chapter 2 pp 85-90

[10] Hidayat S, Suwandi T and Qomarudin M B 2015 The Analysis of Factors Related to Unsafe Acts on Welders in XYZ Ltd Inter. *Journal of Advanced Engineering, Management and Science* Vol. **2** chapter 6 pp 760-766.

[11] SPSS Indonesia 2016 *Cara Uji Paired Sample t-test* (http://www.spssindonesia.com).

[12] Aksorn T and Hadikusumo B H W 2007 The Unsafe Acts and the Decision-to-Err Factors of Thai Construction Workers *Journal of Construction in Developing Countries* Vol. **12** chapter 1 pp 1-25

[13] Zulliyanti S 2010 *Pengaruh Pekerja terhadap Penerapan Manajemen Keselamatan dan Kesehatan Kerja di PT. Gold Coin Indonesia* (Medan : Program Magister Jurusan Teknik Industri USU)

[14] Hart, S.G. and Staveland, L.E., 1988. Development of NASA-TLX (Task Load Index): Results of empirical and theoretical research. *Advances in psychology* Vol. **52**, pp.139-183.

[15] Pratiwi D A 2012 *Analisis Faktor-faktor yang Mempengaruhi Tindakan Tidak Aman* (Unsafe Act) *pada Pekerja di PT. X* (Jakarta: Program Magister Jurusan Teknik Industri Universitas Indonesia).

Inclusive business model in tapioca starch industry in Lake Toba area: a case study

S Tampubolon[1] and Y Manik[1*]
Engineering Management Study Program, Faculty of Industrial Technology, Institut Teknologi Del, Laguboti 22381, North Sumatera, Indonesia

*yosef.manik@del.ac.id

Abstract. The notion of inclusive business calls for additional focus and innovation in the way companies do business which seeks to contribute to poverty alleviation by including Bottom of the Pyramids (BoP) communities within its value chain while not losing sight of the ultimate goal of business. Lake Toba Area has potentials in providing chances for doing businesses. On the other hand, the growth of market size is rather slow and demographically still dominated by BoP. This is a case study which seeks to investigate to what extent the Inclusive Business Model (IBM) is adopted in the strategic planning and applied in the operational management of companies that operate in Lake Toba Area. The study was conducted in qualitative basis. The observation was conducted by gathering data and information through a series of interviews with the top management and desk study of the business plan in a tapioca starch industry in Toba Samosir Regency. The collected data and information were then analyzed qualitatively by comparing them with criteria and parameters of IBM suggested in a vast body of literature. The reference by which the IBM is referred in this study is a series of criteria which is synthesized from a literature review on a vast body of literature about IBM. From data analysis, it is evident that IBM has been incorporated in the strategic plan and applied in the operational activities of the object of this study. However, we also found some rooms for improvement such as expanding the involvement of BoP in their value chain as consumers, by which some innovation in the product diversification is required.

1. Introduction

Inclusive Business Model (IBM) is becoming a trend. The growing enthusiasm for the IBM is adopted by philanthropic organizations, by the newly emerging impact investment industry, and by donors and other organizations, is helping to enhance their visibility and also prompting the creation of new models [1]. That is why IBM becomes an interesting topic to research. Promoting inclusive business is a relatively new, but rapidly growing field of development cooperation and being responded seriously even by the United Nations (UN). The Growing Inclusive Markets Initiative responds to a need for better understanding of how the private sector can contribute to human development and to the Millennium Development Goals [2], where human development is everybody's concern.

Lake Toba region is a natural ecosystem that provides strategic functions of the basic human needs and economic development. Recently, the Government of Indonesia is making every effort to accelerate the development of Lake Toba region into a main tourist destination in the country. Even though tourism has traditionally been considered the main driver for social and economic development for the region, the development of Lake Toba region has is also marked by the rise of industry sectors, particularly the agro-processing industry. Tapioca industry is one of the agricultural processing industries sector located in the area of Lake Toba. The integration of IBM applied in strategic decisions and operational practices in tapioca industry in Lake Toba area is essential considering demographical condition, in which the poor who lives at the base of the global income pyramid (the so-called 'base of the pyramid' or BoP) in the region are still dominant.

Embarking from this reality, the authors consider it important to conduct an assessment on the application of the IBM in the tapioca starch production system in Lake Toba region. The objective of this study is to investigate to what extend the IBM is adopted in the strategic management of the company and to disclose various aspects that have the opportunity for system improvement. This study is a part of a larger agenda in the research roadmap of the Faculty of Industrial Technology at Institut Teknologi Del, which contributes in policy design and development of Lake Toba Area.

2. Methodology

This study begins with literature review about IBM through several studies undertaken by previous researchers and organizations. This literature review is performed to find the parameter or criteria of IBM by revisiting theories, backgrounds, and reports about the practices of IBM in the companies worldwide. After the literature review, a series of data collection and interview was then conducted in a tapioca starch industry located in Toba Samosir Regency (hereafter "the company"). This data collection is performed in order to get the data needed in this study. The data is gathered from the key personnel or the executives in the company in question. After gathering data about how the company operates and the theoretical basis of IBM, author then continue to compare the field data and theory. The comparison will be shown in the form of table. From the table, author states about the conformity of the practical finding with the theory.

Finally, to complete the objective of this study, author then analyze the conformity. It means that the comparison table will be analyzed deeper than merely about conformity. The comparison is discussed to find to what extend the IBM is adopted in the strategic management of the company.

3. Inclusive Business Model Revisited

In this section, an excerpt of literature review of a vast body of literature about Inclusive Business Model (IBM) is presented. IBM is a type of business model that includes the poor on the demand side as clients/ customers and on the supply side as employees, producers and business owners at various points in the value chain [2]. In IBM, the poor is the stakeholder of the business. The poor are those who lives at the base of the global income pyramid (the so-called 'base of the pyramid' or BoP), lacking the access to markets providing essential goods and services [1]. There are many reason and background to the recent unlucky condition of some areas and one of them is the failure of public authorities. Inclusive Business models integrate the poor in situations where markets or public authorities have failed as:

1. They introduce innovation to business models, processes, products and services, such that they become more affordable and/or better tailored to the needs of the poor.
2. They undertake targeted measures to integrate the BoP, e.g. providing training, extending credit to customers, suppliers and sales agents, etc.

To cope with such unlucky conditions, businesses have some roles in promoting IBM which reflects the characteristics of IBM. Author also has stated that doing business in the inclusive way is taking the attention of world organizations. For that, FAO also has its own criteria of IBM [3]. After reviewing relevant literature about IBM, the author decided to characterize the IBM based on Rösler et al. [1] and FAO [3] as a benchmark to conduct the assessment of the object of this case study (Table 1).

IOP Conf. Series: Materials Science and Engineering **337** (2018) 012063 doi:10.1088/1757-899X/337/1/012063

Table 1. Criteria of Inclusive Business Model

No.	Rösler et al. (1)	FAO (3)
1.	Create capacities and space for innovation & research	Provides a living wage for vulnerable group
2.	Deepen community engagement	Uses flexible trading arrangements
3.	Create awareness among clients and other stakeholders	Supports farmers and small enterprises to establish a stronger negotiation position
4.	Build capacities, e.g. training of suppliers	Builds on the skills and expertise of existing market players
5.	Engage in policy dialogue	Is scalable in the medium-term
6.	Organize end-to-end solutions	Allows for diversified income streams
7.	Improve own management skills (as Small Medium Enterprise)	
8.	Commit financing, time and patience to carry IB model to break-even	

In addition to the criteria, Kubzansky et al. [4] argues there are some steps for inclusive enterprise to deliver their social impact at scale as follow:

1. Innovation and grow
   To scale the business, the enterprise need time, investment, and innovation, moreover, when the BoP is the only supplier and customer without any subsidiary entity.
2. Replicate and disseminate
   Creating more enterprise must be seen from many perspectives. One may see it from the existing business' view as the new entrants copy their model. Other may see it from investor's view which wants to invest to new companies with almost similar successful business model.
3. Leverage and improve
   Bringing updates to other sector which are already at scale, this step is asking for improvement. The improvement needed in this step is about formalizing the informal and giving legalization or certification. After that, more investor may come because the enterprise has become more established and acknowledgeable.

Later, these steps will also be discussed for the conformity analysis. The previous interviews from London and Hart [5] reflects the presence of biases associated with top-of-the-pyramid market situation and customer condition may result in a wrong treatment in the base of the pyramid environment. Agribusiness and biotechnology companies can work with small farmer to improve rural productivity [6]. However, the cooperation between agro-processing and BoP may face these two problems:

1. The dispersion laces of small-scale producer's locations which can affect the transportation cost and lead time
2. How appropriate is the supply provided by small-scale producer?

Two big challenges are evident when seeking to apply inclusive models to developing country economies dominated by small scale producers, either for domestic retailing and processing, or for exporting. The first is organizing and upgrading supply from a dispersed producer base. The second is traceability and quality assurance. Through attention to the 'partner network' in the business model framework, the value proposition of modern agro-processing business and cost structure can be maintained or even strengthened by building in inclusion of small-scale producers and suppliers.

4. Data and Analysis

A series of structured interviews have been conducted in the period between February and May 2017 in a tapioca starch industry in Toba Samosir Regency. The data are then compared to the theoretical criteria of IBM. Table 1 compares the roles of business in promoting IBM according to Rösler et al.

[1] and the practices of the company. Table 2 compares the criteria for IBM from FAO and practice in the company.

Table 2. Comparison between Rösler's IBM criteria and practices in the company

No.	Parameter	Practices in the company	Conformity
1	Create capacities and space for innovation & research	The company is open for any educational institution which want to do research or to study about the company in its factory in Lake Toba Area. However, almost all of these researchers cannot be classified as BoP.	Conforming
2	Deepen community engagement	The advent of this company initiated and promoted the cultivation of cassava in the neighborhoods (Lake Toba area), as the community saw the opportunity to become suppliers. Most of the community engaged in cassava planting is considered BoP.	Conforming
3	Create awareness among clients and other stakeholders	To strengthen their supply independency and certainty, the company also has their own field. Basically, the field is not owned by the company, but The company has their own worker and machines to do work on the field. Until 2016, the company still does not own those fields. The field is owned by local people, but was given as rent or contract to the company. The reasons why those fields were given to the company are not only because nobody was working on the field, but also because those fields are not easy enough to reach or to use. The company has made local people (probably as BoP) aware about the opportunity of the factory presence.	Conforming
4	Build capacities, e.g. training of suppliers	The company gives some support in the shape of machinery for local farmer. For that, the company conducts small training about the use of machine.	Conforming
5	Engage in policy dialogue	The company work hard to engage in free market policy and local enterprise protection. Since 2016, the company has felt the competition effect of the free trade in between ASEAN countries. The company recognized that sooner or later, local farmer will then be affected too as their harvesting is not received by the factory.	Conforming
6	Organize end-to-end solutions	There were times when some of the suppliers came in the wrong time, when the factory cannot receive any more cassava even for the next 3 days for certain reasons. It was a dilemma for the supplier who has come from far away. It is almost impossible for them to go home and keep their cassava as bad as if they stay far away from home for days inside of truck. The company provided some solutions to this problem by publishing leaflets and spread posters containing information about their forecasted activity and their contact person information. This solution is effective and also a win-win solution for both the company and local farmer.	Conforming
7	Improve own management skills (as SME)	The company is not classified as SME.	Non-conforming

No.	Parameter	Practices in the company	Conformity
8	Commit financing, time and patience to carry IB model to break-even	According to the company's financial statement, even until 2016 the company has not earned significant profit. However, since the founders commits with their mission in creating opportunities for local farmers, the operation still goes on by depending on the supports from its holding company. However, the managements are optimistic that it still can achieve break-even.	Conforming

From Table 2, it is evident that in most of the criteria, the company conforms to the IBM criteria. The company is, by design, committed to engage local people into their. They believe that when the local farmer's income raises, they can educate their children well and have a better health for their life. There is possibility that those who were in BoP will move up from BoP status. Even if they are not, at least, their condition will be better. This is how the business plan of the company applies IBM in a way to give local people job and work for their unfarmed fields. The company does not give subsides or direct financing support but it gives help or funding support through receiving funding proposal. An influencing factor that enables the company to survive is the support from its holding company. Otherwise, the company would have been closed since some years ago.

Table 3. Comparison between FAO's IBM criteria and the practices in the company

No.	Criteria	Practices in the company	Conformity
1	Provides a living wage for vulnerable group	The worker of the company came from local society. As the local farmers are still assumed as BoP by the company, it makes the worker of the company who came from farmer family are BoP too. The culture which they belief makes them vulnerable to income instability, specially the party or celebration traditions which can be held for days. After they work for the company, those workers come into an arranged and strictly scheduled industrial environment. This condition provides them a better and stable income.	Conforming
2	Uses flexible trading arrangements	The company has deployed their preventive solutions to local farmers so that local farmer will not bring their cassava when the factory closes their supply door. As an industry, it shall strict to this way. But, when it comes to a condition that a local farmer come from far away to sell their cassava when the factory has no demand, the company will still manage to receive the cassava. Of course, the factory will check carefully to ensure that the farmer honestly come from far away. Even though this way is flexible, the company will be careful to that farmer and inform that farmer later to avoid the similar condition happen in the future.	Conforming
3	Supports farmers and small enterprises to establish a stronger negotiation position	As Indonesia enter the free market of ASEAN, government's protection and support on local product is a concern to many companies, including the company. The company has promoted local farmers to use their fields for cassava. This growing number of local cassava farmer may go down together with the factory if there is no place for the company's product in the market. Because The company itself is still not profitable, the company cannot give intense training, skill development, or massive machinery. Both the company and local farmer have	Conforming

No.	Criteria	Practices in the company	Conformity
		known this problem.	
4	Builds on the skills and expertise of existing market players	The company's factory in Lake Toba Area was established by the same people who have experience in other area like Lampung, and other part of Sumatra Island. So, what they wanted to do is to include and persuade people of Lake Toba Area and around to get into their business. The collaboration between local people and the company is clear and transparent. They also risk the same thing if the company does not gain profit and is lost or failed in the free market era.	Conforming
5	Is scalable in the medium-term	The company receives not only massive tonnage of cassava from big players. It also receives medium or small amount of cassava from little player with small field.	Conforming
6	Allows for diversified income streams	The company has ever shipped their product to Surabaya, Java. Thanks to good transportation via sea. The company is used to search for market or buyer far away from home because in Lake Toba Area itself, there is no large-scale industry with tapioca flour as the raw material. So far, the company is used to send their product in its province, North Sumatera, but in the other regency.	Conforming

From Table 3, it is evident that the company is full of people from BoP from the supplier, manufacturing, to customer. It gives many things to do for local people. Its presence motivates local people with unfarmed fields to let theirs being worked by the company. This company also gives chance for small farmer to sell their outputs. In its operation, the company uses skilled and experienced people from all over Sumatra Island as the supervisor or head of departments. Neither local people who are neither scholar nor certified skilled people will be the worker who works under the lead of those who are the heads of departments. Actually, the company spread the job vacancy even in the internet. It also receives candidate from everywhere. Of course, when it comes to two candidates with similar performance, the company will choose the local one.

Analyzing Table 2, it can be concluded that if the company did not exist at its time, the condition will be worse. Even if people in BoP is still in BoP financially, but the condition of education and income stability has been better. One may conclude that the company adopts IBM in spite of its exclusivity in searching for experienced and skilled managers for head of department positions.

As stated clearly in its business plan, the company built to improve the living of people around Lake Toba area. Those people in Lake Toba area are lack of possibility to get higher education, have low income, and around seven kilometers away from the main market to get everything they need. The poor who lives at the base of the global income pyramid (the so-called 'base of the pyramid' or BoP) is lacking access to markets providing essential goods and services.

People living in poverty are unable to sell their product at the right time, in the quality, or to right time. That's why the company, in its operation, gives a lot of information about the market, the price, and the availability to receive cassava demand at a time. The reason why the inclusivity of the company does not include BoP as its main customer is mainly because of the small market size in the Toba Samosir regency, the area where they are operating.

In order to protect and to promote the adoption of IBM by other company, the government side should show their support, especially in the initial years (start-up). Policy makers can support inclusive business development by establishing an innovation system that facilitates exchange and cooperation between academia, the private sector and responsible government actors. Better business and industry environment will exist if the output of the company is used by other kinds of manufacturing located not too far away, in the same regency.

5. Conclusion

This study provide evident on whether and how the company has applied IBM, not only in its business plan, but also further into its practices and operational. Despite most of the practice shows the conformity to the IBM's parameters, the company is still lacking in building new enterprises which also applies IBM as SME. As the company has strong patience for its break-even, and maintaining close relationship with BoP as supplier, we conclude that the company is applying IBM in its operational as well as in its strategic management. The main recommendation of this study is that community and government, in rural area where farming is the main activity, must support companies that adopt IBM. This study also recommends that the government should protect and support companies using IBM like the company.

References

[1] Rösler et al. 2014 *Inclusive Business Models* Deutsche Gesellschaft fur Internationale Zusammenarbeit (GIZ) GmbH.
[2] United Nations Development Program 2008 *Creating Value for all: Strategies for Doing Business with the Poor*
[3] FAO 2015 *Inclusive business models – Guidelines for improving linkages between producer groups and buyers of agricultural produce* by Kelly, S., Vergara, N. & Bammann, H. Rome, Italy
[4] Kubzansky M, Cooper A and Barbary V 2011 *Promise and Progress market-based solutions to poverty in Africa.*
[5] London T and Hart S 2004 Reinventing strategies for emerging markets: beyond the transnational model *Journal of International Business Studies* **35**(5) pp 350-370
[6] Nelson J and Prescott D 2008 *Business and the Millennium Development Goals: A Framework for Action* (2nd edition) International Business Leaders Forum

Determining factors that contribute to financial literacy for small and medium enterprises

M S Hakim[1*], V Oktavianti[1], I K Gunarta[2]
[1]Department of Business Management, Institut Teknologi Sepuluh Nopember, Kampus ITS Sukolilo-Surabaya 60111, Indonesia
[2]Department of Industrial Engineering, Institut Teknologi Sepuluh Nopember, Kampus ITS Sukolilo-Surabaya 60111, Indonesia

*muh.saiful.hakim@gmail.com

Abstract. Financial literacy on the SME owner/ manager will subsequently improve the profitability of SME. Although, numerous research had existed in financial literacy subject, there are still limited research that examined the factor related to financial literacy. In this paper, we are examining demographic and SME factor that contribute to financial literacy using linear regression. Other objective of this paper is to confirm the benefit of financial literacy on access to credit. The finding shown that for demographic factor, level of education is contributed to financial literacy. The second finding is that level of financial literacy is related with greater degree of access to credit

1. Introduction

High variety of financial product had emerged in the era of modern financial. These rapid changes have given opportunities for investor or financial product consumer to select wider range of product that suitable with their investment preferences. In the other hand, this evolution poses a problem for a new investor or unexperienced investor to select the suitable one. To resolve with this matter investor needs specific knowledge and capabilities that called financial literacy [1].

Financial literacy defined as investor or person capability to understand and make use of financial concepts [2]. The component of financial literacy is ranging from the money management, investment and leverage [1]. The rapid development of financial industry has put financial literacy to the spotlight among scholar. Because the existences of financial literacy will have an effect to behavior of investor in finish their financial task.

Financial literacy on investor will highly contribute to investor behavior on debt management [3], Investor wealth [4], and their investment behavior [5]. In the small business area, financial literacy of the owner or manager have an effect on SME performance [6], business growth [7] and a better access to credit [8]. Despite of the financial literacy benefit, recent literature have revealed that majority of people still have a low level on financial literacy [9] [10].

The puzzles left in the table is to identify the factors that contribute to financial literacy. Few literatures have emanated to answer the question. There are researcher that contribute with the indicator to measuring financial literacy [11]. Some of researcher have put an effort on understanding the determinant of financial literacy. Other scholars had studied the influence of psychosocial factors on financial literacy [12]. While other stream of research had put their focus on investigating the effect of demographic properties such as gender and age on financial literacy [13] [14].

The literature on the factors contribute to financial literacy is immense and mainly refers to personal finance scope namely personal investment. In contrast, there are still limited financial literacy factors research on SME owner or manager. Notably, no study has been published that has investigated the financial literacy related factor with a specific factor on small and medium enterprises context such as the age of their business or amount of SME Assets. Thus, this study is attempted to close a gap in the literature

The primary objective of this paper is to investigate factors that related with financial literacy using demographic factors and specific SME factors. Namely age, education, gender, SME ages. This objective tried to test generic demographic factors on different context of SME mixed with a new factor that related to SME factors. The secondary objective is to understand the effect of financial literacy on financial behavior of the owner/ manager of SME in their SME financial capability. In this second objective, this paper attempt to comprehend the benefit of financial literacy on SME financial capability, namely their access to a loan. Financial literacy supposed to contribute to SME access of loan [8]. SME that supported with a good financial literacy will have a better financial report. It is obvious with a better financial report surely will give a SME with greater financing opportunities. Thus in this secondary objective our hypothesis is "SME with a better financial literacy will subsequently have a better access to loan".

2. Literature review

Financial literacy could be defined as investor or person capability to understand and make use of financial concepts [2]. Other alternative definition is that financial literacy comprise with an ability to considering relevant information in make a logical choice among diverse financial choices [15]. From the definition, it's clear that financial literacy comprises of two properties. First properties are knowledge related to financial decision (thought) and second properties related to behavior (decision). In the past literature, the terms financial literacy is often used interchangeably with financial knowledge and financial education. However, few scholars renege that financial literacy is differ with other term. Financial literacy consist of behavioral element, which not existed in other term [11]. The factors of financial literacy are already existed in the past literature. Summary of research in financial literacy factors is listed in table 1 below.

Table 1. Financial Literacy Factors, List of Recent Literature.

Author	Year	Factors	Sample
United States Government Accountability Office	2010	English language, culture	Non-English-speaking population (US)
Chen H, Volpe R.P	1998	Gender, work experience, age	College Student (US)
Arif K	2015	Age, gender, marital status, work activity	Investor (Pakistan)
Bashir T,Arshad A, Nazir A, Afzal N	2013	Psychosocial factor	Bank Consumer (Pakistan)
Sucuachi	2013	Gender	SME (Philippines)

3. Method

Sample in this research is consist of SME owner or manager in Surabaya. The sampling method in this research is using purposive sampling. Requirement of sampling process is the research subject should already be an owner or manager of SME at least 6 months. This requirement created with the expectation that research subject understand the operational the SME. The financial literacy measurement is followed measurement by Nkundabanyanga & Kasozi [16]. Measurement scale employing 5-point Likert scale. Financial literacy variable used 2 dimension that align with SME specifically characteristic namely, financial knowledge and financial skills. The measurement on access to credit using 2 dimensions namely, loan history and source of financing, the complete questionnaire question item in appendix. In this paper, we will employ regression analysis using Ordinary Least Square (OLS).

4. Result and Discussion

In this paper, we utilize 80 sample of SME in Surabaya. 76 of the sample is collected from offline survey, where 4 of the sample collected via email. All sample answered questionnaire completely.

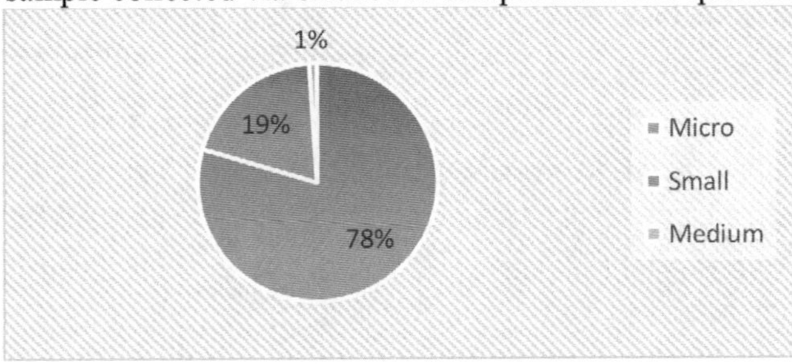

Figure 1. Sample on scale of enterprises.

Scale of sample is illustrated in figure 1. The majority of sample is in micro category enterprises. This category subsequently have a revenue turnover below 300 million rupiah. However, there are only 1 % in medium category. That mean this paper will represent mostly the micro and small enterprises.

Table 2. Regression Result (Financial Literacy).

Independent Variable	Beta	Sig
Gender	-0.116	0.471
Age	-0.088	0.445
SME Age	0.071	0.345
Education	0.288	*0.007

*Significant at 0.05

Table 2 demonstrate the result of Regression analysis. The regression model shown that for demographic variable of gender, age and SME age doesn't not have relate with financial literacy. In the other hand for education of SME owner / Manager shown a significant positive relationship. This finding indicated that higher level of education will contribute to higher level of financial literacy.

Demographic variable isn't related with financial literacy. This result is differ from what we expect based on the past literature [13] [17]. We argue that the differences are existed because of the sample in this paper consist of owner or manager with an experience. To meet on this requirement, the sample should have a minimum level of financial literacy. This argument is clarified by the mean financial literacy variable. Mean of financial literacy is on slightly higher than the midpoint of financial literacy (3.56 out of 3) this means that majority of sample have higher than average financial literacy level.

Level of education have indicated a positive relationship with financial literacy. Individual with a higher level of education will have an ability to learn about financial literacy, therefore will have a consequence on greater degree of financial literacy. This result is aligned with the research from Sucuahi, (2013). For SME context, the education could predict financial literacy better compared with other demographic variable.

Table 3 Descriptive statistic.

Independent Variable	Mode	Mean	Notes
Gender	0		Female
Age	3		40-50 Years
SME Age	2		1-3 Years
. Education	2		Middle Education
Financial Literacy		3.3625	Higher than mid-point
Credit Access		3. 805	Higher than mid-point

Table 4. Regression Result (Access to Credit).

Independent Variable	Beta	Sig
Financial Literacy	0.361	0.000

*Significant at 0,05

The secondary objective in this paper is to investigate the relationships between financial literacy to access of credit. Regression analysis on this matter is listed in table 4. Regression analysis have indicated that there is positive relationship between financial literacy and credit access. This mean that we accepted hypothesis 1. This implied that greater financial literacy for SME will provide benefit for the SME in the form of greater credit access. This result is similar with Nkundabanyanga & Kasozi, (2014) research in Uganda, Africa.

This paper indicated that the owner of SME should have some certain level of literacy because of their experience. This result could be a main point that differentiate between financial literacy in context of regular investor and household and financial literacy for SME context. Related with the issue, scholars usually differentiate between financial literacy on household context with the financial literacy on SME context. This also could pose a problem to compare between groups. In the future, there should be a conformity between SME and household regarding financial literacy that will facilitate the comparison process.

Financial literacy had shown their role in maintaining the success of firm. In this paper, we explore this roles by examining financial literacy role for greater access of credit. Higher financial literacy will contribute to greater degree of access to finance. We interpret the result; financial literacy will provide greater access of credit because this knowledge will provide a stable and good record on financial register. Other issue that should be underpinning the relationships is that people with greater financial literacy should have a clear understanding about debt management. They should have maintained excellent history of debt. This knowledge should be have contributed to relationships between financial literacy and access to credit.

5. Conclusion and Future Research

Positive relationships in demographic variable only met in between level of education on SME owner/manager with a level of financial literacy. Furthermore, other demographic level wasn't significantly related with financial literacy. This result differs from the existing literature because diverge research subject. Financial literacy proof to had a positive relationship with the access to credit. This result is confirming existing literature in the subject.

In this paper behavior of owner/ manager measured using perceptions scale from the sample. In the future, it's a major opportunity to perform analysis based on non-perception scale such as number of credit or variability source of funding. On related factors to financial literacy, there should be more research that assign other demographic or SME factors to financial literacy.

6. References

[1] van Raaij W F 2016 *Understanding Consumer Financial Behavior.* New York: Palgrave Macmillan US

[2] Servon L J and Kaestner R 2008 Consumer Financial Literacy and the Impact of Online Banking on the Financial *J. Consum. Aff. Summer* vol 42 no 2 pp 271–305

[3] Ibrahim M E and Alqaydi F R 2013 Financial literacy, personal financial attitude, and forms of personal debt among residents of the UAE *Int. J. Econ. Financ.* vol **5** no 7 pp 126–138

[4] Lusardi A and Mitchell O S 2006 Baby Boomer Retirement Security: The Roles of Planning, Financial Literacy, and Housing Wealth

[5] Nalini R, Alamelu R, Amudha R, and Motha L C S 2016 Financial Literacy and its Contributing Factors in Investment Decisions among Urban Populace *Indian J. Sci. Technol.*, vol. **9**, no. 27

[6] Eniola A A and Entebang H 2016 Financial literacy and SME firm performance *Int. J. Res. Stud. Manag.* vol 5 no 1 pp 31–43

[7] Eresia-Eke C E and Raath C 2013 SMME Owners' Financial Literacy and Business Growth *Mediterr. J. Soc. Sci.* vol **4** no 13 pp 397–406

[8] Wachira M and Kihiu E 2012 Impact of financial literacy on access to financial services in Kenya *Int. J. Bus. Soc.* vol **3** no 19 pp 42–50

[9] Lusardi A and Mitchell O S 2014 The Economic Importancce of Financial Literacy: Theory and Evidence *J. Econ. Lit.* vol **52** no 1 pp 5–44 2014

[10] Agyei K N 2014 Small Scale Businesses Practice of Basic Financial Management With Regards To Liquidity *Int. J. Res. Soc. Sci.* vol **4** no 5

[11] Huston S J 2010 Measuring Financial Literacy *J. Consum. Aff.* vol **44** no 2 pp 296–316

[12] Bashir T, Arshad A, Nazir A, and Afzal N 2013 Financial literacy and influence of psychosocial factors *Eur. Sci. J.* vol **9** no 28 pp 384–404

[13] Chen H and Volpe R P 2002 Gender Difference in Personal Financial Literacy Among College Students *Financial Services Review* vol **11** pp 289–307

[14] Das S 2016 Financial literacy : measurement and determinants *EPRA Int. J. Econ. Bus. Rev* vol **509** pp 88–93

[15] Mouna A and Jarboui A 2015 Financial literacy and portfolio diversification: an observation from the Tunisian stock market *Int. J. Bank Mark.* vol **33** no 6 pp 808–822

[16] Nkundabanyanga S K and Kasozi D 2014 Lending terms , financial literacy and formal credit accessibility *Int. J. Soc. Econ.* vol **41** no 5 pp 342–361

[17] Arif K 2015 Financial Literacy and other Factors Influencing Individuals' Investment Decision : Evidence from a Developing Economy (Pakistan) *J. Poverty, Invest. Dev.* vol **12** pp 74–85

[18] Sucuahi W T 2013 Determinants of Financial Literacy of Micro Entrepreneurs in Davao City *Int. J. Account. Res.* vol **1** no 1 pp 44–51

Acknowledgement

This paper is a result from "Pengabdian Masyarakat Berbasis Penelitian, Institut Teknologi Sepuluh Nopember - Kampus ITS Sukolilo - Surabaya 60111, Indonesia"

IOP Conf. Series: Materials Science and Engineering **337** (2018) 012064 doi:10.1088/1757-899X/337/1/012064

Appendix

Table A. Financial Literacy.

No	Dimension	Question
1	Financial Knowledge	I know about managing personal finance
2		Am aware of financial products and services provided by banks
3		My firm participates less in the formal financial system
4		I know the different sources of financing from which my firm can use
5		My firm makes effective financial choices
6		I have a working knowledge of financial institutions
7	Financial Skills	I have adequate skills in financial management
8		Am capable of discussing money and financial issues with comfort
9		Am very confident when making financing decisions
10		My firm has control over personal finance

Nkundabanyanga and D. Kasozi 2014

Table B. Credit Access

No	Dimension	Question
1	Loan History	My firm has ever acquired a loan from commercial banks
2		My firm received the full amounts of loan facility applied for
3		The loan amounts received satisfied my firm's credit needs
4	Financing Source	My firm finds it easy to access credit from commercial banks
5		My firm obtains loans more often from commercial bank

Nkundabanyanga and D. Kasozi 2014

IOP Conf. Series: Materials Science and Engineering **337** (2018) 012064 doi:10.1088/1757-899X/337/1/012064

Table C. Questions List

Financial Literacy

No	Dimension	Question
1	Financial Knowledge	I know about managing personal finance
2		Am aware of financial products and services provided by banks
3		My firm participates less in the formal financial system
4		I know the different sources of financing from which my firm can use
5		My firm makes effective financial choices
6		I have a working knowledge of financial institutions
7	Financial Skills	I have adequate skills in financial management
8		Am capable of discussing money and financial issues with comfort
9		Am very confident when making financing decisions
10		My firm has control over personal finance

Nkundabanyanga and D. Kasozi 2014

Credit Access

No	Dimension	Question
1	Loan History	My firm has ever acquired a loan from commercial banks
2		My firm received the full amounts of loan facility applied for
3		The loan amounts received satisfied my firm's credit needs
4	Financing Source	My firm finds it easy to access credit from commercial banks
5		My firm obtains loans more often from commercial bank

Nkundabanyanga and D. Kasozi 2014

Critical operations capabilities in a high cost environment: a multiple case study

C Sansone[1*], P Hilletofth[1], and D Eriksson[1]
[1]Department of Industrial Engineering and Management, School of Engineering, Jönköping University, Jönköping, Sweden

*cinzia.sansone@ju.se

Abstract. Operations capabilities have been a popular research area for many years and several frameworks have been proposed in the literature. The current frameworks do not take specific contexts into consideration, for instance a high cost environment. This research gap is of particular interest since a manufacturing relocation process has been ongoing the last decades, leading to a huge amount of manufacturing being moved from high to low cost environments. The purpose of this study is to identify critical operations capabilities in a high cost environment. The two research questions were: What are the critical operations capabilities *dimensions* in a high cost environment? What are the critical *operations capabilities* in a high cost environment? A multiple case study was conducted and three Swedish manufacturing firms were selected. The study was based on the investigation of an existing framework of operations capabilities. The main dimensions of operations capabilities included in the framework were: cost, quality, delivery, flexibility, service, innovation and environment. Each of the dimensions included two or more operations capabilities. The findings confirmed the validity of the framework and its usefulness in a high cost environment and a new operations capability was revealed (employee flexibility).

1. Introduction

Firms need to develop operations capabilities to maintain a long term competitive advantage [1] and achieve highest level of performance [2] Operations capabilities have been a popular research area for many years, and several frameworks have been proposed in the literature. However, the current frameworks do not take specific contexts into considerations, i.e. high cost environment.

The environment in which a firm compete affects the formulation of the operations strategy [3]and, subsequently the development of different operations capabilities. Considering the manufacturing relocation process that has been ongoing the last decades, from high to low cost environments [4] it is fundamental to investigate the topic of operations capabilities within the specific context of high cost environment.

The purpose of this study is to identify critical operations capabilities in a high cost environment. In order to satisfy this purpose, the following research questions were formulated: What are the critical operations capabilities *dimensions* in a high cost environment? What are the critical *operations capabilities* in a high cost environment? In order to answer these research questions and fulfil the purpose, a multiple case study was conducted and three Swedish manufacturing firms were selected.

2. Theoretical framework

This study was based on the investigation of an existing framework of operations capabilities [5]. The framework consists of seven dimensions of operations capabilities: cost, quality, delivery, flexibility, service, innovation, and environment (Table 1); these dimensions include two or more specific operations capabilities, for a total of 21 operations capabilities. The framework is based on a systematic literature review, including more than 150 papers.

Table 1. Framework of operations capabilities [5].

Dimension	Capabilities	Definition
Cost	Total cost	Ability to reduce production and distribution costs.
	Productivity	Ability to optimize the utilization of manufacturing resources (machines, equipment, and labor) and increase their output.
Quality	Performance	Ability to provide products and processes at the desired high level of performance.
	Conformance	Ability to offer products and manufacturing processes that correspond to the specifications, which help to ensure defect free products.
	Durability	Ability to offer products that withstand hard use over an extended period of time.
Delivery	Dependability	Ability to provide reliable delivery by meeting schedules or keeping promises.
	Speed	Ability to provide fast delivery and respond quickly to customer orders.
Flexibility	Volume flexibility	Ability to change production volume and respond rapidly to volume changes.
	Production mix flexibility	Ability to change the range of products in the production and respond rapidly to changes.
	Customization flexibility	Ability to adjust the product according to customer requirements and needs.
	Broad product line	Ability to offer a wide range of products, with a large number of features.
Service	Customer service	Ability to add value to the product by providing product information and making the product easily available and obtainable.
	After sale service	Ability to add value to the product after the purchase by providing effective after sale services, and delivering appropriate technical assistance and product support.
	Advertising	Ability to market and promote the product, and improve the company's image.
	Broad distribution	Ability to make the product available to a larger group of customers.
Innovation	New product	Ability to develop and introduce updated or novel products to the market.
	New technology	Ability to develop and implement updated and novel technologies.
	New service	Ability to develop and present updated and novel services to the customers.
	New market	Ability to create, expand and develop products and services, as to reach additional groups of customers.
Environment	Environmental friendly products	Ability to produce products with a reduced negative or even positive environmental impact.
	Environmental friendly processes	Ability to have processes with a reduced negative or even positive environmental impact.

3. Research methodology

In order to analyse the framework in a real setting and specific context [6], a multiple case study was conducted. Three Swedish manufacturing firms were selected through a theoretical sampling. The unit of analysis was limited to operations capabilities developed in the case companies.

The data collection was carried out through more than 30 semi structured interviews, approximately ten interviews per case company. The interviews were carried out by three investigators at the respective companies, and lasted for approximately 60 minutes each. Different managers from different departments were interviewed, this helped to gain a more holistic perspective. The semi structured interviews represent the primary data; however, secondary data was also collected, i.e. documents and observations.

The data analysis was carried out through a content analysis within each case study and through a cross-case analysis. Hence, the data collected in the case companies was compared with the existing framework of operations capabilities presented in Section 2. For each of the case company, a case study description was developed; this helped to gain a general overview of the dimensions and operations capabilities developed in the specific case company. Later on, a cross-case analysis was initiated and the link between the cross-case analysis and the selected framework of operations capabilities was further investigated.

The research quality was assured by assessing the criteria of validity and reliability. The first one was achieved by adopting data triangulation and investigator triangulation. The second one was achieved by developing a case study description for each one of the case companies, which included all relevant information for this study.

4. Findings

The findings revealed that all the dimensions included in the specific framework were considered critical by all the case companies involved in this study.

Table 2. Findings company Gamma.

Capabilities	Case Gamma	Source
Total cost	Cost is an important factor for the company. A lot of effort is put on product improvement processes (PIP), the goal is to reduce the cost of a manufacturing product.	I2, I3, I4, I5, I7, I8, I12
Productivity	In order to increase productivity and improve efficiency, the company is focusing on automation.	I7
Performance	In order to increase the performance of a product, the company adopts PIPs. In this way, it is possible to make small changes and improvements per time to increase the performance and quality.	I2, I3
Conformance	The company uses lean as a method to increase sustainable quality and offer to their customers conformed and high-quality products.	I4, I5
Durability	In order to offer durable quality products, the company is adopting lean methods.	I4
Dependability	The goal of the company is to achieve high delivery precision and dependability.	I7, I9, I11
Speed	The company put a lot of effort on offering a fast delivery. The company, indeed, is present in more that 20 countries. This helps the company to increase its dependability and speed.	I4, I9
Volume and production mix flexibility	Volume and production mix flexibility have been improved by shortening the lead time and moving the company's facilities closer to the customers.	I3, I4
Customization flexibility	The manufacturing processes are designed to simplify customization, the customer have a big impact on the product design outcome.	I8
Broad product line	In order to satisfy the customers' demand, the company is working with the customer to improve its product line.	I8
Employee flexibility	The company believes that by investing on employees training and knowledge, it is possible to achieve high level of flexibility.	I2, I3
Customer service	For the company, it is important to always be service-minded in order to be more competitive.	I3, I7
After sale service	Provide after sale service to the customer is a good way to earn customer trust.	I5, I7
Broad distribution	In order to be able to distribute the product broadly in a fast way, the company has moved its facilities closer to the customers.	I3, I7
New product and technology	The company is involving different employees in brainstorming activities, in order to identify different improvements areas. The company believes that new products or new technologies can be created by small changes and improvements to already existing products or technologies. This is part of the Lean philosophy adopted by the company. In general, more than 130 engineers work every day with identifying innovative products and technologies. This is a key factor in order to be competitive in Sweden.	I3, I5, I9, I10
Environmental friendly products and processes	The company is working with continuous improvements to offer environmental friendly products and to reduce energy consumption in the manufacturing processes.	I3

In case company *Gamma*, 18 of the operations capabilities, included in the framework, were considered critical. The operations capabilities not discussed in the interviews were: ‘advertising’; ‘new service’ and ‘new market’ (Table 2). In case company *Delta*, 18 of the operations capabilities, included in the framework, were considered critical. The operations capabilities not discussed in the interviews were: ‘durability’, ‘broad product line’, and ‘broad distribution’ (Table 3). In case company *Epsilon*, 19 of the operations capabilities, included in the framework, were considered critical. The operations capabilities not discussed in the interviews were: ‘after sale service’ and ‘broad distribution’ (Table 4).

From an overall perspective, all dimensions and capabilities included in the framework were discussed in the case companies. One new capability was identified in all the case companies involved in this study, which was ‘employee flexibility’. During the interviews, this capability was mentioned and discussed by the respondents. A commonality among the three case companies was to focus on increasing knowledge and skills of the employees. According to the respondents, in the long term, firms can increase the quality of their products and processes by focusing on the knowledge and skills of their employees. Employees are a fundamental resource for any firm, investing in their education is of vital importance to be competitive in a high cost environment. To achieve a high employee flexibility is part of an effective learning organization.

Table 3. Findings company Delta.

Capabilities	Case Delta	Source
Total cost	The company focus on reduction the total cost, the company believes that in order to reduce cost, it is important to focus on automation.	I2, I3, I4, I5, I7, I9
Productivity	The company is working on reducing wastes in the production and increasing efficiency by using less material bust still offer good products.	I7
Performance	The company is working in offering high performance products, the company has, indeed invested in updating existing machines and technologies to increase the quality level.	I5, I6, I8, I10, I11
Conformance	Have good quality product and conformed to the specifications, it is important in order to win the competition.	I3, I4
Dependability	Delivery precision is one of the most important factor in the company. In order to have high dependability, the company is working with forecasting. The delivery precision is now over 90%, delivery on time is important to be competitive.	I2, I4
Speed	In order to offer a fast delivery, the company is working on reducing the lead time.	I6, I9
Volume and production mix flexibility	The company believes to be quite good in responding to customers' demands, regarding both volume flexibility and production mix flexibility.	I2, I4, I6
Customization flexibility	The company works in producing customized products that satisfy customers' requirements.	I6
Employee flexibility	The company constantly invest on its employees' knowledge. They believe that having well trained and multi skilled employees is a key aspect for achieving high quality and flexibility.	I3, I5
Customer service	Offering good service to the customers is an important factor to gain customers' trust and win the competition.	I2, I4
After sale service	The company plans different meetings to always improve their after-sale service.	I2, I4
Advertising	The company has a really strong brand reputation; this helps them to gain more trust from the customers.	I3
New product and technology	Nowadays competition is strongly related to innovation; the company works in connection with universities, machine manufacturers, to be able to develop new products and new technologies.	I9, I10
New service and market	Service is a key factor to win the competition in high cost environment. Hence, the company is working on understanding the customers' needs and offer more suitable services for both current customers and new customers.	I2, I4
Environmental friendly products and processes	The company thinks that having environmental friendly products and processes are important factors to win the competition in a high cost environment. However, it becomes of high priority only when it supports cost reduction.	I2

Table 4. Findings company Epsilon.

Capabilities	Case Epsilon	Source
Total cost	The company works on reducing total cost by reducing, for example, the lead time and the setup time. Automation is very important.	I3, I4, I9
Productivity	The company works on improving productivity by increasing the efficiency in the manufacturing process. The company sets a goal of increasing productivity 10% more each year.	I2, I3, I4, I7
Performance	The company focus on offering high performance and quality products.	I2
Conformance	The company always works on offering products which correspond to the customers' specifications. This is a really challenging task, this because the customers' specifications change throughout the process.	I2, I3, I5, I6, I7, I8, I9, I10
Durability	The company focus on offering durable products, its products have a long-life time.	I3, I4, I5, I6, I7
Dependability and speed	The company needs to be able to deliver the product fast and according to the agreements in order to be competitive. Speed and dependability in the delivery are highly important, indeed the company works on reducing lead times, setup time and inventories in order to be able to provide short and dependable deliveries.	I2, I4, I5, I6, I7, I8
Volume and production mix flexibility	The company needs to improve its volume and production mix flexibility to cope with rapid changes in the market and survive the competition. This is done by reducing lead time and adopting continuous improvement techniques.	I2, I4, I9
Customization flexibility	It is important to always satisfy the customers. Therefore, the company is working on offering customized products.	I2
Broad product line	In order to satisfy different customers, the company offers a wide range of products	I4, I6
Employee flexibility	The case company focus on achieving high flexibility. It is important to train the employees to achieve different skills. Having more knowledge and skills helps the employees to respond and adapt faster to different types of situations.	I7
Customer service	The company offers good customer support and service. This is presented in different countries in order to satisfy different customers' areas.	I7
Advertising	The company puts a lot of effort in advertising and improving its brand reputation. Advertising is based on the company's high-quality products.	I2, I4, I5
New product, technology, and service	In order to be competitive, the company works in providing new products, technologies and services in the future.	I3, I6. I10
New market	The company is currently working in identifying new market opportunities.	I3
Environmental friendly products and processes	The company is actively working in reducing the energy consumption in the process, but also offers environmental friendly products to satisfy the customers' requests.	I10

IOP Conf. Series: Materials Science and Engineering **337** (2018) 012065 doi:10.1088/1757-899X/337/1/012065

5. Concluding remarks

In this study, the data analysed through a cross-case analysis was linked to the framework of operations capabilities presented in the literature. The findings are summarized in Table 5. From an overall level, all the dimensions and operations capabilities included in the existing framework are considered critical by the case companies involved in this study. The findings revealed that an additional operations capability was identified in all the case companies, employee flexibility.

Table 5. Analysis of the empirical findings in relation to the literature.

Dimensions	Capabilities	Definitions	Gamma	Delta	Epsilon	Literature
Cost	Total cost	Ability to reduce production and distribution costs.	✔	✔	✔	✔
	Productivity	Ability to optimize the utilization of manufacturing resources (machines, equipment, labours) and increase their output.	✔	✔	✔	✔
Quality	Performance	Ability to provide products and processes at a desired high level of performance.	✔	✔	✔	✔
	Conformance	Ability to offer products and manufacturing processes that correspond to the specifications, which help to ensure defects free products.	✔	✔	✔	✔
	Durability	Ability to offer durable products that withstand hard use over an extended period of time.	✔		✔	✔
Delivery	Dependability	Ability to provide reliable delivery by meeting schedules or keeping promises.	✔	✔	✔	✔
	Speed	Ability to provide fast delivery and respond quickly to customer orders.	✔	✔	✔	✔
Flexibility	Volume flexibility	Ability to change production volume and respond rapidly to volume change.	✔	✔	✔	✔
	Production mix flexibility	Ability to change the range of products in the production and respond rapidly to changes.	✔	✔	✔	✔
	Customization flexibility	Ability to adjust the product according to customer requirements and needs.	✔	✔	✔	✔
	Broad product line	Ability to offer a wide range of products, with a large number of features.	✔		✔	✔
	Employee flexibility	Ability to which employees possess skills, which enable the firm to pursue strategic decisions	✔	✔	✔	
Service	Customer service	Ability to add value to the product by providing product information and making the product easily available and obtainable.	✔	✔	✔	✔
	After sale service	Ability to add value to the product after the purchase by providing effective after sale services, and delivering appropriate technical assistance and product support.	✔	✔		✔
	Advertising	Ability to market and promote the product, and improve the company's image.		✔	✔	✔
	Broad distribution	Ability to make the product available to a larger group of customers.	✔			✔
Innovation	New product	Ability to develop and introduce updated or novel products to the market.	✔	✔	✔	✔
	New technology	Ability to develop and implement updated and novel technologies.	✔	✔	✔	✔
	New service	Ability to develop and present updated and novel services to the customers.		✔	✔	✔
	New market	Ability to create, expand and develop products and services, as to reach additional groups of customers.		✔	✔	✔
Environment	Environmental friendly products	Ability to produce products with a reduced negative or even positive environmental impact.	✔	✔	✔	✔
	Environmental friendly processes	Ability to have processes with a reduced negative or even positive environmental impact.	✔	✔	✔	✔

This study contributes in building a complete framework of operations capabilities within the specific context of a high cost environment; this framework will benefit both researchers and practitioners in a better understanding of operations capabilities' development and formulation of winning operations strategies. However, the research findings were only based on three case companies in the manufacturing sector in a high cost environment, future research can focus in investigating the topic of operations capabilities by taking into consideration different industries or contexts.

6. Reference

[1] Ward P T, Bickford D J and Leong G K 1996 Configurations of manufacturing strategy, business strategy, environment and structure *Journal of Management* **22** pp 597-626

[2] Boyer K K and Lewis M W 2002 Competitive priorities: investigating the need for trade-offs in operations strategy *Production & Operations Management* **11** pp 9-20

[3] Bradi M B, Donald D and Donna D 2000 Operations strategy, environmental uncertainty and performance: a path analytic model of industries in developing countries *Omega* **28** pp 155-173

[4] Wiesmann B, Snoei J R, Hilletofth P and Eriksson D 2017 Drivers and barriers to reshoring: a literature review on offshoring in reverse *European Business Review* **29 pp** 15-42

[5] Sansone C, Hilletofth P and Eriksson D 2017 Critical operations capabilities for competitive manufacturing: a systematic review *Industrial Management & Data Systems* **117 pp** 801-837

[6] Yin R K 2014 *Case study research: Design and methods* (SAGE Publications Thousand Oaks)

Determining optimal preventive maintenance interval for component of Well Barrier Element in an Oil & Gas Company

A Siswanto[1*], N Kurniati[2]
[1]Department of Management Technology, Institut Teknologi Sepuluh Nopember, Surabaya, Indonesia
[2]Department of Industrial Engineering, Institut Teknologi Sepuluh Nopember, Kampus ITS Sukolilo-Surabaya 60111, Indonesia

*asiselits@gmail.com

Abstract. An oil and gas company has 2,268 oil and gas wells. Well Barrier Element (WBE) is installed in a well to protect human, prevent asset damage and minimize harm to the environment. The primary WBE component is Surface Controlled Subsurface Safety Valve (SCSSV). The secondary WBE component is Christmas Tree Valves that consist of four valves i.e. Lower Master Valve (LMV), Upper Master Valve (UMV), Swab Valve (SV) and Wing Valve (WV). Current practice on WBE Preventive Maintenance (PM) program is conducted by considering the suggested schedule as stated on manual. Corrective Maintenance (CM) program is conducted when the component fails unexpectedly. Both PM and CM need cost and may cause production loss. This paper attempts to analyze the failure data and reliability based on historical data. Optimal PM interval is determined in order to minimize the total cost of maintenance per unit time. The optimal PM interval for SCSSV is 730 days, LMV is 985 days, UMV is 910 days, SV is 900 days and WV is 780 days. In average of all components, the cost reduction by implementing the suggested interval is 52%, while the reliability is improved by 4% and the availability is increased by 5%.

1. Introduction

An oil and gas company has been established for more than 48 years and operated 2,268 wells. One third of the wells are in production, while two third of the wells are plugged and shut-in due to depleted reservoir, corrosion (leak), and other integrity issues.

Every well has been designed to have well barrier element (WBE) as per company and international standard. WBE consists of primary and secondary elements [1]. The components of WBE directly become a flow path of the hydrocarbon and independent each other **Error! Reference source not found.**.

Every well shall be tested and maintained regularly in order to ensure their integrity for preventing any leak of oil or gas to the surface which can endanger human, environment and asset. Well integrity related to containment and the prevention of the escape of fluids (i.e. liquids or gases) surface [1]. The failure of WBE is mainly related to stuck open or stuck close and leak above the acceptable limit. If there is failure of WBE, the company cannot produce the oil and gas and may cause production loss.

Failure of WBE component can be observed in SCSSV and Christmas Tree Valves following the result of function test, leak test, or gas detection system which leads to the activation of the automatic shutdown system on the well. If there is any unexpected failure during leak test and/or operation, Well Service (WLS) Department will perform Corrective Maintenance (CM), i.e., replacement of the failure component. It requires cost rent of water transportation or slick line unit, spare part material (gate valve for Christmas Tree Valves or retrieved the SCSSV with the new one), and worker. The historical failure and maintenance data are recorded in Well Chronology Database.

The Preventive Maintenance (PM) program is preventive replacement of the component as per schedule stated on the manual, i.e., every 1,095 days. The PM requires rent unit of water transportation or slickline unit, spare part material (seal, seat and API grease for Christmas Tree Valves and retrieved the SCSSV with the new one), and worker. The PM may cause production loss as well.

Currently, there is no systematic evaluation on the current PM program. The historical maintenance data has not been used for any analysis including reliability of WBE. This paper attempts to analyze the failure data and reliability based on historical data. Due to decreasing price of oil and gas between 2015 - 2016, the company need to initiate cost reduction programs. The optimal PM interval will be recommended to the company in order to minimize the total cost of maintenance per unit time.

The WBE is considered as non-repairable item since it has to be replaced with the other equipment. The duration of failure for component of WBE is used to determine Time To Failure (TTF). The Weibull analysis helps to determine the reliability as time function and its parameters [3]. Optimal preventive replacement interval for excavator critical spare part can be determined in order to minimize total cost of maintenance [4].

2. Methods

2.1. Reliability

Reliability is the probability that a system will perform its intended function for a specified period of time under a given set of conditions [5]. Reliability of equipment can be used as key success factor for inspection and maintenance.

In the preventive maintenance, Time To Failure (TTF) is continuous data. In order to calculate the reliability of equipment / component, statistical distribution should be determined. In the reliability analysis, there are some statistical distributions used such as: Weibull, Normal, Lognormal and Exponential distribution. Weibull distribution fits a large number of failure characteristics of equipment. The reliability for three-parameter of Weibull distribution is formulated as follows [6]:

$$R(t) = \exp\left[-\left(\frac{t-\gamma}{\eta}\right)^{\beta}\right], \tag{1}$$

where:
η = scale parameter; $\eta > 0$,
β = shape / slope parameter; $\beta > 0$,
γ = location parameter.

Reliability is usually stated as lifetime expectation of the system, noted in $E[T]$ or Mean Time To Failure (MTTF). MTTF is only used on component or equipment which classified as non-repairable items, once fail, it must be replaced by the new and good one [7]. By applying equation (1), the MTTF for three-parameter Weibull distribution could be written as [6]:

$$MTTF = \gamma + \eta\Gamma\left(1+\frac{1}{\beta}\right), \tag{2}$$

where Γ is gamma function taken from the function table or by Excel software.

2.2. Maintainability

Maintainability is the probability of component which fails to be repaired and work effectively in the period of time given as per procedure [8]. When TTF data is changed with Time To Repair (TTR) data, the CDF or *F*(*t*) will be transformed into maintainability function [9]. Based on equation (1) where CDF is complement of reliability, the maintainability for two-parameter Weibull distribution could be written as follows:

$$M(t) = F(t) = 1 - \exp\left[-\left(\frac{t}{\eta}\right)^{\beta}\right]. \tag{3}$$

Mean Time To Repair (MTTR) is mean time to do maintenance on the component or equipment in order to restore its function into initial state [8]. The MTTF in equation (2) can be transformed as well into MTTR [9]. The MTTR of two-parameter Weibull distribution would be:

$$MTTR = \eta\Gamma\left(1 + \frac{1}{\beta}\right). \tag{4}$$

2.3. Availability

Availability of component or system in a steady state condition is the comparison between up time (T_{Up}) with total operation time (T_{Up}+ T_{Down}). It is a useful parameter that describes the proportion of time for which an item is not failed. The availability can be modeled with mathematical equation as follows [7]:

$$Availability = \frac{T_{Up}}{T_{Up} + T_{Down}} = \frac{MTTF}{MTTF + MTTR}, \tag{5}$$

where MTTF and MTTR refers to equation (2) and (4).

2.4. Optimal preventive maintenance interval with reliability approach

There are three type of maintenance, i.e., inspection, preventive maintenance and predictive maintenance [7]. Preventive maintenance can be performed if the failure rate becomes higher, where in the bathtub curve is shown with β > 1 at Weibull distribution. It is called as wear out area. Component of equipment failure can be prevented in this region by performing preventive maintenance [6].

Equipment sometimes called a replaceable unit or part is subject to sudden failure, and when failure occurs, the item has to be replaced. In order to reduce the number of failures, preventive replacement is scheduled to occur at specified intervals [6]. A balance is required between the amount spent on the preventive replacement compared with the failure replacement.

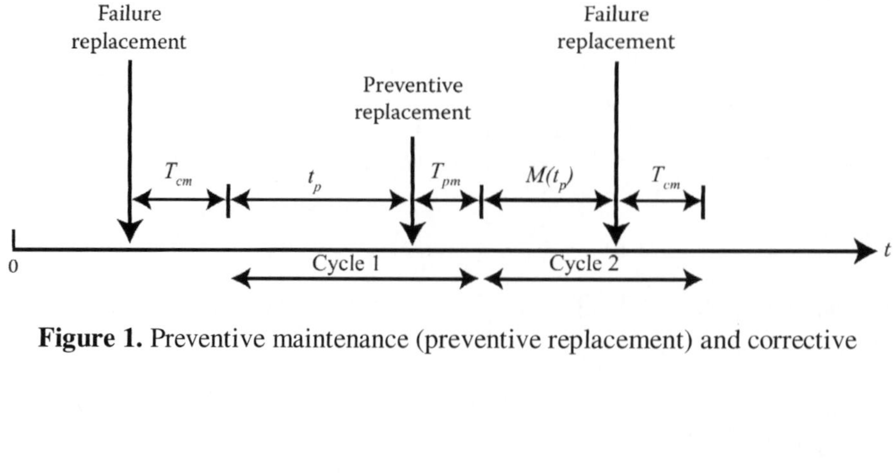

Figure 1. Preventive maintenance (preventive replacement) and corrective

maintenance (failure replacement) cycles, including replacement duration [6].

As seen in Figure 1, there are two possible cycles of operation, i.e., preventive maintenance and corrective maintenance. The objective is to determine the optimal Preventive Maintenance (PM) interval of the component in order to minimize the total cost per unit time. The optimal interval is determined by utilizing the cost model that solved by analytical method. Based on the two cycles, total cost of maintenance per unit time can be formulated as per below [6]:

$$C(t_p) = \frac{C_{pm} \times R(t_p) + C_{cm} \times \left(1 - R(t_p)\right)}{\left(t_p + T_{pm}\right) \times R(t_p) + \int_0^{t_p} tf(t)dt + T_{cm} \times \left[1 - R(t_p)\right]} \tag{6}$$

where:

$C(t_p)$ = total cost per unit time,
C_{pm} = cost of preventive maintenance / preventive replacement,
C_{cm} = cost of corrective maintenance / failure replacement,
$R(t_p)$ = Reliability at period of t_p, refers to equation (1),
T_{pm} = mean time required to perform a preventive maintenance,
T_{cm} = mean time required to perform corrective maintenance, refers to equation (4),
t_p = optimal time interval for preventive replacement once the item has reached a specific age,
t = Mean Time To Failure (MTTF), refers to equation (2)

3. Result

Data processing is used in order to achieve the research objectives. Time To Failure (TTF) data distribution is tested by Kolmogorov Smirnov (K-S) goodness of fit test. The TTF data distribution fits with the three-parameter Weibull. The reliability parameters on the components can be seen in Table 1. The lowest MTTF is SCSSV i.e. 868.927 days. It means the component has the most frequent failure compared with the other component. SCSSV is the first component in contact with the reservoir oil and gas in order to produce the well.

The shape (β) parameter of WBE components is more than 1 means the components were in the wear out region. The maintenance strategy is preventive maintenance / replacement at fixed intervals.

Table 1. Reliability parameters of WBE.

No	Component	Distribution	Beta (β)	Eta (η)	Gamma (γ)	MTTF (day)
1	SCSSV	Weibull 3	1.5734	156.3088	728.5600	868.927
2	LMV	Weibull 3	3.8643	967.0521	699.4250	1574.256
3	UMV	Weibull 3	1.6881	201.7563	907.1350	1087.239
4	SV	Weibull 3	7.7396	1052.3703	273.4400	1262.896
5	WV	Weibull 3	1.9504	209.2257	767.0800	952.601

Time To Repair (TTR) history data is tested as well by using K-S goodness of fit test. The TTR data does not fit with three-parameter Weibull but fit with two-parameter Weibull distribution. The maintainability parameters of WBE are provided in Table 2. The longest time for corrective maintenance (T_{cm}) is LMV i.e. 2.045 days. Because the well need to be plugged in order to have redundant safety barrier (i.e. tubing plug and SCSSV) prior replacing the LMV. The Well Service Supervisor also need to assess the well condition before replacing the component.

Table 2. Maintainability parameters of WBE.

No	Component	Distribution	Beta (β)	Eta (η)	MTTR
1	SCSSV	Weibull 2	3.3339	2.1243	1.907
2	LMV	Weibull 2	3.1701	2.2840	2.045
3	UMV	Weibull 2	3.5556	1.0313	0.929
4	SV	Weibull 2	2.8779	1.0909	0.972
5	WV	Weibull 2	2.9228	1.2515	1.116

Following the failure of WBE, the cost for corrective maintenance is available in Table 3. The biggest cost of corrective maintenance (C_{cm}) is SCSSV i.e. US$ 129,718.693 since the material cost is much higher compared with the other component. Time to repair (T_{cm}) has big impact to production loss thus directly impact on cost corrective maintenance (C_{cm}).

Table 3. Cost of corrective maintenance (C_{cm}) for every component.

Equipment	Type / Component	Cost of Corrective Maintenance (Failure Replacement)					
		Rent Unit (US$/d)	Material / job	Worker (US$/d)	Production Loss (US$/d)	T_{cm} (day)	C_{cm} (US$)
SCSSV	WR-SCSSV	13,247	40,000	350	31,793	1.907	126,536.010
Christmas Tree	LMV	13,247	21,545	350	17,200	2.045	84,516.420
	UMV	1,816	31,950	350	17,200	0.929	49,934.534
	SV	1,816	15,775	350	17,200	0.972	34,606.788
	WV	1,816	24,856	350	17,200	1.116	46,473.954

The company perform component PM every 3 years (1095 days). In average, the time required for performing preventive maintenance (T_{pm}) of WBE component is 0.25 day. Cost for PM is provided in Table 4. The biggest total cost of PM (C_{pm}) is SCSSV i.e. US$ 13,912.375 because the material cost is much higher compared with the other component.

Table 4. Cost of preventive maintenance (C_{pm}) for every component.

Equipment	Type / Component	Cost of Preventive Maintenance (Preventive Replacement)						
		Rent Unit (US$/d)	Material / job	Worker (US$/d)	Production Loss (US$/d)	Total Cost (US$/ d)	T_{pm} (day)	C_{pm} (US$)
SCSSV	WR-SCSSV	1,816	10,000	171	13,663	25,650	0.25	13,912.375
Christmas Tree	LMV	13,247	2,174	171	10,735	15,592	0.25	8,212.167
	UMV	1,816	2,174	171	10,735	3,990	0.25	5,354.375
	SV	1,816	2,174	171	10,735	3,990	0.25	5,354.375
	WV	1,816	2,174	171	10,735	3,990	0.25	5,354.375

In order to determine optimal PM interval, mathematical model is used to minimize total expected cost per unit time. The total cost consists of PM cost (C_{pm}) and corrective maintenance cost (C_{cm}) per cycle. The cost consists of rent unit, material/component, worker, and production loss. Based on Table 3 and Table 4 applied into equation (6), the optimal PM interval could be seen in Table 5. The optimal interval (t_p) is below the MTTF of each component. Means, the component has to be maintained / replaced prior any unexpected failure. Based on the optimal interval (t_p), the reliability and availability can be calculated by using equation (1) and (5).

Table 5. Optimal preventive maintenance interval.

Equipment	Type / Component	Optimal PM Interval			
		t_p (day)	$C(t_p)$ (US$/d)	Reliability	Availability
SCSSV	WR-SCSSV	730	99.743	0.999	0.998
Christmas Tree	LMV	985	9.367	0.992	0.999
	UMV	910	29.489	0.999	0.999
	SV	900	6.539	0.982	0.999
	WV	780	27.860	0.996	0.999

Based on MTTF in Table 1 and existing PM schedule, the total cost per unit time could be calculated as well by applying equation (6). The existing total cost for both corrective and preventive maintenance is provided in Table 6. Based on Table 6, the total cost for performing optimal PM on every component is lower than let the component to fail nor performing existing PM.

Based on existing PM interval (T_p), the reliability and availability on the current PM policy can be calculated as well by using equation (1) and (5). Refer to Table 4 and 5, the optimal and existing PM result is compared. In average, the reliability of every component is improved by 4% compared with the company requirement at 0.95. The availability improvement of every component in average is 5% higher compared with the company requirement.

Table 6. Existing total cost of corrective and preventive maintenance.

Equipment	Type / Component	Existing CM		Existing PM	
		$MTTF$ (day)	$C(MTTF)$ (US$/d)	T_p (day)	$C(T_p)$ (US$/d)
SCSSV	WR-SCSSV	868.927	351.683	1095	488.660
Christmas Tree	LMV	1574.255	31.863	1095	10.010
	UMV	1087.239	99.986	1095	102.722
	SV	1262.896	15.690	1095	8.648
	WV	952.601	83.905	1095	116.710

4. Conclusion

In order to obtain the lowest total cost per unit time, the replacement should be performed at interval t_p i.e. 730 days for SCSSV, 985 days for LMV, 910 days for UMV, 900 days for SV and 780 days for WV. It is sooner than the company's preventive maintenance policy for replacing the component every 1,095 days. In average for all components, the reduction of total cost per unit time is 52% by implementing the suggested optimal preventive maintenance interval.

References

[1] ISO/TS 16530-2:2014 *Well Integrity – Part 2: Well Integrity for the Operational Phase*

[2] Perrin D, Caron M and Gaillot G 1999 *Well completion and servicing: oil and gas field development techniques* (Paris: Editions Technip)

[3] Sheikh A K, Younas M and Al-Anazi D M 2002 Weibull Analysis of Time Between Failures of Pumps Used in an Oil Refinery *The 6th Saudi Engineering Conference* **4** pp 475-491 (Dahran, Saudi Arabia: KFUPM)

[4] Putra C A 2016 Spare Part Inventory Policy with Reliability Approach (Case Study: PT XYZ) *Master Thesis* (Surabaya, Indonesia: MMT ITS)

[5] Lewis E E 1994 *Introduction to Reliability Engineering* 2nd edition (New York, USA: John Wiley & Sons, Inc.)

[6] Jardine A K S and Tsang A H C 2013 *Maintenance, Replacement and Reliability – Theory and Applications* 2nd edition (Florida, USA: Taylor & Francis Group, LLC)
[7] Smith D J 2011 *Reliability, Maintainability and Risk – Practical Method for Engineer* 8th edition (Oxford, UK: Elsevier Ltd)
[8] Ebeling C E 1997 *Reliability and Maintainability Engineering* International Edition (New York, USA: McGraw-Hill)
[9] Dhillon B S 2006 *Maintainability, Maintenance and Reliability for Engineers* (Florida, USA: Taylor & Francis Group, LLC)

Acknowledgments

The authors would like to convey the gratitude for every people who have given their kind help and support in performing the research especially the Company, my Supervisor, colleague, MMT lecturer and staff. Also my family who always give me the patience and their best supports.

Lean waste classification model to support the sustainable operational practice

A Sutrisno[1*], I Vanany[2], I Gunawan[3] and M Asjad[4]
[1]Department of Mechanical Engineering, Sam Ratulangi University, Manado, Indonesia
[2]Department of Industrial Engineering, Institut Teknologi Sepuluh Nopember, Kampus ITS Sukolilo-Surabaya 60111, Indonesia
[3]Faculty of Professions, the University of Adelaide, Adelaide, Australia
[4]Department of Mechanical Engineering, Jamia Militia Islamia University, New Delhi, India

*agungsutrisno@unsrat.ac.id

Abstract. Driven by growing pressure for a more sustainable operational practice, improvement on the classification of non-value added (waste) is one of the prerequisites to realize sustainability of a firm. While the use of the 7 (seven) types of the Ohno model now becoming a versatile tool to reveal the lean waste occurrence. In many recent investigations, the use of the Seven Waste model of Ohno is insufficient to cope with the types of waste occurred in industrial practices at various application levels. Intended to a narrowing down this limitation, this paper presented an improved waste classification model based on survey to recent studies discussing on waste at various operational stages. Implications on the waste classification model to the body of knowledge and industrial practices are provided.

1. Introduction

In an attempt increasing profitability and minimize operating cost, companies globally are implementing endeavours to reduce/eliminate the occurrence of any activity which has no value addition but contributing into an operating cost. One of the viable way to achieve above goal is by implementing a lean operation [1] that not only reduces the wastes but also enhances the effectiveness and efficiency of any process leading to a sustainable operation. Attempting being lean means minimizing the occurrence of non-value-added activity by increasing capability to uncover the modes of waste. Initial endeavour to implement a leaner operation is introduced by Taiichi Ohno by presenting the Toyota Production System that introduces the concept of Seven Waste. Although the use of the Seven waste mode of Ohno is becoming versatile in its application, it is only covering waste during in process (in process type waste) and excluding other types of waste such as human and environmental -related waste. In addition, the applicability of the Seven Waste of Ohno model is limited to manufacturing oriented waste only and at a companywide setting. Motivated by this opportunity, the goal of this paper is to propose an improved model for waste typology for a leaner operation by using survey to the recent references discussing on waste classifications. The structure of the paper is in the followings, in section 2, overview and typology of waste is presented and followed with survey to recent references discussing on waste categorizations. Section 3 relates to the proposal on improvement of the waste taxonomy with inclusion on additional waste not covered by Ohno [2], and Simboli et al. [3]. Discussions on the taxonomy of the improved waste classification waste model and its implication is provided in section 4. Conclusions and new research opportunities for further investigation are presented in section 5.

2. Waste in Lean Philosophy- Overview and Typology

As the name imply, waste is defined as anything that do not provide value to customer. The existence of waste may lead to the increasing increasing the cost and time to the customers pertaining to their demands/goals/objectives. In a production context, there are three kinds of value related activities implemented. The first is Value Added (VA) activity concerning on transforming product and services to customer. The second one is the Necessary Non-value-added activity (NNVA) relates to the wasteful unavoidable activity. The third is a Non-Value-Added activity, a kind of activity that truly waste and should be eliminated [4] at the earliest. Depending on its industry types, the modes of waste may vary from industry to industry. The modes of waste in manufacturing may slightly differ on the modes of waste in a product development stages. In similar situation, in the service industrial settings, the original seven waste modes of Ohno that strongly correlated with manufacturing types can't be transferred directly into service settings due to the characteristics of service. The existence of waste causing system unable to achieve its operational excellence, thus the root cause of waste must be identified and alleviated.

3. Lean Waste Identification and Classification Progress

Becoming the driver to the resource consumption, the occurrence of waste is negatively contributing to sustainability. The waste creating many negative consequences such as productivity and profit losses, customers' dissatisfactions, threatening human life and so on. Originally coming from the birth of the Toyota Production System in 1980s by the introduction of the Seven waste; the classification and the mode(s) of the waste is continually evolving considering the versatility of the application of Toyota Production System or Lean Operation System in various sectors such as service, product development and software development and at both of companywide and supply chain level. Historically introduced by Taichi Ohno in 1980s, there are 7 kinds of waste; Overproduction, Waiting, Unnecessary transport, Incorrect Processing, and Defects. As the time elapsed since its application, due to versatility of application in non-manufacturing settings the original Seven waste of Ohno is insufficient to cope with waste elimination efforts. For example, the seven-original waste of Ohno are unable to capture the occurrence of waste due to behavioural and environmental issues. In addition, time dimension related to waste occurred in communication is not included in the Ohno Model. Therefore, the modes or categories of waste may be expanded to cope with other type of waste escaped from the coverage of the Ohno classification model. The modes of waste are then increased by various scholars by adding the additional waste concerning to the environmental, behavioural and communicational waste. Considering on the new views on waste determination for a leaner operation, the classification of the new modes of waste is then given as in Table 1.

Table 1. Classification of the new modes of waste adding to the Ohno Model.

No	Waste Mode	References	Waste Applicability Setting
1	Overproduction		
2	Waiting		
3	Unnecessary Transport		
4	Incorrect Processing	[5][2][6]	
5	Excess Inventory		Companywide
6	Unnecessary Movement		
7	Defect		
8	Unused people creativity		
9	Environmental Waste	[7][8]	

No	Waste Mode	References	Waste Applicability Setting
10	Behavioral Waste	[9][10]	
11	Failure Demand, Flow Demand, Flow Excess, Flawed Flow	[11]	
11	Outdated Information, Centralized Decision Making, Over specification, Lack of Customer Involvement, Inappropriate assumption	[12]	
12	Avoiding Decision Making, Information Distortion, Limited Information Access, Information Uncertainty	[13]	Companywide
13	Lack of stakeholders' involvement, Lack of Shared Understanding, restricted information access, Scattered Information	[14]	
14	Digital Waste, duplicating efforts, overdesigning, Excessive Overhead, Poor Planning, Poor Measurement, Poor Coordination	[15]	Supply Chain

According to the additional waste classification model as depicted in Table 1, based on defining waste as non-value adding activity, the Seven Waste model of Ohno is added by considering environmental impact [3]. Woscheke et al. [10] and Hick [11] add the additional waste "behavioural waste" concerning to the absence on the waste category dealing with human behavioural characteristics causing the occurrence of non-value-added activity. Considering the application of lean principle in information and communication sectors, several references expanding the waste categories including on communication and information waste [11][12][13][14]. Using waste categorization using application setting as basis, Su [15] provides category of waste under supply chain context.

4. Observations and Implications of the progress of Waste Categorization Models

Since introduced in the last two decades, the concept of waste as non-value adding activity has been used as foundation of study and practices in alleviating waste toward a leaner operation. However, a critical evaluation of waste concept as introduced by Thurer et al. [16] argues that foundation on defining waste as non-value-added activity is reducing visibility to waste differentiation since according to their study, there is a kind of waste that can't be eliminate without creating additional waste. In their waste classification model, the concept of Buffer Waste is proposed. In an attempt to improve definition of waste toward universally acceptable and consistent definition of waste, Lean Waste then redefined as "Any system input (transformed resources, transforming resources) that is not transformed into a system output that is valued by customers (fulfilled customer demand, this is neither unfulfilled nor exceeded) just-in-time". Following this definition, the new classification of Waste is proposed. Those are Waste type I related to the type of waste that can be eliminated without creating another waste, and waste type II, another kind of waste that can't be eliminate without creating another waste. Waste type I is called Obvious Waste and Waste type II is named as Buffer Waste. By using this new waste categorization model, typology of new waste dimension is introduced extending to the structural model of the Shingo Waste Classification Model. Apart of their work, in relation of findings of Thurer et al. [16], we observe that the categorization of waste of both of Ohno [2] and Thurer et al. [16] models overlooked on human behavioural dimensional aspect in waste classification and locus of waste occurrence settings whether a certain waste is occurred during in process stage or out of process stage. This new waste categorical dimensions are added considering to the situation that a particular product or service causing the occurrence of non-value-added events consuming extra amount of resources when used or consumed. Accordingly, the improved waste classification model is then displayed as in Table 2.

Table 2. An improved waste classification model (source own research).

No	Dimensional Basis of Waste Category	Waste typology
1	Process Basis	Over processing, Transportation, inventory, quality defects
2	Waste Typology	Obvious, Buffer
3	Operational Basis	Overproduction, waiting, unnecessary motion
4	Behavioural Basis	Negative Behavioural Waste
5	Locus of Waste Occurrence	In process waste and out of process waste

The improved waste classification of waste as proposed in Table 2 implies that current methodology for evaluating, accessing critically of waste using modification of the risk appraisal tool, the FMEA (Failure Mode and Effect Analysis) must be revisited again toward becoming an effective and efficient waste criticality assessment and mitigation tools. Considering the locus of the waste occurrence that can be out of process implies on our thinking for not only considering on environmental but also societal impact of the wastage during product and or service usage in the form of *loss to society*; a term borrowed from Taguchi Robust Design philosophy. In addition, the existence of method to measure criticality of behavioural waste for a more holistic waste criticality assessment is becoming imperative. Linked with endeavour to improve performance by eliminating waste, there is also becoming important to develop a kind of waste -based performance measurement tool using available performance measurement tool such as Balance Scorecard and Performance Scorecard.

5. Conclusions

Preventing and mitigation adverse effect of the occurrence of lean waste is one of the pre-request for realizing sustainable operation. The existence of waste is disadvantageous since it is causing negative consequences to the organizations. In this paper, a survey to recent studies of waste categorization based on value adding concept is presented, and followed by critical evaluation. Based on survey to previous waste classification category and modes, our views suggest to the need on improving waste classification model using behavioural dimension and locus of waste occurrence. In our proposal, additional waste dimensions are proposed and followed by implications on developing improved tools and methodology considering characteristics of waste for waste alleviation. Opportunities are existing demanding for future investigations. For example, establishment of complex Failure Mode and Effect analysis (FMEA) model considering interrelationship among obvious and buffer waste extending the work of previous studies [17][18][19] are still unknown in literature. Furthermore, development of a metric as means to measure a composite risk impact model which consider operational, behavioural, environmental and societal losses as surrogate of the Risk Priority Number (RPN) in conventional FMEA is vacant in references. Another opportunity is concerning to the development of lean waste risk-based performance evaluation tool in compliment with the work of [20].

References

[1] Benson R and Kulkarni N 2011 *Pharm. Eng.* **6** pp 74 -82
[2] Ohno T 1988 *Toyota Production Systems: Beyond Large Scale Production,* Productivity Press, Cambridge
[3] Simboli A, Taddeo R and Mogrante A 2014 *Adm.Sci.*4 pp 73-91
[4] Mostafa S and Dumrak J 2015 *Int. Conf. Mat. Man. Ind. and Man. Eng* Bali
[5] Shingo S 1989 *A Study of Toyota Production System from An Industrial Viewpoint* (Cambridge: Productivity Press)

[6] Womack J P and Jones D T 2003 *Lean Thinking: Banish Waste and Create Wealth in Your Corporation* Free Press Simon and Schuster
[7] Khan M, Jaber M Y and Glock C H 2012 Impact of Learning on the Environmental Performance of a Two Supply Chain, *Procd. 17-th Int. Sym. Inventories*, Budapest, Hungary
[8] Fallan S L, Han B, Zueller L H, Tarn M J and Roach D M 2011 *Int. J Heath. Inf. Sys.* **6** pp 1-22
[9] Lopez E A, Requena G and Lobera A S 2015 *Proc. Mfg.* **132** pp 23 -30
[10] Woscheke T, Hanse H and Laftenschager A 2016 *Int. J. Prod. Perf. Man.* **65** pp 32-553
[11] Hick B J 2007 *Int. J Inf. Man.* **27** 4 pp 1-22
[12] Al-Baik O and Miller J 2014 *Emp. Soft. Eng.* **19** 29-61
[13] Mandic V, Ono M, Rodriguez P, Kuvaja P, Kaikonen H and Turhan B 2010 *Proc. Of LESS* Helsinki Finland pp 72-84
[14] Korkala M and Mauren F 2014 *J.Syst. Soft.* **95** pp 22-40
[15] Su T C 2012 *MSc Thesis* Chalmers. Univ. Tech
[16] Thurer M, Tomasevuc I and Stevenson M 2016 *Prod. Plan. Inv. Cont.* early view edition pp 1-12
[17] Souza R V B and Carpinetti L C R 2014 *Int. J Qual. Rel. Man.* **31** pp 46 – 66
[18] Sutrisno A, Gunawan I, Tangkuman S and Khorsidhi H A 2015 *Int. J. Qual. Eng. Tech.* **5** pp 27-37
[19] Sutrisno A, Vanany I and Gunawan I 2016 *Proc. IEEE Int. Conf. Ind. Eng. Eng. Man.* Bali
[20] Vernadat F, Shah L, Etienne A and Siadat A 2013 *Int. J. Prod. Res.* **51** pp 20 – 38

Acknowledgements
The authors thanks to Indonesian Ministry of Higher Education, Research and Development for funding this study under research scheme *International Collaborative Research and Joint Publications 2017.*

IOP Conf. Series: Materials Science and Engineering **337** (2018) 012068 doi:10.1088/1757-899X/337/1/012068

Policy for equipment's leasing period extension with minimum cost of maintenance

C Lestari[1] and N Kurniati[1*]
[1]Department of Industrial Engineering, Institut Teknologi Sepuluh Nopember, Kampus ITS Sukolilo-Surabaya 60111, Indonesia

*nanikur@gmail.com

Abstract. The cost structure for equipment investment including purchase cost and maintenance cost is getting more expensive. The company considers to lease the equipment instead of purchase it under a contractual agreement. Offering to extend the lease period, following to the base lease period, will provide more benefits for both the lessor (owner) and the lessee (user). Whenever the lease period extension offered at the beginning of the contract, there are some risks in finance e.g. uncertainty of the equipment performance and lessor responsibility. Therefore, this research attempts to model the optimal maintenance policy for lease period extension offered at the end of the contract. Minimal repair is performed to rectify a failed equipment, while imperfect preventive maintenance is conducted to improve the operational state of the equipment when reaches a certain control limit to avoid failures. The mathematical model is constructed to determine the optimal control limit, the number and degree of preventive maintenance, and the multiplication number of the lease period extension. Finally, numerical examples are given to illustrate the influences of the optimal length of the extended lease and the maintenance policy to minimize the maintenance cost.

1. Introduction

Most companies need various types of equipment for running their business processes, either to produce goods or provide services to consumers. The purchase price and maintenance cost of the equipment is getting more expensive, it may not be economical for the company to purchase the expensive equipment. The company considers to lease the equipment instead of purchase it [1][2].

For repairable leased equipment, there are two types of maintenance actions, they are Corrective Maintenance (CM) and Preventive Maintenance (PM). CM is used to rectify failed equipment back to its operational state and PM is performed to improve the operational state of the equipment to avoid failures. In the CM action, minimal repair is the commonly used to restore failed equipment. After minimal repair, the equipment is under normal operation but the failure rate remains unchanged [3][4]. PM is classified into two major categories, perfect PM and imperfect PM [5]. There are two methods for describing PM degree, Age Reduction Method (ARM) and Failure Rate Reduction Method (FRRM). ARM is used to restore the age of the equipment younger than the current age, FRRM is used to reduce the failure rate of the equipment [6]. Pongpech & Murthy [7] utilized FRRM to describe PM degree and derived the optimal PM policy for leased equipment. Some other issues associated with FRRM can be found in the literature [8][9][10][11]. Yeh et al. [12] utilized ARM to describe PM degree and derived the optimal PM policy for leased equipment with considering lease period extension at the beginning of the contract. Some other issues associated with ARM can be found in the literatures [13][14]. Most studies mentioned focus on determining the optimal PM policy for leased equipment with a specified lease period by minimizing cost or an extension of lease period that was offered at the begin of the contract by maximizing profit. On the other hand, Chang & Lin [15] used ARM to derive the optimal PM policy and length of extended warranty after the base warranty period expires. Thus, extension of the lease period after base lease period ends may be a critical issue to get more benefit for the lessor and the lessee.

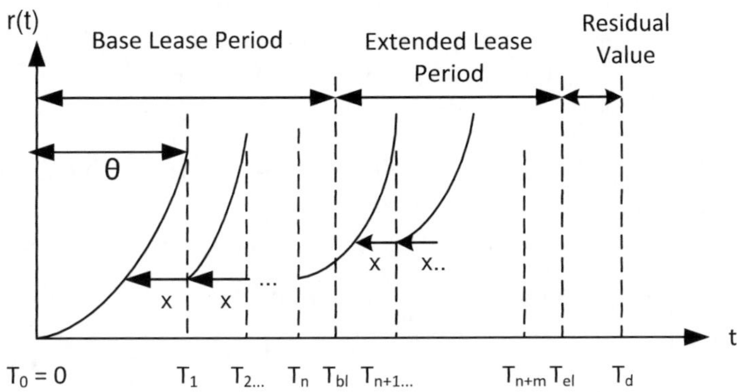

Figure 1. ARM of PM within life cycle of the equipment.

Hence, this paper will adopt ARM to describe PM degree and determine the optimal length of extended lease and the corresponding optimal maintenance policy for the equipment with minimizing the expected total maintenance cost of the lessor. Organization of this paper is as follows. The mathematical formulation is shown in Section 2. Section 3 provides the properties of the optimal maintenance policy and length of extended lease and an effective algorithm. The performance of the maintenance policy and length of extended lease are evaluated and illustrated through numerical examples in Section 4. Finally, some conclutions are illustrated in Section 5.

2. Mathematical formulation

The following notations are used to construct the mathematical model in this paper:

L	Unit length of lease period
k	Number of extended lease periods of a fixed time length unit L
$r(t)$	Failure rate function of the leased equipment
$R(t)$	Cumulative failure rate function of the leased equipment
C_m	Minimal repair cost
C_f	Penalty cost for each failure
τ	Pre-specified time limit for minimal repair action
θ	Controlled-limit of age for performing a PM action
x	Maintenance degree of a PM action
$C_p(x)$	PM cost function with maintenance degree x
n	Total number of PM action within base lease period $(0, t_{bl}]$
m	Total number of PM action within extended lease period $(t_{bl}, t_{el}]$

T_i Time epoch to perform the i-th PM action
t_{bl} Base lease period of leasing equipment
t_{el} Extended lease period of leasing equipment, where $t_{el}=t_{bl}+kL$
t_d Life cycle of the equipment
$E[TC]$ Expected total maintenance cost within the lease period

In Figure 1, there are three phases within life cycle of the equipment. They are base lease period $(0,t_{bl}]$, extended lease period $(t_{bl},t_{el}]$, and residual value at time t_d. When the leased equipment fails, there is a penalty cost $C_f>0$ to the lessor. Each equipment failure is rectified by a minimal repair. The minimal repair time required is less than or equal to pre-specified time limit τ. An imperfect PM action is performed with the same degree x when the age of the equipment reaches a controlled-limit θ over the lease period $(0,t_{bl}+kL)$. T_i, $i=1,2,...,n,n+1,...,n+m$ is the time epochs to perform PM action. Thus, $T_i=\theta$ can be assumed. The number of PM action within the base lease period and extended lease period are n and m.

In addition, the boundary conditions $x\le\theta\le\min\{t_{bl},kL\}$ and $\max\{(t_{bl}-\theta)/n,(kL-\theta)/m\}\le x\le\min\{t_{bl},kL\}$ create an equation $T_i=T_{i-1}+x,i=1,2,...,n+m$. This result shows that $T_i=T_1+(i-x),i=1,2,...,n+m$. After substituting $T_i=\theta$ into $T_i=T_1+(i-1)x$, the equation becomes $T_i=\theta+(i-1)x,i=1,2,...,n+m$.

2.1. Repair cost to the lessor

With the failure rate of the product $r(t)$ when $r(0)=0$, the expected total number of failures within the interval $[0,t]$ is $R(t)=\int_0^t r(u)du$. When using the two-parameter Weibull distribution, the equation becomes $R(t)=(\lambda t)^\beta$ where $\alpha>0$ and $\beta>1$. The failure process of the equipment in the interval $[T_{i,}T_{i+1}]$ is a non-homogeneous Poisson process with intensity $r(t-ix)$ when product failures are rectified using minimal repairs. A minimal repair is performed with a fixed cost $C_m>0$ within the lease period. From Figure 1, the expected total numbers of failures are $\sum_{i=0}^{n-1}\int_{T_i}^{T_{i+1}} r(t-ix)dt+\int_{T_n}^{t_{bl}} r(t-nx)dt$, where $T_0=0$ and $T_i=\theta$ within the base lease period $(0,t_{bl}]$ and $\int_{t_{bl}}^{T_{n+1}} r(t-nx)dt+\sum_{i=1}^{m}\int_{T_{n+i}}^{T_{n+i+1}} r(t-(n+i)x)dt$, where $T_{n+m+1}=t_{el}$ and $t_{el}=t_{bl}+kL,k=0,1,2,...$ within the extended lease period $(t_{bl},t_{el}]$, the repair cost of the equipment over the lease period $(0,t_{el})$ is

$$\begin{aligned}&[C_m+C_f]\left\{\sum_{i=0}^{n-1}\int_{T_i}^{T_{i+1}} r(t-ix)dt+\int_{T_n}^{t_{bl}} r(t-nx)dt+\int_{t_{bl}}^{T_{n+1}} r(t-nx)dt+\sum_{i=1}^{m}\int_{T_{n+1}}^{T_{n+i+1}} r(t-(n+i)x)dt\right\}\\&=[C_m+C_f]\{(n+m)[R(\theta)-R(\theta-x)]+R(t_{bl}+kL-(n+m)x)\}\end{aligned} \quad (1)$$

2.2. PM cost to the lessor

After an imperfect PM action is performed, the age of the equipment becomes x unit of time younger than before and each PM cost is $C_p(x)$. It is a non-negative and non-decreasing function of maintenance degree x, i.e., $C_p(x) \geq 0$ and $C_p^{'}(x) \geq 0$ for all $x \geq 0$. Assuming that the time required to perform an imperfect PM is negligible. PM costs are $\sum_{i=0}^{n} C_p(x) = nC_p(x)$ and $\sum_{i=0}^{m} C_p(x) = mC_p(x)$ within the base and extended lease periode. Therefore, the total PM cost over the lease period $(0, t_{bl} + kL)$ is

$$\sum_{i=0}^{n} C_p(x) + \sum_{i=0}^{m} C_p(x) = (n+m)C_p(x) \tag{2}$$

The repair and PM costs as in equation (1) and (2) are combined yield the expected total maintenance cost over the lease period $(0, t_{bl} + kL)$, which is

$$E[TC] = [C_m + C_f]\{(n+m)[R(\theta) - R(\theta - x)] + R(t_{bl} + kL - (n+m)x)\} + (n+m)C_p(x) \tag{3}$$

3. Optimal policy

Based on the objective function equation (3), we will find an optimal maintenance policy $(n^*, m^*, x^*, \theta^*)$ and number of extended lease period k^* that manimize the expected total maintenance cost of the lessor. The properties of them are investigated as followed. Taking the first and second partial derivatives of equation (3) with respect to k, we have

$$\frac{\partial E[TC]}{\partial k} = -L[C_m + C_f]r(t_{bl} + kL - (n+m)x) \tag{4}$$

and

$$\frac{\partial^2 E[TC]}{\partial k^2} = -[L^2[C_m + C_f]r^{'}(t_{bl} + kL - (n+m)x)] \tag{5}$$

Equation (4) and (5) are observed by the following theorems.

Theorem 1. Given any $n, m, x, \theta > 0$ when $r'(t) > 0, \forall t > 0$, the results can be written as followed.

(i) If $k^* = 0$, then $L[C_m + C_f]r(t_{bl} - (n+m)x) > 0$

(ii) If $k^* = \frac{t_d - t_{bl}}{L}$, then $L[C_m + C_f]r(t_d - (n+m)x) < 0$

(iii) If $-L[C_m + C_f]r(t_d - (n+m)x) > -L[C_m + C_f]r(t_{bl} - (n+m)x)$

then there is a unique solution $k^* \in \left(0, \frac{t_d - t_{bl}}{L}\right)$ such that equation (4) equals zero.

Substituting k^* into equation (3) yields the expected total total maintenance cost, which is

$$E[TC] = [C_m + C_f]\{(n+m)[R(\theta) - R(\theta - x)] + R(t_{bl} + k^*L - (n+m)x)\} - (n+m)C_p(x) \tag{6}$$

Theorem 2. Given any $n, m, x, > 0$, when $r'(t) > 0, \forall t > 0$, the optimal controlled-limit $\theta^* = x$ .The boundary condition becomes $\max\{t_{bl}/(n+1), k^*L/(m+1)\} \leq x \leq \min\{t_{bl}, k^*L\}$ after substituting $\theta^* = x$ into the boundary condition $\max\{(t_{bl} - \theta)/n, (kL - \theta)/m\} \leq x \leq \min\{t_{bl}, kL\}$ and equation (6). Thus, the expected total maintenance cost can be rewritten as

$$E[TC] = [C_m + C_f]\{(n+m)R(x) + R(t_{bl} + k^*L - (n+m)x)\} - (n+m)C_p(x) \tag{7}$$

Theorem 3. Given any $n, m > 0$, when $r'(t) > 0, \forall t > 0$, dan $C'_p(x) \geq 0, \forall x > 0$, the optimal PM degree is $x^* = \max\left\{\frac{t_{bl}}{n+1}, \frac{k^* L}{m+1}\right\}$ Substituting x^* into equation (7) yields the expected total maintenance cost, which is

$$E[TC] = [C_m + C_f]\{(n+m)R(x^*) + R(t_{bl} + k^* L - (n+m)x^*)\} - (n+m)C_p(x^*) \quad (8)$$

Finally, n and m are the desicion variables in equation (8). Assuming that the time required to perform a minimal repair is less than or equal to pre-specified time limit τ. Therefore, we have the boundary condition $0 \leq n\tau \leq t_{bl}$ and $0 \leq m\tau \leq k^* L$. The trivial upper bounds are t_{bl} / τ and $k^* L / \tau$ for n and m. These upper bounds are used to find the optimal n^* in the interval $[0, t_{bl} / \tau]$ and m^* in the interval $[0, k^* L / \tau]$ because n and m are both intergers. The following algorithm can be used to find the optimal maintenance policy $(n^*, m^*, x^*, \theta^*)$ and number of extended lease period k^*.

Algorithm
1. Set $k = 1$, $\min E[TC] = 999999999$, compute $\bar{k} = (t_d - t_{bl}) / L$, $\bar{n} = t_{bl} / \tau$.
2. Set $n = 1$ and compute $\bar{m} = kL / \tau$.
3. Set $m = 1$.
4. Compute $\theta = x = \max\{t_{bl} / (n+1), kL / (m+1)\}$ and
   $E[TC] = [C_m + C_f]\{(n+m)R(x^*) + R(t_{bl} + k^* L - (n+m)x^*)\} - (n+m)C_p(x^*)$
5. If $\min E[TC] > E[TC]$, then set $(k^*, n^*, m^*, x^*, \theta^*) = (k, n, m, x, \theta)$, thus $\min E[TC] = E[TC]$. Otherwise, set $m = m + 1$. If $m > \bar{m}$, then go to 6. Otherwise, go to 4.
6. Set $n = n + 1$. If $n > \bar{n}$, then go to 7. Otherwise, go to 3.
7. Set $k = k + 1$. Jika $k > \bar{k}$, then stop. Otherwise, go to 2.

4. Numerical example

Based on using a two-parameter Weibull distribution as the lifetime distribution of the equipment, $r(t) = \lambda\beta(\lambda\beta)^{\beta-1}$, where $\alpha > 0$ and $\beta > 1$. The PM cost is $C_p(x) = 30 + bx$ (\$). The following values are considered for the model parameter.

$L = 6$ (months)	$t_d = 100$ (months)	$C_m = 220$ (\$)	$C_f = 180$ (\$)	$b = 0, 10, 30, 50, 70, 90$ (\$)
$t_{bl} = 48$ (months)	$\alpha = 5$	$\beta = 1.5, 2, 2.3$	$\tau = 0.1$ (months)	

The optimal maintenance policy and number of extended lease period for the equipment are determined corresponding the minimal expected total maintenance cost of the lessor. One combination of values parameter has one result for the expected total maintenance cost of the lessor. The summary of all the results is shown at Table 1.

Table 1. Optimal policy and number of extended lease period.

b	$\beta=1.5$					$\beta=2$					$\beta=2.5$				
	k^*	n^*	m^*	$\theta^*=x^*$	$\min E[TC]$	k^*	n^*	m^*	$\theta^*=x^*$	$\min E[TC]$	k^*	n^*	m^*	$\theta^*=x^*$	$\min E[TC]$
10	1	33	4	1.4118	3937	1	34	4	1.3714	2862	1	32	3	1.5000	2285
30	1	31	3	1.5000	4971	1	35	4	1.3333	3903	1	30	3	1.5484	3330
50	1	31	3	1.5000	5991	1	31	3	1.5000	4938	1	31	3	1.4118	4352
70	1	31	3	1.5000	7011	1	31	3	1.5000	5958	1	31	3	1.5000	5372
90	1	31	3	1.5000	8031	1	31	3	1.5000	6978	1	31	3	1.5000	6392

Table 1 summarizes the optimal number of extended lease period k^* and maintenance policy (n^*,m^*,x^*,θ^*) with $\lambda=0.2$. For example, when $(\beta,b)=(1.5,50)$ yields $k^*=1$, $(n^*,m^*,x^*,\theta^*)=(31,3,1.5)$, and $E[TC]=5991(\$)$. This means that the optimal number of PM are $n^*=31$ and $m^*=3$ for the base and extended lease periode. In this maintenance policy, PM degree are 1.5 months, thus the time epochs to perform PM are $T_1\approx 1.5, T_2\approx 3, T_3\approx 4.5, T_4\approx 6, T_5\approx 7.5, \ldots, T_{30}\approx 45, T_{31}\approx 46.5$ within the base lease period and $T_1\approx 49.5, T_2\approx 51, T_3\approx 52.5$ within the extended lease period. Furthermore, we have some results from Table 1.

1. When the shape parameter ($\beta>2$) increases, the maintenance degree (x) increase, the number of PM actions (n^*+m^*) and the maintenance cost decrease.
2. When the shape parameter ($\beta<2$) increases, the maintenance cost and the maintenance degree (x) decrease, but the number of PM actions (n^*+m^*) increase.
3. When the marginal PM cost (b) increases, the maintenance cost and the maintenance degree (x) increase, but the number of PM actions decrease.

5. Conclusions

In this paper, PM degree is described by ARM to determine the optimal length of extended lease and the corresponding optimal maintenance policy for the equipment with minimizing the expected total maintenance cost of the lessor. Based on Table 1, the optimal length of extended lease period are one period for every variations of shape parameter and marginal PM cost.

6. References

[1] Glickman T S and Berger P D 1976 *Management Science* **22** L1381

[2] Nisbet A and Ward A 2001 *The British Journal of Radiology* **74** L735

[3] Barlow R E and Hunter L C 1960 *Operations Research* **8** L90

[4] Nakagawa T and Kowada M 1983 *European Journal of Operational Research* **12** L176

[5] Jack N and Murthy D N P 2007 *Journal of the Operational Research Society* **58** L1612

[6] Nakagawa T 1981 *Journal of the Operational Research Society of Japan* **24** L213

[7] Pongpech J and Murthy D N P 2006 *Reliability Engineering and System Safety* **91** L772

[8] Jaturonnatee J, Murthy D N P and Boondiskulchok R 2006 *European Journal of Operational Research* **174** L201–215.

[9] Ben Mabrouk A, Chelbi A and Radhoui M 2016) *Int. J. Production Economics* **178** L57

[10] Yeh R H and Chang W L 2007 *Mathematical and Computer Modelling* **46** L730

[11] Yeh R H, Kao K and Chang W L 2009 *Computers and Industrial Engineering* **57** L304

[12] Yeh R H, Chang W L and Lo H 2011 *Mathematical and Computer Modelling* **54** L2014

[13] Chang W L and Lo H 2011 *Computers and Industrial Engineering* **61** L489

[14] Zhou X, Li Y, Xi L and Lee J 2014 Multi-Phase Preventive Maintenance Policy for Leased Equipment *Int. J. Prod. Res.* L4528

[15] Chang W L and Lin J 2012 *Computers and Mathematics with Applications* **63** L144

Prediction accident triangle in maintenance of underground mine facilities using Poisson distribution analysis

M H Khuluqi[1*], R R Prapdito[1] and F P Sambodo[2]
[1]Underground Mine Maintenance Division, PT. Freeport Indonesia, Tembagapura, Papua, Indonesia
[2]SHE and Loss Prevention Mine Maintenance Division, PT. Freeport Indonesia, Tembagapura, Papua, Indonesia

*husnul.khuluqi@gmail.com

Abstract. In Indonesia, mining is categorized as a hazardous industry. In recent years, a dramatic increase of mining equipment and technological complexities had resulted in higher maintenance expectations that accompanied by the changes in the working conditions, especially on safety. Ensuring safety during the process of conducting maintenance works in underground mine is important as an integral part of accident prevention programs. Accident triangle has provided a support to safety practitioner to draw a road map in preventing accidents. Poisson distribution is appropriate for the analysis of accidents at a specific site in a given time period. Based on the analysis of accident statistics in the underground mine maintenance of PT. Freeport Indonesia from 2011 through 2016, it is found that 12 minor accidents for 1 major accident and 66 equipment damages for 1 major accident as a new value of accident triangle. The result can be used for the future need for improving the accident prevention programs.

1. Introduction

In Indonesia, mining activities are categorized as hazardous, especially for underground mining [1]. In recent years, a dramatic increase in the mining equipments and technological complexities have resulted in higher maintenance expectations that accompanied by changes in the working conditions, and one of these conditions is safety [2]. It has become an important issue because each year a vast number of people die and get seriously injured due to various types of accidents [3]. Ensuring safety during the process of conducting maintenance works in underground mining is important as an integral part of accident prevention programs.

When the results of incidents are analyzed over a period of time and possibly, aggregated by region or industry, the distribution of their severities may look similar to the classical model of an accident pyramid or safety triangle, introduced by H.W.Heinrich in 1931 [4]. Figure 1 presents this model, and also reflects suggestions by Bird and Germain [5]. It supports to safety practitioner to draw a road map in preventing accidents [6].

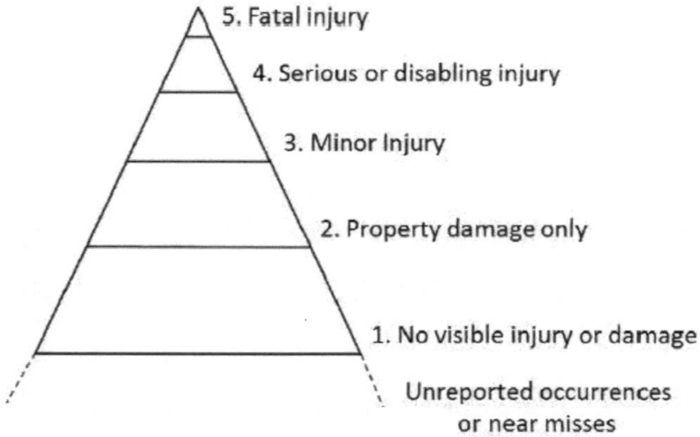

Figure 1. A Safety Triangle Distribution of Incident [7].

The purpose of this paper is to determine the new accident triangle based on accident data statistics in the underground mine maintenance of PT. Freeport Indonesia. Using a Poisson distribution as a statistical method, this analysis will show how information derived from these accident summaries can be used to predict potential future events.

2. Literature Review

The criteria of the mining accidents in Indonesia must involve five elements according to the ministry of mining and energy decree of the *Kepmen 555.K/26/M.PE/1995*, in article no. 39. The mining accident is an unplanned, unexpected, undesired, uncontrolled, and unintentional event which occurs suddenly and caused injury to the workers in connection with a work relationship of the mining operations. The five elements of the mining accident are the actual accident occurred, resulting injured to the mine worker or people who get a permit from the technical mine manager, consequences of the mining business activities, the accident occurred during working hours to injure of mine worker or people who get permit, and the accident occurred within the area of the mining activity or operation or in the project area [8].

In 1931, Herbert William Heinrich, one of the founding fathers of industrial safety and accident prevention, published the results of a study he performed while working for Traveler's Insurance [9]. His original work revealed that, for every "major injury" resulting from a single accident, there were 29 "minor injuries" resulting from accidents and 300 no-injury accidents. The figure produced by this analysis was depicting these ratios, came to be known as Heinrich's Triangle.

According to the ministry of mining and energy decree of the *Kepmen 555.K/26/M.PE/1995*, article no. 40, mining accidents shall be recorded and classified into the following categories:

- Minor, if any injury resulting from a mining accident which causes the mine worker to no longer be able to carry out his/her normal duties for more than one day and less than three weeks, including Sundays and holidays.
- Major, minor accident that any injury which causes the mine worker to become an invalid and who can no longer carry out his/her original duties and irrespective of any duration of absence, any mining accident causing one or more of the following injuries, such as fractures of skull, spine, pelvis, lower arm, upper arm, leg or foot; Internal hemorrhage or unconsciousness caused by oxygen deficiency; Severe wounds or lacerations likely to cause permanent disability; and Any dislocation of any joint which had never previously occurred.
- Fatal, if any accident causing the death of a mine worker within 24 hours calculated from the time the accident occurred.

PT. Freeport Indonesia on Internal FRESH Standards no. 4.11, incident classification will be as follow:

- Fatality, if any injury/ill health or incident directly resulting in death by working at the mine area irrespective of time
- Lost Time (LT) means, injury/ill health in which the employee is not able to return to work on the next regularly scheduled work shift or any subsequently scheduled work shift as a result of the injury /ill health
- Restricted Duty (RD) means, injury or ill health in which the employee returns to work, but is unable to perform all the duties normally connected with the job or where the employee is assigned to another job due to the injury or ill health on a temporary basis.
- Medical Treatment (MT), if any reportable injuries or ill health not classified as an LT or RD but requires medical treatment. Treatment by a medical doctor does not constitute an MT case if the doctor merely performs first-aid treatment, conducts diagnostic procedures (i.e. lab work, x-rays, tetanus injections, etc.) or hospitalizes the employee for observation. Examples of medical treatment include: fracture, removal of embedded foreign objects, debridement of wounds or burns, application of sutures, and administering prescribed medication for eye injuries.
- First Aid (FA), if any one time treatment or follow-up of minor scratches, cuts, burns, splinters, etc. which do not ordinarily require professional medical care or treatment.
- Property damage, if any damage to company property as a result of accident; incidents which are within the control or realm of operational control, supervisory control and established preventive maintenance or other management practices, classified as light vehicle damage, heavy equipment damage, and stationary equipment damage [10].

LT and RD categorized as major accident, in line with MT and FA categorized as minor incident.

3. Methodology

3.1. Accident Data Statistic

The data that shown on table 1 is captured from Safety Department in Mine Maintenance Division of PT. Freeport Indonesia. It provides information about accident that happened in Underground Mine Maintenance Division from 2011 through 2016. Accident is categorized as property damage, major accident, and minor accident.

Table 1. Accident Data in Underground Mine Maintenance Division.

Category		2011	2012	2013	2014	2015	2016	Total
Property Damage	HE Damage	3	7	5	8	8	8	39
	LV Damage	11	16	12	16	14	10	79
	ST Damage	1	3	0	3	2	4	13
Major Accident	LT	1	0	0	0	0	0	1
	RD	0	0	0	0	1	0	1
Minor Accident	ND	1	2	0	1	2	0	6
	FA	0	2	3	4	4	4	17

The data trends were analyzed and used to determine the probability of accident (property damage, major accident, minor accident) that will be occurred. If we assume that accident prevention program will not change or affect the working safety of Underground Mine Maintenance Division, the data accident from past can be used to predict future trends.

3.2. Data Analysis

The number of rare events occurring within a fixed period of time has Poisson distribution [11]. It is appropriate for the analysis of accidents at a specific site in a given time period [12]. Probability using a Poisson distribution can be calculated using the following formula:

$$\text{Probability} = \frac{e^{-\mu}.\,\mu^{k}}{k!} \qquad (1)$$

Where:
μ = Mean of the Poisson distribution = λ. t;
t = Selected time interval;
k = Total number of successes (or failures) or, in this case, one or more major accidents.
Microsoft Excel was used for all of the statistical data analyses presented in this paper.

4. Result and Discussion

Trends during 2011-2016 were analyzed to determine the annualized and overall ratios of minor accident to major accident and property damage cases to major accident. Parameter Poisson that used for minor accident is μ = 4, for major accident μ = 1, and for property damage μ = 22. The result for probability accident occurrence in underground mine maintenance is shown in Figure 2, while the cumulative probability accident occurrence in Figure 3. In these graph, the green line show probability of having n successful event for major accident, while the red line is for property damage, and the blue line is for minor accident. From graph in Figure 2, we can conclude that probability of having 1 major accident is 0.36 (36%), probability of having 4 minor accidents is 0.19 (19%), and probability of having 22 property damages is 0.08 (8%).

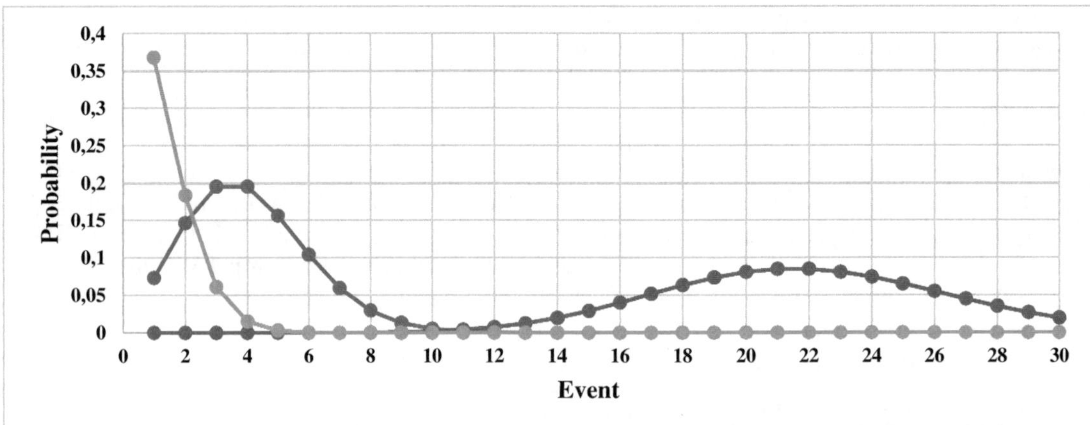

Figure 2. Probability of Accident Occurrences in Underground Mine Maintenance

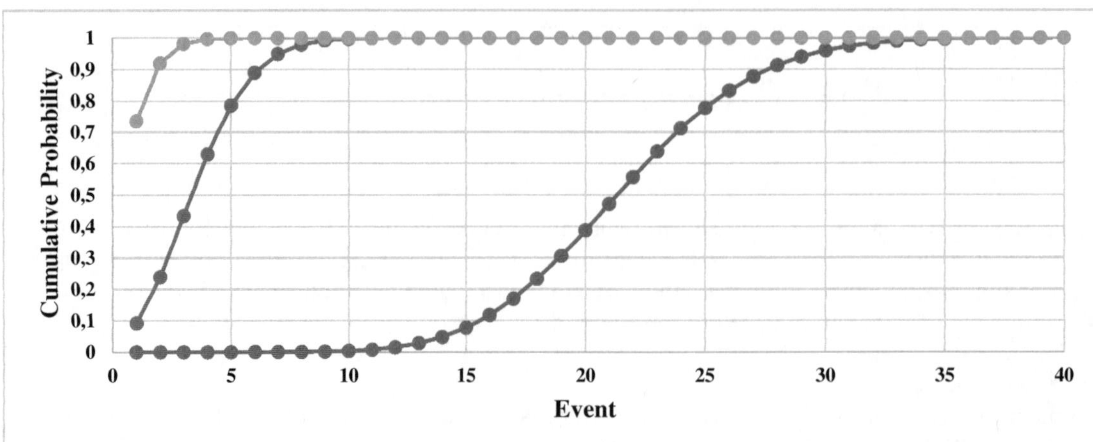

Figure 3. Cumulative Probability of Accident Occurrences in Underground Mine Maintenance

From the ratio result of cumulative data shown in table 1, 12 minor accidents for 1 major accident and 66 equipment damages for 1 major accident were derived as a new value of accident triangle for Underground Mine Maintenance Division PT. Freeport Indonesia appear as seen in figure 4.

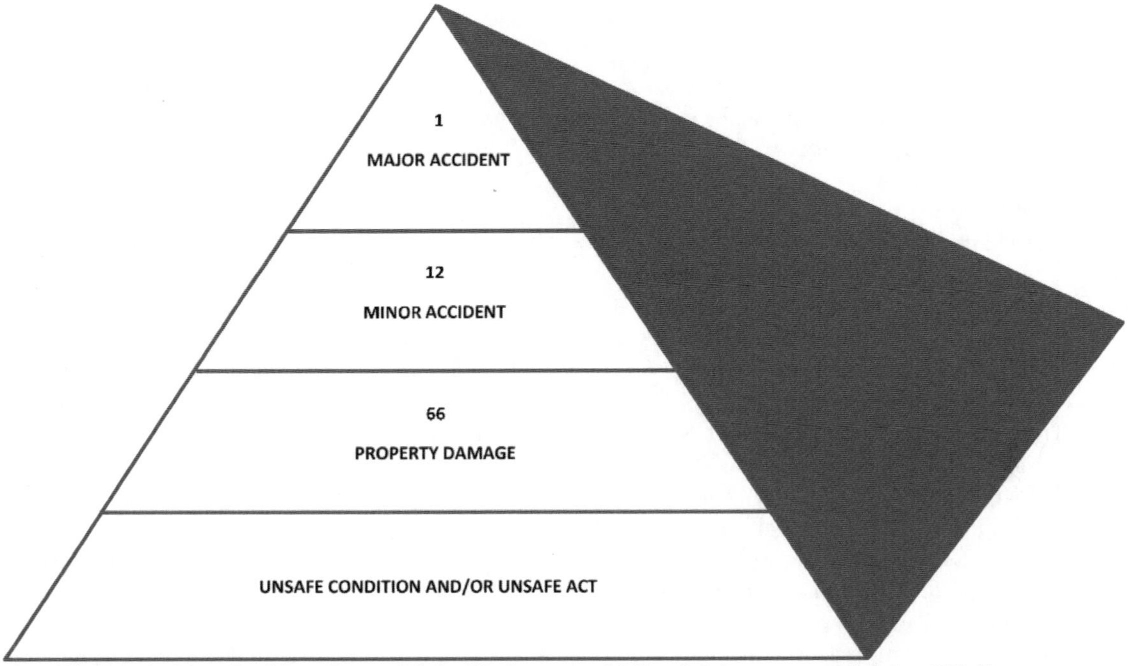

Figure 4. New Incident Pyramid for Underground Mine Maintenance Division PT. Freeport Indonesia

One of the fundamental underlying principles of Heinrich's original triangle is that major accident as the top of pyramid cannot occur without a foundation of less severe incidents. In other words, increasing numbers of minor accident and equipment damage eventually support the existence of more serious and fatal incidents. Number of unsafe condition and/or unsafe act as the direct cause of an accident will be the base layer of the accident triangle.

Realizing that accidents in the mining industry are inevitable does not stop the company for believing that they are also preventable. Based on the triangle, PT. Freeport Indonesia implemented program called as Master Observation Program to reduce the number of unsafe condition and/or unsafe act which was found in the area of work. From this program, we capture the data of unsafe act and/or unsafe condition as shown at table 2.

Table 2. Data Master Observation Program.

Category	2011	2012	2013	2014	2015	2016
PPE	973	690	1056	1298	1396	1353
Housekeeping	4316	2856	4350	4671	4561	4785
Procedure	4307	2624	4194	3918	3710	3846
Position of people	581	385	552	564	568	608
Tools and Equipment	3419	1866	3658	4220	3900	4126
Total	**13596**	**8421**	**13810**	**14671**	**14135**	**14718**

The unsafe condition and/or unsafe act found in area of work should be stopped and followed up. Through the commitment of all line in the organization, especially leaders and supervisors, for the elimination of it, PT. Freeport Indonesia will be able to improve the safety performance.

5. Conclusion

One of the fundamental underlying principles of accident triangle is that major accident cannot occur without a foundation of less severe incidents. In other words, increasing numbers of minor incidents and property damage eventually support the existence of major incidents. The new value of accident triangle from the data is 1 major accident; 12 minor accident; and 66 property damage.

6. References

[1] Markkanen P K 2004 *Working paper no 9, Occupational safety and health in Indonesia* (Philippines: Int. Labor Organization)
[2] Mwanza B G and Mbohwa C 2017 Safety in Maintenance: An Improvement Framework. *Procedia. Manufacturing* **8** pp 657–664
[3] Dhillon B S 2008. *Mining equipment reliability, maintainability, and safety (Springer series in reliability engineering)*
[4] Heinrich H W, Petersen, D and Roos N 1980 *Industrial accident prevention: A scientific approach, 5th edition,* (New York: McGraw-Hill Book Company)
[5] Bird F E and Germain G L 1986 *Practical loss control leadership. Loganville* (Ga: Institute Publishing; International Loss Control Institute)
[6] Penkey S P and Siddiqui N A 2015 A Review on Accident Pyramid and Its Empirical Interpretation in Oil & Gas Industry (Upstream) *Int. Journal of Scientific and Research Publications* **5**(1) pp 1-3
[7] Goodman W M 2014 Measuring Changes in the Distribution of Incident-Outcome Severities: A Tool for Safety Management *CS-BIGS* **5**(1) pp 32-43
[8] 1995. Ministry for Mining and Energy (Decree no. 555.K/26/M.PE/1995)
[9] William H H 1959 *Industrial Accident Prevention: A Scientific Approach* 4th Edition (Boston: McGraw-Hill Book Company)
[10] 2014 Freeport Safety and Health Standard article no. 4.11 *OH&S Incident Reporting, Classification, and Recording*
[11] Baron M 2014 *Probability and Statistics for Computer Scientists, 2nd Edition* (CRC Press)
[12] Nicholson A and Wong Y D 1993 Are accidents poisson distributed? A statistical test. *Accident Analysis & Prevention* **25**(1) pp 91–97

Institute of Physics
Dirac House, Temple Back
Bristol BS1 6BE UK

ISSN: 1757-8981
ISBN 978-1-5108-6158-9

3rd International Conference on Materials and Manufacturing Engineering (ICMME-2018)

IOP Conference Series: Materials Science and Engineering
Volume 390

Tamilnadu, India
8 – 9 March 2018

IOP
Institute of Physics